Lecture Notes in Physics

Springer-Verlag Berlin Heidelberg GmbH

The Editorial Policy for Proceedings

The series Lecture Notes in Physics reports new developments in physical research and teaching – quickly, informally, and at a high level. The proceedings to be considered for publication in this series should be limited to only a few areas of research, and these should be closely related to each other. The contributions should be of a high standard and should avoid lengthy redraftings of papers already published or about to be published elsewhere. As a whole, the proceedings should aim for a balanced presentation of the theme of the conference including a description of the techniques used and enough motivation for a broad readership. It should not be assumed that the published proceedings must reflect the conference in its entirety. (A listing or abstracts of papers presented at the meeting but not included in the proceedings could be added as an appendix.)
When applying for publication in the series Lecture Notes in Physics the volume's editor(s) should submit sufficient material to enable the series editors and their referees to make a fairly accurate evaluation (e.g. a complete list of speakers and titles of papers to be presented and abstracts). If, based on this information, the proceedings are (tentatively) accepted, the volume's editor(s), whose name(s) will appear on the title pages, should select the papers suitable for publication and have them refereed (as for a journal) when appropriate. As a rule discussions will not be accepted. The series editors and Springer-Verlag will normally not interfere with the detailed editing except in fairly obvious cases or on technical matters.
Final acceptance is expressed by the series editor in charge, in consultation with Springer-Verlag only after receiving the complete manuscript. It might help to send a copy of the authors' manuscripts in advance to the editor in charge to discuss possible revisions with him. As a general rule, the series editor will confirm his tentative acceptance if the final manuscript corresponds to the original concept discussed, if the quality of the contribution meets the requirements of the series, and if the final size of the manuscript does not greatly exceed the number of pages originally agreed upon. The manuscript should be forwarded to Springer-Verlag shortly after the meeting. In cases of extreme delay (more than six months after the conference) the series editors will check once more the timeliness of the papers. Therefore, the volume's editor(s) should establish strict deadlines, or collect the articles during the conference and have them revised on the spot. If a delay is unavoidable, one should encourage the authors to update their contributions if appropriate. The editors of proceedings are strongly advised to inform contributors about these points at an early stage.
The final manuscript should contain a table of contents and an informative introduction accessible also to readers not particularly familiar with the topic of the conference. The contributions should be in English. The volume's editor(s) should check the contributions for the correct use of language. At Springer-Verlag only the prefaces will be checked by a copy-editor for language and style. Grave linguistic or technical shortcomings may lead to the rejection of contributions by the series editors. A conference report should not exceed a total of 500 pages. Keeping the size within this bound should be achieved by a stricter selection of articles and not by imposing an upper limit to the length of the individual papers. Editors receive jointly 30 complimentary copies of their book. They are entitled to purchase further copies of their book at a reduced rate. As a rule no reprints of individual contributions can be supplied. No royalty is paid on Lecture Notes in Physics volumes. Commitment to publish is made by letter of interest rather than by signing a formal contract. Springer-Verlag secures the copyright for each volume.

The Production Process

The books are hardbound, and the publisher will select quality paper appropriate to the needs of the author(s). Publication time is about ten weeks. More than twenty years of experience guarantee authors the best possible service. To reach the goal of rapid publication at a low price the technique of photographic reproduction from a camera-ready manuscript was chosen. This process shifts the main responsibility for the technical quality considerably from the publisher to the authors. We therefore urge all authors and editors of proceedings to observe very carefully the essentials for the preparation of camera-ready manuscripts, which we will supply on request. This applies especially to the quality of figures and halftones submitted for publication. In addition, it might be useful to look at some of the volumes already published. As a special service, we offer free of charge LaTeX and TeX macro packages to format the text according to Springer-Verlag's quality requirements. We strongly recommend that you make use of this offer, since the result will be a book of considerably improved technical quality. To avoid mistakes and time-consuming correspondence during the production period the conference editors should request special instructions from the publisher well before the beginning of the conference. Manuscripts not meeting the technical standard of the series will have to be returned for improvement.

For further information please contact Springer-Verlag, Physics Editorial Department II, Tiergartenstrasse 17, D-69121 Heidelberg, Germany

H. Latal W. Schweiger (Eds.)

Perturbative and Nonperturbative Aspects of Quantum Field Theory

Proceedings of the
35. Internationale Universitätswochen
für Kern- und Teilchenphysik,
Schladming, Austria, March 2–9, 1996

Springer

Editors

H. Latal
W. Schweiger
Institut für Theoretische Physik
Karl-Franzens-Universität Graz
Universitätsplatz 5
A-8010 Graz, Austria

Supported by the Österreichische Bundesministerium für Wissenschaft, Verkehr und Kunst, Vienna, Austria

Cataloging-in-Publication Data applied for.

Die Deutsche Bibliothek - CIP-Einheitsaufnahme

Perturbative and nonperturbative aspects of quantum field theory : proceedings of the 35. Internationale Universitätswochen für Kern- und Teilchenphysik, Schladming, Austria, March 2 - 9, 1996 / H. Latal ; W. Schweiger (ed.).
(Lecture notes in physics ; 479)
ISBN 978-3-662-14126-7 ISBN 978-3-540-49735-6 (eBook)
DOI 10.1007/978-3-540-49735-6

NE: Latal, Heimo [Hrsg.]; Internationale Universitätswochen für Kern- und Teilchenphysik <35, 1996, Schladming>; GT

ISSN 0075-8450
ISBN 978-3-662-14126-7

Originally published by Springer-Verlag Berlin Heidelberg New York in 1997
Softcover reprint of the hardcover 1st edition 1997

Typesetting: Camera-ready by the authors/editors
Cover design: *design & production* GmbH, Heidelberg
SPIN: 10550497 55/3144-543210 - Printed on acid-free paper

Preface

This volume contains the written versions of the invited lectures and two of the seminars presented at the "35. Internationale Universitätswochen für Kern- und Teilchenphysik" in Schladming, Austria, which took place from March 2nd to 9th, 1996. The title of the School was "Perturbative and Non-perturbative Aspects of Quantum Field Theory". The choice of this topic was motivated by the fact that the presently most successful theory of elementary particles and their interactions, the so-called "Standard Model", turns out to be still too complicated to be solved explicitly in most of its applications. The lectures given at the school address three of the main techniques for relating the Standard Model, or more generally quantum field theories, to experimentally measurable quantities. These are computer simulations on space-time lattices, perturbative methods, and the reduction to a simpler effective theory by freezing out superficious degrees of freedom.

One of the most intriguing problems is to solve the strong-interaction sector of the Standard Model, i.e. quantum chromodynamics (QCD), at low energies. Major progress in calculating the hadron spectrum and even the structure of hadrons has been achieved since the late 1970s by formulating QCD on a space-time lattice and treating it numerically by means of extremely powerful computers. The tremendous computational effort is connected with the fact that, in order to recover the continuum limit, the lattice spacing has to become small and thus the number of lattice points large. Recent developments that allow a simulation of QCD on coarse lattices – even with personal computers – are discussed in the lecture by G.P. Lepage.

Whereas the objectives of lattice studies are essentially long-distance properties of the strong interaction, perturbative methods can be applied if short-distance phenomena are investigated. This is the case in so-called "hard" high-energy processes which involve large (transverse) momentum transfers. A third regime of QCD, which is neither determined by pure long-distance, nor by pure short-distance physics, can be identified in "soft" hadron-hadron collisions at high energy, in which the transferred momentum is small. This kind of reactions is the main subject of O. Nachtmann's lecture. The description he is proposing for soft hadron-hadron collisions relies on a particular model for the highly non-trivial QCD vacuum, the so-called "stochastic vacuum model". Soft hadron-hadron collisions are then understood via the scattering of the hadronic constituents in a fixed gluon field, which represents a particular vacuum configuration. Finally an appropriate sum over all possible configurations of the gluonic field has to be taken. The problem of developing a perturbative formalism in the presence of a non-perturbative QCD background is addressed in Yu.A. Simonv's lecture. The hope is that such an approach can improve the convergence properties of the perturbation series. Simonov shows that complications arising in ordinary perturbation theory (with a trivial vacuum), e.g. the Landau pole occurring in the running coupling constant, or the renormalon problem, can be avoided by taking into account the non-trivial vacuum structure. Also ordinary QCD perturbation theory has to deal with non-perturbative dynamics. Due to the confinement property of QCD, i.e. the fact that the asymptotic states of QCD are hadrons rather than quarks and gluons, long-distance effects are always present, even in the hard-scattering region. Only the existence of factorization theorems ensures that the long-distance contributions can be lumped together into process-independent functions (structure functions, fragmentation functions, distribution amplitudes). The techniques for separating long- and short-distance contributions to hard inclusive cross sections in a systematic way and the treatment of multiple perturbative scales by means of resummation (in particular, of Sudakov logarithms) is explained in the lecture by G. Sterman. Going beyond leading order perturbation theory is, apart from a few simple cases, usually a very demanding task. The precision of present-day experiments, especially in the electroweak sector of the Standard Model, however, makes such refinements of the perturbative expansion often necessary. The application of computer algebra systems to perform the analytical part of this job is the topic of J. Vermaseren's lecture. His explanations are nicely exemplified in E. Remmiddi's seminar on the analytical evaluation of the 6th-order electron-(g-2) factor in QED.

In many physical applications it appears to be more expedient to work rather with an effective model than with the underlying fundamental theory. The idea behind effective theories is that dynamics at low energies should not depend on the details of the dynamics at high energies. A.V. Manohar's lecture contains a quantitative consideration of this rather qualitative statement. His methods are exemplified, on the one hand, for the weak interaction, where the effective Lagrangian can be directly deduced from the full Lagrangian by means of a systematic perturbative expansion, and on the other hand for the strong interaction, where the relation to QCD is less direct, but rather based on general properties such as the spontaneous breaking of chiral symmetry. An effective model for the low-energy strong interaction, based on constituent quarks which interact via exchange of the octet of pseudoscalar mesons, and its consequences for the baryon spectrum are discussed in the seminar by L.Ya. Glozman. Considerable theoretical progress has been achieved since the beginning of the 1990s for systems that contain a heavy quark. Under these circumstances a systematic approximation to full QCD may be constructed by performing an expansion in the inverse mass of the heavy quark. The formulation of this "heavy quark effective theory" and its application to exclusive and inclusive decays of heavy hadrons is described in the lecture by T. Mannel.

The school was complemented by many excellent seminars which, due to space limitations, could not be included fully in these proceedings. The list of seminar speakers and the topics addressed by them can be found at the end of this volume. The interested reader is requested to contact the speakers directly for detailed information or pertinent material.

Finally, we would like to express our thanks to the lecturers for all their efforts, to the sponsors of the school, above all the Austrian Ministry of Science, Traffic, and Arts, and the Government of Styria, for providing generous support, to our colleagues in the organizing committee for their assistance, and to Mrs. E. Neuhold for her help in bringing the files prepared by the authors in TEX or LATEX into their final form.

Graz, October 1996

H. Latal (Director of the School)
W. Schweiger (Scientific Secretary)

Table of Contents

Redesigning Lattice QCD

G. Peter Lepage

Newman Laboratory of Nuclear Studies, Cornell University,
Ithaca NY 14853, USA

Abstract. There has been major progress in recent years in the development of improved discretizations of the QCD action, current operators, etc for use in numerical simulations that employ very coarse lattices. These lectures review the field theoretic techniques used to design these discretizations, techniques for testing and tuning the new formalisms that result, and recent simulation results employing these formalisms.

1 Introduction

Lattice quantum chromodynamics is *the* fundamental theory of low-energy strong interactions. In principle, lattice QCD should tell us everything we want to know about hadronic spectra and structure, including such phenomenologically useful things as weak-interaction form factors, decay rates and deep-inelastic structure functions. In these lectures I discuss a revolutionary development that makes the techniques of lattice QCD much more widely accessible and greatly extends the range of problems that are tractable.

The basic approximation in lattice QCD is the replacement of continuous space and time by a discrete grid. The nodes or "sites" of the grid are separated by lattice spacing a, and the length of a side of the grid is L:

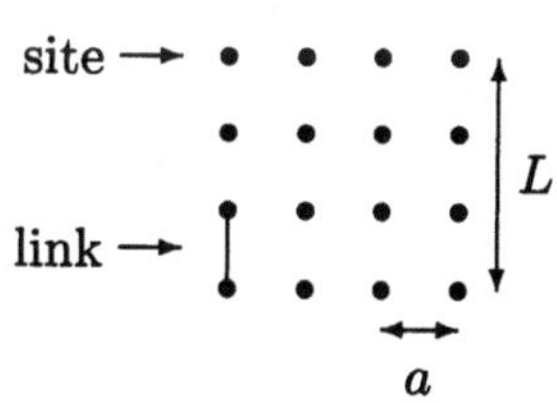

The quark and gluon fields from which the theory is built are specified only on the sites of the grid, or on the "links" joining adjacent sites; interpolation is used to find the fields between the sites. In this lattice approximation, the path integral, from which all quantum mechanical properties of the theory can be extracted, becomes an ordinary multidimensional integral where the integration variables are the values of the fields at each of the grid sites:

$$\int \mathcal{D}A_\mu \ldots \mathrm{e}^{-\int L \mathrm{d}t} \longrightarrow \int \prod_{x_j \epsilon \, \mathrm{grid}} \mathrm{d}A_\mu(x_j) \ldots \mathrm{e}^{-a \sum L_j}. \tag{1}$$

Thus the problem of nonperturbative relativistic quantum field theory is reduced to one of numerical integration. The integral is over a large number of variables and so Monte Carlo methods are generally used in its evaluation. Note that the path integral uses euclidean time rather than ordinary minkowski time, where $t_{\mathrm{eucl}} = \mathrm{i}\, t_{\mathrm{mink}}$; this removes a factor of i from the exponent, getting rid of high-frequency oscillations in the integrand that are hard to integrate.

Early enthusiasm for this approach to nonperturbative QCD gradually gave way to the sobering realization that very large computers would be needed for the numerical integration of the path integral — computers much larger than those that existed in the mid 1970's, when lattice QCD was invented. Much of a lattice theorist's effort in the first twenty years of lattice QCD was spent in accumulating computing time on the world's largest supercomputers, or in designing and building computers far larger than the largest commercially available computers. By the early 1990's, it was widely felt that teraflop computers costing tens of million of dollars would be essential to the final numerical solution of full QCD.

The revolutionary development I discuss here was the introduction of new techniques that allow one to do realistic numerical simulations of QCD on ordinary desktop workstations or even personal computers. To understand this development one must understand the factors that govern the cost of a full QCD simulation. This cost is governed by a formula like

$$\mathrm{cost} \propto \left(\frac{L}{a}\right)^4 \left(\frac{1}{a}\right) \left(\frac{1}{m_\pi^2 \, a}\right) \tag{2}$$

where the first factor is just the number of lattice sites in the grid, and the remaining factors account for the "critical slowing down" of the algorithms used in the numerical integration. This formula shows that the single most important determinant of cost is the lattice spacing. The cost varies as the sixth power of $1/a$, suggesting that one ought to keep the lattice spacing as large as possible. Until very recently it was thought that lattice spacings as small as .05–.1 fm would be essential for reliable simulations of QCD. As I describe in these lectures, new simulation results based on new techniques indicate that spacings as large as .4 fm give accurate results. Assuming that the cost is proportional to $(1/a)^6$, these new simulations using coarse lattices should cost 10^3–10^6 times less than traditional simulations on fine lattices.

The computational advantage of coarse lattices is enormous and will certainly redefine numerical QCD: the simplest calculations can be done on a personal computer, while problems of unprecedented difficulty and precision can be tackled with large supercomputers. In these lectures I explain why we now think coarse lattices can be made to work well. I review the techniques from quantum field theory needed to redesign lattice QCD for coarse lattices. These techniques are based on ideas from renormalization theory and effective field theory. I begin in Sect. 2 discussing the field-theoretic implications of discretizing space and time. Then in Sects. 3, 4 and 5 I discuss in detail how to discretize the dynamics of gluons, light quarks and heavy quarks, respectively. I also discuss the crucial issue of testing and tuning improved discretizations. These sections present a mixture of very old and very new results. I do not discuss in any detail the numerical techniques used in the simulations; these are described in standard texts [1], [2], [3]. As these are lectures from a school, I include a number of exercises that illustrate the concepts developed in the lectures. Finally I summarize the current situation and prospects for the future in Sect. 6.

2 Field Theory on a Lattice

In this section I discuss the factors that limit the maximum size of the lattice spacing. Replacing space-time by a discrete lattice is an approximation. If we make the lattice spacing too large, our answers will not be sufficiently accurate. For the purpose of these lectures, I define a "sufficiently accurate" simulation to be one that reproduces the low-energy properties of hadrons to within a few percent, which is very precise for low-energy strong-interaction physics. The issue then is: How large can we make a while keeping errors of order a few percent or less?

A nonzero lattice spacing results in two types of error: the error that arises when we replace derivatives in the field equations by finite-difference approximations, and the error due to the ultraviolet cutoff imposed by the lattice. I now discuss each of these in turn, and then focus on the key role played by perturbation theory in correcting for finite-a errors.

2.1 Approximate Derivatives

In the lattice approximation, field values are known only at the sites on the lattice. Consequently we approximate derivatives in the field equations by finite-differences that use only field values at the sites. This is completely conventional, and very familiar, numerical analysis. For example, the derivative of a field ψ evaluated at lattice site x_j is approximated by

$$\frac{\partial\psi(x_j)}{\partial x} \approx \Delta_x \psi(x_j) \tag{3}$$

where

$$\Delta_x \psi(x) \equiv \frac{\psi(x+a) - \psi(x-a)}{2a}. \tag{4}$$

It is easy to analyze the error associated with this approximation. Taylor's Theorem implies that

$$\begin{aligned} 2a\,\Delta_x\psi(x) &\equiv \psi(x+a) - \psi(x-a) \\ &= \left(e^{a\partial_x} - e^{-a\partial_x}\right)\psi(x) \end{aligned} \tag{5}$$

and therefore

$$\Delta_x\psi = \left(\partial_x + \frac{a^2}{6}\partial_x^3 + \mathcal{O}(a^4)\right)\psi. \tag{6}$$

Thus the relative error in $\Delta_x\psi$ is of order $(a/\lambda)^2$ where λ is the typical length scale in $\psi(x)$.

On coarse lattices we generally need more accurate discretizations than this one. These are easily constructed. For example from (6) it is obvious that

$$\frac{\partial\psi}{\partial x} = \Delta_x\psi - \frac{a^2}{6}\Delta_x^3\psi + \mathcal{O}(a^4), \tag{7}$$

which is a more accurate discretization. When one wishes to reduce the finite-a errors in a simulation, it is usually far more efficient to improve the discretization of the derivatives than to reduce the lattice spacing. For example, with just the first term in the approximation to $\partial_x\psi$, cutting the lattice spacing in half would reduce a 20% error to 5%; but the cost would increase by a factor of $2^6 = 64$ in a simulation where cost goes like $1/a^6$. On the other hand, including the a^2 correction to the derivative, while working at the larger lattice spacing, achieves the same reduction in error but with a cost increase of only a factor of 2.

Equation (7) shows the first two terms of a systematic expansion of the continuum derivative in powers of a^2. In principle, higher-order terms can be included to obtain greater accuracy, but in practice the first couple of terms are sufficiently accurate for most purposes. Simple numerical experiments with a^3-accurate discretizations like this one (see below) show that only three or four lattice sites per bump in ψ are needed to achieve accuracies of a few percent or less. Since ordinary hadrons are approximately 1.8 fm in diameter, these experiments suggest that a lattice spacing of .4 fm would suffice for simulating these hadrons. However QCD is a quantum theory, and, as I discuss in the next section, quantum effects can change everything.

Exercise: Show that

$$\frac{\partial^2\psi}{\partial x^2} = \Delta_x^{(2)}\psi - \frac{a^2}{12}\left(\Delta_x^{(2)}\right)^2\psi + \mathcal{O}(a^4) \tag{8}$$

where

$$\Delta_x^{(2)}\psi(x) \equiv \frac{\psi(x+a) - 2\psi(x) + \psi(x-a)}{a^2}. \tag{9}$$

Note that Δ_x and $\Delta_x^{(2)}$ can be used to construct lattice approximations for derivatives of any order: for example,

$$\begin{aligned}\Delta_x^{(3)}\psi &\equiv \Delta_x \Delta_x^{(2)}\psi = \Delta^{(2)}\Delta_x\psi \\ &= \partial_x^3\psi + \mathcal{O}(a^2) \end{aligned} \tag{10}$$

$$\begin{aligned}\Delta_x^{(4)}\psi &\equiv \left(\Delta_x^{(2)}\right)^2\psi \\ &= \partial_x^4\psi + \mathcal{O}(a^2) \end{aligned} \tag{11}$$

$$\vdots$$

Exercise: Use

$$\psi(x) = \mathrm{e}^{-x^2/2\sigma^2} \tag{12}$$

with $\sigma = 2a$ as a sample function to test our discretizations of $\partial_x\psi$ and $\partial_x^2\psi$. Evaluate the a^2 errors of the simplest discretization in each case, and the a^4 errors of the more accurate discretizations. Show that the a^2 errors are of order 10–20%, while the a^4 errors are only a few percent or less. Note that $\Delta^{(2)}$ is more substantially more accurate than $\Delta^{(1)}$.

2.2 Ultraviolet Cutoff

The shortest wavelength oscillation that can be modeled on a lattice is one with wavelength $\lambda_{\mathrm{min}} = 2a$; for example, the function $\psi(x) = +1, -1, +1\ldots$ for $x = 0, a, 2a\ldots$ oscillates with this wavelength. Thus gluons and quarks with momenta $k = 2\pi/\lambda$ larger than π/a are excluded from the lattice theory by the lattice; that is, the lattice functions as an ultraviolet cutoff. In simple classical field theories this is often irrelevant: short-wavelength ultraviolet modes are either unexcited or decouple from the long-wavelength infrared modes of interest. However, in a noisy nonlinear theory, like an interacting quantum field theory, ultraviolet modes strongly affect infrared modes by renormalizing masses and interactions. Thus we cannot simply discard all particles with momenta larger than π/a; we must somehow mimick their effects on infrared states. This we can do by modifying our discretized lagrangian.[1]

To see how we might mimick the effects of $k > \pi/a$ states on low momentum states, consider the scattering amplitude T for quark-quark scattering in one-loop perturbation theory (Fig. 1a). The difference between the correct amplitude T in the continuum theory and the cut-off amplitude $T^{(a)}$ in our lattice theory involves internal gluons with momentum k of order π/a or larger. This is because the classical theories agree at low momenta, and therefore all propagators and vertices agree there as well. Thus the loop contributions from low momenta will be the same in T and $T^{(a)}$, and cancel

[1] The idea of modifying the lagrangian to compensate for a finite UV cutoff is central to chiral field theories for pions, nonrelativistic QED/QCD and all other effective field theories. The application to lattice field theories was pioneered in the form discussed here by Symanzik and is referred to as "Symanzik improvement" of lattice operators [4].

in the difference. Given that the external quarks have momenta p_i that are small relative to π/a, we can expand the difference $T - T^{(a)}$ in a Taylor series in $a\,p_i$ to obtain

$$\begin{aligned} T - T^{(a)} &= a^2\, c(a)\, \overline{u}(p_2)\gamma_\mu u(p_1)\, \overline{u}(p_4)\gamma^\mu u(p_3) \\ &+ a^2\, c_A(a)\, \overline{u}\gamma_\mu\gamma_5 u\, \overline{u}\gamma^\mu\gamma_5 u \\ &+ a^4\, d(a)\, (p_1 - p_2)^2\, \overline{u}\gamma_\mu u\, \overline{u}\gamma^\mu u \\ &+ \cdots . \end{aligned} \tag{13}$$

where the couplings c, $c_A \ldots$ are dimensionless functions of the cutoff. This difference is what is missing from the lattice theory; it is the contribution of the $k > \pi/a$ states excluded by the lattice.[2] The key observation is that we can reintroduce this high-k contribution into the lattice theory by adding new interactions to the lattice lagrangian:

$$\begin{aligned} \delta\mathcal{L}^{(a)}_{4\mathrm{q}} &= \tfrac{1}{2}\, a^2\, c(a)\, \overline{\psi}\gamma_\mu\psi\, \overline{\psi}\gamma^\mu\psi \\ &+ \tfrac{1}{2}\, a^2\, c_A(a)\, \overline{\psi}\gamma_\mu\gamma_5\psi\, \overline{\psi}\gamma^\mu\gamma_5\psi \\ &+ a^4\, d(a)\, \overline{\psi}D^2\gamma_\mu\psi\, \overline{\psi}\gamma^\mu\psi \\ &+ \cdots . \end{aligned} \tag{14}$$

These new local interactions give the same contribution to $T(qq \to qq)$ in the lattice theory as the $k > \pi/a$ states do in the continuum theory. Although these correction terms are nonrenormalizable, they do not cause problems in our lattice theory because it is cut off at π/a. On the contrary, they bring our lattice theory closer to the continuum without reducing the lattice spacing.

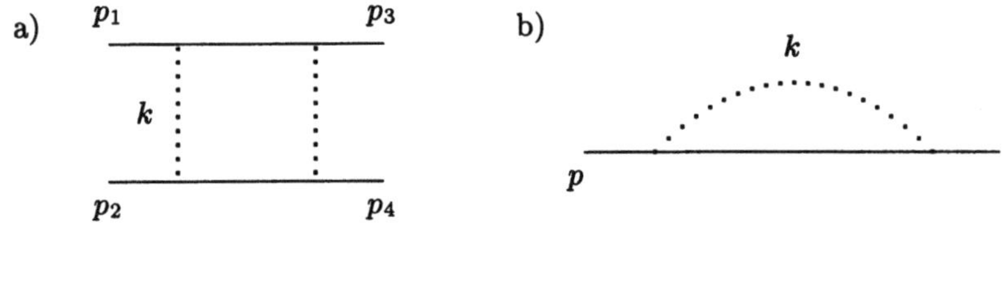

Fig. 1. One-loop amplitudes contributing to: a) $qq \to qq$, and b) the quark self energy.

This simple analysis illustrates a general result of renormalization theory: one can mimick the effects of states excluded by a cutoff with extra local

[2] More precisely, this contribution is due to states above the cutoff, and to corrections needed to fix up the states below but sufficiently near the cutoff that they suffer severe lattice distortion.

interactions in the cut-off lagrangian. The correction terms in $\delta\mathcal{L}^{(a)}$ are all local — that is, they are polynomials in the fields and in derivatives of the fields — since contributions omitted as a consequence of the cutoff can be Taylor-expanded in powers of ap_i. This is intuitively reasonable since these terms correct for contributions from intermediate states that are highly virtual and thus quite local in extent; the locality is a consequence of the uncertainty principle. The relative importance of the different interactions is determined by the number of powers of a in their coefficients, and that is determined by the dimensions of the operators involved: each term in $\delta\mathcal{L}^{(a)}$ must have total (energy) dimension four, and so if the operator in a particular term has dimension $n+4$ then the coefficient must contain a factor a^n. In principle there are infinitely many correction terms, but we need only include interactions with operators of dimension $n+4$ or less to achieve accuracy through order $(ap_i)^n$. In practice we only need precision to a given order n in ap_i, and there are only a finite number of local operators with dimension less than or equal to any given $n+4$.

Initially the $k > \pi/a$ contributions appear to be bad news. As we argued in the previous section, a^2 corrections are necessary for precision when a is large. However the quantum nature of our theory implies that there are a^2 terms, due to $k > \pi/a$ states, which depend in detail on the nature of the theory. Thus, for example, when we discretize the derivative in the QCD quark action we replace

$$\overline{\psi}\partial\cdot\gamma\psi \to \overline{\psi}\Delta\cdot\gamma\psi + a^2\, d(a)\, \overline{\psi}\Delta^{(3)}\cdot\gamma\psi + \cdots \tag{15}$$

where now $d(a)$ has a part, $-1/6$, from numerical analysis (equation (7)) plus a contribution that mimicks the $k > \pi/a$ part of the quark self energy (Fig. 1b). The problem is that the latter contribution is, by its nature, completely theory specific; we cannot look it up in a book on numerical analysis. We must somehow solve the $k > \pi/a$ part of the field theory in order to calculate it; otherwise we are unable to correct our lagrangian and unable to use large lattice spacings.

The good news, for lattice QCD, is that $k > \pi/a$ QCD is perturbative provided a is small enough (because of asymptotic freedom). Then correction coefficients like $d(a)$ can be computed perturbatively, using Feynman diagrams, in an expansion in powers of $\alpha_s(\pi/a)$. Thus, for example, our corrected lattice action for quarks becomes

$$\begin{aligned}\mathcal{L}^{(a)} &= \overline{\psi}\left(\Delta\cdot\gamma + m(a)\right)\psi \\ &\quad + d(a)\, a^2\, \overline{\psi}\Delta^{(3)}\cdot\gamma\psi \\ &\quad + c(a)\, a^2\, \overline{\psi}\gamma_\mu\psi\, \overline{\psi}\gamma^\mu\psi + \ldots\end{aligned} \tag{16}$$

where

$$d(a) = -\frac{1}{6} + d_1\, \alpha_s(\pi/a) + \cdots \tag{17}$$

$$c(a) = c_2\, \alpha_s^2(\pi/a) + \cdots \tag{18}$$

$$\cdots$$

are computed to whatever order in perturbation theory is necessary. In this way we use perturbation theory to, in effect, fill in the gaps between lattice points, allowing us to obtain continuum results without taking the lattice spacing to zero.

2.3 Perturbation Theory and Tadpole Improvement[3]

Improved discretizations and large lattice spacings are old ideas, pioneered by Wilson, Symanzik and others [6]. However, perturbation theory is essential; the lattice spacing a must be small enough so that $k \approx \pi/a$ QCD is perturbative. This was the requirement that drove lattice QCD towards very costly simulations with tiny lattice spacings. Traditional perturbation theory for lattice QCD begins to fail at distances of order 1/20 to 1/10 fm, and therefore lattice spacings must be at least this small before improved actions are useful. This seems very odd since phenomenological applications of continuum perturbative QCD suggest that perturbation theory works well down to energies of order 1 GeV, which corresponds to a lattice spacing of 0.6 fm. The breakthrough, in the early 1990's, was the discovery of a trivial modification of lattice QCD, called "tadpole improvement," that allows perturbation theory to work even at distances as large as 1/2 fm [5], [7].

One can readily derive Feynman diagram rules for lattice QCD using the same techniques as in the continuum, but applied to the lattice lagrangian [8]. The particle propagators and interaction vertices are usually complicated functions of the momenta that become identical to their continuum analogues in the low-momentum limit. All loop momenta are cut off at $k_\mu = \pm\pi/a$.

Testing perturbation theory is also straightforward. One designs short-distance quantities that can be computed easily in a simulation (i.e., in a Monte Carlo evaluation of the lattice path integral). The Monte Carlo gives the exact value which can then be compared with the perturbative expansion for the same quantity. An example of such a quantity is the expectation value of the Wilson loop operator,

$$W(\mathcal{C}) \equiv \langle 0 | \tfrac{1}{3}\,\mathrm{Re\,Tr\,P}\,\mathrm{e}^{-ig\oint_{\mathcal{C}} A\cdot dx} | 0 \rangle, \tag{19}$$

where A is the QCD vector potential, P denotes path ordering, and $\mathcal{C}$ is any small, closed path or loop on the lattice. $W(\mathcal{C})$ is perturbative for sufficiently small loops $\mathcal{C}$. We can test the utility of perturbation theory over any range of distances by varying the loop size while comparing numerical Monte Carlo results for $W(\mathcal{C})$ with perturbation theory.

Fig. 2 illustrates the highly unsatisfactory state of traditional lattice-QCD perturbation theory. There I show the "Creutz ratio" of $2a \times 2a$, $2a \times a$ and $a \times a$ Wilson loops,

$$\chi_{2,2} \equiv -\ln\left(\frac{W(2a\times 2a)\,W(a\times a)}{W^2(2a\times a)}\right), \tag{20}$$

[3] This section is based upon work with Paul Mackenzie that is described in [5].

plotted versus the size $2a$ of the largest loop. Traditional perturbation theory (dotted lines) underestimates the exact result by factors of three or four for loops of order 1/2 fm; only when the loops are smaller than 1/20 fm does perturbation theory begin to give accurate results.

The problem with traditional lattice-QCD perturbation theory is that the coupling it uses is much too small. The standard practice was to express perturbative expansions of short-distance lattice quantities in terms of the bare coupling α_{lat} used in the lattice lagrangian. This practice followed from the notion that the bare coupling in a cutoff theory is approximately equal to the running coupling evaluated at the cutoff scale, here $\alpha_s(\pi/a)$, and therefore that it is the appropriate coupling for quantities dominated by momenta near the cutoff. In fact the bare coupling in traditional lattice QCD is much smaller than true effective coupling at large lattice spacings: for example,

$$\alpha_{\text{lat}} = \alpha_V(\pi/a) - 4.7\,\alpha_V^2 + \cdots \tag{21}$$

$$\leq \tfrac{1}{2}\alpha_V(\pi/a) \qquad \text{for } a > .1\,\text{fm} \tag{22}$$

where $\alpha_V(q)$ is a continuum coupling defined by the static-quark potential,

$$V_{Q\overline{Q}}(q) \equiv -4\,\pi\,C_F\,\frac{\alpha_V(q)}{q^2}. \tag{23}$$

Consequently α_{lat} expansions, though formally correct, badly underestimate perturbative effects, and converge poorly.

The anomalously small bare coupling in the traditional lattice theory is a symptom of the "tadpole problem". As we discuss later, all gluonic operators in lattice QCD are built from the link operator

$$U_\mu(x) \equiv \mathrm{P}\,\mathrm{e}^{-\mathrm{i}\int_x^{x+a\hat{\mu}} gA\cdot dx} \approx \mathrm{e}^{-\mathrm{i}agA_\mu} \tag{24}$$

rather than from the vector potential A_μ. Thus, for example, the leading term in the lagrangian that couples quarks and gluons is $\overline{\psi}U_\mu\gamma_\mu\psi/a$. Such a term contains the usual $\overline{\psi}gA\cdot\gamma\psi$ vertex, but, in addition, it contains vertices with any number of additional powers of agA_μ. These extra vertices are irrelevant for classical fields since they are suppressed by powers of the lattice spacing. For quantum fields, however, the situation is quite different since pairs of A_μ's, if contracted with each other, generate ultraviolet divergent factors of $1/a^2$ that precisely cancel the extra a's. Consequently the contributions generated by the extra vertices are suppressed by powers of g^2 (not a), and turn out to be uncomfortably large. These are the tadpole contributions.

The tadpoles result in large renormalizations — often as large as a factor of two or three — that spoil naive perturbation theory, and with it our intuition about the connection between lattice operators and the continuum. However tadpole contributions are generically process independent and so it is possible to measure their contribution in one quantity and then correct for them in all other quantities.

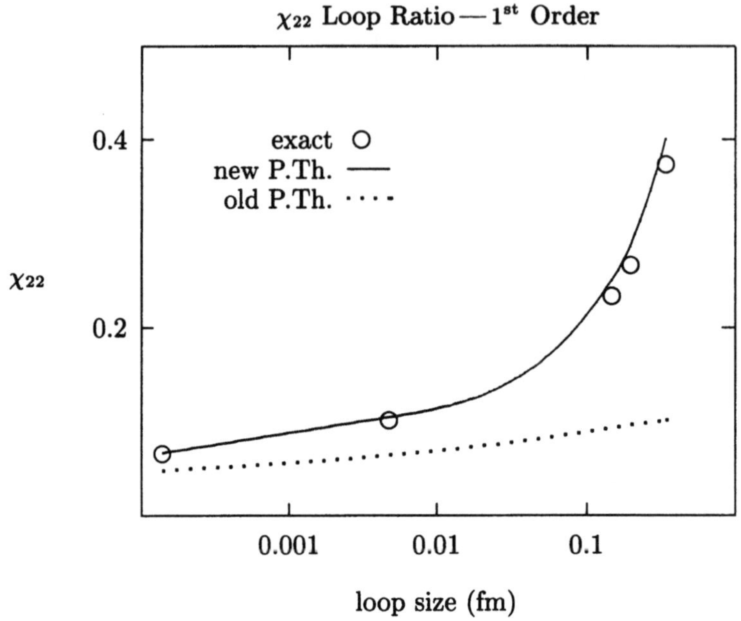

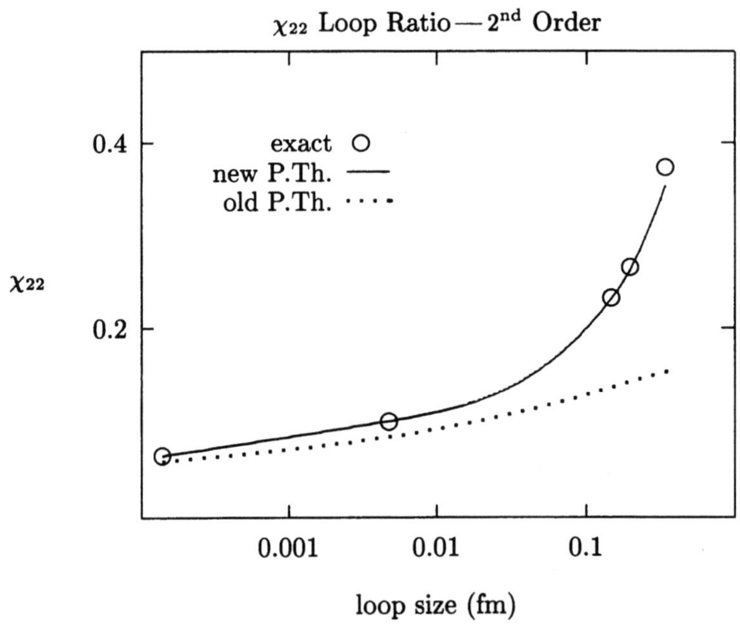

Fig. 2. The χ_{22} Creutz ratio of Wilson loops versus loop size. Results from Monte Carlo simulations (exact), and from tadpole-improved (new) and traditional (old) lattice perturbation theory are shown.

The simplest way to do this is to cancel them out. The mean value u_0 of $\frac{1}{3}\,\mathrm{Re}\,\mathrm{Tr}\,U_\mu$ consists of only tadpoles and so we can largely cancel the tadpole contributions by dividing every link operator by u_0. That is, in every lattice operator we replace

$$U_\mu(x) \to \frac{U_\mu(x)}{u_0} \tag{25}$$

where u_0 is computed numerically in a simulation.

The u_0's cancel tadpole contributions, making lattice operators and perturbation theory far more continuum-like in their behavior. Thus, for example, the only change in the standard gluon action when it is tadpole-improved is that the new bare coupling α_{TI} is enhanced by a factor of $1/u_0^4$ relative to the coupling α_{lat} in the unimproved theory:

$$\alpha_{\mathrm{TI}} = \frac{\alpha_{\mathrm{lat}}}{u_0^4}. \tag{26}$$

Since $u_0^4 < .6$ when $a > .1\,\mathrm{fm}$, the tadpole-improved coupling is typically more than twice as large for coarse lattices. Expressing α_{TI} in terms of the continuum coupling α_V, we find that now our intuition is satisfied:

$$\alpha_{\mathrm{TI}} = \alpha_V(\pi/a) - .5\,\alpha_V^2 + \cdots \tag{27}$$

$$\approx \alpha_V(\pi/a). \tag{28}$$

Perturbation theory for the Creutz ratio (20) converges rapidly to the correct answer when it is reexpressed in terms of α_{TI}. An even better result is obtained if the expansion is reexpressed as a series in a coupling constant defined in terms of a physical quantity, like the static-quark potential, where that coupling constant is measured in a simulation. By measuring the coupling we automatically include any large renormalizations of the coupling due to tadpoles. It is important that the scale q^* at which the running coupling constant is evaluated be chosen appropriately for the quantity being studied [5], [7]. When these refinements are added, perturbation theory is dramatically improved, and, as illustrated in Fig. 2, is still quite accurate for loops as large as $1/2\,\mathrm{fm}$.

This same conclusion follows from Fig. 3 which shows the value of the bare quark mass needed to obtain zero-mass pions using Wilson's lattice action for quarks. This quantity diverges linearly as the lattice spacing vanishes, and so should be quite perturbative. Here we see dramatic improvements as the tadpoles are removed first from the gluon action, through use of an improved coupling, and then also from the quark action.

The Creutz ratio and the critical quark mass are both very similar to the couplings we need to compute for improved lagrangians. Tadpole improvement has been very successful in a wide range of applications.

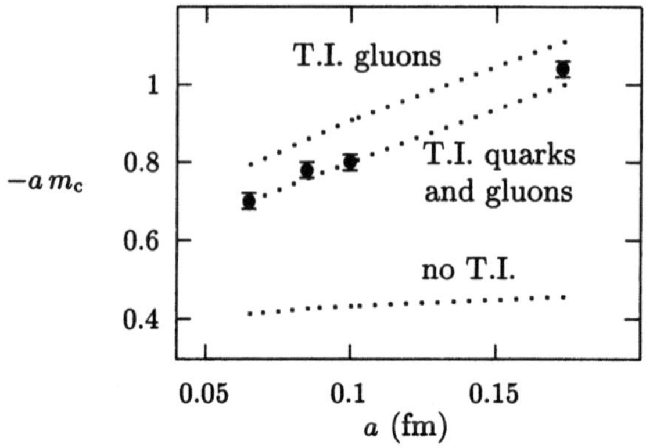

Fig. 3. The critical bare quark mass for Wilson's lattice quark action versus lattice spacing. Monte Carlo data points are compared with perturbation theories in a theory with no tadpole improvement (T.I), tadpole-improved gluon dynamics, and tadpole-improved quark and gluon dynamics.

Exercise: Derive the Feynman rules for lattice ϕ^4 theory where

$$S = \sum_x a^4 \left\{ \frac{-\phi\Delta^{(2)}\phi + m^2\phi^2}{2} + \frac{\lambda\,\phi^4}{4!} \right\}. \tag{29}$$

In particular show that the ϕ propagator is

$$G \propto \frac{a^2}{\sum_\mu \left(2\sin(ap_\mu/2)\right)^2 + a^2 m^2} \tag{30}$$

2.4 An Aside on "Perfect Actions"

Although not necessary in practice, it is possible to sum all the terms in the a^2 expansion of a continuum derivative using Fourier transforms: for example, we replace

$$\partial_x^2 \psi(x_i) \rightarrow -\sum_j d_{\rm pf}^{(2)}(x_i - x_j)\, \psi(x_j) \tag{31}$$

where

$$d_{\rm pf}^{(2)}(x_i - x_j) \equiv \int \frac{dp}{2\pi}\, p^2\, {\rm e}^{{\rm i}p(x_i - x_j)}\, \theta(|p| < \pi/a). \tag{32}$$

We do not do this because $d_{\rm pf}^{(2)}(x_i - x_j)$ falls off only as $1/|x_i - x_j|^2$ for large $|x_i - x_j|$, resulting in highly nonlocal actions and operators that are very costly to simulate, particularly when gauge fields are involved. So it is generally far better to truncate the a^2 expansion than to sum to all orders. However this analysis suggests a different strategy for discretization that we now examine.

The slow fall-off in the all-orders or perfect derivative $d^{(2)}_{\rm pf}$ is caused by the abruptness of the lattice cutoff (the θ function in (32)). Wilson noticed that $d^{(2)}$ can be made to vanish faster than any power of $1/|x_i - x_j|$ if a smoother cutoff is introduced. In his approach the lattice field at node x_i is *not* equal to the continuum field there, but rather equals the continuum field averaged over a small volume centered at that node: for example, the lattice field at site x_i in one dimension is[4]

$$\Psi_i = \frac{1}{a}\int_{x_i-a/2}^{x_i+a/2} \psi(x)\,\mathrm{d}x \tag{33}$$

$$= \int_{-\infty}^{\infty} \frac{dp}{2\pi}\,\mathrm{e}^{\mathrm{i}px_i}\,\Pi(p)\,\psi(p) \tag{34}$$

where, in the transform, smearing function

$$\Pi(p) \equiv \frac{2\sin(pa/2)}{pa} \tag{35}$$

provides a smooth cutoff at large p. The lattice action for the smeared fields is designed so that tree-level Green's functions built from these fields are *exactly* equal to the corresponding Green's functions in the continuum theory, with the smearing function applied at the ends: for example,

$$\langle \Psi_i \Psi_j \rangle = \int_{x_i-a/2}^{x_i+a/2} \mathrm{d}x \int_{x_j-a/2}^{x_j+a/2} \mathrm{d}y\ \langle \psi(x)\,\psi(y)\rangle \tag{36}$$

$$= \int_{-\infty}^{\infty} \frac{\mathrm{d}p}{2\pi}\,\mathrm{e}^{\mathrm{i}p(x_i-x_j)}\,\frac{\Pi^2(p)}{p^2+m^2}. \tag{37}$$

By rewriting the propagator $\langle \Psi_i \Psi_j \rangle$ in the form

$$\int_{-\pi/a}^{\pi/a} \frac{\mathrm{d}p}{2\pi}\,\mathrm{e}^{\mathrm{i}p(x_i-x_j)} \sum_n \frac{\Pi^2(p+2\pi n/a)}{(p+2\pi n/a)^2+m^2}, \tag{38}$$

we find that the quadratic terms of this lattice action are

$$S^{(2)} = \tfrac{1}{2}\sum_{i,j} \Psi_i\, d^{(2)}_{\rm pa}(x_i-x_j,m)\,\Psi_j, \tag{39}$$

where $d^{(2)}_{\rm pa}(x_i-x_j,m)$ is

$$\int_{-\pi/a}^{\pi/a} \frac{\mathrm{d}p}{2\pi}\,\mathrm{e}^{\mathrm{i}p(x_i-x_j)}\left(\sum_n \frac{\Pi^2(p+2\pi n/a)}{(p+2\pi n/a)^2+m^2}\right)^{-1}. \tag{40}$$

[4] The smearing discussed here is conceptually simple, but better smearings, resulting in improved locality, are possible. In [9] the authors use stochastic smearing in which Ψ_i equals the blocked continuum field only on average. This introduces a new parameter that can be tuned to optimize the action.

It is easily shown that, remarkably, $d^{(2)}_{\rm pa}(x_i-x_j,m)$ vanishes faster than any power of $1/|x_i-x_j|$ as the separation increases. So the derivative terms in such an action are both exact to all orders in a and quite local — that is, the classical action is "perfect." Again, this is possible because the lattice field Ψ_i is obtained by smearing the continuum field, which introduces a smooth UV cutoff in momentum space rather than the abrupt cutoff of (32).

The complication with this approach is that it is difficult to reconstruct the continuum fields $\psi(x)$ from the smeared lattice fields Ψ_i. Consequently nonlinear interactions, currents, and other operators are generally complicated functions of the Ψ_i, and much more difficult to design than with the previous approach, where the lattice field is trivially related to the continuum field. This is an important issue in QCD where phenomenological studies involve a wide range of currents and operators. It also means that in practice the "perfect action" used for an interacting theory is not really perfect since the arbitrarily complex functions of Ψ_i that arise in real perfect actions must be simplified for simulations. Although these difficulties gradually are being overcome for QCD [10], [11], I will not discuss this interesting approach further in these lectures.

2.5 Summary

Asymptotic freedom implies that short-distance QCD is simple (perturbative) while long-distance QCD is difficult (nonperturbative). The lattice separates short from long distances, allowing us to exploit this dichotomy to create highly efficient algorithms for solving the entire theory: $k>\pi/a$ QCD is included via corrections $\delta\mathcal{L}$ to the lattice lagrangian that are computed using perturbation theory; $k<\pi/a$ QCD is handled nonperturbatively using Monte Carlo integration. Thus, while we wish to make the lattice spacing a as large as possible, we are constrained by two requirements. First a must be sufficiently small that our finite-difference approximations for derivatives in the lagrangian and field equations are sufficiently accurate. Second a must be sufficiently small that π/a is a perturbative momentum. Numerical experiments indicate that both constraints can be satisfied when $a\approx 1/2\,\mathrm{fm}$ or smaller, provided all lattice operators are tadpole improved.

It is important to remember that most any perturbative analysis in QCD is contaminated by nonperturbative effects at some level. This will certainly be the case for the couplings in $\delta\mathcal{L}$. However we now have extensive experience showing that such nonperturbative effects are rarely significant for physical quantities at the distances relevant to our discussion. Should a situation arise where this is not the case (and one surely will, some day) we may have to supplement the perturbative approach outlined in this section with nonperturbative techniques. Such situations will not pose a problem if they are relatively rare, as now seems likely. Furthermore, as we discuss later, simple nonperturbative techniques already exist, should they be needed, for tuning the leading correction terms in both the gluon and quark actions.

3 Gluon Dynamics on Coarse Lattices

In this section I discuss first the construction of accurate discretizations of the classical theory of gluon dynamics. I then discuss the changes needed to make a quantum theory, and illustrate the discussion with Monte Carlo results. Finally I discuss actions for anisotropic lattices, and the tuning and testing of gluon actions.

3.1 Classical Gluons

The continuum action for QCD is

$$S = \int d^4x \, \tfrac{1}{2} \sum_{\mu,\nu} \mathrm{Tr}\, F_{\mu\nu}^2(x) \tag{41}$$

where

$$F_{\mu\nu} \equiv \partial_\mu A_\nu - \partial_\nu A_\mu + ig[A_\mu, A_\nu] \tag{42}$$

is the field tensor, a traceless 3×3 hermitian matrix. The defining characteristic of the theory is its invariance with respect to gauge transformations where

$$F_{\mu\nu} \to \Omega(x)\, F_{\mu\nu}\, \Omega(x)^\dagger \tag{43}$$

and $\Omega(x)$ is an arbitrary x-dependent SU_3 matrix.

The standard discretization of this theory seems perverse at first sight. Rather than specifying the gauge field by the values of $A_\mu(x)$ at the sites of the lattice, the field is specified by variables on the links joining the sites. In the classical theory, the "link variable" on the link joining a site at x to one at $x + a\hat{\mu}$ is determined by the line integral of A_μ along the link:

$$U_\mu(x) \equiv \mathrm{P} \exp\left(-i \int_x^{x+a\hat{\mu}} gA \cdot \mathrm{d}y\right) \tag{44}$$

where the P-operator path-orders the A_μ's along the integration path. We use U_μ's in place of A_μ's on the lattice, because it is impossible to formulate a lattice version of QCD directly in terms of A_μ's that has exact gauge invariance. The U_μ's, on the other hand, transform very simply under a gauge transformation:

$$U_\mu(x) \to \Omega(x)\, U_\mu(x)\, \Omega(x + a\hat{\mu})^\dagger . \tag{45}$$

This makes it easy to build a discrete theory with exact gauge invariance.

A link variable $U_\mu(x)$ is represented pictorially by a directed line from x to $x + \hat{\mu}$, where this line is the integration path for the line integral in the exponent of $U_\mu(x)$:

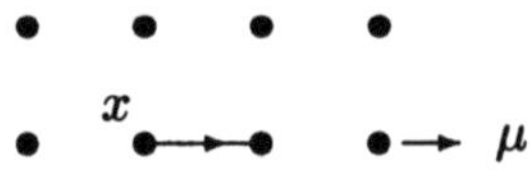

$U_\mu(x)$

In the conjugate matrix $U_\mu^\dagger(x)$ the direction of the line integral is flipped and so we represent $U_\mu^\dagger(x)$ by a line going backwards from $x + \hat\mu$ to x:

$U_\mu^\dagger(x)$

A Wilson loop function,

$$W(C) \equiv \tfrac{1}{3}\,\mathrm{Tr}\,\mathrm{P}\,\mathrm{e}^{-\mathrm{i}\oint_C gA\cdot\mathrm{d}x}, \tag{46}$$

for any closed path C built of links on thc latticc can be computed from the path-ordered product of the U_μ's and $U_\mu^\dagger$'s associated with each link. For example, if C is the loop

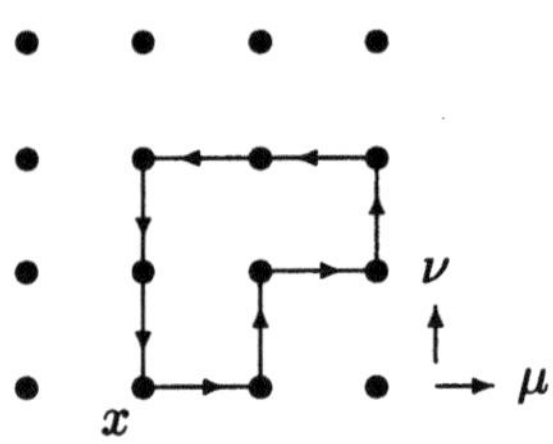

then

$$W(C) = \tfrac{1}{3}\,\mathrm{Tr}\,\big(U_\mu(x)U_\nu(x + a\hat\mu)\ldots U_\nu^\dagger(x)\big)\,. \tag{47}$$

Such quantities are obviously invariant under arbitrary gauge transformations (45).

You might wonder why we go to so much trouble to preserve gauge invariance when we quite willing give up Lorentz invariance, rotation invariance, etc. The reason is quite practical. With gauge invariance, the quark-gluon, three-gluon, and four-gluon couplings in QCD are all equal, and the bare gluon mass is zero. Without gauge invariance, each of these couplings must be tuned independently and a gluon mass introduced if one is to recover QCD. Tuning this many parameters in a numerical simulation is very expensive. This is not much of a problem in the classical theory, where approximate gauge invariance keeps the couplings approximately equal; but it is serious in the quantum theory because quantum fluctuations (loop-effects) renormalize the various couplings differently in the absence of exact gauge invariance. So while it is quite possible to formulate lattice QCD directly in terms of A_μ's, the resulting theory would have only approximate gauge invariance, and thus would be prohibitively expensive to simulate. Symmetries like Lorentz invariance can be given up with little cost because the symmetries of the lattice, though far less restrictive, are still sufficient to prevent the introduction of new interactions with new couplings (at least to lowest order in a).

We must now build a lattice lagrangian from the link operators. We require that the lagrangian be gauge invariant, local, and symmetric with respect to axis interchanges (which is all that is left of Lorentz invariance). The most local nontrivial gauge invariant object one can build from the link operators is the "plaquette operator," which involves the product of link variables around the smallest square at site x in the $\mu\nu$ plane:

$$P_{\mu\nu}(x) \equiv \tfrac{1}{3}\,\mathrm{Re\ Tr}\left(U_\mu(x)U_\nu(x+a\hat{\mu})U_\mu^\dagger(x+a\hat{\mu}+a\hat{\nu})U_\nu^\dagger(x)\right). \tag{48}$$

To see what this object is, consider evaluating the plaquette centered about a point x_0 for a very smooth weak classical A_μ field. In this limit,

$$P_{\mu\nu} \approx 1 \tag{49}$$

since

$$U_\mu \approx \mathrm{e}^{-\mathrm{i}gaA_\mu} \approx 1. \tag{50}$$

Given that A_μ is slowly varying, its value anywhere on the plaquette should be accurately specified by its value and derivatives at x_0. Thus the corrections to (49) should be a polynomial in a with coefficients formed from gauge-invariant combinations of $A_\mu(x_0)$ and its derivatives: that is,

$$\begin{aligned} P_{\mu\nu} \;=\; & 1 - c_1\, a^4\ \mathrm{Tr}\,(gF_{\mu\nu}(x_0))^2 \\ & - c_2\, a^6\ \mathrm{Tr}\left(gF_{\mu\nu}(x_0)(\mathrm{D}_\mu^2+\mathrm{D}_\nu^2)gF_{\mu\nu}(x_0)\right) \\ & + \mathcal{O}(a^8) \end{aligned} \tag{51}$$

where c_1 and c_2 are constants, and D_μ is the gauge-covariant derivative. The leading correction is order a^4 because $F_{\mu\nu}^2$ is the lowest-dimension gauge-invariant combination of derivatives of A_μ, and it has dimension 4. (There are no F^3 terms because $P_{\mu\nu}$ is invariant under $U_\mu \to U_\mu^\dagger$ or, equivalently, $F \to -F$.)

It is straightforward to find the coefficients c_1 and c_2. We need only examine terms in the expansion of $P_{\mu\nu}$ that are quadratic in A_μ; the cubic and quartic parts of $F_{\mu\mu}^2$ then follow automatically, by gauge invariance. Because of the trace, the path ordering is irrelevant to this order. Thus

$$\begin{aligned} P_{\mu\nu} &= \tfrac{1}{3}\,\mathrm{Re\ Tr\ P}\,\mathrm{e}^{-\mathrm{i}\oint_\square gA\cdot\mathrm{d}x} \\ &= \tfrac{1}{3}\,\mathrm{Re\ Tr}\left[1 - i\oint_\square gA\cdot\mathrm{d}x - \tfrac{1}{2}\left(\oint_\square gA\cdot\mathrm{d}x\right)^2 + \mathcal{O}(A^3)\right] \end{aligned} \tag{52}$$

where, by Stoke's Theorem,

$$\begin{aligned}\oint_{\square} A \cdot \mathrm{d}x &= \int_{-a/2}^{a/2} \mathrm{d}x_\mu \mathrm{d}x_\nu \left[\partial_\mu A_\nu(x_0+x) - \partial_\nu A_\mu(x_0+x)\right] \\ &= \int_{-a/2}^{a/2} \mathrm{d}x_\mu \mathrm{d}x_\nu \left[F_{\mu\nu}(x_0) + (x_\mu \mathrm{D}_\mu + x_\nu \mathrm{D}_\mu) F_{\mu\nu}(x_0) + \cdots\right] \\ &= a^2\, F_{\mu\nu}(x_0) + \frac{a^4}{24}\,(\mathrm{D}_\mu^2 + \mathrm{D}_\nu^2) F_{\mu\nu}(x_0) + \mathcal{O}(a^6, A^2). \end{aligned} \tag{53}$$

Thus $c_1 = 1/6$ and $c_2 = 1/72$ in (51).

The expansion in (51) is the classical analogue of an operator product expansion. Using this expansion, we find that the traditional "Wilson action" for gluons on a lattice,

$$S_{\mathrm{Wil}} = \beta \sum_{x,\mu>\nu} (1 - P_{\mu\nu}(x)) \tag{54}$$

where $\beta = 6/g^2$, has the correct limit for small lattice spacing up to corrections of order a^2:

$$S_{\mathrm{Wil}} = \int d^4x \sum_{\mu,\nu} \left\{ \tfrac{1}{2} \,\mathrm{Tr}\, F_{\mu\nu}^2 + \frac{a^2}{24}\, \mathrm{Tr}\, F_{\mu\nu} (\mathrm{D}_\mu^2 + \mathrm{D}_\nu^2) F_{\mu\nu} + \cdots \right\}. \tag{55}$$

We can cancel the a^2 error in the Wilson action by adding other Wilson loops. For example, the $2a \times a$ "rectangle operator"

$$R_{\mu\nu} = \tfrac{1}{3}\, \mathrm{Re}\, \mathrm{Tr} \tag{56}$$

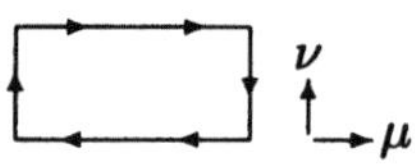

has expansion

$$R_{\mu\nu} = 1 - \frac{4}{6}\, a^4\, \mathrm{Tr}\, (gF_{\mu\nu})^2 - \frac{4}{72}\, a^6\, \mathrm{Tr}\, \left(gF_{\mu\nu}(4\,\mathrm{D}_\mu^2 + \mathrm{D}_\nu^2) gF_{\mu\nu}\right) - \cdots. \tag{57}$$

The mix of a^4 terms and a^6 terms in the rectangle is different from that in the plaquette. Therefore we can combine the two operators to obtain an improved classical lattice action that is accurate up to $\mathcal{O}(a^4)$ [12], [13]:

$$S_{\mathrm{classical}} \equiv -\beta \sum_{x,\mu>\nu} \left\{ \frac{5P_{\mu\nu}}{3} - \frac{R_{\mu\nu} + R_{\nu\mu}}{12} \right\} + \mathrm{const} \tag{58}$$

$$= \int d^4x \sum_{\mu,\nu} \tfrac{1}{2}\, \mathrm{Tr}\, F_{\mu\nu}^2 + \mathcal{O}(a^4). \tag{59}$$

This process is the analogue of improving the derivatives in discretizations of non-gauge theories.[5]

[5] An important step that I have not discussed is to show that the gluon action is positive for any configuration of link variables. This guarantees that the classical ground state of the lattice action corresponds to $F_{\mu\nu} = 0$. See [13] for a detailed discussion.

Exercise: The euclidean Green's function or propagator G for a nonrelativistic quark in a background gauge field A_μ satisfies the equation

$$\left(\mathrm{D}_t - \frac{\mathbf{D}^2}{2M}\right) G(x) = \delta^4(x) \tag{60}$$

where $\mathrm{D}_\mu = \partial_\mu - igA_\mu(x)$ is the gauge-covariant derivative. Show that the static ($M = \infty$) quark propagator is

$$G_\infty(\mathbf{x},t) = \left[\mathrm{P}\, \mathrm{e}^{-\mathrm{i}\int_0^t gA_0(\mathbf{x},t)\mathrm{d}t}\right]^\dagger \delta^3(\mathbf{x}) \tag{61}$$

which on the lattice becomes

$$G_\infty(\mathbf{x},t) = U_t^\dagger(\mathbf{x},t-a)\, U_t^\dagger(\mathbf{x},t-2a)\, \ldots\, U_t^\dagger(\mathbf{x},0). \tag{62}$$

Propagation of a static antiquark is described by $G_\infty^\dagger$ and therefore the "static potential" $V(r)$, which is the energy of a static quark and antiquark a distance r apart, is obtained from

$W(r,t) \equiv \langle 0|\, \frac{1}{3}\,\mathrm{Tr}$ 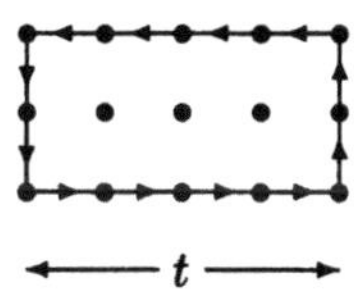

$|0\rangle$ 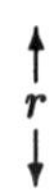

(63)

where for large t

$$W(r,t) \to \mathrm{const}\; \mathrm{e}^{-V(r)\,t}. \tag{64}$$

Exercise: Defining the "twisted-rectangle operator"

$T_{\mu\nu} = \frac{1}{3}\,\mathrm{Re}\,\mathrm{Tr}$ 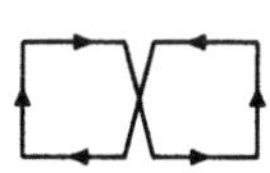 . (65)

show that

$$S_{\text{classical}} \equiv -\beta \sum_{x,\mu>\nu} \left\{P_{\mu\nu} + \frac{T_{\mu\nu} + T_{\nu\mu}}{12}\right\} + \text{const} \tag{66}$$

$$= \int d^4x \sum_{\mu,\nu} \tfrac{1}{2}\,\mathrm{Tr}\, F_{\mu\nu}^2 + \mathcal{O}(a^4). \tag{67}$$

This is an alternative to the improved gluon action derived in the previous exercise.

3.2 Quantum Gluons[6]

In the previous section we derived improved classical actions for gluons that are accurate through order a^4. We now turn these into quantum actions. The most important step is to tadpole improve the action by dividing each link operator U_μ by the mean link u_0: for example, the action built of plaquette and rectangle operators becomes

$$S = -\beta \sum_{x,\mu>\nu} \left\{ \frac{5}{3} \frac{P_{\mu\nu}}{u_0^4} - \frac{R_{\mu\nu} + R_{\nu\mu}}{12\, u_0^6} \right\}. \tag{68}$$

The u_0's cancel lattice tadpole contributions that otherwise would spoil weak-coupling perturbation theory in the lattice theory and undermine our procedure for improving the lattice discretization. Note that $u_0 \approx 3/4$ when $a = .4\,\mathrm{fm}$, and therefore the relative importance of the $R_{\mu\nu}$'s is larger by a factor of $1/u_0^2 \approx 2$ than without tadpole improvement. Without tadpole improvement, we cancel only half of the a^2 error.

The mean link u_0 is computed numerically by guessing a value for use in the action, measuring the mean link in a simulation, and then readjusting the value used in the action accordingly. This tuning cycle converges rapidly to selfconsistent values, and can be done very quickly using small lattice volumes. The u_0's depend only on lattice spacing, and become equal to one as the lattice spacing vanishes.

The expectation value of the link operator is gauge dependent. Thus to minimize gauge artifacts, u_0 is commonly defined as the Landau-gauge expectation value, $\langle 0|\frac{1}{3}\,\mathrm{Tr}\, U_\mu|0\rangle_{\mathrm{LG}}$. Landau gauge is the axis-symmetric gauge that maximizes u_0, thereby minimizing the tadpole contribution; any tadpole contribution that is left in Landau gauge cannot be a gauge artifact. An alternative procedure is to define u_0 as the fourth root of the plaquette expectation value,

$$u_0 = \langle 0|P_{\mu\nu}|0\rangle^{1/4}. \tag{69}$$

This definition gives almost identical results and is more convenient for numerical work since gauge fixing is unnecessary.

Tadpole improvement is the first step in a systematic procedure for improving the action. The next step is to add in renormalizations due to contributions from $k > \pi/a$ physics not already included in the tadpole improvement. These renormalizations induce $a^2\,\alpha_s(\pi/a)$ corrections,

$$\begin{aligned} \delta\mathcal{L} &= \alpha_s\, r_1\, a^2 \sum_{\mu,\nu} \mathrm{Tr}(F_{\mu\nu} \mathrm{D}_\mu^2 F_{\mu\nu}) \\ &+ \alpha_s\, r_2\, a^2 \sum_{\mu,\nu} \mathrm{Tr}(\mathrm{D}_\mu F_{\nu\sigma} \mathrm{D}_\mu F_{\nu\sigma}) \end{aligned}$$

[6] This section and the next are based on work with M. Alford, W. Dimm, G. Hockney and P. Mackenzie that is described in [14].

$$+ \alpha_s r_3 a^2 \sum_{\mu,\nu} \mathrm{Tr}(\mathrm{D}_\mu F_{\mu\sigma} \mathrm{D}_\nu F_{\nu\sigma})$$
$$+ \cdots, \tag{70}$$

that must be removed. The last term is harmless; its coefficient can be set to zero by a change of field variable (in the path integral) of the form

$$A_\mu \to A_\mu + a^2 \alpha_s f(\alpha_s) \sum_\nu \mathrm{D}_\nu F_{\nu\mu}. \tag{71}$$

Since changing integration variables does not change the value of an integral, such field transformations must leave the physics unchanged.[7] Operators that can be removed by a field transformation are called "redundant." The other corrections are removed by renormalizing the coefficient of the rectangle operator $R_{\mu\nu}$ in the action, and by adding an additional operator. One choice for the extra operator is

$$C_{\mu\nu\sigma} \equiv \tfrac{1}{3} \,\mathrm{Re\,Tr}$$ 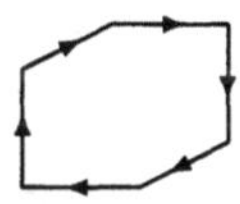 (72)

Then the action, correct up to $\mathcal{O}(a^2\alpha_s^2, a^4)$, is [15]

$$S = -\beta \sum_{x,\mu>\nu} \left\{ \frac{5}{3} \frac{P_{\mu\nu}}{u_0^4} - r_g \frac{R_{\mu\nu} + R_{\nu\mu}}{12\, u_0^6} \right\} + c_g \beta \sum_{x,\mu>\nu>\sigma} \frac{C_{\mu\nu\sigma}}{u_0^6}, \tag{73}$$

where

$$r_g = 1 + .48\, \alpha_s(\pi/a) \tag{74}$$
$$c_g = .055\, \alpha_s(\pi/a). \tag{75}$$

The coefficients r_g and c_g are computed by "matching" physical quantities, like low-energy scattering amplitudes, computed using perturbation theory in the lattice theory with the analogous quantity in the continuum theory. The lattice result depends upon r_g and c_g; these parameters are tuned until the lattice amplitude agrees with the continuum amplitude to the order in a and α_s required:

$$T_{\mathrm{lat}}(r_g, c_g) = T_{\mathrm{contin}}. \tag{76}$$

Note that tadpole improvement has a big effect on these coefficients. Without tadpole improvement, $r_g = 1 + 2\alpha_s$; that is, the coefficient of the radiative correction is four times larger. Tadpole improvement automatically supplies 75% of the one-loop contribution needed without improvement. Since $\alpha_s \approx 0.3$, the unimproved expansion for r_g is not particularly convergent. However, with tadpole improvement, the one-loop correction is only about 10–20%

[7] One must, of course, include the jacobian for the transformation in the transformed path integral. This contributes only in one-loop order and higher; it has no effect on tree-level calculations.

of r_g. Indeed, for most current applications, one-loop corrections to tadpole-improved actions are negligible.

Finally note that the twisted-rectangle action (66) must also be tadpole improved:

$$S_{\rm trt} \equiv -\beta \sum_{x,\mu>\nu} \left\{ \frac{P_{\mu\nu}}{u_0^4} + \frac{T_{\mu\nu}+T_{\nu\mu}}{12\,u_0^8} \right\}. \tag{77}$$

The correction term in this action has two more u_0's than that in our other action. This makes $S_{\rm trt}$ much more dependent upon tadpole improvement. Comparing results generated by our two actions provides a sensitive test of tadpole improvement. The α_s corrections to this action have not yet been computed, but are probably unnecessary.

Exercise: Show that the gauge that maximizes $\langle 0|\frac{1}{3}\,{\rm Tr}\,U_\mu|0\rangle$ becomes Landau gauge ($\partial \cdot A = 0$) in the $a \to 0$ limit.

Exercise: Tadpole diagrams are only important because they are ultraviolet divergent; the loop integrations generate factors of $1/a^2$ that cancel the explicit a's from the link operator U_μ. Consequently such contributions can be significantly reduced if the effective ultraviolet cutoff for the gluons is lowered. Lowering the cutoff by a factor of two, for example, would reduce tadpole contributions by a factor of four, rendering them harmless. This reduction is easily accomplished by smearing, where, for example,

$$A_\mu(x) \to \left(1 - \frac{\epsilon\, a^4 {\rm D}^4}{n}\right)^n A_\mu(x). \tag{78}$$

The low-momentum components of the smeared A_μ field are identical, up to $\mathcal{O}(a^4)$ corrections, with the unsmeared field; but at high momentum the smearing factor acts as an ultraviolet cutoff

$$\left(1 - \frac{\epsilon\, a^4\, {\rm D}^4}{n}\right)^n \to \left(1 - \frac{\epsilon\, a^4\, p^4}{n}\right)^n \approx {\rm e}^{-\epsilon p^4 a^4}. \tag{79}$$

Tadpole improvement should be unnecessary when the smeared field is used in quark actions, etc., in place of the original field. Parameter ϵ is chosen to give the appropriate cutoff, while n is chosen sufficiently large that smearing factor approximately exponentiates as indicated. Design and test an implementation of this smearing scheme for the link variables U_μ.

Exercise: Lattice ϕ^4 theory has a^2 corrections of the form

$$\delta\mathcal{L} = c_1\, a^2 \sum_\mu \left(\partial_\mu^2 \phi\right)^2 + c_2\, a^2\, \phi^6 + c_3\, a^2\, \left(\partial^2\phi\right)^2 + c_4\, a^2\, \phi^3 \partial^2 \phi. \tag{80}$$

Find a field transformation

$$\phi \to \tilde{\phi} = \phi + a^2\, f(\phi, \partial_\mu^2 \phi \ldots) \tag{81}$$

that removes the c_3 and c_4 terms (that is, the redundant operators). Note that $\phi \to \phi + \delta\phi$ implies

$$S[\phi] \to S + \delta\phi\, \frac{\delta S}{\delta \phi}. \tag{82}$$

Thus the leading redundant operators can be rearranged so that they are proportional to $\delta S/\delta\phi$. Since the classical field equation is $\delta S/\delta\phi = 0$, redundant operators are often said to have no effect "because of the equations of motion." More precisely it is because they can be removed by a change of integration variable in the path integral.

Show that the jacobian $|\delta\phi/\delta\tilde{\phi}|$ has no effect at tree-level, but must be included when quantum loop corrections are important. (Hint: Try computing a simple scattering amplitude using perturbation theory in the original and in the transformed theory.)

3.3 Tests of the Gluon Action

Having a procedure for systematically improving the gluon action, we need to establish whether the improvements really do improve simulations.

The effects of the $\mathcal{O}(a^2)$ improvements in the gluon action are immediately apparent if one compares the static-quark potential computed, using (64), with the Wilson action (54) and with the improved action (73) [14]. Monte Carlo simulation results for these potentials are plotted in Fig. 4, together with the continuum potential. The Wilson action has errors as large as 40%. These are reduced to only 4–5% when the improved action is used instead.

The dominant errors in the uncorrected simulation of $V(r)$ reflect a failure of rotational invariance, which is expected since the a^2 error in the Wilson action (54) is neither Lorentz nor rotationally invariant. The points at $r = a, 2a, 3a\ldots$ are for separations that are parallel to an axis of the lattice, while the points at $r = \sqrt{2}a, \sqrt{3}a\ldots$ are for diagonal separations between the static quark and antiquark. In Table 1, I tabulate the error ΔV in the potential at $r = \sqrt{3}a$ for a variety gluon actions, with and without tadpole improvement and one-loop radiative corrections. As expected, the correction term in the action is significantly underestimated without tadpole improvement. The tadpole-improved action is very accurate both with and without one-loop corrections, suggesting that $\mathcal{O}(a^2\alpha_s)$ corrections are comparable to those of $\mathcal{O}(a^4)$ and therefore unimportant for most current applications.

I also include in Table 1 results obtained using the twisted-rectangle action $S_{\rm trt}$ in (77). This action gives a potential that is essentially identical to that obtained with the other improved action. The a^4 errors are very different in the two actions; their agreement suggests that a^4 errors are small. It also is striking confirmation of the importance of tadpole improvement since the u_0's more than double the size of the correction term in the twisted-rectangle action when $a = 0.4\,$fm.

Another test of our improved action is to compute physical quantities on lattices with several different lattice spacings, looking for evidence of finite-a errors. These are consistently small. For example, the temperature at which gluonic QCD has a deconfining phase transition can be computed using the

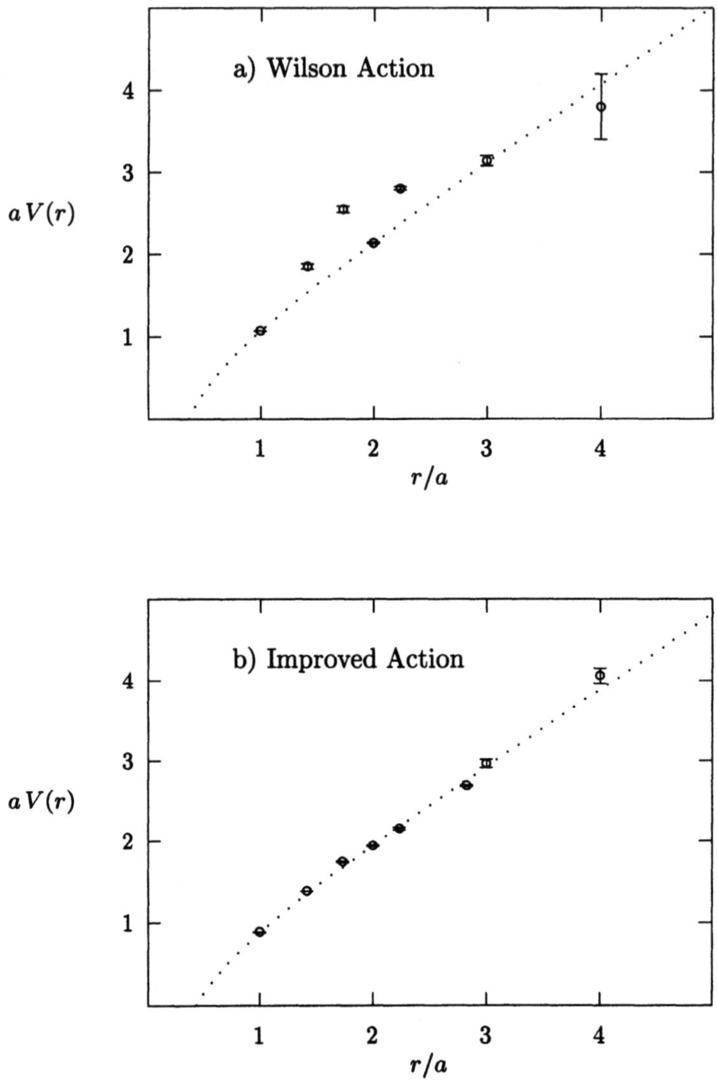

Fig. 4. Static-quark potential computed on 6^4 lattices with $a \approx 0.4\,\mathrm{fm}$ using the Wilson action and the improved action. The dotted line is the standard infrared parameterization for the continuum potential, $V(r) = Kr - \pi/12r + c$, adjusted to fit the on-axis values of the potential.

Table 1. Error in the static quark potential at $V(\sqrt{3}a)$ for a variety of gluon actions. The lattice spacing in each case is $a \approx .4\,\mathrm{fm}$; K is the slope of the linear part of the static potential.

Action	$\Delta V(\sqrt{3}a)/K\sqrt{3}a$
unimproved (Wilson)	.41 (2)
tree-level improved, no tadpole improvement	.15 (1)
one-loop improved, no tadpole improvement	.12 (2)
tree-level improved, with tadpole improvement	.05 (1)
one-loop improved, with tadpole improvement	.04 (1)
twisted-rectangle correction, with tadpole improvement	.04 (2)

improved action. One finds [16]:

$$T_c = \begin{cases} 298\,(6)\,\mathrm{MeV} & \text{for } a = .32\,\mathrm{fm} \\ 309(6)\,(6)\,\mathrm{MeV} & \text{for } a = .21\,\mathrm{fm.} \\ 303\,(8)\,\mathrm{MeV} & \text{for } a = .16\,\mathrm{fm.} \end{cases} \tag{83}$$

The results all agree to within statistical errors of a few percent.

3.4 Anisotropic Lattices and Tuned Gluon Actions[8]

As a final illustration of improved gluon actions with tadpole-improved operators, I now show results of simulations employing anisotropic lattices with temporal lattice spacings a_t that are smaller than the spatial lattice spacings a_s. Such anisotropic lattices lead to greatly improved signal-to-noise in Monte Carlo simulations of hadrons; they are also useful for designing improved quark actions, as I discuss later. On an anisotropic lattice, our rectangle-improved action becomes

$$\begin{aligned} S = -\beta \sum_{x,\,s>s'} \frac{a_t}{a_s} \left\{ \frac{5}{3} \frac{P_{ss'}}{u_s^4} - \frac{1}{12} \frac{R_{ss'}}{u_s^6} - \frac{1}{12} \frac{R_{s's}}{u_s^6} \right\} \\ -\beta \sum_{x,\,s} \frac{a_s}{a_t} \left\{ \frac{5}{3} \frac{P_{st}}{u_s^2 u_t^2} - \frac{1}{12} \frac{R_{st}}{u_s^4 u_t^2} - \frac{1}{12} \frac{R_{ts}}{u_t^4 u_s^2} \right\}, \end{aligned} \tag{84}$$

where s and s' are spatial directions. The mean links u_t and u_s are different on anisotropic lattices. When $a_t \leq a_s/2$, u_t is very close to its continuum

[8] This section is based on work with M. Alford, T. Klassen, C. Morningstar, M. Peardon and H. Trottier.

value of one. Thus we take

$$u_t = 1 \tag{85}$$

$$u_s = \langle P_{ss'} \rangle^{1/4}. \tag{86}$$

The same u_s, to within a few percent, can be obtained from the link expectation value in Coulomb gauge, which is the natural gauge choice for anisotropic lattices with small a_t's.

The anisotropic simulation can be tested by comparing the static quark potential computed from Wilson loops in different orientations: loops in the x–t plane, loops in the x–y plane, and loops in the t–x plane (with x playing the role of time). If the anisotropic theory is tadpole-improved all three potentials agree (Fig. 5). Without tadpole-improvement the slopes disagree by 20% (Fig. 6).

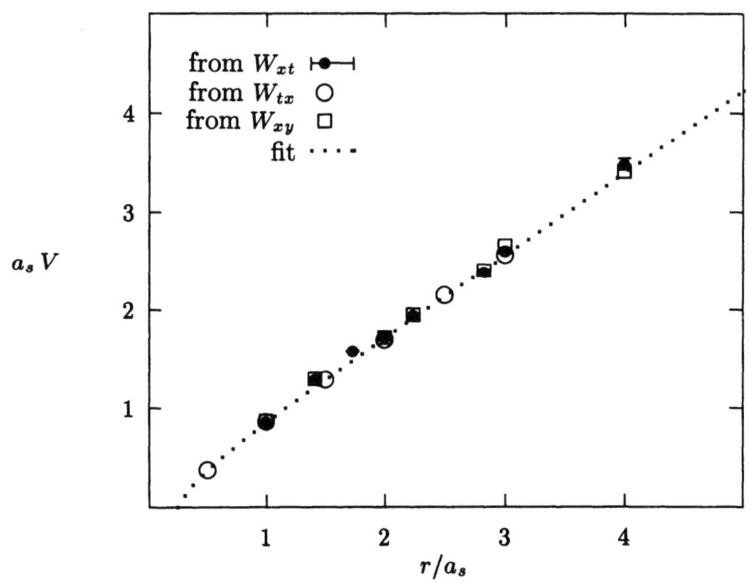

Fig. 5. The static-quark potential computed on an anisotropic lattice in different orientations. Results are shown for a_t/a_s equal 1/2. Except as shown, analysis errors are smaller than the plot symbols. The fit in each plot is to the open circles; the fitting function is $V(r) = Kr - b/r + c$.

The coefficients of the correction terms in the gluon action for anisotropic lattices are known only to leading order in α_s. While it is clear from the simulations that tadpole improvement captures the bulk of the renormalizations,

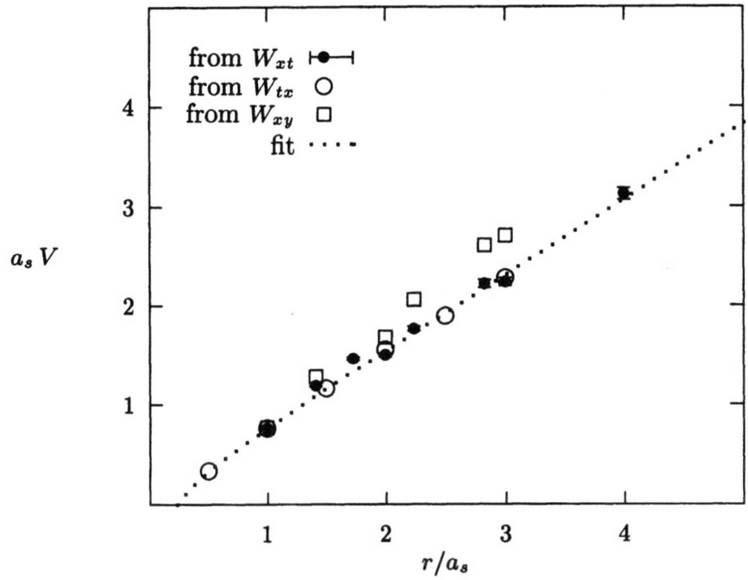

Fig. 6. The static-quark potential computed on an anisotropic lattice in different orientations, as in the previous figure. The action used here was not tadpole improved. The input value for a_t/a_s is 1/2.

these same simulations provide a handle on further radiative corrections. This is because the leading corrections in the gluon action affect the extent to which continuum symmetries, like rotation invariance, are restored. We can use symmetry restoration as a criterion for tuning the couplings beyond their tadpole-improved tree-level values. Thus, for example, the static potential at $\mathbf{r} = (a_s, 2a_s, 2a_s)$ should agree exactly with that at $\mathbf{r} = (3a_s, 0, 0)$. One might tune the value of the spatial mean link u_s used in the action until this is the case. Similarly one would like the anisotropy a_t/a_s built into the action to agree with the renormalized anisotropy coming out of the simulations. The ratio u_t/u_s can easily be tuned to make the two agree exactly. Such tuning does not capture all of the $\mathcal{O}(\alpha_s)$ corrections except in the simplest of situations. However it may be a useful intermediate step between tree-level tadpole improvement and a full perturbative analysis of the radiative corrections.

Exercise: One defect of our improved gluon action is that the gluon spectrum has ghosts. These are caused by the R_{ts} terms in the action, which extend two steps in the time direction. (Numerical ghosts, like these, are discussed in greater detail in the section on light quarks.) These gluon ghosts, being at high energies (of order $2.6/a_t$), should have little effect on simulation results, particularly for smaller a_t's; this is confirmed by simulations. Nevertheless one

might wish to remove them completely. This is easily done, when a_t is small compared with a_s, by dropping the R_{ts} terms in the action. This introduces new errors of order a_t^2 but these will be negligible relative to the a_s^4 errors provided a_t is small enough. Modify the st terms in the action so that all a_s^2 errors are still canceled in the absence of the R_{ts} contribution.

3.5 Summary

In this section I have outlined how one designs actions for simulating gluon dynamics on coarse lattices. The simulation results show that accurate simulations are possible even on lattices as coarse as $a = .4$ fm using very simple actions. Tadpole improvement is essential, but one-loop radiative corrections to the action are not too important. Rotational invariance of the potential and, for anisotropic lattices, space-time interchange symmetry provide sensitive tests of improved gluon actions, and can be used to tune the leading correction terms.

The coarse lattices make simulations very fast. Most of the simulation results shown here were obtained using an IBM RS6000/250 desktop workstation, which is powered by a personal-computer CPU (66MHz PowerPC). Any of these results could be generated easily on a high-end personal computer.

4 Light Quarks on Coarse Lattices[9]

In this section I develop techniques for designing very accurate discretizations of the Dirac equation for quarks in a gluonic field. I begin by reviewing various resolutions of the "doubling problem" which plagues even the simplest discretization. I then discuss two currently popular discretizations, and finally new discretizations that are significantly more accurate.

4.1 Naive Discretization and Doubling

The euclidean Dirac lagrangian in the continuum is

$$\mathcal{L} = \overline{\psi}\,(\mathrm{D}\cdot\gamma + m)\,\psi \tag{87}$$

where

$$\{\gamma_\mu, \gamma_\nu\} = 2\,\delta_{\mu\nu} \tag{88}$$

and I take

$$\gamma_t = \begin{pmatrix} 1 & 0 \\ 0 & -1 \end{pmatrix} \qquad \gamma_s = \begin{pmatrix} 0 & \sigma_s \\ \sigma_s & 0 \end{pmatrix} \qquad \gamma_5 \equiv \gamma_t\gamma_x\gamma_y\gamma_z = i\begin{pmatrix} 0 & 1 \\ -1 & 0 \end{pmatrix} \tag{89}$$

[9] This section is based on work with M. Alford and T. Klassen.

where, again, s and t as subscripts signify spatial and temporal indices respectively. The obvious discretization is

$$\mathcal{L}_{\rm lat} = \overline{\psi}\left(\Delta\cdot\gamma + m + \mathcal{O}(a^2)\right)\psi \tag{90}$$

where now Δ_μ is a (tadpole-improved) gauge-covariant derivative:

$$\Delta_\mu\psi(x) \equiv \frac{U_\mu(x)\,\psi(x+a\hat{\mu}) - U_\mu^\dagger(x-a\hat{\mu})\,\psi(x-a\hat{\mu})}{2\,a\,u_0}. \tag{91}$$

The other finite-difference operators, $\Delta_\mu^{(2)}$, $\Delta_\mu^{(3)}\ldots$, can all be made gauge-covariant by adding U_μ's as in Δ_μ; in what follows I will always use these symbols to refer to the gauge-covariant differences.

Unfortunately this simple lattice Dirac action has a very serious pathology. To see this consider the $\mathbf{p}=0$ spin-up solution of the corresponding Dirac equation without gluon fields:

$$\psi_\uparrow(t) = \mathrm{e}^{-Et}\begin{bmatrix}1\\0\\0\\0\end{bmatrix} \tag{92}$$

(There is no i in the exponent because t is euclidean; E is the minkowski energy.) Substituting into the Dirac equation,

$$(\Delta\cdot\gamma + m)\;\psi_\uparrow(t) = 0, \tag{93}$$

we obtain

$$(\Delta_t + m)\,\mathrm{e}^{-Et} = 0 \qquad\Longrightarrow\qquad \frac{\mathrm{e}^{-aE} - \mathrm{e}^{aE}}{2\,a} + m = 0. \tag{94}$$

This equation is quadratic in $z\equiv\exp(aE)$ and therefore it has two solutions for E. We only want one! The two solutions are

$$E = \begin{cases} m + \mathcal{O}(a^2) & \text{(normal)} \\ -m + \mathrm{i}\pi/a + \mathcal{O}(a^2) & \text{(doubler)} \end{cases}. \tag{95}$$

The first is the solution we want; the second is known as a "doubler."

The possibility of a second solution for the energy is obvious from inspection of the lattice Dirac equation. The problem lies in the $\Delta_t\psi$ term. This links ψ's two lattice spacings apart which means that the dispersion relation is second order in $z\equiv\exp(aE)$. In general one expects n roots if the ψ's are spread over n lattice spacings — a potential problem when we go to improve the discretization. The doubler cannot be ignored. It is related to the normal solution by a symmetry of the gluon-free lattice theory: if $\psi(\mathbf{x},t)$ is a solution of the lattice Dirac equation then so is

$$\tilde{\psi}(\mathbf{x},t) \equiv i\gamma_5\gamma_t\,(-1)^{t/a}\,\psi(\mathbf{x},t). \tag{96}$$

This symmetry operation turns the normal solution into the doubler solution. Consequently the doubler acts as an extra flavor of quark, which is degenerate with the normal quark (when there are no gauge fields). Furthermore, in euclidean space there is nothing unique about the time direction. Thus there is an analogous doubling symmetry for each direction, and successive transformations involving two or more directions also are symmetries. This implies that there are fifteen doublers in all.

The root of the doubler problem lies in the finite-difference approximation for first-order derivatives. Fourier transforming ∂ and Δ we obtain

$$\partial \to \mathrm{i}\, p \tag{97}$$

$$\Delta \to \mathrm{i}\, \sin(pa)/a. \tag{98}$$

These agree well for low momenta, but, as p approaches the ultraviolet cutoff π/a, the transform of Δ vanishes rather than becoming large. As a consequence, for example, high-momentum states can have very low or even vanishing energies. This is disastrous. Note that the same problem does not arise for second-order derivatives:

$$\partial^2 \to -p^2 \tag{99}$$

$$\Delta^{(2)} \to -\left(\frac{2}{a}\sin(pa/2)\right)^2 . \tag{100}$$

Although the transform of $\Delta^{(2)}$ becomes quite different from p^2 at large p, it nevertheless remains large, going to its maximum of $4/a^2$ at the ultraviolet cutoff. Second-order (and higher even-order) derivatives do not cause doubling problems.

I will discuss two approaches to the doubling problem. One involves adding operators to the quark action that are almost redundant but remove the doublers or drive them to high energies. The other involves "staggering" the quark degrees of freedom on the lattice.

Exercise: Show that gauge transformation

$$\psi(x) \to \Omega(x)\,\psi(x) \qquad U_\mu(x) \to \Omega(x)\,U_\mu(x)\,\Omega(x+a\hat{\mu})^\dagger. \tag{101}$$

implies that

$$\Delta_\mu\psi(x) \to \Omega(x)\,\Delta_\mu\psi(x). \tag{102}$$

Also show that (setting $u_0 = 1$)

$$\Delta_\mu = \frac{\mathrm{e}^{a\mathrm{D}_\mu} - \mathrm{e}^{-a\mathrm{D}_\mu}}{2\,a}. \tag{103}$$

(Hint: Why can't there be $F_{\mu\nu}$'s on the right-hand side?) This relation is crucial to what follows; find a different definition of the covariant difference operator Δ for which this relation is incorrect in $\mathcal{O}(a)$.

4.2 Field Transformations to Remove Doublers

The first approach to removing the doublers is to introduce into the action an operator that is redundant up to $\mathcal{O}(a^2)$ and has higher order non-redundant parts that drive the doubler to high energies. A redundant operator remember, is one that can be eliminated from the action by a change of variables on the quark fields in the path integral, and hence has no effect on any physical observables. For example, consider the field transformation from the continuum field ψ_c to a new field ψ where

$$\begin{aligned}\psi_c &= \left[1 - \frac{r\,a}{4}(\Delta\cdot\gamma - m_c)\right]\psi \\ &\equiv \Omega\psi,\end{aligned} \tag{104}$$

and

$$\overline{\psi}_c = \overline{\psi}\,\Omega \tag{105}$$

With this transformation our naive lattice action becomes

$$\begin{aligned}\overline{\psi}_c\,(\Delta\cdot\gamma + m_c)\,\psi_c &= \overline{\psi}\,\Omega\;(\Delta\cdot\gamma + m_c)\;\Omega\,\psi &(106)\\ &= \overline{\psi}\left[\Delta\cdot\gamma + m - \frac{r\,a}{2}\left(\Delta\cdot\Delta + \frac{\sigma\cdot gF}{2}\right)\right]\psi + \mathcal{O}(a^2) &(107)\end{aligned}$$

where $m \equiv m_c + ram_c^2/2$. In deriving the final formula we used the lattice version of the identity

$$(\mathrm{D}\cdot\gamma)^2 = \mathrm{D}\cdot\mathrm{D} + \frac{\sigma\cdot gF}{2} \tag{108}$$

where

$$\sigma_{\mu\nu} \equiv -\frac{i}{2}\left[\gamma_\mu, \gamma_\nu\right]. \tag{109}$$

The new $\mathcal{O}(a)$ term in the transformed lagrangian does not introduce new $\mathcal{O}(a)$ errors because it results from a field transformation; it is redundant in the sense discussed above for gluon operators. In its present form this term also has no effect on the doubling problem. The trick is to replace the operator $\Delta\cdot\Delta$, which vanishes at the ultraviolet cutoff, with $\Delta^{(2)}$, which is the same up to $\mathcal{O}(a^2)$ errors in the infrared but large at the ultraviolet cutoff. The $\Delta^{(2)}$ pushes the doublers off to high energies, of order $1/a$, where they are harmless. Thus an $\mathcal{O}(a)$-accurate quark action that is doubler free is

$$\begin{aligned}\mathcal{L}_{\mathrm{SW}} &= \overline{\psi}\left[\Delta\cdot\gamma + m - \frac{r\,a}{2}\left(\Delta^{(2)} + \frac{\sigma\cdot gF}{2}\right)\right]\psi &(110)\\ &= \overline{\psi}_c\,(\mathrm{D}\cdot\gamma + m_c)\,\psi_c + \mathcal{O}(a^2), &(111)\end{aligned}$$

where, again,

$$\psi_c = \Omega\,\psi \qquad m = m_c + ram_c^2/2. \tag{112}$$

This is the Sheikholeslami-Wohlert (SW) quark action [17]. It becomes Wilson's original action if the $\sigma \cdot gF$ term is dropped (resulting in $\mathcal{O}(a^1)$ errors).

We can check that the doublers are removed by again substituting the $\mathbf{p}=0$ solution into the corresponding Dirac equation. We obtain

$$\left(\Delta_t + m - \frac{ra}{2}\,\Delta^{(2)}\right) \mathrm{e}^{-Et} = 0$$
$$\Longrightarrow \quad (1-r)\,\mathrm{e}^{-aE} - (1+r)\,\mathrm{e}^{aE} + 2r + 2\,a\,m = 0. \qquad (113)$$

which has roots

$$E = \begin{cases} m_c + \mathcal{O}(a^2) & \text{(normal)} \\ (1/a)\,\ln\left[(r-1)/(r+1)\right] + \cdots & \text{(ghost)} \end{cases}. \qquad (114)$$

The doubler has been replaced by a high-energy "ghost" particle, as desired. The ghost goes away completely for $r=1$.

Exercise: Create a lattice version of $F_{\mu\nu}$ for use in the SW quark action. Build your operator out of tadpole-improved link operators. Try designing a discretization that is accurate up to corrections of $\mathcal{O}(a^4)$.

4.3 a^3-Accurate Lattice Dirac Theory

Our numerical experiments in the first Section suggest that the a^2 errors in the SW quark action might still be quite large when the lattice spacing is of order .4 fm. Thus it is worthwhile developing a quark action that has errors only of order a^4 and higher.

A attempt at such an improved Dirac theory is the action

$$\overline{\psi}\left[\Delta\cdot\gamma - \frac{a^2}{6}\,\Delta^{(3)}\cdot\gamma + m_c\right]\psi \qquad (115)$$

where the a^2 correction cancels the leading errors induced by the $\Delta\cdot\gamma$ term. Not surprisingly this action is plagued with doublers and ghosts. Since the temporal derivatives extend two steps in each direction, the energy dispersion relation has four branches instead of one.

There are several approaches one can take to solving this doubler disaster. The "D234" quark actions rely upon three ingredients [18]:

1. anisotropic lattices, with $a_t \leq a_s/2$, which push ghost states to higher energies, where they decouple. This works because ghost energies are governed by the temporal lattice spacing and grow like $1/a_t$ when a_t is reduced. Reducing the temporal lattice spacing also makes Monte Carlo measurements of hadron propagators much more efficient.
2. redundant operators like
$$-\frac{r\,a_t}{2}\left(\Delta^{(2)} + \frac{\sigma\cdot gF}{2} + \cdots\right) \qquad (116)$$
in the quark action, which can be tuned to remove some of the doublers and ghosts.

3. "irrelevant" operators like

$$s\, a_t^3\, \Delta_t^{(4)} \tag{117}$$

in the quark action, which have negligible effect on low-momentum physics (if $a_t \leq a_2/2$) but are useful at high energies for removing doublers and ghosts. Terms involving higher-order spatial derivatives, like $a_s^4 \Delta_s^{(6)}$, can also be added.

A variety of actions can be created by combining differing amounts of redundant and irrelevant operators. In general one tries to tune these additions to minimize the number and importance of ghost states, to give the best low-momentum behavior, and to minimize pathologies at high momentum.

A simple example of a D234-type action is obtained by generalizing our derivation of the SW action. We begin with a field transformation $\psi_c \to \psi$ where

$$\psi_c = \Omega\, \psi \tag{118}$$

with

$$\Omega \equiv 1 - X - \frac{X^2}{2} - \frac{X^3}{2} \tag{119}$$

$$= \sqrt{1-2X} + \mathcal{O}(X^4) \tag{120}$$

and

$$X \equiv \frac{r\, a_t}{4} \left(\Delta \cdot \gamma - \sum_\mu \frac{a_\mu^2}{6}\, \Delta_\mu^{(3)}\, \gamma_\mu - m_c \right) \tag{121}$$

$$= \frac{r\, a_t}{4} \left(D \cdot \gamma - m_c + \mathcal{O}(a^4) \right). \tag{122}$$

Applying this transformation to the naive improved action (115) and adding an irrelevant a_t^3 operator, we obtain[10]

$$\begin{aligned} \mathcal{L}_{\mathrm{D234x}} = \overline{\psi} \Bigg[& \Delta \cdot \gamma - \sum_\mu \frac{a_\mu^2}{6}\, \Delta_\mu^{(3)}\, \gamma_\mu + m \\ & - \frac{r\, a_t}{2} \left(\Delta^{(2)} - \sum_\mu \frac{a_\mu^2}{12}\, \Delta_\mu^{(4)} + \frac{\sigma \cdot gF}{2} \right) \\ & + s\, a_t^3\, \Delta_t^{(4)} \Bigg] \psi. \end{aligned} \tag{123}$$

[10] The designers of this action, having examined a large number of similar actions, all very effective, are currently grappling with a nomenclature crisis. I refer to the particular action shown here as the "D234x" action, but this is certainly not its final name. Most of what I say about this action is true of almost all other D234 actions, and so "D234x" can usually be taken to refer to a generic D234 action.

This action is the same as the continuum action up to corrections of order a_t^3 and a_s^4. It is easy to show that the choice

$$s = \frac{1}{12} - \frac{r}{24} \tag{124}$$

removes one of the ghost/doubler states, leaving two ghosts that have high energies for any value of r near 1. If in addition $r = 2/3$ is chosen, a second ghost disappears and only one remains (with an energy of order $2/a_t$.) The free-quark dispersion relation for this set of parameters is shown in Fig. 7 for a lattice with $a_s = 2\,a_t = .4\,\mathrm{fm}$; the ghost is roughly 2 GeV above the real quark and so harmless in most applications.

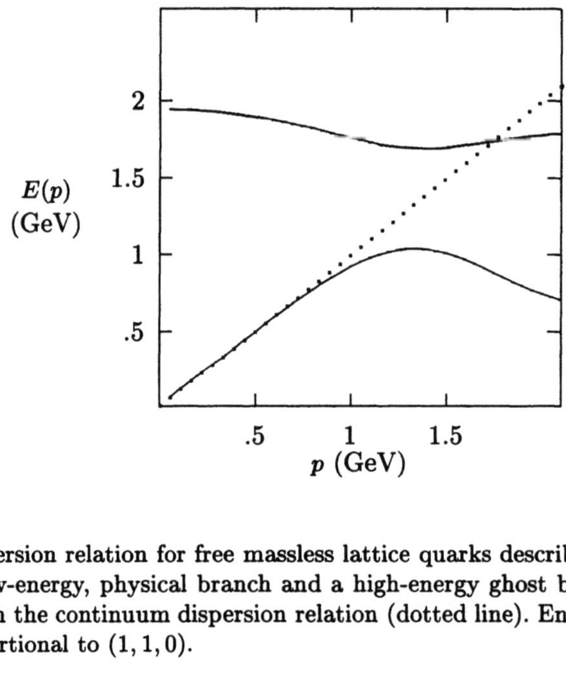

Fig. 7. Dispersion relation for free massless lattice quarks described by the D234x action. A low-energy, physical branch and a high-energy ghost branch are shown, together with the continuum dispersion relation (dotted line). Energies are for momenta proportional to $(1, 1, 0)$.

4.4 Tests of Quark Action

Highly improved actions like the D234x action (123) have a large number of bare coupling constants that must be computed in perturbation theory — one for each term in the action. In the previous section we worked only to lowest order (tree-level) in perturbation theory. However our discussion of the gluon action suggests that this is sufficiently accurate provided all operators are tadpole improved; the renormalizations due to $k > \pi/a$ physics affect physical quantities by only a few percent. How can we ascertain whether the same is true of our quark actions?

One powerful approach is to verify that continuum symmetries are restored by the corrections. For example, the leading error in the discretized Dirac operator, $\Delta \cdot \gamma + m$, breaks Lorentz invariance. The $a_\mu^2 \Delta_\mu^{(3)} \gamma_\mu$ term in the D234x action is meant to cancel this error. We can check that this term has done its job by comparing the dispersion relation of a meson (or baryon) computed in a simulation with the Lorentz-invariant dispersion relation of the continuum. In a Lorentz-invariant theory, a hadron's energy $E_{\rm h}(\mathbf{p})$ and its three momentum $\mathbf{p}$ are such that the combination

$$c^2(\mathbf{p}) \equiv \frac{E_{\rm h}^2(\mathbf{p}) - m_{\rm h}^2}{\mathbf{p}^2} \tag{125}$$

is independent of $\mathbf{p}$ and equal to the square of the speed of light ($= 1$). In Fig. 8 I show D234x simulation results for meson's $c(\mathbf{p})$. The lattice for this simulation is very coarse, with $a_s = 2\,a_t = .4\,\mathrm{fm}$, and yet $c(\mathbf{p})$ is within a few percent of one all the way out to momenta of order $1\,\mathrm{GeV}$ ($\approx 2/a_s$). For comparison I include the same quantity calculated using the SW action (on an isotropic lattice), which has no correction for the a^2 error; it has 25% errors by $p = 1\,\mathrm{GeV}$, and the momentum dependence of the discrepancy is consistent with the quadratic errors expected. This comparison shows that the coefficient of the of the a^2 correction in the D234x action is not renormalized by more than 10–20%, and that this renormalization will have no more than a few percent effect on low-momentum physics. The momenta used for the Monte Carlo points shown in this plot include momenta that are parallel to a lattice axis, and momenta that point along lattice diagonals; the fact that different momenta agree to within a few percent shows that rotation invariance is also restored by the correction term.

The $c(\mathbf{p})$ plot provides a second important check on the D234x action. On anisotropic lattices, the symmetry between space and time is not exact. Consequently the spatial and temporal parts of the leading Dirac operator $\Delta \cdot \gamma$ are renormalized differently:

$$D \cdot \gamma \to \Delta_t \, \gamma_t + c_0 \, \boldsymbol{\Delta} \cdot \boldsymbol{\gamma}, \tag{126}$$

where c_0 is a bare speed-of-light,

$$c_0 = 1 + \mathrm{const}\, \alpha_s + \cdots. \tag{127}$$

Such a renormalization would shift the low-momentum values of $c(\mathbf{p})$ away from one. In fact, with only our tree-level action, $c(\mathbf{p}) = 1.025(20)$ at low momenta; parameter c_0 must be within a percent of one. Note that this success depends crucially on tadpole improvement; without tadpole improvement $c(\mathbf{p}) = 1.2$ at low momenta on our $a_s = .4\,\mathrm{fm}$ lattice.

The only other potentially large renormalization is in the $\mathcal{O}(r\,a_t)$ terms induced by the field transformation. Again there is a symmetry associated

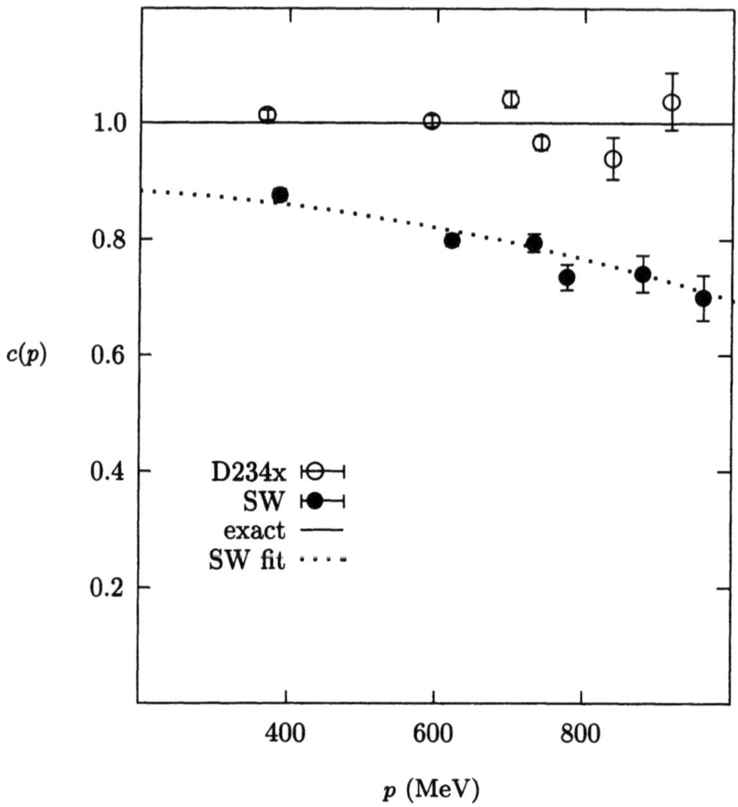

Fig. 8. The speed of light computed using the D234x and SW quark actions from the energy and three momenta of a meson. The fit to the SW data is $\tilde{c}_0 + k\,(ap)^2$ where $\tilde{c}_0$ and k are constants.

with this term: field-redefinition or "r symmetry". The operator multiplied by r is redundant and so should have no effect on the low-momentum physics. Consequently the low-momentum physics should be invariant under changes in the value of r. If the renormalizations of the $\Delta^{(2)}$ and $\sigma \cdot gF$ terms in the action are very different, then the tree-level operator is no longer redundant and results must change as r is varied; on the other hand, if the renormalizations are roughly the same then the net effect of the renormalizations is a harmless shift in r.[11] The results of such an r-test for a close relative of the D234x action are shown in Table 2. These show that the pion, rho, nucleon

[11] I am simplifying the discussion here. On an anisotropic lattice, where space–time interchange is not an exact symmetry, the temporal and spatial parts of $\Delta^{(2)}$ and $\sigma \cdot gF$ have different radiative corrections. Thus the r-test is checking on the relative coefficients of four operators, as opposed to two in the isotropic case.

and delta masses all shift by less than a percent when r is changed from 2/3 to 1. Since omitting the $\sigma \cdot gF$ term shifts masses by 20%, our r-test implies that the relative renormalization given by tree-level tadpole improvement is correct to within roughly 10–20%. A 10% change in the coefficient of $\sigma \cdot gF$ would shift the meson masses by only 2%.

Table 2. Mass differences between two simulations, one with $r=1$ and the other with $r=2/3$, for different quark masses. The bare quark masses in the two simulations were tuned to give the same m_π/m_ρ.

m_π/m_ρ	Δm_π	Δm_ρ	$\Delta m_{\rm N}$	Δm_Δ
.65	0.6 (2) %	-0.13 (13) %	-0.3 (2) %	0.4 (3) %
.57	0.6 (4) %	-0.6 (1.0) %		

In [19] a different test of the $\sigma \cdot gF$ coefficient is proposed. This is based on the fact that the r terms in the action break the quark's chiral symmetry even for massless quarks. If the $\sigma \cdot gF$ coefficient is properly tuned, the chiral symmetry breaking terms are redundant and the chiral symmetry Ward identity should be restored. Using Monte Carlo evaluations of the various terms involved, the authors tune the coefficient until the identity is satisfied. Working only with the SW action and relatively small lattice spacings, they find renormalizations that are larger than those above for D234: the coefficient of $\sigma \cdot gF$ in the SW action is a little more 20% larger than the tadpole-improved tree-level coefficient at $a \approx .1$ fm. Such a shift at such small lattice spacings has only a small effect on hadron masses; but in the same study the authors encountered pathological behavior in the SW action, with this correction, at large lattice spacings. Also the correction to tadpole improvement, though small throughout the range of their study, grows much more rapidly than naively expected as a increases from .03 fm to .1 fm. It would be very interesting to verify these results using the r-test, which provides an independent method of testing the same correction.

Our tests of the D234 action confirm that tree-level improvement of the quark action, using tadpole-improved operators, is quite accurate. However the work I just described concerning the SW action suggests that nonperturbative tuning of parameters in the action may be important for some actions and operators. Such tuning is straightforward for D234: the same tests I described above can be used to tune the leading corrections in the action. In particular, the c_0 parameter, which is potentially the most important, can be easily adjusted so that $c(\mathbf{p}=0) = 1$.

Finally, in Fig. 9, I show simulation results for the rho mass as a function of lattice spacing. I give results for the D234x action [18], which should have a_s^4 errors, for the SW action [20], which should have a^2 errors, and for the

Wilson action [21], which should have a^1 errors. The D234x and SW simulations use an improved gluon action; the Wilson results are for an uncorrected gluon action. Assuming that the a dependence of the errors is as expected, all three actions agree quite well on the continuum limit $(a \to 0)$ mass of the rho, but the more improved the quark action, the sooner it gets there. This figure demonstrates very clearly that we know how to systematically improve quark actions for use on very coarse lattices. Furthermore it is clear that simulations that are accurate to within a few percent are possible on lattices with spatial lattice spacings between .3 and .4 fm.

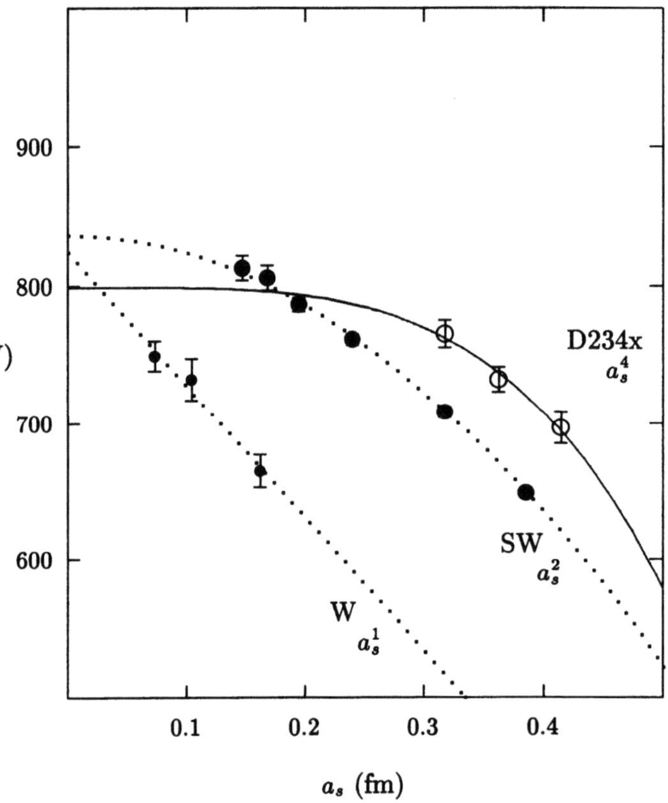

Fig. 9. Rho mass versus lattice spacing computed using the Wilson quark action (W), the Sheikholeslami-Wohlert action (SW) and the D234x action. The charmonium S-P splitting was used to determine a_s. These simulations have no quark vacuum polarization and therefore the rho mass is expected to differ from its experimental value.

4.5 *t*-Staggered Quarks

A second approach to removing doublers is to "stagger" the quark degrees of freedom on the lattice. If we rewrite the Dirac field in terms of two two-component fields, ϕ and χ,

$$\psi = (\gamma_t)^{t/a_t} \begin{bmatrix} \phi + \chi \\ \phi - \chi \end{bmatrix} \tag{128}$$

the naive lattice Dirac equation (90) falls apart into two decoupled two-component equations:

$$\left(m_c + \Delta_t + (-1)^{t/a_t} \boldsymbol{\Delta} \cdot \boldsymbol{\sigma}\right) \phi = 0 \tag{129}$$

$$\left(m_c + \Delta_t + (-1)^{1+t/a_t} \boldsymbol{\Delta} \cdot \boldsymbol{\sigma}\right) \chi = 0 \tag{130}$$

The second equation is identical to the first equation, but shifted one lattice spacing in the t direction; it leads to identical quark physics. One of these equations describes the quark while the other describes the degenerate doubler. We get rid of the (temporal) doubler simply by dropping one of the two equations, say the second one. The remaining equation describes two quark polarizations and two antiquark polarizations, but no doublers. This approach is called "staggered" because dropping the χ field is equivalent to specifying only half the spinor components of the original Dirac field on time slices of the lattice with even t/a_t, while only the other half are specified on time slices with odd t/a_t. The spinor components are staggered between time slices.

This discretization still has a^2 errors and the spatial derivatives have poor behavior at high momentum. These problems are easily remedied by replacing

$$\boldsymbol{\Delta} \to \boldsymbol{\Delta} - \frac{a_s^2}{6} \boldsymbol{\Delta}^{(3)} \tag{131}$$

$$m \to m - s\, a_s^5 \left(\boldsymbol{\Delta}^{(2)}\right)^3 \tag{132}$$

in (129), where the correction terms involve spatial derivatives only. The parameter s is set at some small value ($\approx .02$) that is large enough to keep the high-momentum energy large but small enough that is has little effect on the low-momentum physics. The corrected equations still have a_t^2 errors, but these are negligible on anisotropic lattices with $a_t \leq a_s/2$.

This action has never been tried. It should be as accurate as the D234-type actions. It has half as many fields, no field transformation, no ghosts, and is somewhat simpler than the D234 actions; it may well be competitive.

A more conventional action, the Kogut-Susskind or staggered-quark action, is obtained by staggering the quark degrees of freedom in spatial as well as temporal directions. This reduces the number of doublers to only three (from fifteen in the original naive action). Rather than removing the remaining doublers, in this approach one treats them as three extra flavors of quark. If these extra flavors were degenerate with the original quark, this approach might be computationally very efficient, perhaps giving four times the statistics for the same computational cost as other discretizations. Unfortunately the gluonic interactions lift the degeneracy between the flavors in the simplest version of this theory, resulting in significant mass differences. It would be very interesting to develop improved versions of this theory where the degeneracy is more accurately preserved.

4.6 Summary

Temporal derivatives usually cause problems when improving the discretization of an action. This is because successive improvements to the temporal derivatives involve operators that are increasingly nonlocal in time and therefore that almost inevitably result in ghosts of one sort or another. This problem is unusually serious for quark actions where even the leading-order discretization leads to doublers.

As illustrated in the previous sections, there are two standard techniques for removing the doublers. One is to introduce redundant operators that break the doubling symmetry; this leads to the SW and D234-type actions. The second is to stagger the quark spinor components on the lattice; this leads to actions for t-staggered quarks and Kogut-Susskind quarks.

Having repaired the leading-order operator, we face an even more acute problem in the a_t^2 correction to $\Delta_t\gamma_t$. This can be dealt with, as in the D234 actions, by using anisotropic lattices, with reduced a_t's, to push the ghosts to high energies, and by adding irrelevant operators to cancel some of the ghosts. Alternatively, using anisotropic lattices, one might leave the a_t^2 errors uncorrected; these will be negligible compared with a_s^4 errors provided a_t is small enough compared with a_s. (Numerical experiments suggest that the a_t^2 errors are only a few percent when a_t is less than .2 fm.) A more complicated alternative, which I have not discussed, is to add irrelevant operators that replace the $a_t^2\Delta_t^{(3)}\gamma_t$ correction term with less troublesome operators: for example, one can replace

$$\begin{aligned}
\Delta_t^{(3)}\gamma_t &\to -\tfrac{1}{2}\left[(\boldsymbol{\Delta}\cdot\gamma+m_c)\,(\Delta_t\gamma_t)^2+(\Delta_t\gamma_t)^2\,(\boldsymbol{\Delta}\cdot\gamma+m_c)\right] && (133)\\
&\to (\boldsymbol{\Delta}\cdot\gamma+m_c)\,\Delta_t\gamma_t\,(\boldsymbol{\Delta}\cdot\gamma+m_c) && (134)\\
&\to -\tfrac{1}{2}\left\{\boldsymbol{\Delta}\cdot\gamma+m_c\,,\,m_c-\boldsymbol{\Delta}^{(2)}-\sigma\cdot gF/2\right\} && (135)
\end{aligned}$$

using successive field transformations and obtain, in the last step, an operator that has only spatial derivatives.

The numerical results in this section suggest that these various strategies work about as well as expected, and that accurate simulations are possible with spatial lattice spacings between .3 and .4 fm. They also show that the meson dispersion relation (i.e., $c(\mathbf{p})$) at high and low momenta, the r-test and the ρ mass are all very sensitive to finite-a errors.

5 Heavy Quarks

As a final illustration of improved actions I briefly review techniques for simulating heavy quarks like the c and b quarks. Heavy quarks play a central role in the experimental study of weak interactions, and consequently it is important that we have reliable simulation techniques for their study. However heavy quark dynamics is not easily simulated using the quark actions of the previous section. Those actions only work well for quarks with energies and momenta small compared with the ultraviolet cutoff π/a. But heavy-quark hadrons have large rest energies (masses) and so necessitate small lattice spacings. Simulating an Υ, for example, using such techniques requires a lattice spacing of order $1/9.4\,\mathrm{GeV} = .02\,\mathrm{fm}$ or less — much too costly with today's computers. Fortunately the heavy quarks are generally nonrelativistic in such mesons. Most of such a meson's energy is due to the rest mass energy of its heavy-quark constituents, and rest mass energy plays little role in the dynamics of nonrelativistic particles. Consequently we can develop accurate simulation techniques that work well when the lattice spacing is chosen according to the size, rather than the mass, of the heavy-quark hadron: that is, for example, $a = .1\,\mathrm{fm}$ rather than $.02\,\mathrm{fm}$ for upsilons.[12]

Here I discuss only the NRQCD approach to heavy quark dynamics, where one takes maximum advantage of the fact that the heavy quarks are nonrelativistic [22]. I illustrate the technique by reviewing simulation results for the upsilon and psi families of mesons. These are attractive systems to study because so much is known already about them. The quark potential model, for example, provides an accurate phenomenological model of the internal structure of the mesons. Also there is much experimental data for these mesons. The low-lying states are largely insensitive to light-quark vacuum polarization — for example, the Υ, Υ', $\chi_b \ldots$ are all far below the threshold for decays into B mesons — and therefore can be accurately simulated with $n_f = 0$ light-quark flavors. Furthermore, these mesons are very small; for example, the Υ is about five times smaller than a light hadron. This makes them an excellent testing ground for our ideas concerning large lattice spacings.

Given that the heavy quarks are nonrelativistic, the most important momentum scales governing a heavy-quark meson's dynamics are smaller than the heavy quark's mass. We can take advantage of this fact by choosing an inverse lattice spacing of order the quark mass, thereby excluding relativistic states from the theory. Then it is efficient to analyze the heavy-quark dynamics using a nonrelativistic lagrangian (NRQCD). The lagrangian used to

[12] Actually since it is only the energy that is large, only the temporal lattice spacing need be small. Consequently the actions from the previous section can be used efficiently if one works with anisotropic lattices and small a_t's. Experiments using D234 actions are very encouraging [18].

generate the results I show below was

$$\mathcal{L}_{\rm NRQCD} = \sum_x \left\{ \psi^\dagger(x+a_t\hat{t}) \left(1-\frac{a_t H_0}{2n}\right)^n U_{\hat{t}}^\dagger \left(1-\frac{a_t H_0}{2n}\right)^n (1-a_t\delta H)\,\psi(x) \right\}$$
$$- \sum_x \psi^\dagger(x)\psi(x), \tag{136}$$

where $n = 2$, H_0 is the nonrelativistic kinetic-energy operator,

$$H_0 = -\frac{\boldsymbol{\Delta}^{(2)}}{2M_0}, \tag{137}$$

M_0 is the bare heavy-quark mass, and δH is the leading relativistic and finite-lattice-spacing correction,

$$\begin{aligned} \delta H = & -\frac{(\boldsymbol{\Delta}^{(2)})^2}{8M_0^3}\left(1+\frac{a_t M_0}{2n}\right) + \frac{a_s^2 \boldsymbol{\Delta}^{(4)}}{24M_0} \\ & -\frac{g}{2M_0}\boldsymbol{\sigma}\cdot\mathbf{B} + \frac{ig}{8M_0^2}(\boldsymbol{\Delta}\cdot\mathbf{E}-\mathbf{E}\cdot\boldsymbol{\Delta}) \\ & -\frac{g}{8M_0^2}\boldsymbol{\sigma}\cdot(\boldsymbol{\Delta}\times\mathbf{E}-\mathbf{E}\times\boldsymbol{\Delta}). \end{aligned} \tag{138}$$

Here $\mathbf{E}$ and $\mathbf{B}$ are the chromoelectric and chromomagnetic fields. As usual, the entire action is tadpole improved by dividing every link operator U_μ by u_0. Potential models indicate that corrections beyond δH contribute only of order 5 MeV to Υ energies, and two or three times this for ψ's.

The details of this lagrangian are unimportant to us here. What I want to focus on is the extent to which the improvement program outlined in earlier sections is effective for heavy-quark dynamics. The δH term consists of corrections just like the ones we have been analyzing for the other parts of QCD, the only difference here being that we are correcting both for finite-a and for the absence of relativity (ie, order v^2/c^2 errors). The coefficients of the correction terms were determined with tree-level perturbation theory and tadpole improvement, using precisely the techniques outlined in the earlier sections. The extent to which δH improves the simulation results is a measure of the efficacy of all the techniques discussed in these lectures.

Consider first a set of simulations of the Υ family that used a lattice spacing of about 1/12 fm — roughly half the meson's radius [23]. The spectrum is very well described by the simulations; results for the low-lying spectrum in different channels are shown in Figure 10. These compare well with experimental results (the horizontal lines), as they should since systematic errors are estimated to be less than 10–20 MeV. Two sets of Monte Carlo results are shown: one has no light-quark vacuum polarization, the other includes u and d quarks; as expected quark vacuum polarization has little effect on the spectrum. It is important to realize that these are calculations from first principles. The only inputs are the lagrangians describing gluon and quark dynamics, and the only parameters are the bare coupling constant and the bare quark mass. In particular, these results are not based on a phenomenological quark-potential model. These are among the most accurate lattice results to date.

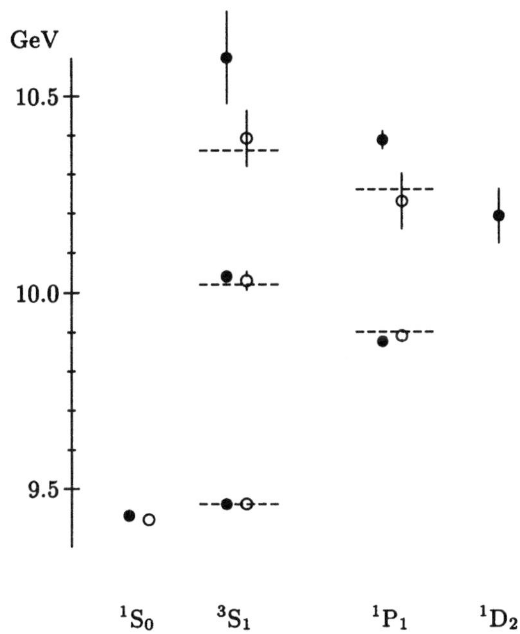

Fig. 10. NRQCD simulation results for the spectrum of the $\Upsilon(^3S_1)$ and $h_b(^1P_1)$ and their radial excitations. Experimental values (dashed lines) are indicated for the S-states, and for the spin-average of the P-states. Simulation results are for $n_f = 0$ (solid circles) and $n_f = 2$ (open circles) light-quark flavors. The energy zero for the simulation results is adjusted to give the correct mass to the Υ.

The corrections in δH have only a small effect on the overall spectrum, but the spin structure is strongly affected. The lagrangian without δH is spin independent, and gives no spin splittings at all. Simulation results for the spin structure of the lowest lying P state are shown in Figure 11. Again these compare very well with the data, giving strong evidence that corrected lagrangians work. Systematic errors here are estimated to be of order 5 MeV. Note that the spin terms in δH all involve either chromoelectric or chromomagnetic field operators. These operators are built from products of four-link operators and so tadpole improvement increases their magnitude by almost a factor of two at the lattice spacing used here. Simulations without tadpole improvement give spin splittings that are much too small.

The correction terms also affect the extent to which the simulation repro-

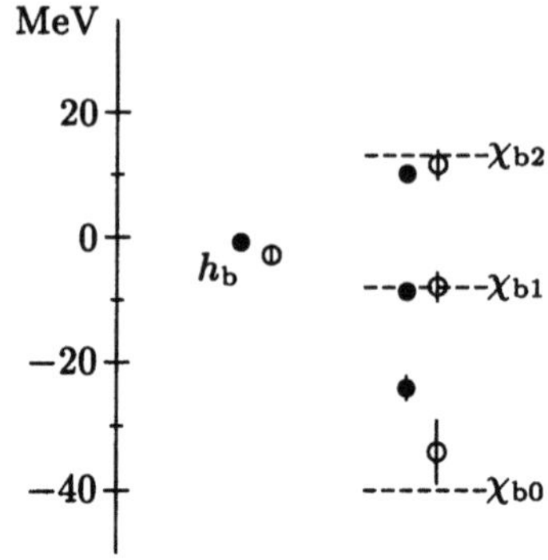

Fig. 11. Simulation results for the spin structure of the lowest lying P-state in the Υ family. The dashed lines are the experimental values for the triplet states. Simulation results are for $n_f = 0$ (solid circles) and $n_f = 2$ (open circles) light-quark flavors.

duces continuum symmetries. For example, the low-momentum dispersion relation of an upsilon in the simulation must have the form

$$E_\Upsilon(\mathbf{p}) = M_1 + \frac{\mathbf{p}^2}{2\,M_2} + \cdots. \tag{139}$$

In a relativistically invariant theory $M_1 = M_2$ is the upsilon mass. However in a purely nonrelativistic theory M_2 equals the sum of the quark masses, which differs from M_1 since M_1 also includes the quarks' binding energy. Only when relativistic corrections are added is M_2 shifted to include the binding energy. Simulations tuned to the correct mass $M_1 = 9.5(1)$ GeV give $M_2 = 8.2(1)$ GeV when δH is omitted. When δH is included, the simulations give $M_2 = 9.5(1)$ GeV, in excellent agreement with M_1. All of the spin-independent pieces of δH contribute to the shift in M_2; once again we have striking evidence that corrected actions work.

This Υ simulation has only two parameters: the bare coupling constant and the bare quark mass. These were tuned until the simulation agreed with experimental data. From the bare parameters we can compute the renormalized coupling and mass. This simulation implies that the renormalized or "pole mass" of the b quark is [24]

$$M_b = 5.0\,(2)\,\mathrm{GeV}. \tag{140}$$

The renormalized coupling that is obtained corresponds to [25]

$$\alpha^{(5)}_{\overline{\mathrm{MS}}}(M_Z) = \begin{cases} .1175\,(25) & \text{from } \chi_b\text{–}\Upsilon \text{ splitting} \\ .1180\,(27) & \text{from } \Upsilon'\text{–}\Upsilon \text{ splitting} \end{cases} \tag{141}$$

depending upon which Υ splitting in the simulation is tuned to agree exactly with experiment. These values agree with results from high-energy phenomenology (but are more accurate). This last result is striking: it shows that the QCD of hadronic structure and the QCD of high-energy quark and gluon jets are really the same theory.

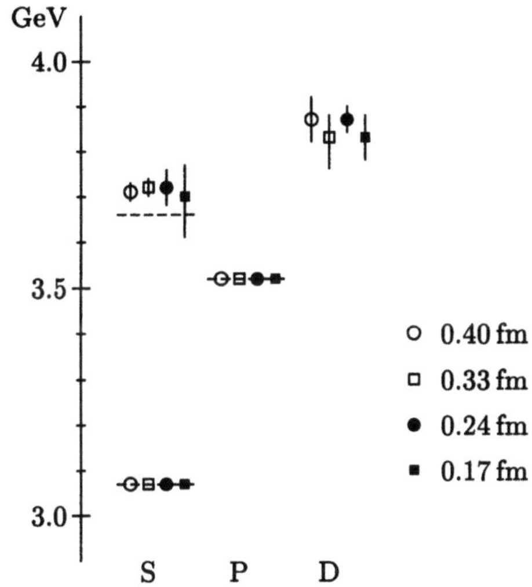

Fig. 12. S, P and D states of charmonium computed on lattices with a range of lattice spacings. Improved quark and gluon actions were used in each case except $a = .17$ fm where only the quark action was improved.

Another check on the corrected action is to show that results for physical quantities are independent of the lattice spacing. In Fig. 12 I show the low-lying spin-averaged spectrum of the ψ mesons computed using very coarse lattices, with a ranging from .17 fm to .40 fm [14]. The 1S and 1P masses are tuned; the 2S and 1D masses are predicted. The latter masses are independent of lattice spacing to within statistical errors of a few percent.

In Fig. 13 I show the 1S and 1P radial wavefunctions for charmonium, computed on lattices with two different lattice spacings using the (improved) NRQCD actions [14]. The most striking observation from these pictures is that the charge radius of the ψ is almost exactly equal to one lattice spacing on the coarser lattice, and yet the results from the coarser lattice are essentially identical to those from the finer lattice. These data confirm that in general one needs only a few lattice points per direction within a hadron to obtain accurate results (few percent) from an a^2-accurate action. Numerical experiments with the NRQCD action, where a^2 corrections are easily turned

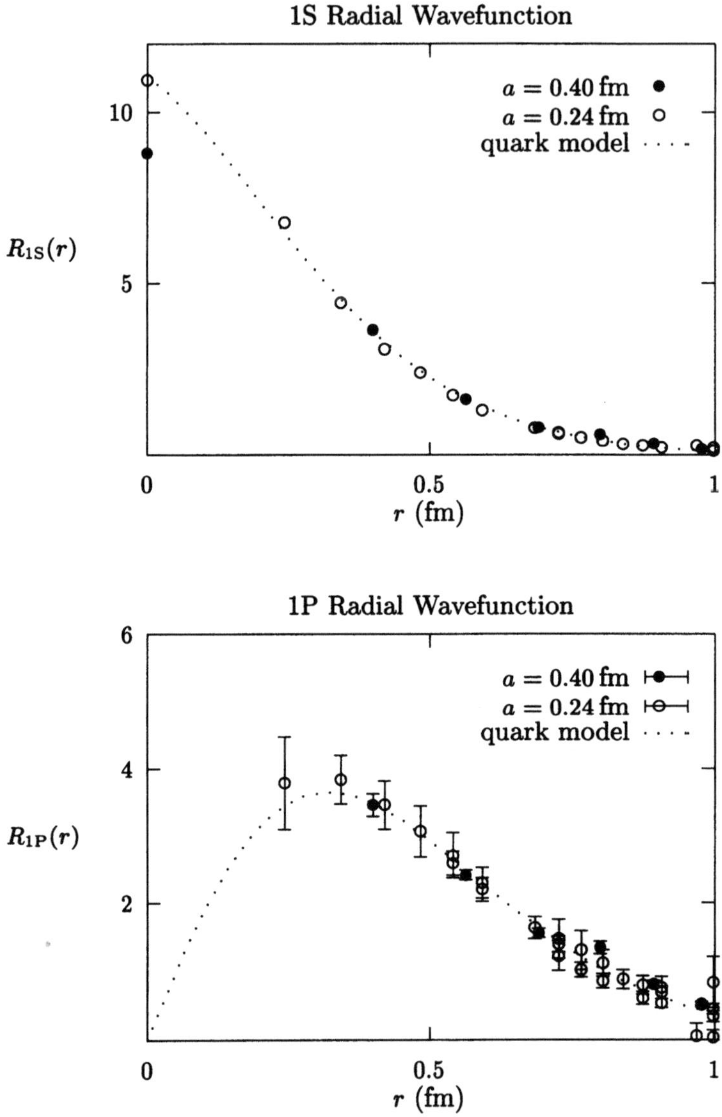

Fig. 13. The radial wavefunctions for the 1S and 1P charmonium computed using improved actions and two different lattice spacings. Wavefunctions from a continuum quark model are also shown. Statistical errors are negligible for the 1S wavefunction.

on and off, suggest that

$$r_{\text{hadron}} \approx a \Rightarrow \begin{cases} \mathcal{O}(a^2) \text{ corrections} \approx 15\text{–}20\% \\ \mathcal{O}(\alpha_s a^2, a^4) \approx 2\text{–}4\% \end{cases} \tag{142}$$

NRQCD simulations of heavy quarks were among the very first in which tadpole improvement played a crucial role. They remain among the best demonstrations of the technique.

Exercise: Verify that the continuum limit of the lattice NRQCD lagrangian (136) is a Schrödinger action with the standard relativistic corrections (chromomagnetic moment, Darwin term, spin-orbit coupling ...).

6 Conclusions

Improved discretizations of QCD dynamics and large lattice spacings work. Tree-level (i.e., classical) improvement of the quark and gluon actions give results that are accurate to within a few to several percent for lattice spacings as large as .3–.4 fm, provided that all operators are tadpole-improved. Tree-level improvement is *not* surprisingly effective; it isn't perfect. Rather, it works about as well as one would naively expect: there are certainly $\mathcal{O}(\alpha_s)$ corrections (and probably also nonperturbative corrections), and these will shift couplings by 10 or 20% and physical quantities by a few percent. That QCD simulations now conform to naive expectations and intuition is a major advance, perhaps *the* major advance. The current level of precision is more than sufficient to have a large impact on QCD phenomenology. And these techniques provide a firm foundation, a starting point, for future high-precision work that will rely upon some mixture of systematic perturbative calculations of the corrections, nonperturbative tuning, nontrivial transformations of the fields like those that lead to "perfect actions", and perhaps analytic techniques like strong-coupling expansions.

The potential impact on QCD simulations of coarse lattices is enormous. This potential will be realized if lots of people become involved in the design, implementation and application this technology.

References

1. M. Creutz, *Quarks, gluons and lattices* (Cambridge University Press, Cambridge, 1985).
2. I. Montvay and G. Münster, *Quantum Fields on a Lattice* (Cambridge University Press, Cambridge, 1994).
3. G.P. Lepage, "The Analysis of Algorithms for Lattice Field Theory" in *From Actions to Answers*, edited by T. DeGrand and D. Toussaint (World Scientific, Singapore, 1989).
4. K. Symanzik, *Nucl. Phys.* **B226** (1983) 187.

5. G.P. Lepage and P.B. Mackenzie, *Phys. Rev.* **D48** (1993) 2250.
6. For an overview see K.G. Wilson, *Rev. Mod. Phys.* **55** (1983) 583.
7. G.P. Lepage, *Lattice QCD for Small Computers* in *The Building Blocks of Creation,* edited by S. Raby and T. Walker (World Scientific Press, Singapore, 1994).
8. See, for example, H. Kawai, R. Nakayama and K. Seo, *Nucl. Phys.* **B189** (1981) 40.
9. P. Hasenfratz and F. Niedermayer, *Nucl. Phys.* **B414** (1994) 785.
10. T. DeGrand, A. Hasenfratz, P. Hasenfratz and F. Niedermayer, *Phys.Lett.* **B365** (1996) 233; *Nucl.Phys.* **B454** (1995) 587; *Nucl.Phys.* **B454** (1995) 615.
11. W. Bietenholz and U.J. Wiese, *Nucl.Phys.* **B464** (1996) 319.
12. G. Curci, P. Menotti, and G. Paffuti, *Phys. Lett.* **130B** (1983) 205; Erratum: *ibid.* **135B** (1984) 516.
13. M. Lüscher and P. Weisz, *Comm. Math. Phys.* **97** (1985) 59.
14. M. Alford, W. Dimm, G.P. Lepage, G. Hockney and P. Mackenzie, *Phys. Lett.* **B361** (1995) 87.
15. M. Lüscher and P. Weisz, *Phys. Lett.* **158B**, (1985) 250, and references therein.
16. D.W. Bliss, K. Hornbostel and G.P. Lepage, heplat-9605041 preprint (1996).
17. B. Sheikholeslami and R. Wohlert, *Nucl.Phys.* **B259** (1985) 572.
18. M. Alford, T. Klassen and G.P. Lepage, *Nucl. Phys.* **B47** (Proc. Suppl.) (1996) 370; and talks by M. Alford and T. Klassen at *Lattice '96* (June 1996, St. Louis).
19. K. Jansen, C. Liu, M. Lüscher, H. Simma, S. Sint, R. Sommer, P. Weisz and U. Wolff, *Nucl.Phys.* **B372** (1996) 275; and R. Sommer's talk at *Lattice '96* (June 1996, St. Louis).
20. S. Collins, R.G. Edwards, U.M. Heller and J. Sloan *Nucl. Phys.* **B47** (Proc. Suppl.) (1996) 366; and R.G. Edward's talk at *Lattice '96* (June 1996, St. Louis).
21. F. Butler, H. Chen, J. Sexton, A. Vaccarino and D. Weingarten, *Nucl.Phys.* **B430** (1994) 179.
22. G.P. Lepage and B. Thacker, *Nucl. Phys.* **B**(Proc. Suppl.)**4** (1988) 199; G.P. Lepage, K. Hornbostel, L. Magnea, U. Magnea and C. Nakhleh, *Phys. Rev.* **D46** (1992) 4052.
23. C. Davies et al., *Phys. Rev.* **D50** (1994) 6963; P. McCallum and J. Shigemitsu, *Nucl. Phys.* **B47** (Proc. Suppl.) (1996) 409.
24. C. Davies et al., *Phys. Rev. Lett.* **73** (1994) 2654.
25. C. Davies et al., *Phys. Lett.* **B345** (1995) 42; J. Shigemitsu's talk at *Lattice '96* (June 1996, St. Louis).

High Energy Collisions and Nonperturbative QCD

Otto Nachtmann

Institut für Theoretische Physik, Universität Heidelberg,
D-69120 Heidelberg, Philosophenweg 16, Germany

Abstract. We discuss various ideas on the nonperturbative vacuum structure in QCD. The stochastic vacuum model of Dosch and Simonov is presented in some detail. We show how this model produces confinement. The model incorporates the idea of the QCD vacuum acting like a dual superconductor due to an effective chromomagnetic monopole condensate. We turn then to high energy, small momentum transfer hadron-hadron scattering. A field-theoretic formalism to treat these reactions is developed, where the basic quantities governing the scattering amplitudes are correlation functions of light-like Wegner-Wilson lines and loops. The evaluation of these correlation functions with the help of the Minkowskian version of the stochastic vacuum model is discussed. A further surprising manifestation of the nontrivial vacuum structure in QCD may be the production of anomalous soft photons in hadron-hadron collisions. We interpret these photons as being due to "synchrotron radiation from the vacuum". A duality argument leads us from there to the expectation of anomalous pieces proportional to $(Q^2)^{1/6}$ in the electric form factors of the nucleons for small Q^2. Finally we sketch the idea that in the Drell-Yan reaction, where a quark-antiquark pair annihilates with the production of a lepton pair, a "chromodynamic Sokolov-Ternov effect" may be at work. This leads to a spin correlation of the $q\bar{q}$ pair, observable through the angular distribution of the lepton pair.

1 Introduction

In these lectures I would like to review some ideas on the way nonperturbative QCD may manifest itself in high energy collisions. Thus we will be concerned with strong interactions where we claim to know the fundamental Lagrangian for a long time now [1]:

$$\mathcal{L}_{\rm QCD}(x) = -\frac{1}{4}G^a_{\lambda\rho}(x)G^{\lambda\rho a}(x) + \sum_q \bar{q}(x)(i\gamma^\lambda D_\lambda - m_q)q(x). \tag{1.1}$$

Here $q(x)$ are the quark fields for the various quark flavours ($q = u, d, s, c, b, t$) with masses m_q. We denote the gluon potentials by $G^a_\lambda(x)$ ($a = 1, ..., 8$) and the gluon field strengths by

$$G^a_{\lambda\rho}(x) = \partial_\lambda G^a_\rho(x) - \partial_\rho G^a_\lambda(x) - gf_{abc}G^b_\lambda(x)G^c_\rho(x), \tag{1.2}$$

where g is the strong coupling constant and f_{abc} are the structure constants of SU(3). The covariant derivative of the quark fields is

$$D_\lambda q(x) = (\partial_\lambda + igG^a_\lambda \frac{\lambda_a}{2})q(x), \tag{1.3}$$

with λ_a the Gell-Mann matrices of the SU(3) group. The gluon potential and field strength matrices are defined as

$$\begin{aligned} G_\lambda(x) &:= G^a_\lambda(x)\frac{\lambda_a}{2}, \\ G_{\lambda\rho}(x) &:= G^a_{\lambda\rho}(x)\frac{\lambda_a}{2}. \end{aligned} \tag{1.4}$$

The Lagrangian (1.1) is invariant under SU(3) gauge transformations. Let $x \to U(x)$ be an arbitrary matrix function, where for fixed x the $U(x)$ are SU(3) matrices:

$$\begin{aligned} U(x)U^\dagger(x) &= 1, \\ \det U(x) &= 1. \end{aligned} \tag{1.5}$$

With the transformation laws:

$$q(x) \to U(x)q(x), \tag{1.6}$$

$$G_\lambda(x) \to U(x)G_\lambda(x)U^\dagger(x) - \frac{i}{g}U(x)\partial_\lambda U^\dagger(x), \tag{1.7}$$

we find

$$G_{\lambda\rho}(x) \to U(x)G_{\lambda\rho}(x)U^\dagger(x),$$

and invariance of $\mathcal{L}_{\text{QCD}}$:

$$\mathcal{L}_{\text{QCD}}(x) \to \mathcal{L}_{\text{QCD}}(x). \tag{1.8}$$

If we want to derive results from the Lagrangian (1.1), we face problems, the most notable being that $\mathcal{L}_{\text{QCD}}$ is expressed in terms of quark and gluon fields whose quanta have not been observed as free particles. In the real world we observe only hadrons, colourless objects, quark and gluons are permanently confined. Nevertheless it has been possible in some cases to derive first principle results which can be compared with experiment, starting from $\mathcal{L}_{\text{QCD}}$ (1.1). These are in essence the following:

(1) Pure short-distance phenomena: Due to asymptotic freedom [2] the QCD coupling constant becomes small in this regime and one can make reliable perturbative calculations. Examples of pure short distance processes are for instance the total cross section for electron-positron annihilation into hadrons and the total hadronic decay rate of the Z-boson.

(2) Pure long-distance phenomena: Here one is in the nonperturbative regime of QCD and one has to use numerical methods to obtain first principle results from $\mathcal{L}_{\text{QCD}}$, or rather from the lattice version of $\mathcal{L}_{\text{QCD}}$ introduced by

Wilson [3]. Today, Monte Carlo simulations of lattice QCD are a big industry among theorists. Typical quantities one can calculate in this way are hadron masses and other low energy hadron properties. (For an up-to-date account of these methods cf. [4]).

There is a third regime of hadronic phenomena, hadron-hadron collisions, which are – apart from very low energy collisions – neither pure long-distance nor pure short-distance phenomena. Thus, none of the above-mentioned theoretical methods apply directly. Traditionally one classifies high-energy hadron-hadron collisions as "hard" and "soft" ones:

(3) High energy hadron-hadron collisions:
   (a) hard collisions,
   (b) soft collisions.

A typical hard reaction is the Drell-Yan process, e.g.

$$\begin{aligned} \pi^- + N \rightarrow\ & \gamma^* + X \\ & \hookrightarrow \ell^+\ell^- \end{aligned} \tag{1.9}$$

where $\ell = e, \mu$. All energies and momentum transfers are assumed to be large. However, the masses of the π^- and N in the initial state stay fixed and thus we are not dealing with a pure short distance phenomenon.

In the reaction (1.9) we claim to see directly the fundamental quanta of the theory, the partons, i.e. the quarks and gluons, in action (cf. Fig. 1). In the usual theoretical framework for hard reactions, the QCD improved parton model (cf. e.g. [5]), one describes the reaction of the partons, in the Drell-Yan case the $q\bar{q}$ annihilation into a virtual photon, by perturbation theory. This should be reliable, since the parton process involves only high energies and high momentum transfers. All the long distance physics due to the bound state nature of the hadrons is then lumped into parton distribution functions of the participating hadrons. This is called the factorization hypothesis, which after early investigations of soft initial and final state interactions [6] was formulated and studied in low orders of QCD perturbation theory in [7]. Subsequently, great theoretical effort has gone into proving factorization in the framework of QCD perturbation theory [8]-[10]. The result seems to be that factorization is most probably correct there (cf. the discussion in [11]). However, it is legitimate to ask if factorization is respected also by nonperturbative effects. To my knowledge this question was first asked in [12] -[14]. In Sect. 4 of these lectures I will come back to this question and will argue that there may be evidence for a breakdown of factorization in the Drell-Yan reaction due to QCD vacuum effects.

Let us consider now soft high energy collisions. A typical reaction in this class is proton-proton elastic scattering:

$$p + p \rightarrow p + p \tag{1.10}$$

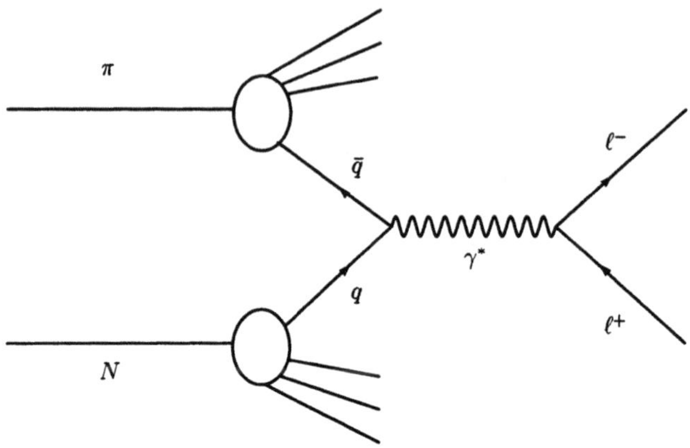

Fig. 1. The lowest order diagram for the Drell-Yan reaction (1.9) in the QCD improved parton model.

at c.m. energies $E_{cm} = \sqrt{s} \gtrsim 5\text{GeV}$ say and small momentum transfers $\sqrt{|t|} = |q| \lesssim 1\text{GeV}$. Here we have two scales, one staying finite, one going to infinity:

$$\begin{aligned} E_{cm} &\to \infty, \\ |q| &\lesssim 1\text{GeV}. \end{aligned} \tag{1.11}$$

Thus, none of the above mentioned calculational methods is directly applicable. Indeed, most theoretical papers dealing with reactions in this class develop and apply models which are partly older than QCD, partly QCD "motivated". Let me list some models for hadron-hadron elastic scattering at high energies:

geometric [15],
eikonal [16],
additive quark model [17],
Regge poles [18],
topological expansions and strings [19],
valons [20],
leading log summations [21],
two-gluon exchange [22],
the Donnachie-Landshoff model for the "soft Pomeron" [23].

It would be a forbidding task to collect all references in this field. The references given above should thus only be considered as representative ones. In addition I would like to mention the inspiring general field theoretic considerations for high energy scattering and particle production by Heisenberg

[24] and the impressive work by Cheng and Wu on high energy behaviour in field theories in the framework of perturbative calculations [25].

I will now argue that the theoretical description of measurable quantities of soft high energy reactions like the total cross sections should involve in an essential way nonperturbative QCD. To see this, consider massless pure gluon theory where all "hadrons" are massive glueballs. Then we know from the renormalization group analysis that the glueball masses must behave as

$$m_{\text{glueball}} \propto M e^{-c/g^2(M)} \tag{1.12}$$

for $M \to \infty$, i.e. for $g(M) \to 0$, due to asymptotic freedom. Here M is the renormalization scale, $g(M)$ is the QCD coupling strength at this scale and c is a constant:

$$\begin{aligned} &\frac{g^2(M)}{4\pi^2} \longrightarrow \frac{12}{33\ln(M^2/\Lambda^2)} \quad \text{for} \quad M \to \infty, \\ &c = \frac{8\pi^2}{11}, \\ &\Lambda: \quad \text{QCD scale parameter.} \end{aligned} \tag{1.13}$$

Masses in massless Yang-Mills theory are a purely nonperturbative phenomenon, due to "dimensional transmutation". Scattering of glueball-hadrons in massless pure gluon theory should look very similar to scattering of hadrons in the real world, with finite total cross sections, amplitudes with analytic t dependence etc. At least, this would be my expectation. If the total cross section σ_{tot} has a finite limit as $s \to \infty$ we must have from the same renormalization group arguments:

$$\lim_{s\to\infty} \sigma_{\text{tot}}(s) \propto M^{-2} e^{2c/g^2(M)} \tag{1.14}$$

for $g(M) \to 0$. In this case, the total cross sections in pure gluon theory are also nonperturbative objects! It is easy to see that this conclusion is not changed if $\sigma_{\text{tot}}(s)$ has a logarithmic behaviour with s for $s \to \infty$, e.g.

$$\sigma_{\text{tot}}(s) \to \text{const} \times (\log s)^2. \tag{1.15}$$

I would then expect that also in full QCD total cross sections are nonperturbative objects, at least as far as hadrons made out of light quarks are concerned.

Some time ago P.V. Landshoff and myself started to think about a possible connection between the nontrivial vacuum structure of QCD – a typical nonperturbative phenomenon – and soft high energy reactions [26]. In the following I will first review some common folklore on the QCD vacuum and discuss in more detail the so-called "stochastic vacuum model". I will then sketch possible consequences of these ideas for high energy collisions.

2 The QCD Vacuum

According to current theoretical prejudice the vacuum state in QCD has a very complicated structure [27]-[37]. It was first noted by Savvidy [27] that by introducing a constant chromomagnetic field

$$\mathbf{B}^a = \mathbf{n}\eta^a B, \quad (a = 1, \ldots, 8), \tag{2.1}$$

into the perturbative vacuum one can lower the vacuum-energy density $\varepsilon(B)$. Here $\mathbf{n}$ and η^a are constant unit vectors in ordinary and colour space. The result of his one-loop calculation was

$$\varepsilon(B) = \frac{1}{2}B^2 + \frac{\beta_0 g^2}{32\pi^2}B^2\left[\ln\frac{B}{M^2} - \frac{1}{2}\right] \tag{2.2}$$

where g is the strong coupling constant, M is again the renormalization scale, and β_0 is given by the lowest order term in the Callan-Symanzik β-function:

$$M\frac{\mathrm{d}g(M)}{\mathrm{d}M} =: \beta(g) = -\frac{\beta_0}{16\pi^2}g^3 + \ldots \tag{2.3}$$

For 3 colours and f flavours:

$$\beta_0 = 11 - \frac{2}{3}f. \tag{2.4}$$

Thus, as long as we have asymptotic freedom, i.e. for $f \leq 16$, the energy density $\varepsilon(B)$ looks as indicated schematically in Fig. 2 and has its minimum for $B = B_{\text{vac}} \neq 0$. Therefore, we should expect the QCD-vacuum to develop spontaneously a chromomagnetic field, the situation being similar to that in a ferromagnet below the Curie temperature where we have spontaneous magnetization.

Of course, the vacuum state in QCD has to be relativistically invariant and cannot have a preferred direction in ordinary space and colour space. What has been considered [33] are states composed of domains with random orientation of the gluon-field strength (Fig. 3). This is analogous to Weiss domains in a ferromagnet. The vacuum state should then be a suitable linear superposition of states with various domains and orientation of the fields inside the domains. This implies that the orientation of the fields in the domains as well as the boundaries of the domains will fluctuate.

A very detailed picture for the QCD vacuum along these lines has been developed in Ref. [33]. I cannot refrain from comparing this modern picture of the QCD vacuum (Fig. 4a) with the "modern picture" of the ether developed by Maxwell more than 100 years ago [38] (Fig. 4b). The analogy is quite striking and suggests to me that with time passing on we may also be able to find simpler views on the QCD vacuum. Remember that Einstein made great progress by eliminating the ether from electrodynamics. In the following we

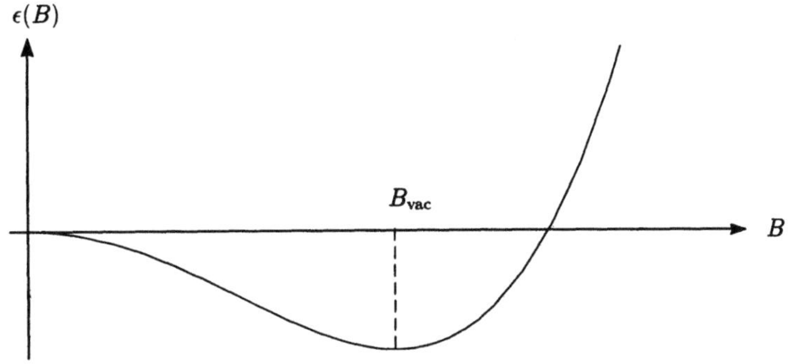

Fig. 2. The schematic behaviour of the vacuum energy density $\varepsilon(B)$ as function of a constant chromomagnetic field B according to Savvidy's calculation (Eq. (2.2)).

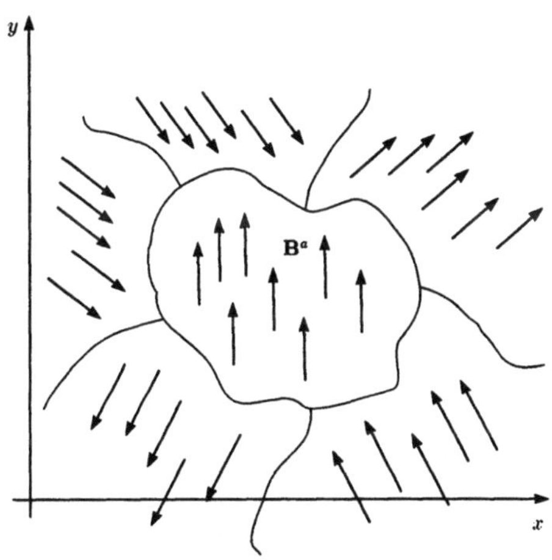

Fig. 3. A "snapshot" of the QCD vacuum showing a domain structure of spontaneously created chromomagnetic fields.

will adopt the picture of the QCD vacuum as developed in Refs. [27]-[34],[36] and outlined above as a working hypothesis.

Let me now come to the values for the field strengths $\mathbf{E}^a$ and $\mathbf{B}^a$ in the vacuum. These must also be determined by Λ, the QCD scale parameter, the only dimensional parameter in QCD if we disregard the quark masses. Therefore, we must have on dimensional grounds for the renormalization

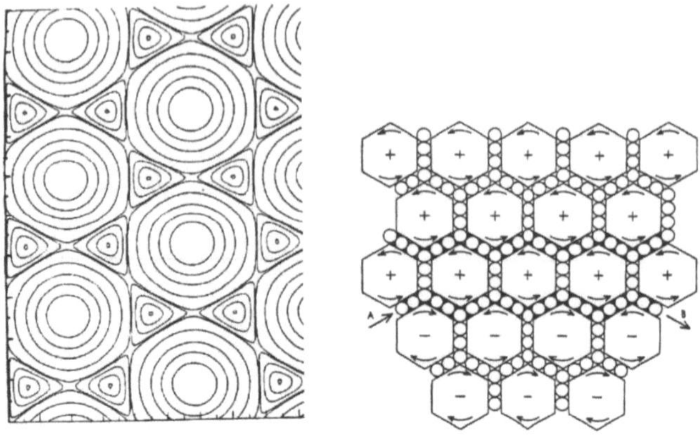

a) b)

Fig. 4. The QCD-vacuum according to Ambjørn and Olesen [33] (a). The ether according to Maxwell [38] (b).

group invariant quantity $(gB)^2$

$$(gB)^2 \propto \Lambda^4. \tag{2.5}$$

But we have much more detailed information on the values of these field strengths due to the work of Shifman, Vainshtein, and Zakharov (SVZ), who introduced the gluon condensate and first estimated its value using sum rules for charmonium states [30]:

$$\begin{aligned} <0|\frac{g^2}{4\pi^2}G^a_{\mu\nu}(x)G^{\mu\nu a}(x)|0> &\equiv <0|\frac{g^2}{2\pi^2}\left(\mathbf{B}^a(x)\mathbf{B}^a(x)-\mathbf{E}^a(x)\mathbf{E}^a(x)\right)|0> \\ &\equiv G_2 = (2.4\pm 1.1)\cdot 10^{-2}\mathrm{GeV}^4 \\ &= (335-430\mathrm{MeV})^4. \end{aligned} \tag{2.6}$$

Here we quote numerical values as given in the review [39]. A simple analysis shows that this implies

$$<0|g^2\mathbf{B}^a(x)\mathbf{B}^a(x)|0> = -<0|g^2\mathbf{E}^a(x)\mathbf{E}^a(x)|0> = \pi^2 G_2 \simeq (700\mathrm{MeV})^4. \tag{2.7}$$

To prove (2.7) we note that Lorentz- and parity-invariance require the vacuum expectation value of the uncontracted product of two gluon field strengths to be of the form

$$<0|\frac{g^2}{4\pi^2}G^a_{\mu\nu}(x)G^b_{\rho\sigma}(x)|0> = (g_{\mu\rho}g_{\nu\sigma}-g_{\mu\sigma}g_{\nu\rho})\delta^{ab}\frac{G_2}{96} \tag{2.8}$$

where G_2 is the same constant as in (2.6). Taking appropriate contractions leads to (2.6) and (2.7).

We find that $< 0|\mathbf{B}^a(x)\ \mathbf{B}^a(x)|0 >$ is positive, $< 0|\mathbf{E}^a(x)\ \mathbf{E}^a(x)|0 >$ negative! This can happen because we are really considering products of field operators, normal-ordered with respect to the perturbative vacuum. The interpretation of (2.7) is, therefore, that the B-field fluctuates with bigger amplitude, the E-field with smaller amplitude than in the perturbative vacuum state.

What about the size a of the colour domains and the fluctuation times τ of the colour fields? On dimensional grounds we must have

$$a \sim \tau \sim \Lambda^{-1}. \tag{2.9}$$

A detailed model for the QCD vacuum incorporating the gluon condensate idea and a fall-off of the correlation of two field strengths with distance was proposed in [40]: the "stochastic vacuum model" (SVM). In the following we will discuss the basic assumptions of the model and then apply it to derive the area law for the Wegner-Wilson loop, i.e. confinement of static quarks. For the rest of this section we will work in Euclidean space-time. To accomplish the analytic continuation from Minkowski to Euclidean space-time we have to make the following replacements for x^λ and G^λ (cf. (1.4)):

$$\begin{aligned} x^0 &\to -iX_4, \\ \mathbf{x} &\to \mathbf{X}, \\ G^0 &\to i\mathcal{G}_4, \\ \mathbf{G} &\to -\boldsymbol{\mathcal{G}}. \end{aligned} \tag{2.10}$$

Here $X = (\mathbf{X}, X_4)$ denotes an Euclidean space-time point and $\mathcal{G}_\alpha$ ($\alpha = 1, ..., 4$) the Euclidean gluon potential. With (2.10) we get

$$\begin{aligned} &(x \cdot y) \to -(X \cdot Y) = -X_\alpha Y_\alpha, \\ &-ig \int \mathrm{d}x^\mu G_\mu(x) \to -ig \int \mathrm{d}X_\alpha \mathcal{G}_\alpha(X), \\ &G^{0j} \to -i\mathcal{G}_{4j}, \\ &G^{jk} \to \mathcal{G}_{jk}, \end{aligned} \tag{2.11}$$

where $1 \le j, k \le 3, \quad 1 \le \alpha, \beta \le 4$ and

$$\mathcal{G}_{\alpha\beta} = \partial_\alpha \mathcal{G}_\beta - \partial_\beta \mathcal{G}_\alpha + ig[\mathcal{G}_\alpha, \mathcal{G}_\beta] \tag{2.12}$$

is the Euclidean gluon field strength tensor.

2.1 Connectors

Consider classical gluon fields in Euclidean space-time. Let X, Y be two points there and C_X a curve from X to Y (Fig. 5). We define the connector, the non-abelian generalization of the "Schwinger string" [41] of QED as

$$V(Y,X;C_X) = \mathrm{P}\{\exp[-ig\int_{C_X} \mathrm{d}Z_\alpha \mathcal{G}_\alpha(Z)]\}. \tag{2.13}$$

where P means path ordering. The connector can be obtained as solution of a differential equation. Let

$$\begin{aligned} &\tau \to Z(\tau), \\ &\tau_1 \leq \tau \leq \tau_2, \\ &Z(\tau_1) = X, \quad Z(\tau_2) = Y, \end{aligned} \tag{2.14}$$

be a parameterization of C_X. Consider the differential equation for a 3×3 matrix function $V(\tau)$:

$$\frac{\mathrm{d}}{\mathrm{d}\tau} V(\tau) = -ig \frac{\mathrm{d}Z_\alpha(\tau)}{\mathrm{d}\tau} \mathcal{G}_\alpha(Z(\tau)) V(\tau), \tag{2.15}$$

with the boundary condition

$$V(\tau_1) = \mathbb{1}. \tag{2.16}$$

The solution of (2.15), (2.16) gives for $\tau = \tau_2$ just the connector (2.13).

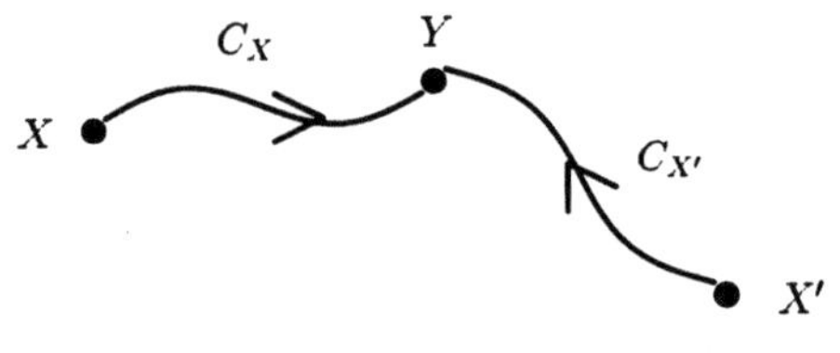

Fig. 5. Points X, X', Y in Euclidean space time and curves $C_X, C_{X'}$ running from X to Y and X' to Y, respectively.

Under a gauge transformation

$$\mathcal{G}_\alpha(X) \to U(X)\mathcal{G}_\alpha(X)U^\dagger(X) - \frac{i}{g} U(X)\partial_\alpha U^\dagger(X) \tag{2.17}$$

where $U(X) \in SU(3)$, we have

$$V(Y,X;C_X) \to U(Y)V(Y,X;C_X)U^{-1}(X). \tag{2.18}$$

The connector can be used to "shift" various objects from one space-time point to another in a gauge-covariant way. We define for instance the field strength tensor shifted from X to Y along C_X as

$$\hat{\mathcal{G}}_{\alpha\beta}(Y,X;C_X) := V(Y,X;C_X)\mathcal{G}_{\alpha\beta}(X)V^{-1}(Y,X;C_X). \qquad (2.19)$$

Under a gauge transformation $\hat{\mathcal{G}}_{\alpha\beta}(Y,X;C_X)$ transforms like a field strength tensor at Y:

$$\hat{\mathcal{G}}_{\alpha\beta}(Y,X;C_X) \to U(Y)\hat{\mathcal{G}}_{\alpha\beta}(Y,X;C_X)U^{-1}(Y). \qquad (2.20)$$

Connectors can, of course, be defined for arbitrary SU(3) representations, not only for the fundamental one used in (2.13). Let $T_a (a = 1, \ldots 8)$ be the generators of SU(3) in some arbitrary unitary representation R where

$$[T_a, T_b] = i f_{abc} T_c. \qquad (2.21)$$

We define the connector for this representation by:

$$V_R(Y,X;C_X) := \mathrm{P}\ \exp[-ig \int_{C_X} \mathrm{d}Z_\alpha \mathcal{G}^a_\alpha(Z) T_a]. \qquad (2.22)$$

We list some basic properties of connectors:
(i) For 2 adjoining curves C_1, C_2 (Fig. 6), the connectors are multiplied:

$$V_R(X_3,X_1;C_2+C_1) = V_R(X_3,X_2;C_2).V_R(X_2,X_1;C_1). \qquad (2.23)$$

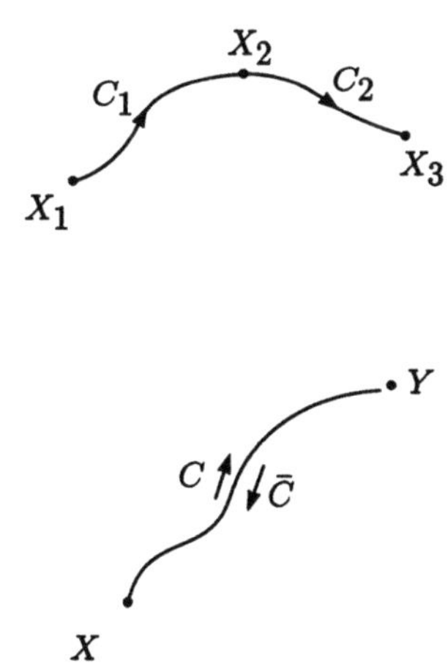

Fig. 6. Curves in Euclidean space time: C_1 going from X_1 to X_2, C_2 from X_2 to X_3, C from X to Y, and $\bar{C}$ from Y to X.

(ii) Let C be a curve from X to Y and $\bar{C}$ the same curve but oriented in inverse direction, running from Y to X (Fig. 6). Then

$$V_R(X,Y;\bar{C})V_R(Y,X;C) = \mathbb{1}, \tag{2.24}$$

$$V_R^\dagger(Y,X;C) = V_R(X,Y;\bar{C}). \tag{2.25}$$

The product of the connectors for the path and the reverse path is equal to the unit matrix. The reversal of the path is equivalent to hermitian conjugation.

We leave the proof of (2.23)-(2.25) as an exercise for the reader.

2.2 The Non-Abelian Stokes Theorem

In this subsection we will derive the non-abelian generalization of the Stokes theorem [42]. Consider a surface S in Euclidean space time with boundary $C = \partial S$. Let X be some point on C as indicated in Fig. 7 and consider the connector (2.22) from X back to X along C:

$$V_R(X,X;C) = \mathrm{P}\ \exp[-ig \int_C \mathrm{d}Z_\alpha \mathcal{G}_\alpha^a(Z)T_a]. \tag{2.26}$$

The problem is to transform this line integral into a surface integral.

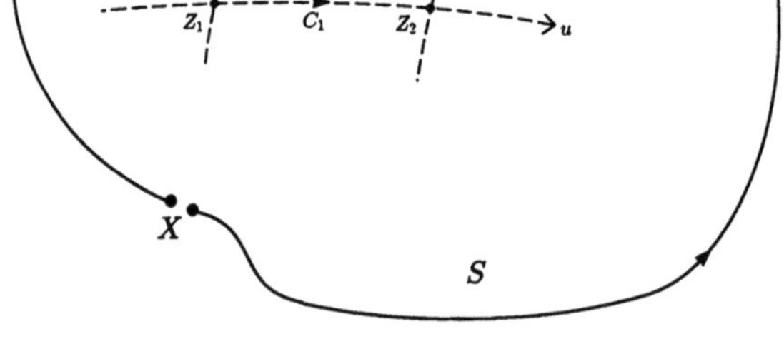

Fig. 7. A surface S with boundary $C = \partial S$. A plaquette with corner points $Z_1 ..., Z_4$ and boundary formed by lines $u =$ const., $v =$ const..

We start by considering a point Z_1 in S and a small plaquette formed by curves $C_1, .., C_4$ where one corner point is Z_1 (Fig. 7). We choose a coordinate system on S in the neighbourhood of Z_1:

$$(u,v) \to Z(u,v) \tag{2.27}$$

such that

$$\begin{aligned} Z_1 &= Z(0,0), \\ Z_2 &= Z(\Delta u, 0), \\ Z_3 &= Z(\Delta u, \Delta v), \\ Z_4 &= Z(0, \Delta v) \end{aligned} \tag{2.28}$$

and the curves $C_1, C_3 (C_2, C_4)$ correspond to lines v=const. (u=const.). The matrix representing the line integral around the small plaquette is

$$U_R(\Delta u, \Delta v) := V_R(C_4)V_R(C_3)V_R(C_2)V_R(C_1). \tag{2.29}$$

We want to make a Taylor expansion of $U_R(\Delta u, \Delta v)$ in Δu and Δv. From (2.24) we find immediately that

$$U_R(0, \Delta v) = U_R(\Delta u, 0) = \mathbb{1}. \tag{2.30}$$

Thus the lowest order term in the expansion after the zeroth order is proportional to $\Delta u \cdot \Delta v$ and we get easily:

$$\begin{aligned} U_R(\Delta u, \Delta v) &= \mathbb{1} - ig\frac{1}{2}\Delta\sigma_{\alpha\beta}G^a_{\alpha\beta}(Z_1)T_a \\ &+O(\Delta u^2 \Delta v, \Delta u \Delta v^2), \end{aligned} \tag{2.31}$$

where

$$\Delta\sigma_{\alpha\beta} = \Delta u \Delta v \frac{\partial(Z_\alpha, Z_\beta)}{\partial(u,v)}. \tag{2.32}$$

In the limit $\Delta u, \Delta v \to 0$ $\Delta\sigma_{\alpha\beta}$ becomes the surface element $d\sigma_{\alpha\beta}$ of the plaquette.

The next step is to choose an arbitrary point Y, the reference point, on the surface S and to draw a fan-type net on S as a spider would do (Fig. 8). The system of curves of the net consists of the curve C_X running from X to Y, then $\bar{C}_{Z_1}$ from Y to Z_1, then around a small plaquette at Z_1, back to Y along C_{Z_1} and so on. The final curve is $\bar{C}_X$ from Y to X. Apart from the initial and final curves C_X and $\bar{C}_X$ we have a system of plaquettes with "handles" connecting them to Y. With the help of (2.23) -(2.25) we see that the connector along the whole net is equivalent to the original connector (2.26).

$$\begin{aligned} V_R(X,X;C) = V_R(X,Y;\bar{C}_X) &\cdot \text{ product of connectors} \\ \text{for the plaquettes with handles } &\cdot V_R(Y,X;C_X). \end{aligned} \tag{2.33}$$

Let us consider one plaquette with handle, for instance the one at Z_n in Fig. 8. For this contribution to (2.33) we get

$$\begin{aligned}&V_R(Y,Z_n;C_{Z_n})V_R(\text{plaquette at } Z_n)V_R(Z_n,Y;\bar{C}_{Z_n})\\&=V_R(Y,Z_n;C_{Z_n})\left[\mathbb{1}-ig\frac{1}{2}\Delta\sigma_{\alpha\beta}\mathcal{G}^a_{\alpha\beta}(Z_n)T_a+\cdots\right]V_R(Z_n,Y;\bar{C}_{Z_n})\\&=\mathbb{1}-ig\frac{1}{2}\Delta\sigma_{\alpha\beta}\hat{\mathcal{G}}^a_{\alpha\beta}(Y,Z_n;C_{Z_n})T_a+\cdots.\end{aligned}\tag{2.34}$$

Here we use (2.31) and the shifted field strengths as defined in (2.19). We leave it as an exercise to the reader to show that (2.19) implies

$$\begin{aligned}&V_R(Y,Z_n;C_{Z_n})\mathcal{G}^a_{\alpha\beta}(Z_n)T_aV_R(Z_n,Y;\bar{C}_{Z_n})\\&=\hat{\mathcal{G}}^a(Y,Z_n;C_{Z_n})T_a\end{aligned}\tag{2.35}$$

for arbitrary representation R of $SU(3)$.

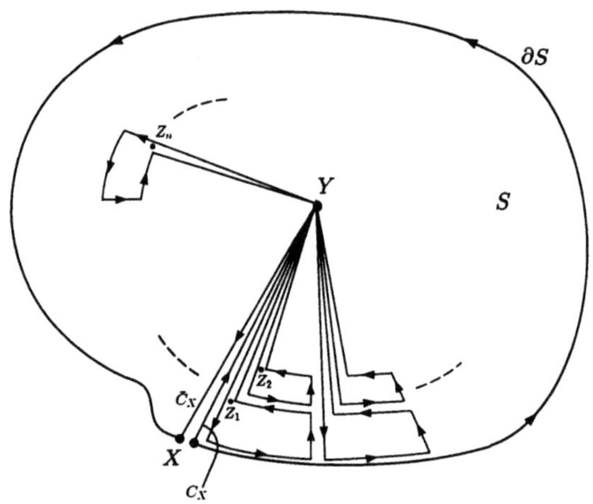

Fig. 8. A reference point Y on S and the fan-type net with center Y spanned over S.

Inserting (2.34) in (2.33) and summing up the contribution of all plaquettes with handles, where we have of course to respect the ordering, we get in the limit that the net is infinitesimally fine:

$$\begin{aligned}&V_R(X,X;C)=V_R(X,Y;\bar{C}_X)\cdot\\&\mathrm{P}\exp\left[-i\frac{g}{2}\int_S \mathrm{d}\sigma_{\alpha\beta}\hat{\mathcal{G}}^a_{\alpha\beta}(Y,Z;C_Z)T_a\right]V_R(Y,X;C_X).\end{aligned}\tag{2.36}$$

Here P denotes the ordering on the whole surface as implied by the net. Usually one takes the trace in (2.36) which leads with (2.24) to

$$\mathrm{Tr}\ V_R(X,X;C) = \mathrm{Tr}\ \mathrm{P}\exp\left[-i\frac{g}{2}\int_S \mathrm{d}\sigma_{\alpha\beta}\hat{\mathcal{G}}^a_{\alpha\beta}(Y,Z;C_Z)T_a\right]. \tag{2.37}$$

This is the desired non-abelian version of Stokes' theorem. We leave it to the reader as an exercise to show that for the abelian case (2.37) reduces to the conventional Stokes theorem.

2.3 The Cumulant Expansion

As a last mathematical tool for making calculations with the SVM we discuss the cumulant expansion [43]. Consider functions

$$\tau \to B(\tau) \tag{2.38}$$

on the interval $O \le \tau \le 1$ where $B(\tau)$ are quadratic matrices. We assume that an averaging procedure over products of the functions $B(\cdot)$ is defined:

$$E(B(\tau_1)), E(B(\tau_1)B(\tau_2)), \ldots .$$

We consider first the case that all averages $E(\cdot)$ are c numbers and that

$$E(1) = 1. \tag{2.39}$$

Let us consider the expectation value of the τ-ordered exponential:

$$f(t) := E(\mathrm{P}\exp\left[t\int_0^1 \mathrm{d}\tau B(\tau)\right]), \tag{2.40}$$

where $t \in \mathcal{C}$. The cumulant expansion asserts that $\ln f(t)$ can be expanded as

$$\ln f(t) = \sum_{n=1}^{\infty}\frac{t^n}{n!}\int_0^1 \mathrm{d}\tau_1 \ldots \int_0^1 \mathrm{d}\tau_n K_n(\tau_1,\ldots,\tau_n), \tag{2.41}$$

where the n-th cumulant $K_n(\tau_1,\ldots,\tau_n)$ is a symmetric function of its arguments. A frequently used notation for the K_n is:

$$K_n(\tau_1,\ldots,\tau_n) \equiv \langle\langle B(\tau_1)\ldots B(\tau_n)\rangle\rangle. \tag{2.42}$$

To prove (2.41) we proceed as follows. Expanding in powers of t on the r.h.s. of (2.40) we get

$$f(t) = 1 + \sum_{n=1}^{\infty}\frac{t^n}{n!}\mathcal{B}_n \tag{2.43}$$

where

$$\mathcal{B}_n = \int_0^1 \mathrm{d}\tau_1 \ldots \int_0^1 \mathrm{d}\tau_n E(\mathrm{P}(B(\tau_1)\ldots B(\tau_n))). \tag{2.44}$$

Now we expand $\ln f(t)$:

$$\ln f(t) = \sum_{n=1}^{\infty} \frac{t^n}{n!} \mathcal{K}_n, \tag{2.45}$$

where the expansion coefficients $\mathcal{K}_n$ are obtained from:

$$\begin{aligned} f(t) &= \exp(\ln f(t)), \\ 1 + \sum_{n=1}^{\infty} \frac{t^n}{n!} \mathcal{B}_n &= \exp\left[\sum_{n=1}^{\infty} \frac{t^n}{n!} \mathcal{K}_n\right]. \end{aligned} \tag{2.46}$$

From this we obtain the $\mathcal{K}_n$ as solution of the following system of equations:

$$\begin{aligned} \mathcal{B}_1 &= \mathcal{K}_1, \\ \mathcal{B}_2 &= \mathcal{K}_2 + \mathcal{K}_1^2, \\ \mathcal{B}_3 &= \mathcal{K}_3 + \frac{3}{2}(\mathcal{K}_2\mathcal{K}_1 + \mathcal{K}_1\mathcal{K}_2) + \mathcal{K}_1^3, \\ \mathcal{B}_4 &= \mathcal{K}_4 + 2(\mathcal{K}_3\mathcal{K}_1 + \mathcal{K}_1\mathcal{K}_3) + 3\mathcal{K}_2^2 + 2(\mathcal{K}_2\mathcal{K}_1^2 + \mathcal{K}_1\mathcal{K}_2\mathcal{K}_1 + \mathcal{K}_1^2\mathcal{K}_2) + \mathcal{K}_1^4, \\ &\dots . \end{aligned} \tag{2.47}$$

Clearly, this system can be inverted and we get $\mathcal{K}_n$ as sum of monomials of the form

$$\mathcal{B}_{i_1} \cdot \mathcal{B}_{i_2} \dots \mathcal{B}_{i_k}, \tag{2.48}$$

where

$$\sum_{j=1}^{k} i_j = n. \tag{2.49}$$

Using (2.44), every monomial (2.48) can be written as n-fold integral over $\tau_1, \dots, \tau_n$ with $0 \le \tau_j \le 1$ ($j = 1, \dots, n$). Since the integration domain is symmetric under arbitrary permutations of $\tau_1, \dots, \tau_n$ we can symmetrize the integrand completely. In this way we get

$$\mathcal{K}_n = \int_0^1 d\tau_1 \dots \int_0^1 d\tau_n K_n(\tau_1, \dots, \tau_n), \tag{2.50}$$

where K_n is a totally symmetric function of its arguments. Inserting (2.50) in (2.45) we get the cumulant expansion (2.41), q.e.d. Explicitly we find for the first few cumulants:

$$\begin{aligned}
K_1(1) &= E(B(1)),\\
K_2(1,2) &= E(\mathrm{P}(B(1)B(2)))\\
&\quad -\frac{1}{2}E(B(1))E(B(2)) - \frac{1}{2}E(B(2))E(B(1)),\\
K_3(1,2,3) &= E(\mathrm{P}(B(1)B(2)B(3)))\\
&\quad -\frac{1}{2}[E(\mathrm{P}(B(1)B(2)))E(B(3))\\
&\quad +E(B(1))E(\mathrm{P}(B(2)B(3)))\\
&\quad +\text{cycl. perm. }]\\
&\quad +\frac{1}{3}[E(B(1))E(B(2))E(B(3)) + \text{perm. }],\\
&\quad \ldots .
\end{aligned} \tag{2.51}$$

Here we write as a shorthand notation $K_1(1) \equiv K_1(\tau_1)$, $B(1) \equiv B(\tau_1)$ etc.

The cumulant expansion (2.41) has the so-called "cluster" property: Let us assume that the expectation values of the P-ordered products factorize

$$E(\mathrm{P}(B(1)\ldots B(n))) = \frac{1}{n!}\{E(B(1))\ldots E(B(n)) + \text{perm.}\} \tag{2.52}$$

for all $n \geq 2$ and all

$$|\tau_i - \tau_j| \geq \tau_{\min} \quad (i \neq j). \tag{2.53}$$

We can then show that the cumulants for $n \geq 2$ vanish:

$$K_n(1,\ldots,n) = 0, \quad (n \geq 2) \tag{2.54}$$

if the τ_i satisfy (2.53).

To prove (2.54) we show first that it is true for $n = 2$. Indeed, from (2.51) and (2.52) we get for $|\tau_1 - \tau_2| \geq \tau_{\min}$

$$\begin{aligned}
K_2(1,2) &= \frac{1}{2!}\{E(B(1))E(B(2)) + \text{perm.}\}\\
&\quad -\frac{1}{2}E(B(1))E(B(2)) - \frac{1}{2}E(B(2))E(B(1))\\
&= 0.
\end{aligned} \tag{2.55}$$

Now we proceed by mathematical induction. Assume that (2.54) has been shown for all k with $2 \leq k \leq n-1$. We have from (2.47):

$$\begin{aligned}
K_n(1,\ldots,n) &= E(\mathrm{P}(B(1)\ldots B(n)))\\
&\quad -\frac{1}{n!}[K_1(1)\ldots K_1(n) + \text{perm.}]\\
&\quad + [\text{symmetrized products of cumulants } K_k(1,\ldots,k)\\
&\qquad \text{with } 1 \leq k \leq n-1]
\end{aligned} \tag{2.56}$$

but at least one factor with $k \geq 2$. With (2.52) we get now immediately $K_n(1,\ldots,n) = 0$ in the region defined by (2.53), q.e.d.

Up to now we have assumed the expectation values $E(\cdot)$ to be c-numbers. In this case many of the formulae (2.47) - (2.56) can be simplified by using

$$E(B(1))E(B(2)) = E(B(2))E(B(1)), \quad \text{etc.} \tag{2.57}$$

We have, on purpose, not used such commutativity relations, since we are now going to generalize the cumulant expansion to the case where

$$E(B(1)), E(B(1)B(2)), \ldots \tag{2.58}$$

are themselves quadratic matrix valued expectation values with

$$E(\mathbb{1}) = \mathbb{1}. \tag{2.59}$$

Then we have, of course, in general, no more the commutativity relations (2.57). But all formulae (2.40) - (2.56) are written in such a way that they remain true also for the case of matrix valued expectation values $E(\cdot)$.

2.4 The Basic Assumptions of the Stochastic Vacuum Model

The basic object of the SVM is the correlator of two field strengths shifted to a common reference point. Let X, X' be two points in Euclidean space-time, Y a reference point and C_X, $C_{X'}$ curves from X to Y and X' to Y, respectively (Fig. 5). We consider the shifted field strengths as defined in (2.19) and the vacuum expectation value of their product in the sense of Euclidean QFT:

$$\langle \frac{g^2}{4\pi^2}\left[\hat{\mathcal{G}}^a_{\mu\nu}(Y,X;C_X)\hat{\mathcal{G}}^b_{\rho\sigma}(Y,X';C_{X'})\right]\rangle =: \frac{1}{4}\delta^{ab}F_{\mu\nu\rho\sigma}(X,X',Y;C_X,C_{X'}). \tag{2.60}$$

Here (2.20) and colour conservation allow us to write the r.h.s. of (2.60) proportional to δ^{ab}. It is easy to see that $F_{\mu\nu\rho\sigma}$ depends only on X, X' and the curve $C_X + \bar{C}_{X'}$ connecting them; i.e. the reference point Y can be freely shifted on the connecting curve. In the SVM one makes now the strong assumption that the correlator (2.60) even does not depend on the connecting curve at all:

- **Ass. 1:** $F_{\mu\nu\rho\sigma}$ is independent of the reference point Y and of the curves C_X and $C_{X'}$.

Translational, $O(4)$- and parity invariance require then the correlator (2.60) to be of the following form:

$$\begin{aligned} F_{\mu\nu\rho\sigma} = F_{\mu\nu\rho\sigma}(Z) = \frac{1}{24}G_2\Big\{&(\delta_{\mu\rho}\delta_{\nu\sigma} - \delta_{\mu\sigma}\delta_{\nu\rho})\,\kappa D(-Z^2) \\ +\frac{1}{2}\Big[\frac{\partial}{\partial Z_\nu}(Z_\sigma\delta_{\mu\rho} - Z_\rho\delta_{\mu\sigma}) &+ \frac{\partial}{\partial Z_\mu}(Z_\rho\delta_{\nu\sigma} - Z_\sigma\delta_{\nu\rho})\Big](1-\kappa)D_1(-Z^2)\Big\}. \end{aligned} \tag{2.61}$$

Here $Z = X - X'$, G_2 is the gluon condensate, D, D_1 are invariant functions normalized to

$$D(0) = D_1(0) = 1 \tag{2.62}$$

and κ is a parameter measuring the non-abelian character of the correlator.

Indeed, if we consider an abelian theory we have to replace the gluon field strengths $\mathcal{G}_{\mu\nu}$ by abelian field strengths $\mathcal{F}_{\mu\nu}$, which satisfy the homogenous Maxwell equation

$$\epsilon_{\mu\nu\rho\sigma}\partial_\nu \mathcal{F}_{\rho\sigma}(X) = 0, \tag{2.63}$$

if there are no magnetic monopoles present. It is easy to see that this implies $\kappa = 0$ in (2.61). Thus, in an abelian theory, the D-term in (2.61) is absent without magnetic monopoles but would be non-zero if the vacuum contained a magnetic monopole condensate. In the non-abelian theory the D-term has no reason to vanish. In fact, we will see that it dominates over the D_1-term. The abelian analogy suggests an interpretation of the D-term as being due to an effective chromomagnetic monopole condensate in the QCD vacuum.

Two further assumptions are made in the SVM:

- Ass. 2: The correlation functions $D(-Z^2)$ and $D_1(-Z^2)$ fall off rapidly for $Z^2 \to \infty$. There exists a characteristic finite correlation length a, which we define as

$$a := \int_0^\infty \mathrm{d}Z \; D(-Z^2). \tag{2.64}$$

A typical ansatz for the function D, incorporating assumption 2, is as follows:

$$D(-Z^2) = \frac{27}{64} a^{-2} \int \mathrm{d}^4 K e^{iKZ} K^2 \left[K^2 + \left(\frac{3\pi}{8a}\right)^2 \right]^{-4}, \tag{2.65}$$

which leads to

$$D(-Z^2) \propto \exp\left(-\frac{3\pi|Z|}{8a}\right) \qquad \text{for } Z^2 \to \infty. \tag{2.66}$$

The function D_1 is chosen such that

$$\left(4 + Z_\mu \frac{\partial}{\partial Z_\mu}\right) D_1(-Z^2) = 4D(-Z^2) \tag{2.67}$$

which leads to

$$D_1(-Z^2) = (Z^2)^{-2} \int_0^{Z^2} \mathrm{d}v 2v D(-v). \tag{2.68}$$

With (2.67) the contracted field strength correlator has the form (cf. (2.61):

$$F_{\mu\nu\mu\nu} = \frac{1}{2} G_2 D(-Z^2). \tag{2.69}$$

The ansatz (2.65), (2.67) can be compared to a lattice gauge theory calculation of the gluon field strength correlator [44] in order to fix the parameters. One finds (cf. [44], [45] and Fig. 9):

$$\begin{aligned} a &= 0.35 \text{ fm}, \\ \kappa &= 0.74, \\ G_2 &= (496 \text{ MeV})^4, \end{aligned} \tag{2.70}$$

with an educated guess for the error of $\approx 10\%$. Of course, from Fig. 9 we get only the product $\kappa \cdot G_2$. These quantities are obtained separately by measuring on the lattice several components of the correlator (2.60). The value for G_2 in (2.70) from the lattice calculations is somewhat larger than from phenomenology. This can be understood as follows. The lattice calculations of [44] are for the pure gluon theory. In the real world light quarks are present. Their effect is estimated to reduce the value of G_2 substantially [46].

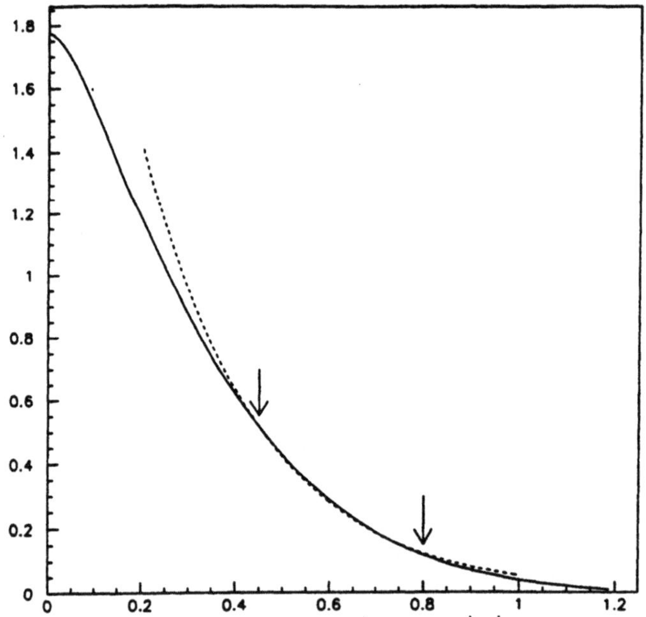

Fig. 9. The correlator function $4\pi^2\kappa G_2 D(-Z^2)$ (cf. (2.61)) as function of $|Z|$. The dashed line is the lattice result of [44] with the arrows indicating the range where these results are considered reliable. The solid line corresponds to the ansatz (2.65) with $a = 0.35$ fm (cf. Fig. 5.1 of [45]).

We note that the correlation length is smaller, albeit not much smaller than a typical radius R of a light hadron (cf. e.g. [47], [48], [49], [50]):

$$R \sim 0.7 - 1 \text{ fm}. \tag{2.71}$$

Still

$$a^2/R^2 \approx 0.2 - 0.3 \tag{2.72}$$

is a reasonably small number and this will be important for us in the following.

We come now to the third assumption made in the SVM:

- __Ass. 3:__ Factorization of higher point gluon field strength correlators.

In detail assumption 3 reads as follows (cf. [45]):

All expectation values of an odd number of products of shifted field strengths vanish:

$$\langle \hat{\mathcal{G}}(1)...\hat{\mathcal{G}}(2n+1)\rangle = 0 \quad \text{for} \quad n = 0, 1, 2, ... \tag{2.73}$$

Here we set as shorthand

$$\hat{\mathcal{G}}(i) \equiv \hat{\mathcal{G}}^{a_i}_{\alpha_i\beta_i}(Y, X_i; C_{X_i}). \tag{2.74}$$

For even number of shifted field strengths we set in the SVM:

$$\langle \hat{\mathcal{G}}(1)...\hat{\mathcal{G}}(2n)\rangle = \sum_{\substack{\text{all pairings} \\ (i_1,j_1)...(i_n,j_n)}} \langle \hat{\mathcal{G}}(i_1)\hat{G}(j_1)\rangle ... \langle \hat{\mathcal{G}}(i_n)\hat{\mathcal{G}}(j_n)\rangle, \tag{2.75}$$

where $n = 1, 2, ...$.

We note that $\langle \hat{\mathcal{G}}(1)\rangle$ must vanish due to colour conservation since the QCD vacuum has no preferred direction in colour space. The vanishing of the other correlators of odd numbers of field strengths, postulated in (2.73), as well as the factorization property (2.75) are strong dynamical assumptions. They mean that the vacuum fluctuations are assumed to be of the simplest type: a Gaussian random process.

For some applications of the SVM, for instance the calculation of the Wegner-Wilson loop described below, assumption 3 is not crucial and can be relaxed. But for the applications of the SVM to high energy scattering (cf. Sect. 3) assumption 3 is crucial. In any case we prefer to specify the model completely, thus giving it maximal predictive power. On the other hand, of course, the model can then more easily run into difficulties in comparison with experiments.

2.5 The Wegner-Wilson Loop in the Stochastic Vacuum Model

We have now specified the SVM completely and can proceed to show how this model produces confinement. We consider a static quark-antiquark pair at distance R from each other and ask for the potential $V(R)$. To calculate $V(R)$ we start with a rectangular Wegner-Wilson loop in the $X_1 - X_4$ plane (Fig. 10): Let C be the loop and S the rectangle, $C = \partial S$. Then

$$W(C) = \frac{1}{3}\langle \mathrm{Tr}\ \mathrm{P}\ \exp[-ig \int_C \mathrm{d}Z_\mu \mathcal{G}_\mu(Z)]\rangle \tag{2.76}$$

and

$$V(R) = -\lim_{T\to\infty} \frac{1}{T} \ln W(C). \tag{2.77}$$

To evaluate $W(C)$ in the SVM we first transform the line integral of the potentials in (2.76) into a surface integral of field strengths, using the non-abelian version of Stokes theorem (cf. Sect. 2.2). For this we choose a reference point Y in S. We get then

$$W(C) = \frac{1}{3}\langle \mathrm{Tr}\ \mathrm{P}\ \exp[-ig \int_S \mathrm{d}X_1 \mathrm{d}X_4 \hat{\mathcal{G}}_{14}(Y, X, C_X)]\rangle \tag{2.78}$$

where $\hat{\mathcal{G}}_{\mu\nu}$ are the field strengths parallel-transported from X to Y along a straight line C_X. The path-ordered way to integrate over S in a fan-type net (cf. Fig. 8) is indicated by P. To evaluate the expectation value of the

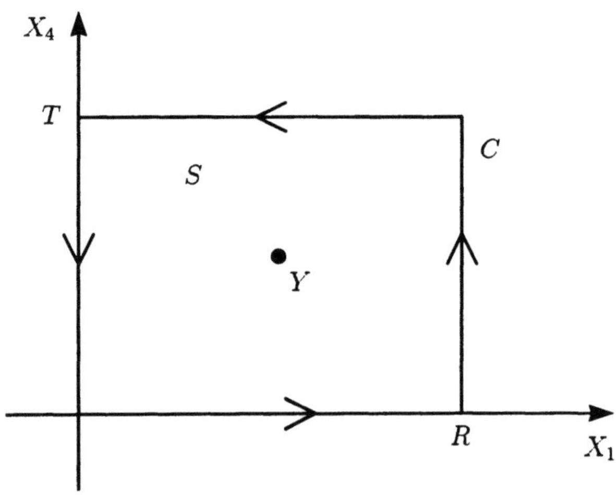

Fig. 10. Rectangular Wegner-Wilson loop in Euclidean space-time in the $X_1 - X_4$ plane. The linear extensions are R in X_1 direction and T in X_4 direction. Y is the reference point used in the application of the non-abelian Stokes theorem.

path-ordered exponential in (2.78), we will use the technique of the cumulant expansion and the assumptions 1-3 of the SVM. We make the following replacements in the formulae (2.38) ff. of Sect. 2.3:

$$\begin{aligned}
&\tau \to (X_1, X_4),\\
&B(\tau_i) \to \hat{\mathcal{G}}_{14}(Y, X^{(i)}, C_{X^{(i)}}),\\
&E(B(\tau_1)..B(\tau_n)) \to \frac{1}{3}\mathrm{Tr}\langle \hat{\mathcal{G}}_{14}(Y, X^{(1)}; C_{X^{(1)}})...\hat{\mathcal{G}}_{14}(Y, X^{(n)}; C_{X^{(n)}})\rangle,\\
&t \to -ig,\\
&f(t) \to W(C).
\end{aligned} \tag{2.79}$$

With this we can express $W(C)$ as an exponential of cumulants as given in (2.41). From Ass. 3 of the SVM we get that all cumulants for odd numbers of gluon field strengths vanish. The lowest nontrivial cumulant is the second one, K_2, and from (2.51) we find

$$K_2(1,2) \to \frac{1}{3}\mathrm{Tr}\langle \mathrm{P}(\hat{\mathcal{G}}_{14}(Y, X^{(1)}; C_{X^{(1)}})\hat{\mathcal{G}}_{14}(Y, X^{(2)}; C_{X^{(2)}})\rangle. \tag{2.80}$$

If we cut off the cumulant expansion (2.41) at $n = 2$ we get then for the Wegner-Wilson loop

$$\begin{aligned}
W(C) &= \exp\Big\{-\frac{g^2}{2}\int_S \mathrm{d}X_1 \mathrm{d}X_4 \int_S \mathrm{d}X_1' \mathrm{d}X_4'\\
&\langle \frac{1}{3}\mathrm{Tr}\ \mathrm{P}\ [\hat{\mathcal{G}}_{14}(Y, X, C_X)\hat{\mathcal{G}}_{14}(Y, X', C_{X'})]\rangle\Big\}\\
&= \exp\left\{-\frac{2\pi^2}{3}\int_S \mathrm{d}X_1 \mathrm{d}X_4 \int_S \mathrm{d}X_1' \mathrm{d}X_4' F_{1414}(X - X')\right\},
\end{aligned} \tag{2.81}$$

where in the last step we used (2.61), i.e. Ass. 1 of the SVM. Next the assumption 2 of short-range correlation for the field strengths enters in a crucial way. This implies that for large Wegner-Wilson loops the integration over X_1', X_4' keeping X_1, X_4 fixed in (2.81) gives essentially a factor a^2. The remaining X_1, X_4 integration gives then the area of $S = RT$. Thus we arrive at an area law for the Wegner-Wilson loop for $R, T \gg a$:

$$W(C) = e^{-\sigma \cdot R \cdot T}, \tag{2.82}$$

where the constant σ is obtained as

$$\sigma = \frac{\pi^3 \kappa G_2}{36}\int_0^\infty \mathrm{d}Z^2 D(-Z^2) = \frac{32\pi\kappa G_2 a^2}{81}. \tag{2.83}$$

We leave the proof of (2.83) as an exercise for the reader. Comparing (2.77) and (2.82) we find that the SVM produces a linearly rising potential

$$V(R) = \sigma \cdot R \quad \text{for} \quad R \gg a, \tag{2.84}$$

where σ is the string tension.

The results (2.82)-(2.84) were derived in the framework of the SVM in [40], where the model was introduced, and they are interesting in quite a number of respects:

Only for $\kappa \neq 0$ does one get an area law and thus confinement. The D_1 term which is the only one present in the abelian theory produces no confinement. In the SVM confinement is related to an effective chromomagnetic monopole condensate in the vacuum.

The short range correlation for the field strengths produces a long range correlation for the potentials if the D term is present in (2.61), i.e. if $\kappa \neq 0$.

The string tension σ is obtained numerically with the input (2.70) as

$$\sigma = (420 \text{ MeV})^2 . \tag{2.85}$$

This is very consistent with the phenomenological determination of $\sigma \simeq (430 \text{ MeV})^2$ from the charmonium spectrum [51].

The SVM has been applied in many other studies of low energy hadronic phenomena (cf. [39] for a review). It was, for instance, possible to calculate flux distributions around a static quark-antiquark pair [52]. The results compare well with lattice gauge theory calculations wherever the latter are available.

Let us come back to the calculation of the Wegner-Wilson loop above. It is legitimate to ask about the contribution of higher cumulants to $W(C)$. How do they modify (2.81)? It turns out that higher cumulants may cause some problems, which we discuss in Appendix A together with a proposal for their remedy.

3 Soft Hadronic Reactions

3.1 General Considerations

In this section we will present a microscopic approach towards hadron-hadron diffractive scattering (cf. [26], [53]). Consider as an example elastic scattering of two hadrons h_1, h_2

$$h_1 + h_2 \to h_1 + h_2 \tag{3.1}$$

at high energies and small momentum transfer. We will look at reaction (3.1) from the point of view of an observer living in the "femto-universe", i.e. we imagine having a microscope with resolution much better than 1 fm for observing what happens during the collision. Of course, we should choose an appropriate resolution for our microscope. If we choose the resolution much too good, we will see too many details of the internal structure of the hadrons which are irrelevant for the reaction considered and we will miss the essential features. The same is true if the resolution is too poor. In [53] we used a series of simple arguments based on the uncertainty relation to estimate this appropriate resolution.

Let $t = 0$ be the nominal collision time of the hadrons in (3.1) in the c.m. system. This is the time when the hadrons $h_{1,2}$ have maximal spatial overlap. Let furthermore be $t_0/2$ the time when, in an inelastic collision, the first produced hadrons appear. We estimate $t_0 \approx 2$ fm from the LUND model of particle production [54]. Then the appropriate resolution, i.e. the cutoff in transverse parton momenta k_T of the hadronic wave functions to be chosen for describing reaction (3.1) in an economical way is

$$k_T^2 \leq \sqrt{s}/(2t_0) \tag{3.2}$$

where $\sqrt{s}$ is the c.m. energy. Modes with higher k_T can be assumed to be integrated out. With this we could argue that over the time interval

$$-\frac{1}{2}t_0 \leq t \leq \frac{1}{2}t_0 \tag{3.3}$$

the following should hold or better: could be assumed:

(a) The parton state of the hadrons does not change qualitatively, i.e. parton annihilation and production processes can be neglected for this time.

(b) Partons travel in essence on straight lightlike world lines.

(c) The partons undergo "soft" elastic scattering.

The strategy is now to study first soft parton-parton scattering in the femto-universe. There, the relevant interaction will turn out to be mediated by the gluonic vacuum fluctuations. We have argued at length in Sect. 2 that these have a highly nonperturbative character. In this way the nonperturbative QCD vacuum structure will enter the picture for high energy soft hadronic reactions. Once we have solved the problem of parton-parton scattering we have to fold the partonic S-matrix with the hadronic wave functions of the appropriate resolution (3.2) to get the hadronic S-matrix elements.

We will now give an outline of the various steps in this program.

3.2 The Functional Integral Approach to Parton-Parton Scattering

Consider first quark-quark scattering:

$$q(p_1) + q(p_2) \to q(p_3) + q(p_4), \tag{3.4}$$

where we set

$$\begin{aligned} s &= (p_1 + p_2)^2 \\ t &= (p_1 - p_3)^2 \\ u &= (p_1 - p_4)^2. \end{aligned} \tag{3.5}$$

Of course, free quarks do not exist in QCD, but let us close our eyes to this at the moment. Now we should calculate the scattering of the quarks over the finite time interval (3.3) of length $t_0 \approx 2$ fm. Let us assume that 2 fm is nearly infinitely long on the scale of the femto universe and use the standard reduction formula, due to Lehmann, Symanzik, and Zimmermann, to relate the S-matrix element for (3.4) to an integral over the 4-point function of the quark fields. We use the following normalization for our quark states

$$\begin{aligned}&\langle q(p_j, s_j, A_j) \mid q(p_k, s_k, A_k)\rangle \\ &\qquad = \delta_{s_j s_k}\delta_{A_j A_k}(2\pi)^3\sqrt{2p_j^0 2p_k^0}\,\delta^3(\mathbf{p}_j - \mathbf{p}_k) \\ &\qquad \equiv \delta(j,k),\end{aligned} \tag{3.6}$$

where s_j, s_k are the spin and A_j, A_k the colour indices. With this we get

$$\begin{aligned}&\langle q(p_3, s_3, A_3) q(p_4, s_4, A_4)|S|q(p_1, s_1, A_1) q(p_2, s_2, A_2)\rangle \\ &\equiv \langle 3,4|S|1,2\rangle \\ &= \langle 3,4|1,2\rangle + Z_\psi^{-2}\Big\{(4|(i\overrightarrow{\not\partial} - m_q') \otimes (3|(i\overrightarrow{\not\partial} - m_q') \\ &\qquad\qquad \langle 0|\mathrm{T}(q(4)q(3)\bar{q}(1)\bar{q}(2))|0\rangle \\ &\qquad\qquad (i\overleftarrow{\not\partial} + m_q')|1) \otimes (i\overleftarrow{\not\partial} + m_q')|2)\Big\}.\end{aligned} \tag{3.7}$$

Here Z_ψ is the quark wave function renormalization constant and m_q' the renormalized quark mass. We use a shorthand notation

$$\begin{aligned}&|j) \equiv u_{s_j, A_j}(p_j) e^{-ip_j x_j}, \\ &(j| = e^{ip_j x_j}\bar{u}_{s_j, A_j}(p_j), \\ &q(j) \equiv q(x_j), \\ &(j = 1, .., 4),\end{aligned} \tag{3.8}$$

where u is the spinor in Dirac and colour space. Two repeated arguments $j, k, ...$ imply a space-time integration, for instance

$$\bar{q}(1)(i\overleftarrow{\not\partial} + m_q')|1) \equiv \int \mathrm{d}x_1 \bar{q}(x_1)(i\overleftarrow{\not\partial} + m_q') e^{-ip_1 x_1} u_{s_1, A_1}(p_1). \tag{3.9}$$

Thus in (3.7) we have four integrations over $x_1, ..., x_4$.

We can represent the 4-point function of the quark fields as a functional integral:

$$\begin{aligned}&\langle 0|\mathrm{T}(q(4)q(3)\bar{q}(1)\bar{q}(2))|0\rangle \\ &= Z^{-1}\int \mathcal{D}(G, q, \bar{q}) \exp\Big\{i\int \mathrm{d}x \mathcal{L}_{\mathrm{QCD}}(x)\Big\} q(4)q(3)\bar{q}(1)\bar{q}(2),\end{aligned} \tag{3.10}$$

where Z is the partition function:

$$Z = \int \mathcal{D}(G, q, \bar{q}) \exp\Big\{i\int \mathrm{d}x \mathcal{L}_{\mathrm{QCD}}(x)\Big\}. \tag{3.11}$$

The QCD Lagrangian (1.1) is bilinear in the quark and antiquark fields. Thus – as is well known – the functional integration over q and $\bar{q}$ can be carried out immediately. After some standard manipulations we arrive at the following expression:

$$\begin{aligned}
&\langle 0|\mathrm{T}(q(4)q(3)\bar{q}(1)\bar{q}(2))|0\rangle \\
&= \frac{1}{Z}\int \mathcal{D}(G)\exp\left\{-i\int \mathrm{d}x\frac{1}{2}\mathrm{Tr}(G_{\lambda\rho}(x)G^{\lambda\rho}(x))\right\} \\
&\det[-i(i\gamma^{\lambda}D_{\lambda}-m_q+i\epsilon)] \\
&\left\{\frac{1}{i}S_F(4,2;G)\frac{1}{i}S_F(3,1;G)-(3\leftrightarrow 4)\right\}.
\end{aligned} \tag{3.12}$$

Here $S_F(j,k;G)\equiv S_F(x_j,x_k;G)$ is the unrenormalized quark propagator in the given gluon potential $G_\lambda(x)$. We have

$$(i\gamma^{\mu}D_{\mu}-m_q)S_F(x,y;G)=-\delta(x-y). \tag{3.13}$$

Functional integrals as in (3.12) will occur frequently further on. Let $F(G)$ be some functional of the gluon potentials. We will denote the functional integral over $F(G)$ by brackets $\langle F(G)\rangle_G$:

$$\begin{aligned}
\langle F(G)\rangle_G &:= \frac{1}{Z}\int \mathcal{D}(G)\exp\left\{-i\int \mathrm{d}x\frac{1}{2}\mathrm{Tr}(G_{\lambda\rho}G^{\lambda\rho})\right\}\cdot \\
&\det[-i(i\gamma^{\lambda}D_{\lambda}-m_q+i\epsilon)]F(G).
\end{aligned} \tag{3.14}$$

Now we insert (3.12) in (3.7) and get

$$\langle 3,4|S|1,2\rangle = \langle 3,4|1,2\rangle - Z_{\psi}^{-2}\langle \mathcal{M}^F_{31}(G)\mathcal{M}^F_{42}(G)-(3\leftrightarrow 4)\rangle_G, \tag{3.15}$$

where

$$\begin{aligned}
&\mathcal{M}^F_{kj}(G)=(k|(i\overrightarrow{\not\partial}-m_q')S_F(i\overleftarrow{\not\partial}+m_q')|j), \\
&(k=3,4;\ j=1,2).
\end{aligned} \tag{3.16}$$

The term $\mathcal{M}^F_{31}\cdot\mathcal{M}^F_{42}$ on the r.h.s. of (3.15) corresponds to the t-channel exchange diagrams, the second term, where the role of the quarks 3 and 4 is interchanged, to the u-channel exchange diagrams (Fig. 11). The latter term should be unimportant for high energy, small $|t|$ scattering. Thus we neglect it in the following. For the scattering of different quark flavours it is absent anyway. We set, therefore:

$$\langle 3,4|S|1,2\rangle \cong \langle 3,4|1,2\rangle - Z_{\psi}^{-2}\langle \mathcal{M}^F_{31}(G)\mathcal{M}^F_{42}(G)\rangle_G. \tag{3.17}$$

We can interpret $\mathcal{M}^F_{kj}(G)$ as scattering amplitude for quark j going to k in the fixed gluon potential $G_\lambda(x)$. To see this, let us define the wave function

$$|\psi^F_{p_j})=S_F(i\overleftarrow{\not\partial}+m_q')|j) \tag{3.18}$$

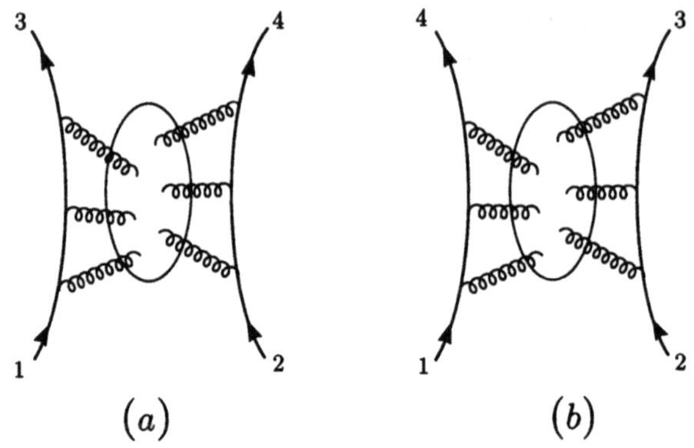

Fig. 11. The t-channel (a) and u-channel (b) exchange topologies for the diagrams describing quark-quark scattering.

which satisfies the Dirac equation with the gluon potential $G_\lambda(x)$:

$$(i\gamma^\lambda D_\lambda - m_q)|\psi^F_{p_j}) = 0. \tag{3.19}$$

Furthermore we use the Lippmann-Schwinger equation for S_F:

$$S_F = S_F^{(0)} - S_F^{(0)}(g\not{G} - \delta m_q)S_F \tag{3.20}$$

where $S_F^{(0)}$ is the free quark propagator for mass m_q' and $\delta m_q = m_q' - m_q$ is the quark mass shift. Inserting (3.20) and (3.18) in (3.16) gives after some simple algebra

$$\mathcal{M}^F_{kj}(G) = (p_k|(g\not{G} - \delta m)|\psi^F_{p_j}). \tag{3.21}$$

This represents $\mathcal{M}^F_{kj}$ in the form a scattering amplitude should have: a complete incoming wave is folded with the potential and the free outgoing wave. However, there is a small problem. The wave function $|\psi^F_{p_j})$ defined in (3.18) does not satisfy the boundary condition which we should have for using it in the scattering amplitude, i.e. it does not go to a free incoming wave for time $t \to -\infty$. The wave function with this boundary condition is obtained by replacing the Feynman propagator S_F in (3.18) by the retarded one, S_r. We have shown in [53] that in the high energy limit this replacement can indeed be justified for the calculation of $\mathcal{M}^F_{kj}(G)$ if the gluon potential $G_\lambda(x)$ contains only a limited range of frequencies:

$$\mathcal{M}^F_{kj}(G) \simeq \mathcal{M}^r_{kj}(G) = (p_k|(g\not{G} - \delta m)|\psi^r_{p_j}), \tag{3.22}$$

where

$$|\psi^r_{p_j}) = S_r(i\overleftarrow{\not{\partial}} + m_q')|j), \tag{3.23}$$

which satisfies

$$(i\gamma^\mu D_\mu - m_q)|\psi^r_{p_j}) = 0, \tag{3.24}$$

$$|\psi^r_{p_j}) \to |j) \quad \text{for} \quad t \to -\infty. \tag{3.25}$$

We summarize the results of this subsection: At high energies and small $|t|$ the quark-quark scattering amplitude can be obtained by calculating first the scattering of quark 1 going to 3 and 2 going to 4 in the same fixed gluon potential $G_\lambda(x)$. Let the corresponding scattering amplitudes be $\mathcal{M}^r_{31}(G)$ and $\mathcal{M}^r_{42}(G)$ (cf. (3.22)). Then the product of these two amplitudes is to be integrated over all gluon potentials with the measure given by the functional integral (3.14) and this gives the quark-quark scattering amplitude (3.17). The point of our further strategy is to continue making suitable high energy approximations in the integrand of this functional integral which will be evaluated finally using the methods of the stochastic vacuum model.

In the following it will be convenient to choose a coordinate system for the description of reaction (3.4) where the quarks 1,3 move with high velocity in essence in positive x^3 direction, the quarks 2,4 in negative x^3 direction. We define the light cone coordinates

$$x_\pm = x^0 \pm x^3 \tag{3.26}$$

and in a similar way the $\pm$ components of any 4-vector. With this we have for the 4-momenta of our quarks:

$$p_j = \begin{pmatrix} \frac{1}{2}p_{j+} & + & \frac{\mathbf{p}^2_{jT}+m'^2_q}{2p_{j+}} \\ & \mathbf{p}_{jT} & \\ \frac{1}{2}p_{j+} & - & \frac{\mathbf{p}^2_{jT}+m^2_{q'}}{2p_{j+}} \end{pmatrix} \tag{3.27}$$

for $j = 1,3$ with $p_{j+} \to \infty$ and

$$p_k = \begin{pmatrix} \frac{1}{2}p_{k-} & + & \frac{\mathbf{p}^2_{kT}+m'^2_{q'}}{2p_{k-}} \\ & \mathbf{p}_{kT} & \\ -\frac{1}{2}p_{k-} & + & \frac{\mathbf{p}^2_{kT}+m^2_{q'}}{2p_{k-}} \end{pmatrix} \tag{3.28}$$

for $k = 2,4$ with $p_{k-} \to \infty$.

3.3 The Eikonal Expansion

The problem is now to solve the Dirac equation (3.24) for arbitrary external gluon potential $G_\lambda(x)$. Of course, we cannot do this exactly. But we are only interested in the high energy, small $|t|$ limit. This suggests to use an eikonal type approach. This works indeed, but it is not as straightforward as one would think at first, since the Dirac equation is of first order in the derivatives, whereas the eikonal expansion is easy to make for a second-order differential equation. What we did in [53] was to make an ansatz for the Dirac field $\psi^r_{p_j}(x)$ in terms of a "potential" $\phi_j(x)$ as follows:

$$\psi^r_{p_j}(x) = (i\gamma^\lambda D_\lambda + m_q)\phi_j(x). \tag{3.29}$$

A suitable boundary condition for $\phi_j(x)$ which is compatible with (3.25) is

$$\phi_j(x) \to \frac{1}{p_j^0 + m_q}\frac{1+\gamma^0}{2}e^{-ip_jx}u(p_j) \tag{3.30}$$

for $t \to -\infty$. Inserting (3.29) into the Dirac equation (3.24) gives:

$$\{i\gamma^\lambda D_\lambda - m_q\}\{i\gamma^\rho D_\rho + m_q\}\phi_j(x) = 0. \tag{3.31}$$

For the case of no gluon field, $G_\lambda(x) = 0$, the covariant derivatives D_λ degenerate to the ordinary ones, ∂_λ, and (3.31) to the Klein-Gordon equation:

$$(\Box + m_q^2)\phi_j(x) = 0. \tag{3.32}$$

Thus the problem for the Dirac field is in essence reduced to a scalar field-type problem which is handled more easily.

Now it is more or less straightforward to turn the theoretical crank and to obtain the eikonal approximations for $\phi_j(x)$ and $\psi^r_{p_j}(x)$, respectively. Take $j = 1$ as an example. We make the ansatz

$$\phi_1(x) = e^{-ip_1x}\tilde{\phi}_1(x). \tag{3.33}$$

Inserting p_1 from (3.27) we see that on the r.h.s. of (3.33) the fast varying phase factor is

$$\exp(-i\frac{1}{2}p_{1+}x_-),$$

since $p_{1+} \to \infty$. Assuming all the remaining factors to vary slowly with x we insert (3.33) in (3.31) and order the resulting terms in powers of $1/p_{1+}$. The solution of (3.31) to leading order in $1/p_{1+}$ is then easily obtained. The final formula for $\psi^r_{p_1}(x)$ reads:

$$\psi^r_{p_1}(x) = V_-(x_+, x_-, \mathbf{x}_T).\left\{1 + O\left(\frac{1}{p_{1+}}\right)\right\}e^{-ip_1x}u(p_1), \tag{3.34}$$

where

$$V_-(x_+, x_-, \mathbf{x}_T) = \mathrm{P}\left\{\exp\left[-\frac{i}{2}g\int_{-\infty}^{x_+} dx'_+ G_-(x'_+, x_-, \mathbf{x}_T)\right]\right\} \tag{3.35}$$

and P means path ordering. When coming in, the quark picks up a non-abelian phase factor, just the ordered integral of G along the path. Of course, V_- is a connector, as studied in Sect. 2.1, but now in Minkowski space and for a straight light-like line running from $-\infty$ to x (Fig. 12).

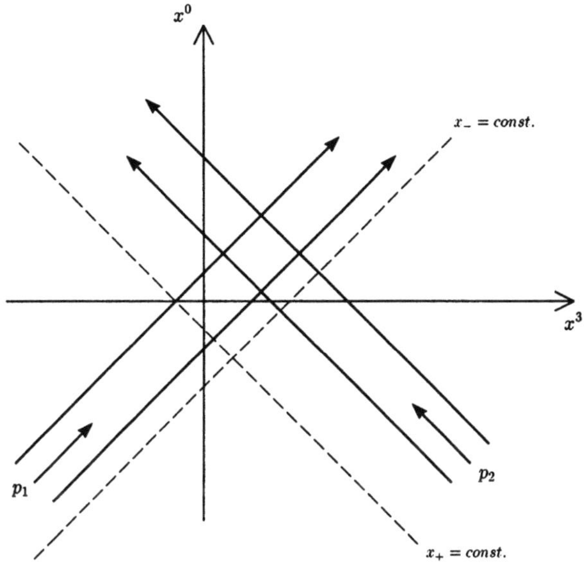

Fig. 12. Projection of the world lines of the quarks 1(2) moving at high velocity in positive (negative) x^3 direction onto the $x^0 - x^3$ plane in Minkowski space. The non-abelian phase factors V_- in (3.35) (V_+ in (3.37)) are the connectors taken along lines $x_-, \mathbf{x}_T = const.$ ($x_+, \mathbf{x}_T = const.$).

In a similar way we obtain for the quark with initial momentum p_2, i.e. the one coming in from the right, for $p_{2-} \to \infty$:

$$\psi^r_{p_2}(x) = V_+(x_+, x_-, \mathbf{x}_T) \left\{ 1 + O\left(\frac{1}{p_{2-}}\right) \right\} e^{-ip_2 x} u(p_2), \tag{3.36}$$

where

$$V_+(x_+, x_-, \mathbf{x}_T) = \mathrm{P}\left\{ \exp\left[-\frac{i}{2} g \int_{-\infty}^{x_-} \mathrm{d}x'_- G_+(x_+, x'_-, \mathbf{x}_T) \right] \right\}. \tag{3.37}$$

Recall that

$$G_\pm(x) = \left(G^{0a}(x) \pm G^{3a}(x)\right) \frac{\lambda_a}{2} \tag{3.38}$$

are matrices in colour space. Thus path ordering in (3.35) and (3.37) is essential.

A solution for $\psi^r_{p_1}$ ($\psi^r_{p_2}$) as a series expansion in powers of $1/p_{1+}$($1/p_{2-}$) was obtained to all orders in [55].

3.4 The Quark-Quark Scattering Amplitude

We can now insert our high energy approximations (3.34), (3.36) for $\psi^r_{p_{1,2}}(x)$ in the expression for $\mathcal{M}^{(r)}_{kj}$ in (3.22). The resulting integrals are easily done and we get for $p_{1+}, p_{3+} \to \infty$:

$$\begin{aligned}
\mathcal{M}^r_{31}(G) \to & \int \mathrm{d}x e^{i(p_3-p_1)x} \\
& \bar{u}(p_3)(g\not{G}(x) - \delta m)V_-(x)u(p_1) \\
& \to \frac{i}{2}\int \mathrm{d}x_+\mathrm{d}x_-\mathrm{d}^2x_T \ \exp\left[\frac{i}{2}(p_3-p_1)_+x_- - i(\mathbf{p}_3-\mathbf{p}_1)_T\cdot\mathbf{x}_T\right] \\
& \bar{u}(p_3)\gamma_+\frac{\partial}{\partial x_+}V_-(x_+,x_-,\mathbf{x}_T)u(p_1) \\
& \to i\sqrt{p_{3+}p_{1+}}\cdot\delta_{s_3,s_1}\int \mathrm{d}x_-\mathrm{d}^2x_T \\
& \exp[\frac{i}{2}(p_3-p_1)_+x_- - i(\mathbf{p}_3-\mathbf{p}_1)_T\cdot\mathbf{x}_T] \\
& [V_-(\infty,x_-,\mathbf{x}_T) - \mathbb{1}]_{A_3,A_1}.
\end{aligned} \tag{3.39}$$

In a similar way we obtain for $p_{2-}, p_{4-} \to \infty$:

$$\begin{aligned}
\mathcal{M}^r_{42}(G) \to & \, i\sqrt{p_{4-}\cdot p_{2-}}\cdot\delta_{s_4,s_2} \\
& \int \mathrm{d}y_+\mathrm{d}^2y_T \exp[\frac{i}{2}(p_4-p_2)_-y_+ - i(\mathbf{p}_4-\mathbf{p}_2)_T\cdot\mathbf{y}_T] \\
& [V_+(y_+,\infty,\mathbf{y}_T) - \mathbb{1}]_{A_4,A_2}.
\end{aligned} \tag{3.40}$$

Here we have written out the spin and colour indices of the in- and outgoing quarks (cf. (3.6), (3.7)). We have used furthermore:

$$\begin{aligned}
\bar{u}_{s_3}(p_3)\gamma^\mu u_{s_1}(p_1) &\to \sqrt{p_{3+}p_{1+}}\cdot\delta_{s_3,s_1}n^\mu_+ \quad \text{for} \quad p_{1,3+}\to\infty, \\
\bar{u}_{s_4}(p_4)\gamma^\mu u_{s_2}(p_2) &\to \sqrt{p_{4-}p_{2-}}\cdot\delta_{s_4,s_2}n^\mu_- \quad \text{for} \quad p_{2,4-}\to\infty,
\end{aligned} \tag{3.41}$$

where

$$n^\mu_\pm = \begin{pmatrix} 1 \\ 0 \\ 0 \\ \pm 1 \end{pmatrix}. \tag{3.42}$$

Finally, the $x_+(x_-)$ integration for $\mathcal{M}^r_{31}(\mathcal{M}^r_{42})$ could be done with the help of (2.15) or rather the analogous equation for connectors in Minkowski space time.

Now we can insert everything in our expression (3.17) for the S-matrix element. This gives:

$$\begin{aligned}&\langle 3,4|S|1,2\rangle = \langle 3,4|1,2\rangle + \\ &\sqrt{p_{3+}p_{1+}p_{4-}p_{2-}} \cdot \delta_{s_3,s_1}\delta_{s_4,s_2} Z_\psi^{-2} \int \mathrm{d}x_- \mathrm{d}^2 x_T \int \mathrm{d}y_+ \mathrm{d}^2 y_T \\ &\exp\left[\frac{i}{2}(p_3-p_1)_+ x_- - i(\mathbf{p}_3-\mathbf{p}_1)_T \cdot \mathbf{x}_T\right] \\ &\exp\left[\frac{i}{2}(p_4-p_2)_- y_+ - i(\mathbf{p}_4-\mathbf{p}_2)_T \cdot \mathbf{y}_T\right] \\ &\langle [V_-(\infty, x_-, \mathbf{x}_T) - 1\!\!1]_{A_3,A_1} [V_+(y_+,\infty,\mathbf{y}_T) - 1\!\!1]_{A_4,A_2}\rangle_G \,. \qquad (3.43)\end{aligned}$$

From translational invariance of the functional integral we have:

$$\begin{aligned}&\langle [V_-(\infty, x_-, x_T) - 1\!\!1]_{A_3,A_1} [V_+(y_+,\infty,\mathbf{y}_T) - 1\!\!1]_{A_4,A_2}\rangle_G \\ &= \langle [V_-(\infty, 0, \mathbf{x}_T - \mathbf{y}_T) - 1\!\!1]_{A_3,A_1} [V_+(0,\infty,0) - 1\!\!1]_{A_4,A_2}\rangle_G. \qquad (3.44)\end{aligned}$$

Inserting (3.44) in (3.43), we can pull out the δ-function for the overall energy-momentum conservation and we get finally:

$$\langle 3,4|S|1,2\rangle = \langle 3,4|1,2\rangle + +i(2\pi)^4 \delta(p_3+p_4-p_1-p_2)\langle 3,4|T|1,2\rangle,$$

$$\begin{aligned}&\langle 3,4|T|1,2\rangle = i2\sqrt{p_{3+}p_{1+}p_{4-}p_{2-}} \cdot \delta_{s_3,s_1}\delta_{s_4,s_2}(-Z_\psi^{-2}) \int \mathrm{d}^2 z_T e^{i\mathbf{q}_T\cdot\mathbf{z}_T} \\ &\langle [V_-(\infty,0,\mathbf{z}_T) - 1\!\!1]_{A_3,A_1} [V_+(0,\infty,0) - 1\!\!1]_{A_4,A_2}\rangle_G \,. \qquad (3.45)\end{aligned}$$

Here q is the momentum transfer:

$$\begin{aligned} q &= p_1 - p_3 = p_4 - p_2, \\ q^2 &= t. \end{aligned} \qquad (3.46)$$

In the high energy limit q is purely transverse

$$q \to \begin{pmatrix} 0 \\ \mathbf{q}_T \\ 0 \end{pmatrix}, \quad q^2 \to -\mathbf{q}_T^2. \qquad (3.47)$$

Using different techniques the type of formula (3.45) was also obtained in [56].

In (3.45) we still have to calculate the wave function renormalization constant Z_ψ. This can be done by considering a suitable matrix element of the baryon number current

$$\frac{1}{3}\bar{q}(x)\gamma^\mu q(x)$$

which is conserved and, therefore, needs no renormalization. The result is (cf. [53]):

$$Z_\psi = \frac{1}{3}\langle \mathrm{Tr} V_-(\infty,0,0)\rangle_G. \qquad (3.48)$$

Let us summarize the results obtained so far:

The quark-quark scattering amplitude (3.45) is diagonal in the spin indices. Thus we get helicity conservation in high energy quark-quark scattering. Using (3.41) we can write the spin factor in (3.45) as

$$2\sqrt{p_{3+}p_{1+}p_{4-}p_{2-}}\delta_{s_3,s_1}\delta_{s_4,s_2} \cong \bar{u}_{s_3}(p_3)\gamma^\mu u_{s_1}(p_1)\bar{u}_{s_4}(p_4)\gamma_\mu u_{s_2}(p_2) \quad (3.49)$$

for $p_{1,3+} \to \infty$ and $p_{2,4-} \to \infty$. This $\gamma^\mu \otimes \gamma_\mu$ structure was postulated for high energy quark-quark scattering in the Donnachie-Landshoff model for the "Pomeron" coupling [23]. However, a study of quark-antiquark scattering (cf. [53] and below) reveals that (3.45) does not allow an interpretation in terms of an effective Lorentz vector exchange between the quarks. The amplitude (3.45) has both, charge conjugation C even and odd contributions. The C even part corresponds to the "Pomeron" and this is not a Lorentz-vector exchange, but the coherent sum of spin 2,4,6,... exchanges (cf. [57], [53]). The C-odd part corresponds to the "Odderon" introduced in [58] and this is indeed a Lorentz vector exchange.

The quark-quark scattering amplitude (3.45) is governed by the correlation function of two connectors or string operators $V_\pm$ associated with two light-like Wegner-Wilson lines (Fig. 13).

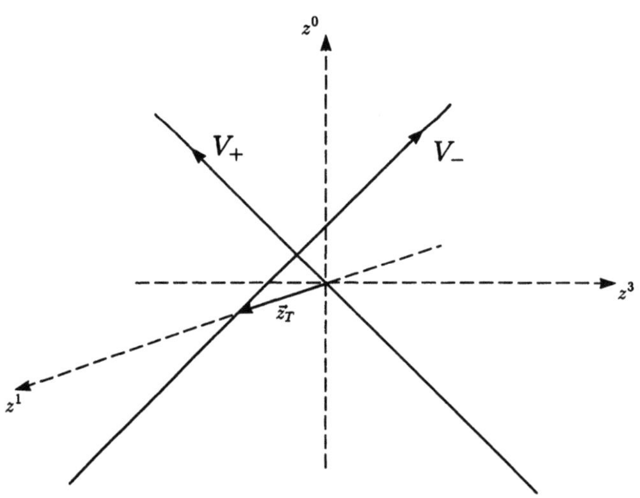

Fig. 13. Two light-like lines on which the associated string operators $V_\pm$ in (3.45) are evaluated. Their correlation function governs quark-quark scattering at high energies.

The first numerical evaluations of (3.45) using the methods of the SVM were done in [59]. However, it turned out that quark-quark scattering was calculable from (3.45) for abelian gluons only. Indeed, we are embarked on a program to reproduce in this way the results obtained in [25] in the framework of perturbation theory for high energy scattering [60]. For the non-abelian gluons difficulties arose having to do with our neglect of quark confinement. This was really a blessing in disguise and the solution proposed in [59] was to consider directly hadron-hadron scattering, representing the hadrons as $q\bar{q}$ and qqq wave packets for mesons and baryons, respectively. We will see below how this is done.

3.5 The Scattering of Systems of Quarks, Antiquarks, and Gluons

Let us consider now the scattering of systems of partons. As an example we study the scattering of two $q\bar{q}$ pairs on each other:

$$q(1) + \bar{q}(1') + q(2) + \bar{q}(2') \to q(3) + \bar{q}(3') + q(4) + \bar{q}(4'), \tag{3.50}$$

where we set $q(i) \equiv q(p_i, s_i, A_i)$, $\bar{q}(i') \equiv \bar{q}(p'_i, s'_i, A'_i)$ $(i = 1, ..., 4)$ with p_i, s_i, A_i (p'_i, s'_i, A'_i) the momentum, spin, and colour labels for quarks (antiquarks). We assume the particles with odd (even) indices i to have very large momentum components in positive (negative) x^3 direction (Fig. 14), i.e. we assume:

$$\begin{aligned} p_{i+}, p'_{i+} &\to \infty \quad \text{for} \quad i \text{ odd}, \\ p_{i-}, p'_{i-} &\to \infty \quad \text{for} \quad i \text{ even}. \end{aligned} \tag{3.51}$$

The transverse momenta are assumed to stay limited.

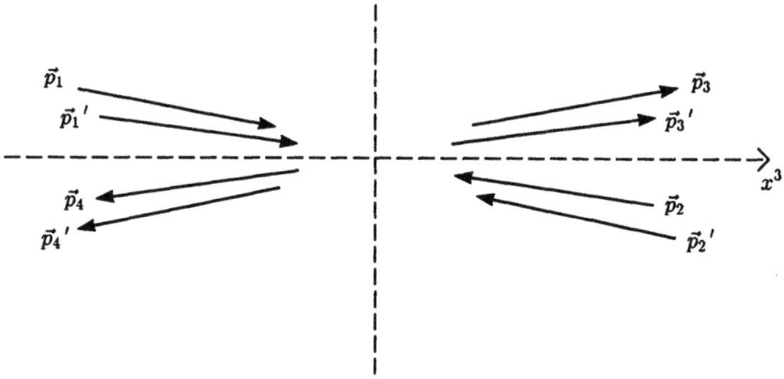

Fig. 14. Sketch of reaction (3.50) in the overall c.m. system

Of course, the reduction formula can also be applied for the reaction (3.50). We have to be careful in keeping disconnected pieces. The further strategy is completely analogous to the one employed in Sect. 3.2 for deriving (3.7)-(3.17). As we dropped the u-channel exchange diagrams in Sect. 3.2, we drop now all terms which are estimated to give a vanishing contribution to high energy small momentum transfer scattering. These terms are characterized by large momenta of order of the c.m. energy $\sqrt{s}$ flowing through gluon lines, leading to suppression factors of order $1/s$. Keeping only the t-channel exchange terms and performing all the steps as done in Sects. 3.2, 3.3 for quark-quark scattering leads finally to a very simple answer for the S-matrix element corresponding to reaction (3.50) in the limit (3.51):

$$\begin{aligned}&\langle 3,3',4,4'|S|1,1',2,2'\rangle \rightarrow\\ &\langle[\delta(3,1) - iZ_{\psi}^{-1}\mathcal{M}_{31}^{r}(G)][\delta(3',1') - iZ_{\psi}^{-1}\mathcal{M}'^{r}_{3'1'}(G)]\\ &[\delta(4,2) - iZ_{\psi}^{-1}\mathcal{M}_{42}^{r}(G)][\delta(4',2') - iZ_{\psi}^{-1}\mathcal{M}'^{r}_{4'2'}(G)]\rangle_G .\end{aligned} \tag{3.52}$$

Here $\delta(i,j)$ is as defined in (3.6), $\mathcal{M}_{31}^{r}(G)$ and $\mathcal{M}_{42}^{r}(G)$ are as in (3.22) and $\mathcal{M}'^{r}_{3'1'}(G)$ and $\mathcal{M}'^{r}_{4'2'}(G)$ are the corresponding amplitudes for the scattering of antiquarks on the gluon potential $G_\lambda(x)$. We define

$$(j'| = \bar{v}_{s'_j,A'_j}(p'_j)e^{-ip'_j x}, \tag{3.53}$$

where $v_{s'_j,A'_j}(p'_j)$ is the Dirac and colour spinor for the antiquark $\bar{q}(j')$. We have then with S_r the retarded Green's function for quarks in the gluon potential $G_\lambda(x)$:

$$\mathcal{M}'^{r}_{k'j'}(G) = -(j'|(i\overrightarrow{\not\partial} - m'_q)S_r(i\overleftarrow{\not\partial} + m'_q)|k'), \tag{3.54}$$

where $(k',j') = (3',1'),(4',2')$.

In the high energy limit (3.51) the scattering amplitudes (3.54) can again be obtained in the eikonal approximation. Indeed, we can just use C-invariance to get:

$$\begin{aligned}&\mathcal{M}'^{r}_{3'1'}(G) \rightarrow i\sqrt{p'_{3+}p'_{1+}}\cdot\delta_{s'_3,s'_1}\\ &\int dx_- d^2x_T \exp\left[\frac{i}{2}(p'_3-p'_1)_+x_- - i(\mathbf{p}'_3-\mathbf{p}'_1)_T\cdot\mathbf{x}_T\right]\\ &[V_-^*(\infty,x_-,\mathbf{x}_T) - \mathbb{1}]_{A'_3,A'_1},\end{aligned} \tag{3.55}$$

$$\begin{aligned}&\mathcal{M}'^{r}_{4'2'}(G) \rightarrow i\sqrt{p'_{4-}p'_{2-}}\cdot\delta_{s'_4,s'_2}\\ &\int dy_+ d^2y_T \exp\left[\frac{i}{2}(p'_4-p'_2)_-y_+ - (\mathbf{p}'_4-\mathbf{p}'_2)_T\cdot\mathbf{y}_T\right]\\ &[V_+^*(y_+,\infty,\mathbf{y}_T) - \mathbb{1}]_{A'_4,A'_2}.\end{aligned} \tag{3.56}$$

In Sect. 3.1 we have argued that over the time interval (3.3) we can neglect parton production and annihilation processes. For the scattering over such a time interval we should have an effective wave function renormalization constant

$$Z_\psi = 1, \tag{3.57}$$

since the deviation of Z_ψ from 1 is just a measure of the strength of quark splitting processes: $q \to q + G$ etc. In the calculation of Z_ψ in the framework of the SVM to be described below, one finds indeed $Z_\psi = 1$, showing the consistency of this approach with the simple physical picture of Sect. 3.1. Anticipating this result we see from (3.6), (3.39), (3.40) and (3.55), (3.56) that in the S-matrix element (3.52) the $\delta(k,j)$ ($\delta(k',j')$) terms cancel with the $\mathbb{1}$ terms in $\mathcal{M}^r_{kj}(G)$ ($\mathcal{M}'^r_{k'j'}(G)$) in the limit (3.51). This leads us to the following simple rules for obtaining the S-matrix element in the high energy limit: For the right-moving quark ($1 \to 3$) we have to insert the factor:

$$\begin{aligned}&\mathcal{S}_{q+}(3,1) = \sqrt{p_{3+}p_{1+}} \cdot \delta_{s_3,s_1} \int \mathrm{d}x_- \mathrm{d}^2 x_T \\ &\exp\left[\frac{i}{2}(p_3 - p_1)_+ x_- - i(\mathbf{p}_3 - \mathbf{p}_1)_T \cdot \mathbf{x}_T\right] \\ &V_-(\infty, x_-, \mathbf{x}_T)_{A_3,A_1} .\end{aligned} \tag{3.58}$$

For the right-moving antiquark ($1' \to 3'$) we have to insert the factor:

$$\begin{aligned}&\mathcal{S}_{\bar{q}+}(3',1') = \sqrt{p'_{3+}p'_{1+}} \cdot \delta_{s'_3,s'_1} \int \mathrm{d}x_- \mathrm{d}^2 x_T \\ &\exp\left[\frac{i}{2}(p'_3 - p'_1)_+ x_- - i(\mathbf{p}'_3 - \mathbf{p}'_1)_T \cdot \mathbf{x}_T\right] \\ &V^*_-(\infty, x_-, \mathbf{x}_T)_{A'_3,A'_1} .\end{aligned} \tag{3.59}$$

For the left-moving quark ($2 \to 4$) and antiquark ($2' \to 4'$) we have to exchange the $+$ and $-$ labels everywhere in (3.58) and (3.59). This gives:

$$\begin{aligned}&\mathcal{S}_{q-}(4,2) = \sqrt{p_{4-}p_{2-}} \cdot \delta_{s_4,s_2} \int \mathrm{d}y_+ \mathrm{d}^2 y_T \\ &\exp\left[\frac{i}{2}(p_4 - p_2)_- y_+ - i(\mathbf{p}_4 - \mathbf{p}_2)_T \cdot \mathbf{y}_T\right] \\ &V_+(y_+, \infty, \mathbf{y}_T)_{A_4,A_2} ,\end{aligned} \tag{3.60}$$

$$\begin{aligned}&\mathcal{S}_{\bar{q}-}(4',2') = \sqrt{p'_{4-}p'_{2-}} \cdot \delta_{s'_4,s'_2} \int \mathrm{d}y_+ \mathrm{d}^2 y_T \\ &\exp\left[\frac{i}{2}(p'_4 - p'_2)_- y_+ - i(\mathbf{p}'_4 - \mathbf{p}'_2)_T \cdot \mathbf{y}_T\right] \\ &V^*_+(y_+, \infty, \mathbf{y}_T)_{A'_4,A'_2} .\end{aligned} \tag{3.61}$$

Finally we have to multiply together the factors $\mathcal{S}_{q^\pm}, \mathcal{S}_{\bar{q}^\pm}$ and integrate over all gluon potentials with the functional integral measure (3.14) to get

$$\langle 3,3',4,4'|S|1,1',2,2'\rangle = \langle \mathcal{S}_{q+}(3,1)\mathcal{S}_{\bar{q}+}(3',1')\mathcal{S}_{q-}(4,2)\mathcal{S}_{\bar{q}-}(4',2')\rangle_G. \quad (3.62)$$

Going from quarks to antiquarks corresponds, of course, just to the change from the fundamental representation (3) of $SU(3)_c$ to the complex conjugate representation (3^*) as we see by comparing (3.58) with (3.59) and (3.60) with (3.61).

It is an easy exercise to show that these rules can be generalized in an obvious way for the scattering of arbitrary systems of quarks and antiquarks on each other. Here we always assume that we have one distinguished collision axis and that one group of partons moves with momenta approaching infinity to the right, the other group to the left. The transverse momenta are assumed to stay limited.

In Appendix B we show that these rules can also be extended to gluons participating in the scattering. We simply have to change the colour representation in (3.58), (3.60) from the fundamental to the adjoint one. In detail we find that for a right-moving gluon

$$G(p_1, j_1, a_1) \rightarrow G(p_3, j_3, a_3) \quad (3.63)$$

the following factor has to be inserted in the S-matrix element (cf. Appendix B):

$$\begin{aligned}
&\mathcal{S}_{G+}(3,1) = \sqrt{p_{3+}p_{1+}} \cdot \delta_{j_3,j_1} \\
&\int \mathrm{d}x_- \mathrm{d}^2x_T \exp\left[\frac{i}{2}(p_3 - p_1)_+ x_- - i(\mathbf{p}_3 - \mathbf{p}_1)_T \cdot \mathbf{x}_T\right] \\
&\mathcal{V}_-(\infty, x_-, \mathbf{x}_T)_{a_3,a_1}. \qquad (3.64)
\end{aligned}$$

Here $j_{1,3}$ are the spin indices which are purely transverse, $1 \leq j_{1,3} \leq 2$. The colour indices are a_1, a_3 with $1 \leq a_{1,3} \leq 8$ and $\mathcal{V}$ is the connector for the adjoint representation of $SU(3)_c$ (cf. (B.23)).

For a left-moving gluon we have again to exchange + and − labels in (3.64).

3.6 The Scattering of Wave Packets of Partons Representing Mesons

In this section we will go from the parton-parton to hadron-hadron scattering. Our strategy will be to represent hadrons by wave packets of partons, where we make simple "Ansätze" for the wave functions. Then the partonic S-matrix element obtained by the rules derived in Sect. 3.5 will be folded with these wave functions to give the hadronic S-matrix elements. Of course, we always work in the limit of high energies and small momentum transfers.

Let us start by considering meson-meson scattering:

$$M_1(P_1) + M_2(P_2) \to M_3(P_3) + M_4(P_4), \tag{3.65}$$

where $M_{1,3}$ are again the right movers, $M_{2,4}$ the left movers. We make simple "Ansätze" for the mesons as $q\bar{q}$ wave packets as follows:

$$\begin{aligned} |M_j(P_j)\rangle = &\int \mathrm{d}^2 p_T \int_0^1 \mathrm{d}\zeta \frac{1}{(2\pi)^{3/2}} h^j_{s_j,s'_j}(\zeta, \mathbf{p}_T) \\ &\frac{1}{\sqrt{3}} \delta_{A_j,A'_j} |q(p_j, s_j, A_j), \bar{q}(P_j - p_j, s'_j, A'_j)\rangle \\ &(j = 1, ..., 4), \end{aligned} \tag{3.66}$$

where for $j = 1, 3$:

$$\begin{aligned} P_{j+} &\to \infty \\ p_{j+} &= \zeta P_{j+}, \\ \mathbf{p}_{jT} &= \frac{1}{2}\mathbf{P}_{jT} + \mathbf{p}_T \end{aligned} \tag{3.67}$$

and for $j = 2, 4$:

$$\begin{aligned} P_{j-} &\to \infty \\ p_{j-} &= \zeta P_{j-}, \\ \mathbf{p}_{jT} &= \frac{1}{2}\mathbf{P}_{jT} + \mathbf{p}_T. \end{aligned} \tag{3.68}$$

Here ζ is the longitudinal momentum fraction of the quark in the meson, $\mathbf{p}_T$ is the relative transverse momentum of q and $\bar{q}$.

We stress that we are not restricting ourselves to spin 0 mesons only. With appropriate functions $h(\zeta, \mathbf{p}_T)$ in (3.66) we can represent states of $q\bar{q}$-mesons of arbitrary spin.

We choose for our states (3.66) the usual continuum normalization:

$$\langle M_j(P')|M_j(P)\rangle = (2\pi)^3 2P^0 \delta^3(\mathbf{P}' - \mathbf{P}). \tag{3.69}$$

With (3.6) this requires:

$$\int \mathrm{d}^2 p_T \int_0^1 \mathrm{d}\zeta \; 2\zeta(1-\zeta) h^{*j}_{s,s'}(\zeta, \mathbf{p}_T) h^j_{s,s'}(\zeta, \mathbf{p}_T) = 1 \tag{3.70}$$

In (3.69) and (3.70) no summation over j is to be taken.

For later use we define the wave functions in transverse position space at fixed longitudinal momentum fraction ζ:

$$\begin{aligned} \varphi^j_{s,s'}(\zeta, \mathbf{x}_T) := \sqrt{2\zeta(1-\zeta)} \frac{1}{2\pi} \int \mathrm{d}^2 p_T \exp(i\mathbf{p}_T \cdot \mathbf{x}_T) h^j_{s,s'}(\zeta, \mathbf{p}_T) \\ (j = 1, ..., 4). \end{aligned} \tag{3.71}$$

With this we define profile functions for the transitions $M_j \to M_k$ for right and left movers as:

$$w_{k,j}(\mathbf{x}_T) := \int_0^1 d\zeta (\varphi^k_{s,s'}(\zeta, \mathbf{x}_T))^* \varphi^j_{s,s'}(\zeta, \mathbf{x}_T), \tag{3.72}$$

where k, j are both odd or even. Clearly we have (cf. (3.70))

$$\begin{aligned} &w_{j,j}(\mathbf{x}_T) \geq 0, \\ &\int d^2x_T w_{j,j}(\mathbf{x}_T) = 1, \\ &(\text{no summation over } j). \end{aligned} \tag{3.73}$$

Let us first study the transition of the right movers alone, i.e. the "reaction":

$$M_1(P_1) \to M_3(P_3). \tag{3.74}$$

For stable mesons $M_{1,3}$ we should find that the corresponding S-matrix elements are identical to the matrix elements of the unit operator. Is this borne out in our approach?

From the rules given in Sect. 3.5 we find easily the S-matrix element for the transition

$$q(1) + \bar{q}(1') \to q(3) + \bar{q}(3') \tag{3.75}$$

in the form

$$\langle 3, 3'|S|1, 1'\rangle = \langle S_{q+}(3,1) S_{\bar{q}+}(3',1')\rangle_G. \tag{3.76}$$

After folding (3.76) with the mesonic wave functions (3.66) we get:

$$\begin{aligned} &\langle M_3(P_3)|S|M_1(P_1)\rangle = (2\pi)^3 2P_1^0 \delta^3(\mathbf{P}_3 - \mathbf{P}_1) \\ &\int d^2z_T \int_0^1 d\zeta_3 \int_0^1 d\zeta_1 \; \varphi^{3^*}_{s,s'}(\zeta_3, \mathbf{z}_T)\varphi^1_{s,s'}(\zeta_1, \mathbf{z}_T) \\ &\frac{1}{2}P_{1+} \int \frac{dz_-}{2\pi} \exp\left[\frac{i}{2}P_{1+}(\zeta_3 - \zeta_1)z_-\right] \\ &\langle \frac{1}{3}\mathrm{Tr}[V_-(\infty, z_-, \mathbf{z}_T)V_-^\dagger(\infty, 0, 0)]\rangle_G. \end{aligned} \tag{3.77}$$

Now we remember that we consider the limit $P_{1+} \to \infty$. Therefore we perform a change of variables in the z_- integral by setting:

$$z'_- := \frac{1}{2}P_{1+}z_-. \tag{3.78}$$

This gives:

$$\begin{aligned}
&\langle M_3(P_3)|S|M_1(P_1)\rangle = (2\pi)^3 2P_1^0 \delta^3(\mathbf{P}_3 - \mathbf{P}_1) \\
&\int \mathrm{d}^2 z_T \int_0^1 \mathrm{d}\zeta_3 \int_0^1 \mathrm{d}\zeta_1 \; \varphi^{3^*}_{s,s'}(\zeta_3, \mathbf{z}_T) \varphi^{1}_{s,s'}(\zeta_1, \mathbf{z}_T) \\
&\int \frac{\mathrm{d}z'_-}{2\pi} \exp[i(\zeta_3 - \zeta_1) z'_-] \\
&\langle \frac{1}{3} \mathrm{Tr}[V_-(\infty, \frac{2}{P_{1+}} z'_-, \mathbf{z}_T) V_-^\dagger(\infty, 0, 0)] \rangle_G \\
&\longrightarrow (2\pi)^3 2P_1^0 \delta^3(\mathbf{P}_3 - \mathbf{P}_1) \int \mathrm{d}^2 z_T \; w_{3,1}(\mathbf{z}_T) \\
&\langle \frac{1}{3} \mathrm{Tr}[V_-(\infty, 0, \mathbf{z}_T) V_-^\dagger(\infty, 0, 0)] \rangle_G \quad \text{for} \quad P_{1+} \to \infty. \qquad (3.79)
\end{aligned}$$

In (3.79) $V_-(\infty, 0, \mathbf{z}_T)(V_-^\dagger(\infty, 0, 0))$ is the quark (antiquark) connector taken along the line $C_q(C_{\bar{q}})$, where:

$$C_q : \tau \to z_q(\tau) = \begin{pmatrix} \tau \\ \mathbf{z}_T \\ \tau \end{pmatrix}, \quad (-\infty < \tau < \infty), \qquad (3.80)$$

$$C_{\bar{q}} : \tau \to z_{\bar{q}}(\tau) = \begin{pmatrix} \tau \\ 0 \\ \tau \end{pmatrix}, \quad (-\infty < \tau < \infty). \qquad (3.81)$$

But from Sect. 2.1 we know that the connector $V_-^\dagger$ along $C_{\bar{q}}$ is equal to the connector V_- taken along the oppositely oriented line $\bar{C}_{\bar{q}}$ (cf. (2.25)). Now we will allow ourselves to join the lines C_q and $\bar{C}_{\bar{q}}$ at large positive and negative times. We can imagine the gluon potentials to be turned off adiabatically there. We obtain then from the product of the connectors in (3.79) a connector taken along a closed lightlike Wegner-Wilson loop

$$\frac{1}{3} \mathrm{Tr}[V_-(\infty, 0, \mathbf{z}_T) V_-^\dagger(\infty, 0, 0)] \longrightarrow \mathcal{W}_+(\frac{1}{2}\mathbf{z}_T, \mathbf{z}_T). \qquad (3.82)$$

Here we define

$$\mathcal{W}_\pm(\mathbf{y}_T, \mathbf{z}_T) = \frac{1}{3} \mathrm{Tr} \mathrm{P} \exp[-ig \int_{C_\pm} \mathrm{d}x_\mu G^\mu(x)] \qquad (3.83)$$

with $C_+(C_-)$ a lightlike Wegner-Wilson loop in the plane $x_- = 0 (x_+ = 0)$, where in the transverse space the centre of the loop is at $\mathbf{y}_T$ and the vector from the antiquark to the quark line is $\mathbf{z}_T$ (Fig. 15).

Inserting now everything in (3.79) we get the simple answer:

$$\begin{aligned}
&\langle M_3(P_3)|S|M_1(P_1)\rangle = (2\pi)^3 2P_1^0 \delta^3(\mathbf{P}_3 - \mathbf{P}_1) \\
&\int \mathrm{d}^2 z_T \; w_{3,1}(z_T) \langle \mathcal{W}_+(\frac{1}{2}\mathbf{z}_T, \mathbf{z}_T) \rangle_G. \qquad (3.84)
\end{aligned}$$

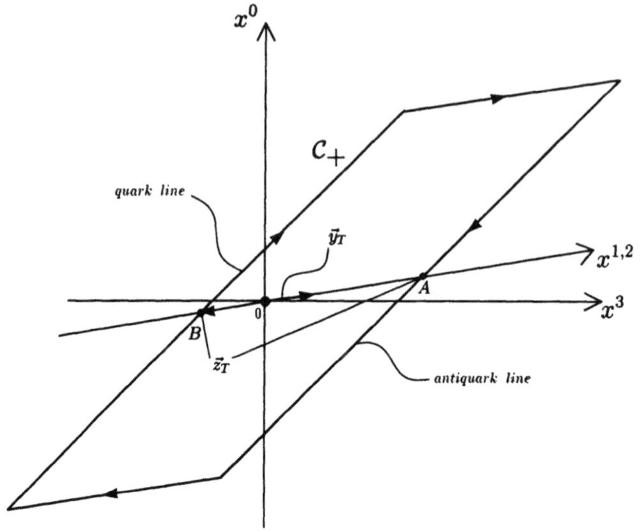

Fig. 15. The light-like Wegner-Wilson loop in Minkowski space-time, C_+ consisting of two light-like lines in the hyperplane $x_- = 0$ and connecting pieces at infinity. In transverse space the centre of the loop is at $\mathbf{y}_T$, the vector from the antiquark to the quark line is $\mathbf{z}_T$ (from A to B).

In the next section we will evaluate the functional integral in (3.84) in the SVM. We will find

$$\langle \mathcal{W}_+(\mathbf{y}_T, \mathbf{z}_T)\rangle_G|_{\mathrm{SVM}} = 1. \tag{3.85}$$

Inserting this in (3.84) leads to the expected result (cf. (3.66)-(3.72)):

$$\langle M_3(P_3)|S|M_1(P_1)\rangle = \langle M_3(P_3)|M_1(P_1)\rangle. \tag{3.86}$$

In our approach the $q\bar{q}$ pair in the right-moving meson $M_1 \to M_3$ does not interact. Of course this is only valid over our finite time interval (3.3)!

The techniques developed thus far are now easily employed for the reaction (3.65). After performing similar steps as above we arrive at the following S-matrix element:

$$\begin{aligned} S_{fi} &= \delta_{fi} + i(2\pi)^4\delta(P_3 + P_4 - P_1 - P_2)T_{fi}, \\ T_{fi} &\equiv \langle M_3(P_3), M_4(P_4)|T|M_1(P_1), M_2(P_2)\rangle \\ &= -2is \int \mathrm{d}^2b_T\mathrm{d}^2x_T\mathrm{d}^2y_T e^{i\mathbf{q}_T\cdot\mathbf{b}_T} w_{3,1}(\mathbf{x}_T)w_{4,2}(\mathbf{y}_T) \\ &\quad \langle \mathcal{W}_+(\tfrac{1}{2}\mathbf{b}_T, \mathbf{x}_T)\mathcal{W}_-(-\tfrac{1}{2}\mathbf{b}_T, \mathbf{y}_T) - 1\rangle_G. \end{aligned} \tag{3.87}$$

Here $s = (P_1 + P_2)^2$ is the c.m. energy squared and $\mathbf{q}_T$ is the momentum transfer, which is purely transverse in the high energy limit:

$$\mathbf{q}_T = (\mathbf{P}_1 - \mathbf{P}_3)_T. \tag{3.88}$$

From (3.87) we see that the amplitude for soft meson-meson scattering at high energies is governed by the correlation function of two lightlike Wegner-Wilson loops, where one is in the hyperplane $x_- = 0$, the other in $x_+ = 0$. The transverse separation between the centres of the two loops is given by $\mathbf{b}_T$, the impact parameter. The vectors $\mathbf{x}_T$ and $\mathbf{y}_T$ give the extensions and orientations of the loops in transverse space (Fig. 16). The loop-loop correlation function has to be integrated over all orientations of the loops in transverse space with the (transition) profile functions of the mesons, $w_{3,1}$ and $w_{4,2}$. Finally a Fourier transform in the impact parameter has to be done.

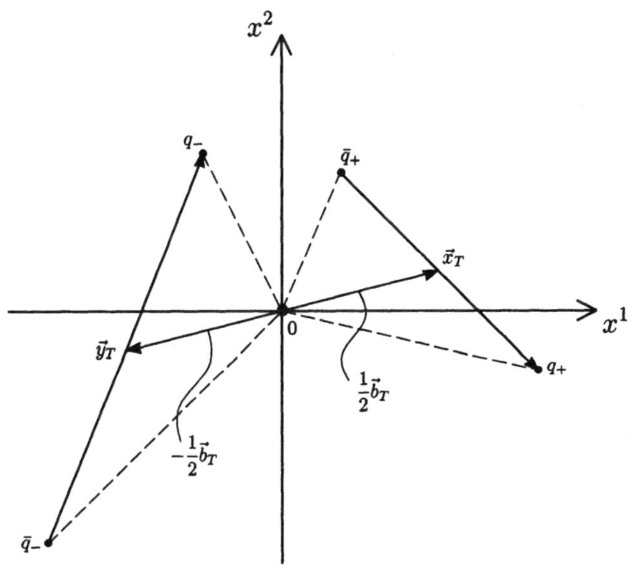

Fig. 16. The projections of the two lightlike Wegner-Wilson loops C_+, C_- occurring in the definition of $W_+(\frac{1}{2}\mathbf{b}_T, \mathbf{x}_T), W_-(-\frac{1}{2}\mathbf{b}_T, \mathbf{y}_T)$ (cf. (3.87), (3.83)) into transverse space. The points marked $q_+, \bar{q}_+$ $(q_-, \bar{q}_-)$ correspond to the projections of the quark and antiquark lines of C_+ (C_-).

The methods presented here for meson-meson scattering can of course also be employed for scattering reactions involving baryons and antibaryons. This is sketched in Appendix C.

3.7 The Evaluation of Scattering Amplitudes in the Minkowskian Version of the Stochastic Vacuum Model

In the previous section we have derived expressions for the amplitudes of soft meson-meson scattering at high energies in terms of correlation functions of light-like Wegner-Wilson loops. The task is now to evaluate the corresponding functional integral $\langle\rangle_G$ in (3.87). Surely we do not want to make a perturbative expansion there, remembering our argument of Sect. 1 (cf. (1.14)). Instead, we will turn to the SVM which did quite well in its applications in Euclidean QCD (Sect. 2). Of course, the generalization of the SVM to Minkowski space-time is a bold step which was done in [59], [45]. The authors of these Refs. proposed to use in Minkowski space-time just the assumptions 1-3 of the SVM (cf. Sect. 2.4), but after having made a suitable analytic continuation. In this way we obtain for instance the Minkowski version of Ass. 1 as (cf. (2.60), (2.61)):

Ass. 1: The correlator of two field strengths, shifted to a common reference point y, is independent of the connecting path and given by:

$$\langle \frac{g^2}{4\pi^2}\hat{G}^a_{\mu\nu}(y,x;C_x)\hat{G}^b_{\rho\sigma}(y,x';C_{x'})\rangle_G = \frac{1}{4}\delta^{ab}F_{\mu\nu\rho\sigma}(z),$$

$$F_{\mu\nu\rho\sigma}(z) = \frac{1}{24}G_2\{(g_{\mu\rho}g_{\nu\sigma} - g_{\mu\sigma}g_{\nu\rho})\kappa D(z^2)$$
$$+\frac{1}{2}[\frac{\partial}{\partial z^\nu}(z_\sigma g_{\mu\rho} - z_\rho g_{\mu\sigma}) + \frac{\partial}{\partial z^\mu}(z_\rho g_{\nu\sigma} - z_\sigma g_{\nu\rho})]\cdot(1-\kappa)D_1(z^2)\}. \quad (3.89)$$

Here $z = x - x'$ and $D(z^2)$ and $D_1(z^2)$ are defined as in (2.65), (2.67) for $z^2 \leq 0$ and by analytic continuation for $z^2 > 0$.

As a first application let us calculate the expectation value of one light-like Wegner-Wilson loop. We have from (3.83) using the non-abelian Stokes theorem:

$$\langle \mathcal{W}_+(\mathbf{y}_T,\mathbf{z}_T)\rangle_G = \langle \frac{1}{3}\mathrm{Tr}\mathrm{P}\exp[-ig\int_{C_+} \mathrm{d}x_\mu G^\mu(x)]\rangle_G$$
$$= \langle \frac{1}{3}\mathrm{Tr}\ \mathrm{P}\exp[-i\frac{g}{2}\int_{S_+} \mathrm{d}u\mathrm{d}v \frac{\partial(x^\mu,x^\nu)}{\partial(u,v)}\hat{G}_{\mu\nu}(R,x;C_x)]\rangle_G. \quad (3.90)$$

Here S_+ is the (planar) surface spanned into C_+ (Fig. 15) and parameterized by

$$x^\mu(u,v) = u\, n_+^\mu + y^\mu - (v - \frac{1}{2})z^\mu,$$

$$-\infty < u < \infty,\ 0 \leq v \leq 1, \quad (3.91)$$

where n_+ is as in (3.42) and

$$y^\mu = \begin{pmatrix} 0 \\ \mathbf{y}_T \\ 0 \end{pmatrix}, \qquad z^\mu = \begin{pmatrix} 0 \\ \mathbf{z}_T \\ 0 \end{pmatrix}. \tag{3.92}$$

The reference point on the surface S_+ is denoted by R and C_x are straight lines running from x to R. From (3.91) we find

$$\frac{\partial(x^\mu, x^\nu)}{\partial(u,v)} = z^\mu n_+^\nu - n_+^\mu z^\nu. \tag{3.93}$$

Now we apply the cumulant expansion formulae (cf. Sect. 2.3) to (3.90) and use assumptions 1-3 of the SVM. This is completely analogous to the calculations done in Sect. 2.5. We get:

$$\begin{aligned} \langle \mathcal{W}_+(\mathbf{y}_T, \mathbf{z}_T)\rangle_G = \exp\Big\{ -\frac{\pi^2}{6}\int_{S_+} \mathrm{d}u\mathrm{d}v \int_{S_+} \mathrm{d}u'\mathrm{d}v' K_2(x-x') \\ + \text{ higher cumulant terms}\Big\}, \end{aligned} \tag{3.94}$$

where

$$\begin{aligned} K_2(x-x') &= \frac{\partial(x^\mu, x^\nu)}{\partial(u,v)} \frac{\partial(x'^\rho, x'^\sigma)}{\partial(u',v')} F_{\mu\nu\rho\sigma}(x-x'), \\ x &\equiv x(u,v), \quad x' \equiv x'(u',v'). \end{aligned} \tag{3.95}$$

It is an easy exercise to evaluate $K_2(x-x')$ using (3.93) and $F_{\mu\nu\rho\sigma}$ from (3.89). The result is

$$K_2(x-x') = 0 \tag{3.96}$$

for all $x, x' \in S_+$. With assumption 3 of the SVM (cf. (2.73)-(2.75)) all higher cumulants are related to the second one. Thus, (3.96) implies also the vanishing of all higher cumulant terms in (3.94) and we get

$$\langle \mathcal{W}_+(\mathbf{y}_T, \mathbf{z}_T)\rangle_G = 1. \tag{3.97}$$

A similar argument leads, of course, also to

$$\langle \mathcal{W}_-(\mathbf{y}_T, \mathbf{z}_T)\rangle_G = 1. \tag{3.98}$$

In Sect. 3.6 we have already used the result (3.97) in the discussion of the transition $M_1 \to M_3$ to obtain (3.86). We can also use it for the calculation of the wave function renormalization constant Z_ψ. The expression we obtained for Z_ψ in (3.48) can be interpreted as the expectation value of the non-abelian phase factor picked up by a very fast right-moving quark. Now, isolated quarks do not exist. The best approximation for it we can think of is a fast, right-moving quark-antiquark pair with the antiquark being very far away from the quark in transverse direction. In this way we obtain for Z_ψ instead of (3.48):

$$\begin{aligned} Z_\psi &= \lim_{|\mathbf{z}_T|\to\infty} \langle \frac{1}{3}\mathrm{Tr}[V_-(\infty,0,0)V_-^\dagger(\infty,0,\mathbf{z}_T)]\rangle_G \\ &= \lim_{|\mathbf{z}_T|\to\infty} \langle \mathcal{W}_+(-\frac{1}{2}\mathbf{z}_T,-\mathbf{z}_T)\rangle_G \\ &= 1. \end{aligned} \tag{3.99}$$

This result was already used in Sect. 3.5, (3.57)ff.

We come now to the evaluation of the loop-loop correlation function of (3.87):

$$\begin{aligned} &\langle \mathcal{W}_+(\frac{1}{2}\mathbf{b}_T,\mathbf{x}_T)\mathcal{W}_-(-\frac{1}{2}\mathbf{b}_T,\mathbf{y}_T) - 1\rangle_G \\ &= \langle [\mathcal{W}_+(\frac{1}{2}\mathbf{b}_T,\mathbf{x}_T) - 1][\mathcal{W}_-(-\frac{1}{2}\mathbf{b}_T,\mathbf{y}_T) - 1]\rangle_G. \end{aligned} \tag{3.100}$$

Here we used (3.97), (3.98). The strategy is as before. We want to transform the line integrals of $\mathcal{W}_\pm$ (cf. (3.83)) into surface integrals using the non-abelian Stokes theorem of Sect. 2.2. Following the authors of [59], [45] we choose as surface with boundary C_+ and C_- a double pyramid with apex at the mid-point of C_+ and C_- which is the origin of our coordinate system (Fig. 17). The mantle of this pyramid is $\mathcal{P}_+ + \mathcal{P}_-$ and we have with suitable orientation

$$\partial(\mathcal{P}_+ + \mathcal{P}_-) = C_+ + C_-. \tag{3.101}$$

In the transverse projection of Fig. 16 the basis surface $S_+(S_-)$ appears as the line $\bar{q}_+q_+$ ($\bar{q}_-q_-$) and the mantle surface $\mathcal{P}_+(\mathcal{P}_-)$ as the triangle $0\bar{q}_+q_+(0\bar{q}_-q_-)$. From the non-abelian Stokes theorem we obtain now for (3.100):

$$\begin{aligned} &\big\langle \big[\mathcal{W}_+(\frac{1}{2}\mathbf{b}_T,\mathbf{x}_T) - 1\big]\big[\mathcal{W}_-(-\frac{1}{2}\mathbf{b}_T,\mathbf{y}_T) - 1\big]\big\rangle_G \\ &= \langle \{\frac{1}{3}\mathrm{Tr P exp}\big[-i\frac{g}{2}\int_{\mathcal{P}_+} \mathrm{d}u\mathrm{d}v\frac{\partial(x^\mu,x^\nu)}{\partial(u,v)}\hat{G}_{\mu\nu}(0,x;C_x)\big] - 1\} \\ &\quad \{\frac{1}{3}\mathrm{Tr\ P\,exp}\big[-i\frac{g}{2}\int_{\mathcal{P}_-} \mathrm{d}u'\mathrm{d}v'\frac{\partial(x'^\rho,x'^\sigma)}{\partial(u',v')}\hat{G}_{\rho\sigma}(0,x';C_{x'})\big] - 1\}\big\rangle_G. \end{aligned} \tag{3.102}$$

So far, it has not been possible to use some version of the cumulant expansion for (3.102). Thus in [59], [45] the path-ordered exponentials on the r.h.s. of (3.102) were expanded directly. The structure of this expansion is as follows:

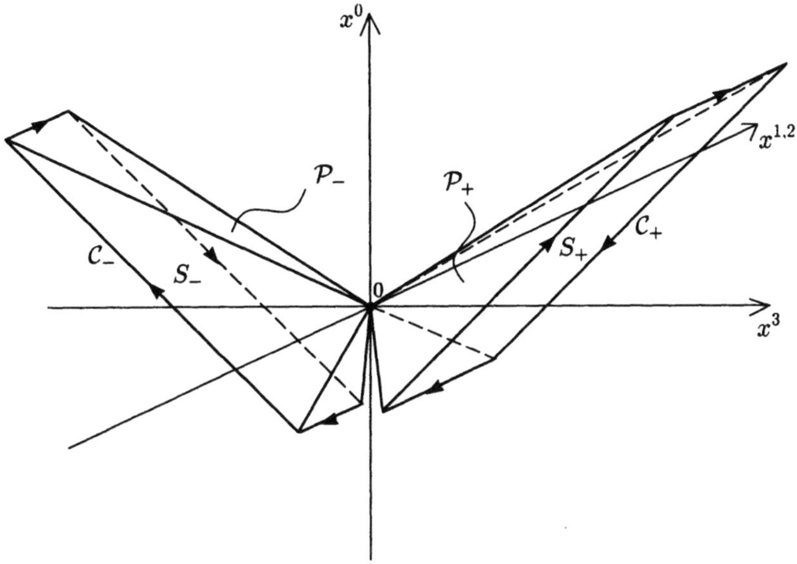

Fig. 17. The curves C_+ and C_- along which the path integrals $\mathcal{W}_\pm$ in (3.100) are taken. The mantle of the pyramid with apex at the origin of the coordinate system and boundary C_+ (C_-) is $\mathcal{P}_+$ ($\mathcal{P}_-$), the basis surface S_+ (S_-).

$$\begin{aligned}&\langle[\mathcal{W}_+ - 1][\mathcal{W}_- - 1]\rangle_G\\ &\sim \langle\frac{1}{3}\mathrm{Tr}[\int_{\mathcal{P}_+}\hat{G} + \int_{\mathcal{P}_+}\int_{\mathcal{P}_+}\hat{G}\hat{G} + ...]\\ &\quad\frac{1}{3}\mathrm{Tr}[\int_{\mathcal{P}_-}\hat{G} + \int_{\mathcal{P}_-}\int_{\mathcal{P}_-}\hat{G}\hat{G} + ...]\rangle_G. \end{aligned} \tag{3.103}$$

The trace of a single shifted field strength vanishes. This means that we cannot exchange a single coloured object in meson-meson scattering (3.65). The first non-trivial contribution in (3.103) comes from the term with two shifted field strengths in each trace. The corresponding correlation function

$$\langle\frac{1}{3}\mathrm{Tr}[\int_{\mathcal{P}_+}\int_{\mathcal{P}_+}\hat{G}\hat{G}]\;\;\frac{1}{3}\mathrm{Tr}[\int_{\mathcal{P}_-}\int_{\mathcal{P}_-}\hat{G}\hat{G}]\rangle_G$$

can now be evaluated using the assumptions 1-3 of the Minkowskian version of the SVM. In doing so it is advantageous to transform the surface integrals over the mantles of the pyramids $\mathcal{P}_\pm$ into surface integrals over $S_\pm$ and integrals over the volumes $V_\pm$ enclosed by $\mathcal{P}_+$ and S_+ and $\mathcal{P}_-$ and S_-, respectively:

$$\partial(V_\pm) = \mathcal{P}_\pm - S_\pm. \tag{3.104}$$

Then one can use the ordinary Gauss theorem to get simpler integrals.

We will not enter into the details of these calculations here, but only note that the integrations along the directions x_+ and x_- can easily be done analytically and that one ends up with integrals over the projections of $S_\pm$ and $V_\pm$ into the transverse space. Thus, one finally needs the correlator (3.89) for space-like separations only:

$$(x - x')^2 = z^2 \leq 0,$$
$$x = \begin{pmatrix} 0 \\ \mathbf{x}_T \\ 0 \end{pmatrix}, \quad x' = \begin{pmatrix} 0 \\ \mathbf{x}'_T \\ 0 \end{pmatrix}, \tag{3.105}$$

where $\mathbf{x}_T$ runs over the triangle $\Delta_+ = 0\bar{q}_+q_+$ and $\mathbf{x'}_T$ over $\Delta_- = 0\bar{q}_-q_-$ in Fig. 16. For space-like separations the correlator functions $D(z^2)$ and $D_1(z^2)$ in (3.89) are as in Euclidean space time. The resulting expressions are then of the following structure

$$\langle[\mathcal{W}_+ - 1][\mathcal{W}_- - 1]\rangle_G \sim$$
$$\Big\{\int_{\Delta_+} \mathrm{d}^2 z_T \int_{\Delta_-} \mathrm{d}^2 z'_T \, G_2[..D(-(\mathbf{z}_T - \mathbf{z}'_T)^2) + ...D_1(-(\mathbf{z}_T - \mathbf{z}'_T)^2)]\Big\}^2. \tag{3.106}$$

These integrals have to be evaluated numerically.

With the methods outlined above, we have obtained an (approximate) expression (3.106) for the functional integral $\langle\rangle_G$ governing the meson-meson scattering amplitude (3.87). Note that the nonperturbative gluon condensate parameter G_2 sets the scale in (3.106) and in the integrals to be performed there the vacuum correlation length a enters through the D and D_1 functions. Thus, on dimensional grounds, we must have in our approximation:

$$\langle(\mathcal{W}_+ - 1)(\mathcal{W}_- - 1)\rangle_G = G_2^2 a^8 f\left(\frac{\mathbf{b}_T}{a}, \frac{\mathbf{x}_T}{a}, \frac{\mathbf{y}_T}{a}\right), \tag{3.107}$$

where f is a dimensionless function. To obtain the meson-meson scattering amplitude (3.87) we still have to integrate over the profile functions $w_{3,1}(\mathbf{x}_T)$ and $w_{4,2}(\mathbf{y}_T)$. Here the transverse extensions of the mesons - i.e. of the wave packets representing them - enter in the results.

For a detailed exposition of the numerical results obtained in the way sketched above we refer to [45]. Here we only discuss the outcome for proton-proton scattering when treated in a similar way (cf. Appendix C). A fit to the numerical results gives for the total cross section and the slope parameter at $t = 0$ of elastic proton-proton scattering the following representation:

$$\sigma_{\mathrm{tot}}(pp) = 0.00881 \left(\frac{R_p}{a}\right)^{3.277} \cdot (3\pi^2 G_2)^2 \cdot a^{10}, \tag{3.108}$$

$$b_{pp} := \frac{\mathrm{d}}{dt} \ln \frac{\mathrm{d}\sigma_{\mathrm{el}}}{\mathrm{d}t}(pp)|_{t=0} = 1.558a^2 + 0.454R_p^2. \tag{3.109}$$

Here R_p is the proton radius and the formulae (3.108), (3.109) are valid for

$$1 \leq R_p/a \leq 3. \tag{3.110}$$

To compare (3.108), (3.109) with experimental results, we can, for instance, consider the c.m. energy $\sqrt{s} = 20$ GeV and take as input the following measured values (cf. [45]):

$$\begin{aligned} \sigma_{\mathrm{tot}}(pp)|_{\mathrm{Pomeron\ part}} &= 35 \text{ mb}, \\ b_{pp} &= 12.5 \text{ GeV}^{-2}, \\ R_p \equiv R_{p,\mathrm{elm}} &= 0.86 \text{ fm}. \end{aligned} \tag{3.111}$$

We obtain then from (3.108) and (3.109):

$$\begin{aligned} a &= 0.31 \text{ fm}, \\ G_2 &= (507 \text{ MeV})^4. \end{aligned} \tag{3.112}$$

The values for the correlation length a and for the gluon condensate G_2 come out in surprising good agreement with the determination of these quantities from the fit to the lattice results (2.70).

But perhaps we were lucky in picking out the right c.m. energy $\sqrt{s}$ and radius for our comparison of theory and experiment. What about the s-dependence of the total cross section σ_{tot} and slope parameter b? The vacuum parameters G_2 and a should be independent of the energy $\sqrt{s}$. On the other hand, from the discussion in Sect. 3.1 leading to (3.2), it seems quite plausible to us that the effective strong interaction radii R of hadrons may depend on $\sqrt{s}$. Let us consider again pp (or $p\bar{p}$) elastic scattering. Once we have fixed G_2 and a from the data at $\sqrt{s} = 20$GeV (3.108) and (3.109) give us $\sigma_{\mathrm{tot}}(pp)$ and b_{pp} in terms of the single parameter R_p, i.e. we obtain as prediction of the model a curve in the plane b_{pp} versus $\sigma_{\mathrm{tot}}(pp)$. This is shown in Fig. 18. It is quite remarkable that the data from $\sqrt{s} = 20$GeV up to Tevatron energies, $\sqrt{s} = 1.8$ TeV follow this curve.

Summarizing this section we can say that explicit calculations for high energy-elastic hadron-hadron scattering near the forward direction have been performed combining the field-theoretic methods of [53] and the Minkowski version of the stochastic vacuum model of [59],[45]. The results are encouraging and support the idea that the vacuum structure of QCD plays an essential role in soft high-energy scattering. We want to point out that these calculations also resolve a possible paradoxon of QCD: On the one hand there are suggestions that the gluon propagator must be highly singular (probably $\propto (Q^2)^{-2}$, cf. e.g. [61]) for momentum transfers $Q^2 \to 0$ in order to produce confinement. On the other hand high energy scattering amplitudes are completely regular for $t \to 0$. A singular gluon propagator will lead in the 2-gluon exchange model to a singularity for $t = 0$ not only for quark-quark

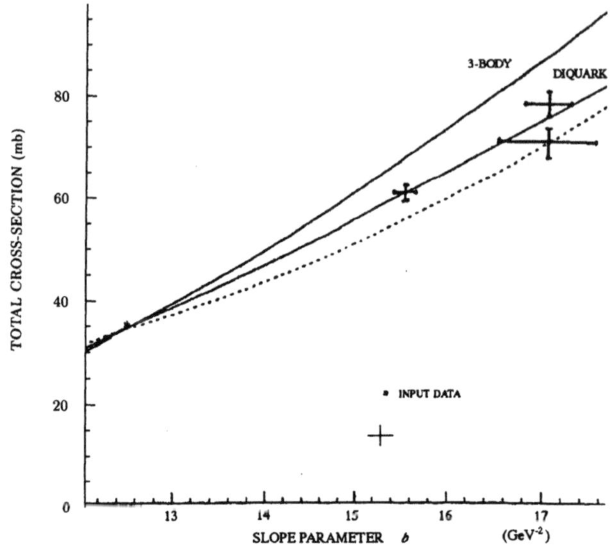

Fig. 18. The relation between the total cross section σ_{tot} and the slope parameter b for proton-proton and proton-antiproton scattering. The dotted line is the prediction from Regge theory. The prediction of the calculation for soft high energy scattering in the stochastic vacuum model is that the data points should lie in the area between the full lines. In essence this is given by (3.108), (3.109) with an uncertainty estimate from different assumptions for the proton wave functions (cf. [45]).

scattering but also for hadron-hadron scattering if the latter are considered e.g. as colour dipoles. The resolution of the paradoxon which we can present is intimately connected with the confinement mechanism which we found in the SVM in Sect. 2. The short range correlation of the gluon field strengths governs the t-dependence of the hadronic scattering amplitudes and gives rise to their regularity for $t = 0$. The gluon propagator on the other hand can be singular for $Q^2 \to 0$, since the gluon potentials have a long range correlation as we have seen in Sect. 2. The result (3.108) for the total cross section depends also on the proton radius. This radius dependence does not saturate for large radii in the calculation with non-abelian gluons but does saturate in an abelian model [45]. Thus with non-abelian gluons we do not get the additive quark model result [17], and thus not the picture of Donnachie and Landshoff [23], where the "soft Pomeron" couples to individual quarks in the hadrons. The strong radius dependence in (3.108) is due to the D-term in the correlator (3.89) which is related to the effective chromomagnetic monopole condensate in the QCD vacuum and which gives rise to the linearly rising quark-antiquark potential, i.e. to string formation (see Sect. 2.5). The calculations reported above suggest that in high energy scattering this same term gives rise to a string-string interaction which leads to the radius dependence in (3.108). Note that one does not have to put in the strings by hand. They enter the picture automatically through our lightlike Wegner-Wilson loops. The radius dependence of the cross section occurs, of course, also for meson-baryon and meson-meson scattering and gives a quantitative understanding of the difference between the Kp and πp total cross sections and slope parameters at high energies. For this and for further results we refer to [45], [62]. The success of the calculation for $pp(p\bar{p})$ scattering describing correctly the relation of the total cross section versus the slope parameter from $\sqrt{s} \simeq 20$ GeV up to $\sqrt{s} = 1.8$ TeV suggests the following simple interpretation: In soft elastic scattering the hadrons act like effective "colour dipoles" with a radius increasing with c.m. energy. The dipole-dipole interaction is governed by the correlation function of two lightlike Wegner-Wilson loops which receives the dominant contribution from the same non-perturbative phenomenon - effective chromomagnetic monopole condensation - which leads to string formation and confinement.

4 "Synchrotron Radiation" from the Vacuum, Electromagnetic Form Factors of Hadrons, and Spin Correlations in the Drell-Yan Reaction

Let us consider for definiteness again a proton-proton collision at high c.m. energy $\sqrt{s} \gg m_p$. We look at this collision in the c.m. system and choose as x^3-axis the collision axis (Fig. 19). According to Feynman's parton dogma [63] the hadrons look like jets of almost non-interacting partons, i.e. quarks and gluons. Accepting our previous views of the QCD vacuum (Sect. 2), these partons travel in a background chromomagnetic field.

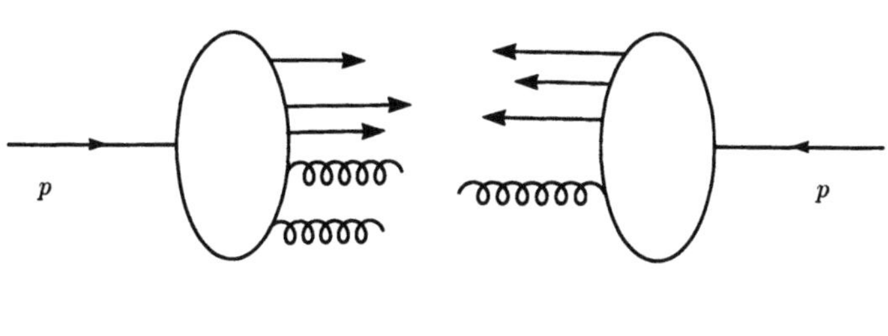

Fig. 19. A proton-proton collision at high energies in the parton picture.

What sort of new effects might we expect to occur in this situation? Consider for instance a quark-antiquark collision in a chromomagnetic field. In our picture this is very similar to an electron-positron collision in a storage ring (Fig. 20). We know that in a storage ring e^- and e^+ are deflected and emit synchrotron radiation. They also get a transverse polarization due to emission of spin-flip synchrotron radiation [64], [65]. Quite similarly we can expect the quark and antiquark to be deflected by the vacuum fields. Since quarks have electric and colour charge, they should then emit both photon and gluon "synchrotron radiation". Of course, as long as we have quarks within a single, isolated proton (or other hadron) traveling through the vacuum no emission of photons can occur, and we should consider such processes as contributing to the cloud of quasi-real photons surrounding a fast-moving proton. (This is similar in spirit to the well-known Weizsäcker-Williams approximation.) But in a collision process the parent quark or antiquark will be scattered away and the photons of the cloud can become real, manifesting themselves as prompt photons in hadron-hadron collisions.

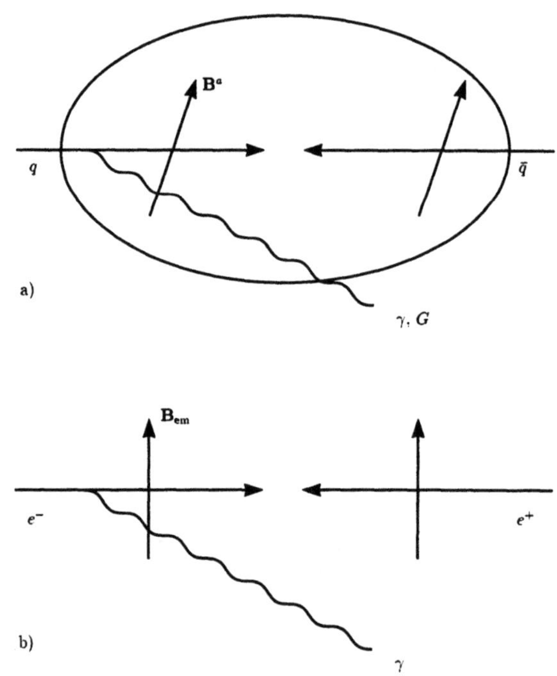

Fig. 20. A quark and antiquark traversing a region of chromomagnetic field (a). An electron and a positron in a storage ring (b). In both cases we expect the emission of synchrotron radiation to occur.

In Ref. [14] we have given an estimate for the rate and the spectrum of such prompt photons using the classical formulae for synchrotron radiation [65]. A more detailed study of soft photon production in hadronic collisions was made in [66]. A sketch of our arguments and calculations is as follows.

In Sect. 2 we discussed the domain picture of the QCD vacuum. In Euclidean space time we have domains in the vacuum of linear size $\simeq a$. Inside one domain the colour fields are highly correlated. The colour field orientations and domain sizes fluctuate, i.e. have a certain distribution. If we translate this picture naively to Minkowski space, we arrive at colour correlations there being characterized by invariant distances of order a. Then the colour fields at the origin of Minkowski space, for instance, should be highly correlated with the fields in the region

$$|x^2| \lesssim a^2 \tag{4.1}$$

(cf. Fig. 21). Consider now a fast hadron passing by with one of its quarks going right through $x = 0$ on a nearly lightlike world line. It is clear that in such a situation the quark will, from the point of view of the observer, spend a long time in a correlated colour background field. An easy exercise shows furthermore that two quarks of the same hadron will have a very good chance to travel in two different colour domains. The argument is in essence

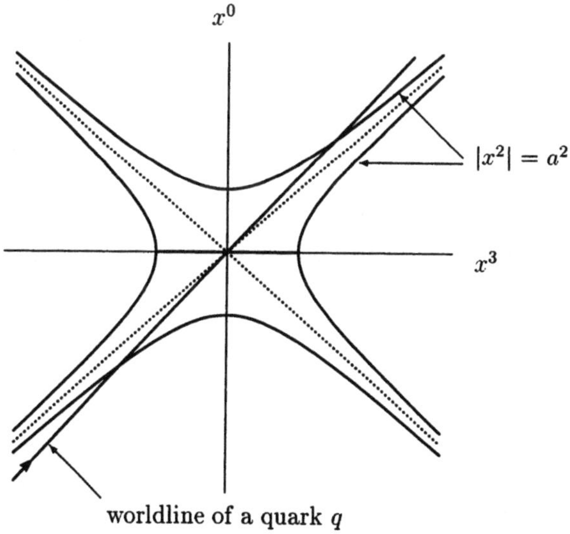

Fig. 21. Sketch of a "colour domain" in Minkowski space and of the worldline of a quark from a fast hadron moving through it.

as follows. The quarks have a transverse separation of the order of the hadron radius R whereas the transverse size of a domain is of order a and we have $a^2/R^2 \ll 1$ (cf. (2.72)). Each quark will then wiggle around due to the deflection by the background colour fields in an uncorrelated fashion. This gives us a justification for adding the synchrotron gluon and photon emission of the quarks incoherently. The result we found can be summarized as follows: In the overall c.m. system of the hadron-hadron collision "synchrotron" photons should appear with energies $\omega < 300 - 500$ MeV, i.e. in the very central region of the rapidity space. The number of photons per collision and their spectrum are — apart from logarithms — independent of the c.m. energy $\sqrt{s}$. The dependence of the number of photons on the energy ω and on the emission angle ϑ^* with respect to the beam axis is obtained as follows for pN collisions:

$$\frac{dn_\gamma}{d\omega d\cos\vartheta^*} = \frac{2\pi\alpha}{\omega^{1/3}}(l_{\text{eff}})^{2/3} \cdot \Sigma(\cos\vartheta^*) \ . \tag{4.2}$$

Here α is the fine structure constant and l_{eff} is the length or time over which the fast quark traveling in the background chromomagnetic field B_c obtains by its deflection a transverse momentum of order $\bar{p}_T \approx 300$ MeV, the mean transverse momentum of quarks in a hadron (Fig. 22):

$$l_{\text{eff}} = \frac{\bar{p}_T}{gB_c}. \tag{4.3}$$

The quantity Σ in (4.2) sums up the contributions from all quarks of the initial and final state hadrons. It involves an integration over the quark distribution functions of these hadrons. In [66] we found

$$\Sigma(\cos\vartheta^*) \simeq \frac{0.21}{(\sin\vartheta^*)^{2/3}} \tag{4.4}$$

for pN collisions at $\sqrt{s} = 29$ GeV.

Our result (4.2) for synchrotron photons should be compared to the innerbremsstrahlungs spectrum

$$\left.\frac{dn_\gamma}{d\omega d\cos\vartheta^*}\right|_{\text{bremsstr.}} \propto \frac{1}{\omega\sin^2\vartheta^*} \tag{4.5}$$

The "synchrotron" radiation from the quarks (4.2) is thus harder than the hadronic bremsstrahlung spectrum. This is welcome, since for $\omega \to 0$ bremsstrahlung should dominate according to Low's theorem [67].

It is amusing to note that in several experiments an excess of soft prompt photons over the bremsstrahlung calculation has been observed [68]-[72]. The gross features and the order of magnitude of this signal make it a candidate for our "synchrotron" process. A detailed comparison with our formulae has been made in [66] for the results from one experiment [72] with encouraging

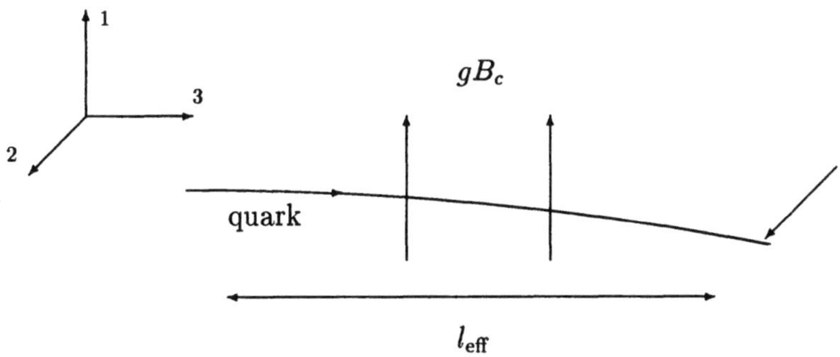

Fig. 22. A quark moving in 3-direction in a transverse (in 1-direction) chromomagnetic field of strength gB_c and picking up a transverse momentum (in 2-direction) of magnitude $\bar{p}_T$ over a length l_{eff}. Here $\bar{p}_T$ is the mean transverse momentum of quarks in the hadron.

results. This is shown in Fig. 23 for the k_T spectrum of photons at $y = 0$, where

$$\begin{aligned} k_T &= \omega \sin \vartheta^*, \\ y &= -\ln \tan(\vartheta^*/2). \end{aligned} \tag{4.6}$$

We see that the addition of synchrotron photons to the bremsstrahlung ones improves the agreement of theory with the data considerably. We deduce from Fig. 23 $l_{\text{eff}} \simeq 20 - 40$ fm. Taking $l_{\text{eff}} = 20$ fm and $\bar{p}_T = 300$ MeV, we find for the effective chromomagnetic deflection field from (4.3)

$$gB_c = \frac{\bar{p}_T}{l_{\text{eff}}} = (55 \text{ MeV})^2. \tag{4.7}$$

This is much smaller than the vacuum field strength (2.7). Our interpretation of this puzzle is as follows: The colour fields in a fast moving hadron must be shielded. Indeed, a chromomagnetic field of the strength (2.7) would lead to a ridiculously small value for the radius of cyclotron motion of a fast quark. The necessary shielding could be done by gluons in a fast hadron. We know from the deep inelastic lepton-nucleon scattering results that a fast nucleon contains many gluons. We may even be brave and turn the argument around: in order for a fast hadron to be able to move through the QCD vacuum, the very strong vacuum chromomagnetic fields must be shielded, making gluons in a fast hadron a necessity. Thus soft photon production in pp collisions may give us a quite unexpected insight into the quark and gluon structure of fast hadrons.

The next topic we want to discuss briefly concerns electromagnetic form factors of hadrons.

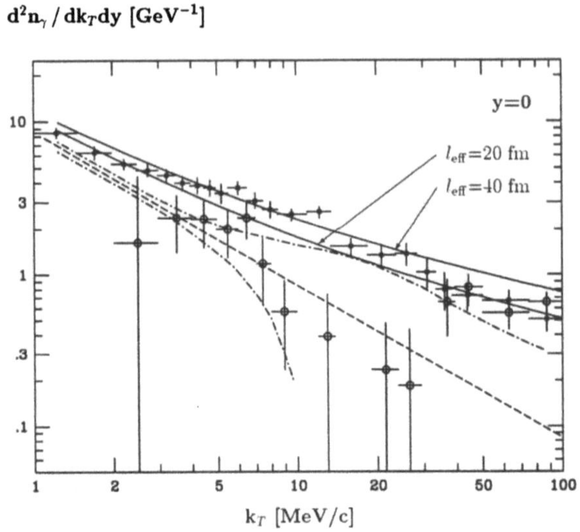

Fig. 23. The $|\mathbf{k}_T|$ distribution for direct photons emitted at c.m. rapidity $y = 0$ in $p - Be$ collisions at 450 GeV incident proton momentum from [72]. The normalization is according to a private communication by H. J. Specht. The background of decay photons is subtracted. The dash-dotted line gives the expected yield of photons from hadronic bremsstrahlung, the dashed lines show the upper and lower limits including the systematic errors in the shape of the decay background and the bremsstrahlung calculation (cf. [72]). The lower (upper) solid line is the result of the calculation for synchrotron photons ((4.2ff.) with $l_{\text{eff}} = 20$ fm ($l_{\text{eff}} = 40$ fm) added to the spectrum of hadronic bremsstrahlung (cf. [66]).

We have argued above that the colour fields in the vacuum should give a contribution to the virtual photon cloud of hadrons and we made an estimate of the distribution of these photons using the synchrotron radiation formulae. Consider now any reaction where a quasi-real photon is emitted from a hadron with the hadron staying intact and the photon interacting subsequently. In Fig. 24 we draw the corresponding diagram for a nucleon N:

$$N(p) \to N(p') + \gamma(q). \tag{4.8}$$

The flux of these quasi-real photons is well known. The first calculations in this context are due to Fermi, Weizsäcker, and Williams [73]. For us the relevant formula is given in equation (D.4) of [74]. Let E be the energy of the initial nucleon, $G_E^N(Q^2)$ its electric Sachs formfactor, and let ω and $q^2 = -Q^2$ be the energy and mass of the virtual photon. Then the distribution of quasi-real photons in the fast-moving nucleon is given by

$$\mathrm{d}n_\gamma^{(\mathrm{excl})} = \frac{\alpha}{\pi}\frac{\mathrm{d}\omega}{\omega}\frac{\mathrm{d}Q^2}{Q^2}\left[G_E^N(Q^2)\right]^2 \tag{4.9}$$

where we neglect terms of order ω/E and Q^2/m_N^2 and assume

$$Q^2 \gg Q^2_{\min} \simeq \frac{m_N^2\omega^2}{E^2}. \tag{4.10}$$

We call (4.9) the exclusive flux since the nucleon stays intact. Now we want to translate (4.9) into a distribution in ω and the angle ϑ^* of emission of the γ (cf. Fig. 24). A simple calculation gives

$$Q^2 \simeq \omega^2 \sin^2\vartheta^*, \tag{4.11}$$

$$\mathrm{d}n_\gamma^{(\mathrm{excl})} \simeq \frac{2\alpha}{\pi}\frac{\mathrm{d}\omega}{\omega}\frac{\mathrm{d}\vartheta^*}{\sin\vartheta^*}\left[G_E^N(\omega^2\sin^2\vartheta^*)\right]^2. \tag{4.12}$$

Now we made an "exclusive-inclusive connection" argument in [66]: We require $\mathrm{d}n_\gamma^{(\mathrm{excl})}/\mathrm{d}\omega\mathrm{d}\cos\vartheta^*$ to behave as $\omega^{-1/3}$ for fixed ϑ^* as we found in (4.2). This implies for the form factor $G_E^N(Q^2)$ a behaviour as $(Q^2)^{1/6}$.

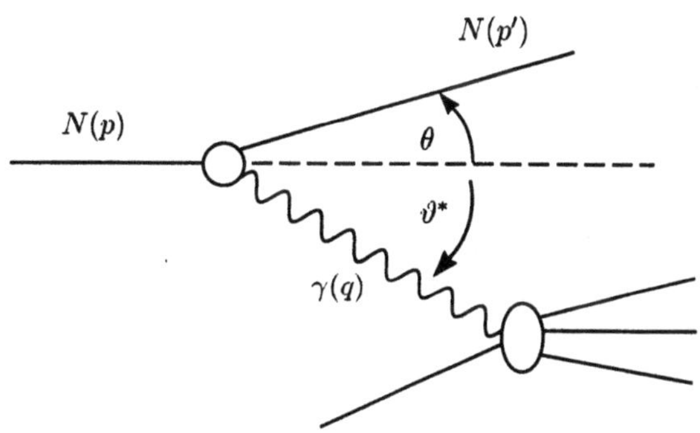

Fig. 24. A nucleon interacting by emission of a quasi-real photon.

Thus we arrive at the following conclusion: The proton form factor G_E^p should contain in addition to a "normal" piece connected with the total charge and the hadronic bremsstrahlung in inelastic collisions a piece $\propto (Q^2)^{1/6}$ connected with "synchrotron" radiation from the QCD vacuum. For the neutron which has total charge zero we would expect the "normal" piece in G_E^n to be quite small and the "anomalous" piece to be quite important for not too large Q^2. Thus the neutron electric formfactor should be an interesting quantity to look for "anomalous" effects $\propto (Q^2)^{1/6}$.

In Fig. 25 we show the data on the electric formfactor of the neutron from [75], [76]. We superimpose the curve

$$G^n_{(\mathrm{syn})}(Q^2) = 3.6 \cdot 10^{-2} \left(\frac{Q^2}{5\,\mathrm{fm}^{-2}}\right)^{1/6} \tag{4.13}$$

which is normalized to the data at $Q^2 = 5\,\mathrm{fm}^{-2}$. We see that except in the very low Q^2 region we get a decent description of the data. For $Q^2 \to 0$ (4.13) has to break down since $G^n_E(Q^2)$ is regular at $Q^2 = 0$. Indeed one knows the slope of $G^n_E(Q^2)$ for $Q^2 = 0$ from the scattering of thermal neutrons on electrons ([77] and references cited therein):

$$\left.\frac{\mathrm{d}G^n_E(Q^2)}{\mathrm{d}Q^2}\right|_{Q^2=0} = 0.019\ \mathrm{fm}^2. \tag{4.14}$$

We see from Fig. 25 that the behaviour of $G^n_E(Q^2)$ has to change rather quickly as we go away from $Q^2 = 0$. We will now make a simple ansatz which takes into account that $G^n_E(Q^2)$ can have singularities in the complex Q^2-plane only for

$$-\infty < Q^2 \leq -4m_\pi^2,$$

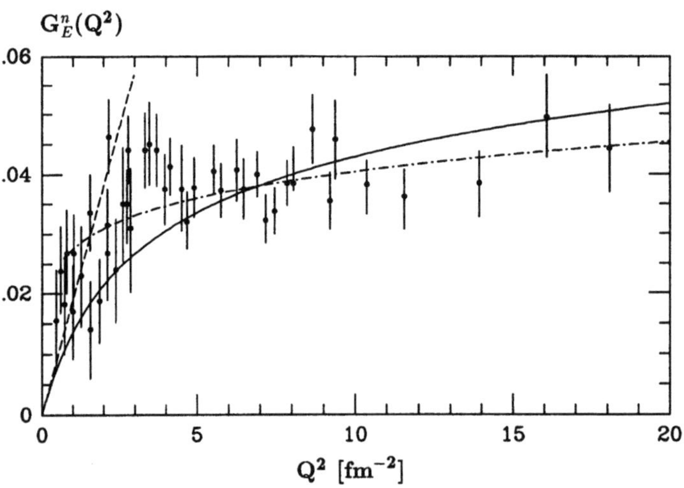

Fig. 25. The data for the electric form factor of the neutron $G^n_E(Q^2)$ from Refs. [75], [76]. Dash-dotted line: our naive "synchrotron" prediction $\propto (Q^2)^{1/6}$ (4.13) normalized to the data at $Q^2 = 5\ \mathrm{fm}^{-2}$. Dashed line: the slope of $G^n_E(Q^2)$ at $Q^2 = 0$ as deduced from thermal neutron-electron scattering [77]. Full line: the ansatz (4.15).

where m_π is the pion mass. We require a $(Q^2)^{1/6}$ behaviour for positive Q^2 and take the slope of $G_E^n(Q^2)$ at $Q^2 = 0$ from experiment (4.14). This leads us to the following functional form for $G_E^n(Q^2)$:

$$G_E^n(Q^2) = 0.019 \text{ fm}^2 \cdot Q^2 \left[1 + \frac{Q^2}{4m_\pi^2}\right]^{-5/6} . \tag{4.15}$$

It is amusing to see that this gives a decent description of the data (Fig. 25).

What about the electric form factor of the proton $G_E^p(Q^2)$? Here, clearly, we have a dominant "normal" piece connected with the total charge. We will assume that this normal contribution is represented by the usual dipole formula

$$\begin{aligned} G_D(Q^2) &= \left(1 + \frac{Q^2}{m_D^2}\right)^{-2} , \\ m_D^2 &= 18.23 \text{ fm}^{-2} \hat{=} 0.710 \text{ GeV}^2 \end{aligned} \tag{4.16}$$

which gives a good representation of the data for $Q^2 = 2-4$ GeV^2 [78]. Let us add to this an anomalous piece for smaller Q^2, connected with synchrotron radiation, and let us assume that this is a purely isovector contribution, consistent with the singularity at $Q^2 = -4m_\pi^2$ in (4.15). We obtain then from (4.15) the following ansatz for $G_E^p(Q^2)$:

$$G_E^p(Q^2) = G_D(Q^2) - G_E^n(Q^2) = G_D(Q^2)(1 - \Delta(Q^2)), \tag{4.17}$$

where

$$\Delta(Q^2) = 0.019 \text{ fm}^2 \cdot Q^2 (1 + \frac{Q^2}{4m_\pi^2})^{-5/6} \cdot (1 + \frac{Q^2}{m_D^2})^2 . \tag{4.18}$$

We predict a deviation of the ratio $G_E^p(Q^2)/G_D(Q^2)$ from unity for small Q^2. It is again amusing to note that such a deviation is indeed observed experimentally [47], [48]. Our ansatz does even quite well quantitatively (Fig. 26). For the electromagnetic radius of the proton we predict from (4.17)

$$\begin{aligned} \langle r_E^2 \rangle : &= -6 \frac{dG_E^p(Q^2)}{dQ^2}\Big|_{Q^2=0} \\ &= \frac{12}{m_D^2} + 6 \cdot 0.019 \text{ fm}^2 \\ &= (0.88 \text{ fm})^2 . \end{aligned} \tag{4.19}$$

This checks well with the experimental values quoted in [47]:

$$\begin{aligned} \langle r_E^2 \rangle^{1/2} &= 0.88 \pm 0.03 \text{ fm}, \text{ or} \\ & \quad 0.92 \pm 0.03 \text{ fm}, \end{aligned} \tag{4.20}$$

depending on the fit used for $G_E^p(Q^2)$ at low Q^2.

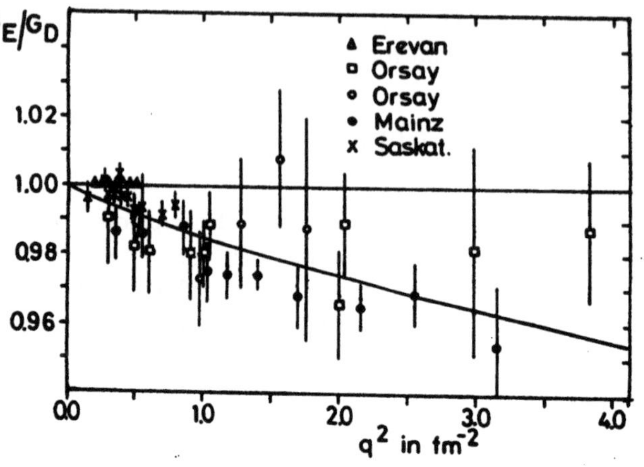

Fig. 26. The ratio G_E^p/G_D of the electric fom factor of the proton to the dipole fit versus Q^2. The data points are from various experiments as summarized in [48]. The solid line corresponds to the ansatz (4.17), (4.18).

To conclude this brief discussion of nucleon form factors, we can summarize our picture as follows: The quarks in the nucleon make a cyclotron-type motion in the chromomagnetic vacuum field. This leads to a spreading out of the charge distribution of the neutron

$$\rho_n(\mathbf{x}) \propto |\mathbf{x}|^{-10/3} \tag{4.21}$$

corresponding to $G_E^n(Q^2) \propto (Q^2)^{1/6}$. The same effect should lead to a deviation of the proton form factor $G_E^p(Q^2)$ from the dipole formula for small Q^2. Concerning the sign of $G_E^n(Q^2)$ we can in essence follow the arguments put forward in [79].

One might think — maybe rightly — that these ideas are a little crazy. But we have also worked out some consequences of them for the Drell-Yan reaction (1.9), which make us optimistic. In the lowest order parton process contributing there, we have a quark-antiquark annihilation giving a virtual photon γ^*, which decays then into a lepton pair (Fig. 1):

$$q + \bar{q} \to \gamma^* \to \ell^+\ell^-. \tag{4.22}$$

In the usual theoretical framework q and $\bar{q}$ are assumed to be uncorrelated and unpolarized in spin and colour if the original hadrons are unpolarized. From our point of view we expect a different situation. Let us go back to Fig. 21 where we sketched the world line of a quark of one fast hadron in a colour domain of extension $|x^2| \lesssim a^2$. Let the quark q and antiquark $\bar{q}$ in

(4.22) annihilate at the point $x = 0$. Here q and $\bar{q}$ come from two different hadrons h_1, h_2 moving with nearly light-like velocity in opposite directions. It is clear that in this situation q and $\bar{q}$ will spend a long time in a highly correlated colour background field (Fig. 27). In [14] we speculated that this may lead to a correlated transverse spin and colour spin polarization of q and $\bar{q}$ due to the chromomagnetic Sokolov-Ternov effect [64], [65]. In [80] we worked out this idea in more detail and found that a transverse $q\bar{q}$ spin correlation influences the $\ell^+\ell^-$ angular distribution in a profound way. Then our colleague H. J. Pirner pointed out to us that data which may be relevant in this connection existed already [81]. And very obligingly these data show a large deviation from the standard perturbative QCD prediction. On the other hand, we can nicely understand the data in terms of our spin correlations and thus vacuum effects in high energy collisions. For more details we refer to [80]. If such spin correlations are confirmed by experiments at higher energies, we would presumably have to reconsider the fundamental factorization hypothesis for hard reactions which we sketched in Sect. 1 and which is discussed in detail in [11].

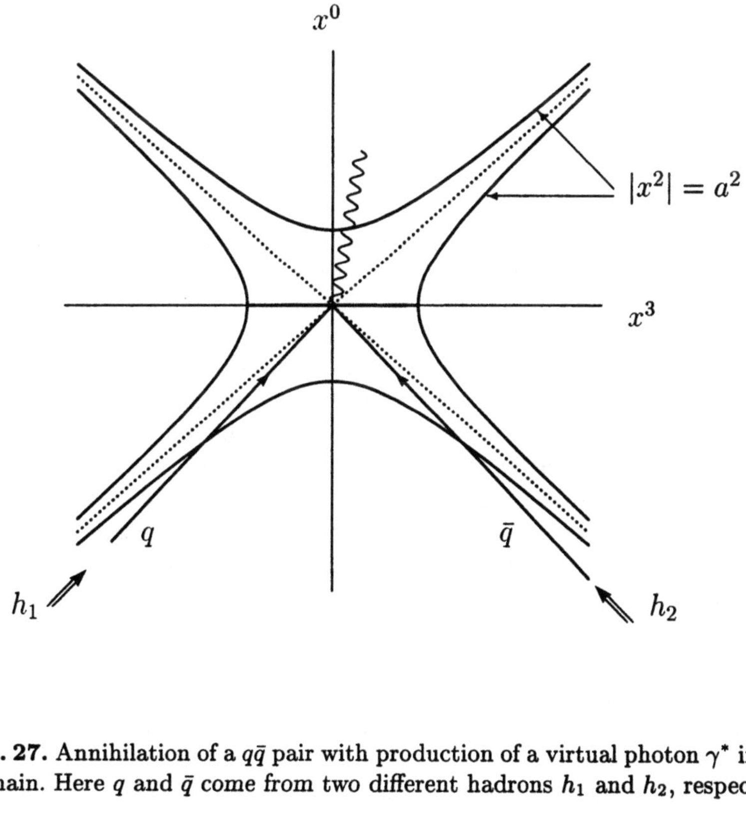

Fig. 27. Annihilation of a $q\bar{q}$ pair with production of a virtual photon γ^* in a colour domain. Here q and $\bar{q}$ come from two different hadrons h_1 and h_2, respectively.

5 Conclusions

In these lectures we have discussed various ideas connected with non-perturbative QCD and in particular with the QCD vacuum structure. In Sect. 2 we introduced connectors, the non-abelian Stokes theorem and the cumulant expansion. Then we presented the domain picture of the QCD vacuum and the stochastic vacuum model (SVM). The latter is consistent with the view of the QCD vacuum acting like a dual superconductor: We found that in the SVM the vacuum contains an effective chromomagnetic monopole condensate, whose effect is parameterized by the D-term in the gluon field strength correlator (2.61). With these tools we could calculate the expectation value of the Wegner-Wilson loop in the SVM. We found a linearly rising potential between a heavy quark-antiquark pair, $Q\bar{Q}$. This is related to the formation of a "string", a chromoelectric flux tube between $Q\bar{Q}$ as can be seen explicitly in the SVM [52]. Thus in this framework confinement is an effect of the nontrivial QCD vacuum structure. All calculations in Sect. 2 were done in Euclidean space time.

In Appendix A we discuss some problems which arise when one considers higher cumulant terms in the SVM. We propose as remedy for these problems to do the calculation of the expectation value of the Wegner-Wilson loop in an iterative way, summing step by step over the contributions of various plaquettes. This also leads us to a proposal for including dynamical fermions in the calculation with the expected result that the linearly rising potential levels off and goes to a constant at large $Q\bar{Q}$ separation, where, of course, we have then mesons $Q\bar{q}, q\bar{Q}$ formed of a heavy quark Q (antiquark $\bar{Q}$) and a light antiquark $\bar{q}$ (quark q).

In Sects. 3, 4 we looked at various possible effects of the nontrivial vacuum structure in the Minkowski world. In Sect. 3 we gave a detailed account of a field-theoretic method for the calculation of scattering amplitudes of high-energy soft hadronic reactions. We started from the functional integral and made high energy approximations in the integrand. This led us to give general rules for writing down scattering amplitudes for high energy soft reactions at the parton level. Then we considered as representation for hadrons wave packets of partons. The corresponding scattering amplitudes for (quasi-)elastic hadron-hadron scattering were found to be governed by the correlation functions of lightlike Wegner-Wilson loops. The evaluation of these correlation functions was possible with the help of the Minkowski version of the stochastic vacuum model. The comparison with experiment gave very good consistency, supporting the view that the QCD vacuum structure plays an essential role in high energy soft reactions. The framework developed in these lectures can be applied directly to elastic and diffractive hadron-hadron scattering at high energies. In principle we should also be able to apply it to non-diffractive reactions, fragmentation processes etc., but this remains to be worked out.

In Sect. 4 we argued that some more startling QCD vacuum effects in high energy collisions may be the appearance of anomalous soft photons in hadron-hadron collisions due to "synchrotron radiation" and spin correlations in the Drell-Yan reaction due to the chromodynamic Sokolov-Ternov effect. Furthermore, we gave an argument that electromagnetic form factors at small Q^2 should reflect the vacuum structure. Finally we would like to mention that in [82] the rapidity gap events observed at HERA are quantitatively described in terms of the parton model but invoking again nonperturbative QCD effects, possibly connected with the vacuum structure. Another place where the QCD vacuum structure may show up is in certain correlations of hadrons in Z^0 decays to two jets [14], [83] for which there is also some experimental evidence [84].

We hope to have convinced the reader that the non-perturbative structure of the QCD vacuum is useful in order to understand confinement of heavy quarks. In our view this vacuum structure manifests itself also in high energy soft and hard reactions. We think it is very worth-while to study such effects both from the theoretical and the experimental point of view.

Acknowledgements

The author is grateful to the organizers of the 1996 Schladming winter school headed by H. Latal for the invitation to lecture there. As basis for the present article the author could use lectures - partly collected by U. Grandel - given at the Workshop on Topics in Field Theory, Graduiertenkolleg Erlangen-Regensburg, Banz (1993). The author also thanks the organizers of that meeting, especially F. Lenz, for the invitation to lecture there. The present article is the contribution of the author to the proceedings of both meetings. For many fruitful discussions on topics of these lecture notes the author extends his gratitude to A. Brandenburg, W. Buchmüller, H. G. Dosch, U. Ellwanger, D. Gromes, P. Haberl, P. V. Landshoff, P. Lepage, H. Leutwyler, Th. Mannel, E. Mirkes, H. J. Pirner, M. Rueter, Yu. A. Simonov, G. Sterman, and W. Wetzel. Special thanks are due to H. G. Dosch for a critical reading of the manuscript and to E. Berger for his help in the preparation of the manuscript. Finally the author thanks Mrs. U. Einecke for her excellent typing of the article.

Appendix A: Higher Cumulant Terms and Dynamic Fermions in the Calculation of the Wegner-Wilson Loop in the Stochastic Vacuum Model

In this appendix we discuss first some problems arising in the calculation of the Wegner-Wilson loop in the SVM (cf. Sect. 2.5) if higher cumulants are taken into account. Then we outline a possible remedy which may also point to a way of including the effects of dynamical fermions in the SVM. We start with the replacements (2.79) which allow us to use the cumulant expansion (2.41) for calculating $W(C)$. The second cumulant is given in (2.80):

$$K_2(1,2) = \frac{4}{3}\frac{\pi^2}{g^2}\mathcal{F}(1,2), \tag{A.1}$$

where we set with $F_{\mu\nu\rho\sigma}$ as defined in (2.61):

$$\mathcal{F}(i,j) \equiv F_{1414}(X^{(i)} - X^{(j)}), \tag{A.2}$$

$$\begin{aligned} F_{1414}(Z) &= \frac{G_2}{24}[\kappa D(-Z^2) \\ &+\frac{1}{2}\left(\frac{\partial}{\partial Z_1}Z_1 + \frac{\partial}{\partial Z_4}Z_4\right)(1-\kappa)D_1(-Z^2)], \\ Z^2 &= Z_1^2 + Z_4^2. \end{aligned} \tag{A.3}$$

With the assumptions 1-3 of the SVM (Sect. 2.4), the next nonvanishing cumulant is K_4, for which we obtain from (2.47), (2.60), (2.73), (2.75):

$$K_4(1,2,3,4) = -2\left(\frac{\pi^2}{g^2}\right)^2 [\mathcal{F}(1,3)\mathcal{F}(2,4)\Theta(1,2,3,4) \ + \ perm.]. \tag{A.4}$$

Here $\Theta(1,2,3,4) = 1$ if $X^{(1)} > X^{(2)} > X^{(3)} > X^{(4)}$ in the sense of the path-ordering on the surface S (cf. Figs. 8, 10) and $\Theta(1,2,3,4) = 0$ otherwise.

We start now again from the expression (2.78) for the expectation value of the Wegner-Wilson loop and use the cumulant expansion (2.41) with the identifications (2.79). We get then:

$$W(C) = \exp\left\{\sum_{n=1}^{\infty}\frac{(-ig)^n}{n!}\int_{S_1}\cdots\int_{S_n}K_n(1,...,n)\right\}, \tag{A.5}$$

where

$$\int_{S_i} \equiv \int_S \mathrm{d}X_1^{(i)}\mathrm{d}X_4^{(i)}. \tag{A.6}$$

In the SVM as formulated in Sect. 2.4 the cumulants for odd n vanish (cf. Ass. 3). The lowest contribution in (A.5) is then from $n = 2$, the next from

$n = 4$. Cutting off the infinite sum in (A.5) at $n = 4$, we get with (A.1) and (A.4):

$$\begin{aligned} W(C) &= \exp\left\{ -\frac{g^2}{2!} \int_{S_1} \int_{S_2} K_2(1,2) \right. \\ &\quad \left. + \frac{(g^2)^2}{4!} \int_{S_1} \cdots \int_{S_4} K_4(1,2,3,4) \right\} \\ &= \exp\left\{ -\frac{1}{2!}\frac{4\pi^2}{3} \int_{S_1} \int_{S_2} \mathcal{F}(1,2) \right. \\ &\quad \left. - \frac{1}{4!} 2\pi^4 \int_{S_1} \cdots \int_{S_4} [\mathcal{F}(1,3)\mathcal{F}(2,4)\Theta(1,2,3,4) \; + \; perm.] \right\} \end{aligned} \tag{A.7}$$

Consider now the contribution of the second cumulant in (A.7) for a large Wegner-Wilson loop (Fig. A1 with $R, T \gg a$):

$$I_2 = \int_{S_1} \int_{S_2} \mathcal{F}(1,2). \tag{A.8}$$

For fixed integration point (1) on S the integration over (2) will give signifi-

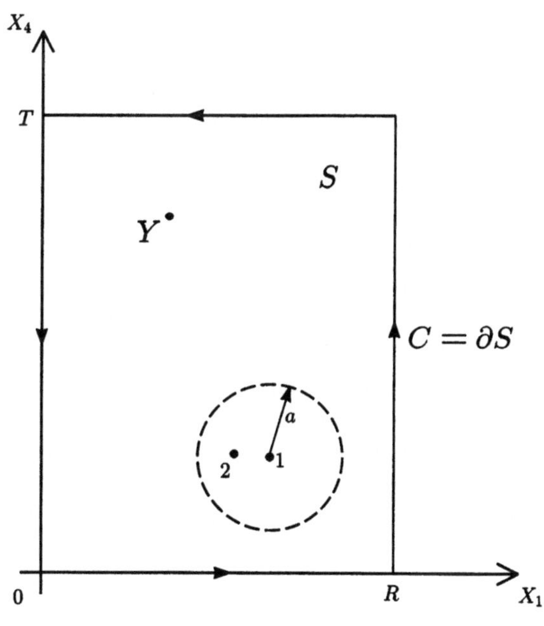

Fig. A.1. The relevant integration region for the integral (A.8). The point (1) runs freely over the surface S, the point (2) is constrained to a distance $\lesssim a$ from (1).

cant contributions only for a region of radius $\sim a$ around (1) since the functions $D(-Z^2)$ and $D_1(-Z^2)$, where $Z = (X^{(1)} - X^{(2)})$, are assumed to fall off rapidly with increasing Z^2. From (A.3) we see that the D_1-term contributes as a total divergence in $\mathcal{F}(1,2)$. Thus we can apply Gauss' law in 2 dimensions for it to transform it to an integral over the boundary $\partial S = C$. In this way we find a contribution from the D_1 term of order a^4/R^2, a^4/RT, a^4/T^2. From the D-term in (A.3) we get a factor $\propto a^2$ from the integration over (2) in (A.8). Then the integration over (1) is unconstrained and gives a factor proportional to the whole area of S:

$$I_2 \propto RTa^2 + O\left(a^4, a^4\frac{T}{R}, a^4\frac{R}{T}\right). \tag{A.9}$$

Putting in all factors from (A.3) and using the explicit form (2.65) for the function $D(-Z^2)$ gives for $T \to \infty$ and $R \gg a$ the result:

$$\frac{1}{2!}\frac{4\pi^2}{3} I_2 = RT\sigma \tag{A.10}$$

with σ as given in (2.83).

We turn next to the contribution of the 4th cumulant in (A.7) and study the integral

$$I_4 = \int_{S_1} \dots \int_{S_4} \mathcal{F}(1,3)\mathcal{F}(2,4)\Theta(1,2,3,4). \tag{A.11}$$

Consider again $R, T \gg a$ and fix the integration point (1) (Fig. A2). Then the short-range correlation of the field strengths requires (3) and (1) as well as (4) and (2) to be near to each other, i.e. at a distance $\lesssim a$. The function $\Theta(1,2,3,4)$ requires $(1) > (2) > (3) > (4)$ in the sense of the path ordering on the surface S. Using the fan-type net as indicated in Fig. 8 for the application of the non-abelian Stokes theorem with straight line handles from the points $X^{(i)}$ in the surface to the reference point Y we see the following. The path-ordering function $\Theta(1,2,3,4)$ restricts (2) to be in the hatched sector of S in Fig. A2. In general this does not restrict (2) to a region close to (1). On the contrary, (2) can vary freely over a triangle of area $\gtrsim a.L$, where L is of order R or T or in between. Thus, the integration over (2) in (A.11) will give at least a factor $\propto (aT)$ if the hatched triangle has its long sides in X_4 direction (cf. Fig. A2). The integrations over (3) and (4) in (A.11) should give factors of a^2 each. The integration over (1) will give a factor RT. Thus we estimate

$$I_4 \propto RT.a^4.aT \propto T^2. \tag{A.12}$$

This is very unpleasant. The 4th cumulant gives a contribution which dominates over the one from the second cumulant for $T \to \infty$. There is no finite limit for $T \to \infty$ in the expression (2.77). The quark-antiquark potential comes out infinite.

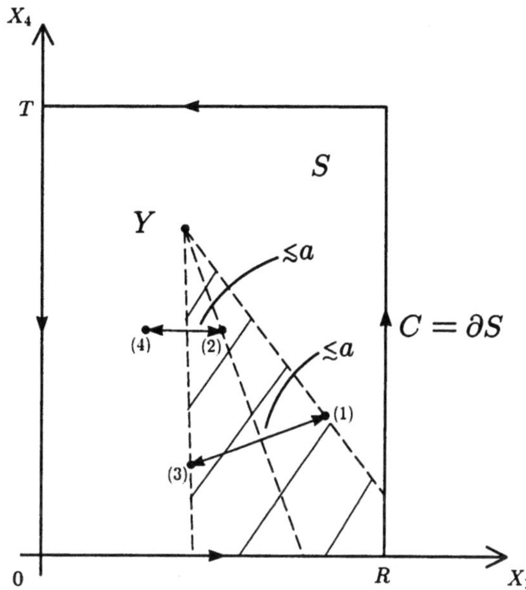

Fig. A.2. The relevant integration region for the integral (A.11). The point (1) runs freely over S. The points (1), (2), (3), (4) are ordered in the angle as seen from the reference point Y due to the path ordering function $\Theta(1,2,3,4)$. The points (1), (3) and (2), (4) must be at a distance $\lesssim a$ to each other.

This problem was recognized in Ref. [52] and eliminated by hand making an additional assumption: that all but the leading powers in the quotient of the correlation length to the extension parameters R, T of the loop could be neglected. Another simple cure of the problem would be to postulate instead of Ass. 3 (cf. (2.73)-(2.75)) a behaviour of the higher point correlation functions of the shifted gluon field strengths which gives precisely zero for the cumulants K_n with $n > 2$ in (A.5). In our opinion this would be an unsatisfactory solution, since the model would then only work for a particular fine-tuned set of correlation functions whereas generically the above problem would remain.

We think that the origin of these difficulties in the SVM is the assumption 1, which states that the correlator should be independent of the reference point Y for arbitrary Y. But why should the field strength correlation (2.60) (cf. Fig. 5) be the same if we use a straight line path $C_X + \bar{C}_{X'}$ to connect X and X' or a path which runs on a loop behind the moon? We will replace Ass. 1 by a milder one:

– **Ass. 1'**: $F_{\mu\nu\rho\sigma}$ in (2.60) is independent of the reference point Y and of the curves C_X and $C_{X'}$ if the reference point Y is close to X and X':

$$|X - Y| \lesssim a', \; |X' - Y| \lesssim a',$$

where a' is of order of a.

Now we try to calculate the expectation value of the Wegner-Wilson loop (2.76) using only this weaker hypothesis. We start from the rectangle S of area RT and insert a smaller rectangular loop C_1 with sides $R-2a', T-2a'$. On C_1 we choose N_1 reference points $Y_1, ..., Y_{N_1}$ and we partition the area between C and C_1 in N_1 plaquettes $P_1, ..., P_{N_1}$ of size $\sim a'^2$ (Fig. A3). We can now apply the non-abelian Stokes theorem. We start from Y_0 on C and construct a path equivalent to C in the following way: From Y_0 to Y_1, then in a fan-type net over the plaquette P_1 with Y_1 as reference point, etc., until we arrive at $Y_{N_1} \equiv Y_1$ from where the path runs back to Y_0. According to (2.36) we get then:

$$W(C) = \frac{1}{3}\mathrm{Tr}\langle V_{0,N_1} V(P_{N_1}) V_{N_1,N_1-1} ... V_{2,1} V(P_1) V_{1,0}\rangle, \tag{A.13}$$

where

$$V(P_j) = \mathrm{P}\ \exp\left[-ig \int_{P_j} \hat{G}_{14}(Y_j, X^{(j)}; C_{X^{(j)}})\right] \tag{A.14}$$

and $V_{i,j}$ are the connectors from Y_j to Y_i on straight lines.

– **Ass. 4**: Now we will make a <u>mean field-type approximation</u> and replace the path-ordered integrals over the plaquettes P_j by the corresponding vacuum expectation values:

$$V(P_j) \rightarrow \langle V(P_j)\rangle \cdot \mathbb{1}. \tag{A.15}$$

This is similar in spirit to the "block spin" transformations considered in [85]. For the r.h.s. of (A.15) we can use the cumulant expansion and assumptions 1', 2 and 3 of the SVM. Here the reference point Y_j is never too far away from $X^{(j)}$. We get with $\mathbb{1}$ the unit matrix in colour space

$$\begin{aligned}\langle V(P_j)\rangle &= \mathbb{1} \cdot \exp\left\{\sum_{n=1}^{\infty} \frac{(-ig)^n}{n!} \int_{P_{j,1}} ... \int_{P_{j,n}} K_n(1, ..., n)\right\} \\ &= \mathbb{1} \cdot \exp\left\{-|P_j|\sigma\left[1 + O\left(\frac{1}{4!}\sigma \cdot a \cdot a'\right)\right]\right\}.\end{aligned} \tag{A.16}$$

Here we cut off the cumulant expansion at $n = 4$ and use the results and estimates (A.7)-(A.12), but now for each plaquette P_j instead of the whole

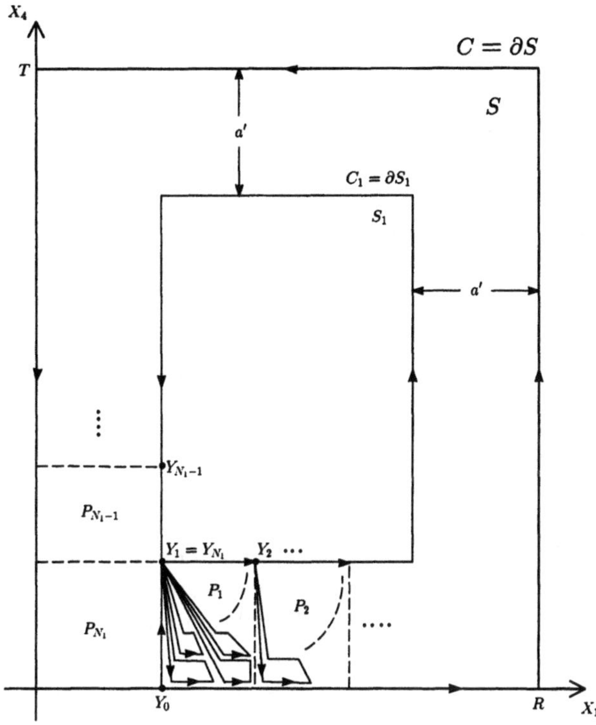

Fig. A.3. The rectangle S, $C = \partial S$ and N_1 reference points $Y_1, ..., Y_{N_1}$ on the curve C_1. The area between C_1 and C is partitioned in N_1 plaquettes $P_1, ..., P_{N_1}$.

surface S. We see that the correction terms from the 4th cumulant are now of manageable size since (cf. (2.70), (2.85))

$$\sigma a a' = \left(\frac{a'}{a}\right) \sigma a^2 = 0.56 \left(\frac{a'}{a}\right) = O(1). \tag{A.17}$$

Then (hopefully) the factorials $1/n!$ in the cumulant expansion will lead to small corrections from higher cumulants.

We can now insert (A.15), (A.16) in (A.13) and get

$$\begin{aligned} W(C) &\simeq \frac{1}{3}\mathrm{Tr}\langle V_{0,N_1}\langle V(P_{N_1})\rangle V_{N_1,N_1-1}...V_{2,1}\langle V(P_1)\rangle V_{1,0}\rangle \\ &= \frac{1}{3}\mathrm{Tr}\langle V_{0,N_1} V_{N_1,N_1-1}...V_{2,1}V_{1,0}\rangle \cdot \exp\Big\{ -\sum_{j=1}^{N_1} |P_j|\sigma\big[1+O(\frac{\sigma a a'}{4!})\big]\Big\} \\ &= \frac{1}{3}\mathrm{Tr}\langle V_{N_1,N_1-1}...V_{2,1}\rangle \exp\Big\{ -(|S|-|S_1|)\sigma\big[1+O(\frac{\sigma a a'}{4!})\big]\Big\}. \end{aligned} \tag{A.18}$$

Here we used also the cyclicity of the trace and $V_{1,0}V_{0,N_1} = 1\!\!1$ (cf. (2.24)). The product of the remaining connectors in (A.18) gives just the Wegner-Wilson loop of the curve $C_1 = \partial S_1$ in Fig. A.3 and we obtain:

$$W(C) = W(C_1)\exp\left\{-(|S| - |S_1|)\sigma\left[1 + O\left(\frac{\sigma a a'}{4!}\right)\right]\right\}. \tag{A.19}$$

The procedure can easily be repeated by inscribing a rectangle S_2 with boundary $C_2 = \partial S_2$ in S_1 etc. The final result of this iterative procedure obviously is:

$$W(C) = \exp\left\{-|S|\sigma\left[1 + O\left(\frac{\sigma a a'}{4!}\right)\right]\right\}. \tag{A.20}$$

From this we deduce for the "true" string tension in the SVM:

$$\sigma_{\text{true}} = \sigma\left[1 + O\left(\frac{\sigma a a'}{4!}\right)\right] = \sigma\left[1 + \text{ terms of order } 5\%\right], \tag{A.21}$$

where σ is given in (2.83). Thus we have justified the result for the string tension in Sect. 2.5, where we used the second cumulant only. We have now relied on the assumption 1' and avoided transportation of field strengths to far away reference points Y. The fourth cumulant is estimated to give only a correction of a few percent and the contribution of the 6th, 8th, etc. cumulants can be estimated in an analogous way to be even smaller[1] Most important, we have avoided the unpleasant result (A.12).

We will now discuss another question which can be raised in connection with the calculation of $W(C)$ in the SVM. Why should we span a minimal surface S into the loop C and not use some other, more complicated surface S' with the same boundary, $\partial S' = C$? From the point of view of applying the non-abelian Stokes theorem (2.37) any wiggly surface S' would be as good as the flat rectangle S which is the surface of minimal area. However, from the point of view of the iterative calculation with the plaquettes, as explained in this appendix, an arbitrary surface S' is clearly not equivalent to S. In (A.15) we made the approximation of replacing $V(P_j)$ ($j = 1, ..., N_1$) by its vacuum expectation value. This should be a good approximation if the various plaquettes are well separated. Indeed, we would expect that then we can perform the functional integral over the variables related to the regions in Euclidean space time where the various plaquettes are located in a separate and independent way. If, however, two plaquettes overlap or are very near to each other, the above approximation will break down. The point is now that on the minimal surface S neighbouring plaquettes will be at maximal distance from each other. For some arbitrary surface S' with wiggles we will inevitably find plaquettes closer together which will make our approximation worse. This

[1] Numerical studies suggest that $a' \approx 3a$ should be large enough for obtaining the area law for the plaquettes P_j in (A.16) (H. G. Dosch, private communication). We obtain then $\sigma a a' \simeq 1.7$ and still correction terms $\lesssim 10$ % in (A.21).

gives us some rationale to use a minimal surface S in the applications of the non-abelian Stokes theorem in the framework of the SVM.

So far our calculations should apply to QCD with dynamical gluons and static quarks only. The quark-antiquark potential $V(R)$ in (2.84) rises linearly for $R \to \infty$. In real life we have, of course, dynamic quarks. If we separate a heavy quark-antiquark pair $Q\bar{Q}$ starting from small R we will first see a linear potential as in (2.84) but at some point the heavy quark Q (antiquark $\bar{Q}$) will pick up a light antiquark $\bar{q}$ (quark q) from the vacuum and form a meson $Q\bar{q}$ ($\bar{Q}q$). The two mesons can escape to infinity, i.e. the force between them vanishes, the potential $V(R)$ must go to a constant as $R \to \infty$:

$$V(R) \to V_\infty \quad \text{for} \quad R \to \infty. \tag{A.22}$$

Can we understand also this feature of nature in an extension of the SVM?

Let us go back to the approximation (A.15), where we have replaced the integral over the plaquette P_j by its vacuum expectation value. More generally we can argue that $V(P_j)$ as defined in (A.14) is an object transforming under a gauge transformation (2.17) as follows (cf. (2.20)):

$$V(P_j) \to U(Y_j)V(P_j)U^{-1}(Y_j). \tag{A.23}$$

Any approximation we make should respect this gauge property and indeed the replacement (A.15) does. Now we can ask for a generalisation of (A.15) in the presence of dynamic light quarks. We have then the quark and antiquark variables at the point Y_j at our disposal and can construct from them the object

$$q(Y_j)\bar{q}(Y_j) \tag{A.24}$$

which has the gauge transformation property (A.23) and is also rotationally invariant. Let us consider only u and d quarks. Then we suggest as generalization of (A.15) the following replacement:

$$V_{AB}(P_j) \to \langle V(P_j)\rangle_0 \delta_{AB} - w(P_j) q_{A,f,\alpha}(Y_j)\bar{q}_{B,f,\alpha}(Y_j), \tag{A.25}$$

where A, B are the colour indices, $f = u, d$ is the flavour index and α the Dirac index. Furthermore, $\langle V(P_j)\rangle_0$ is as in (A.15), (A.16), and $w(P_j)$ is a coefficient depending on the size of the plaquette P_j. It can be thought of as representing the chance of producing a $q\bar{q}$ pair from the vacuum gluon fields over the plaquette P_j. (This is inspired by the discussions of particle production in the LUND model [54]). On dimensional grounds $w(P_j)$ must be proportional to a volume, thus we will set

$$w(P_j) = c|P_j| \cdot a, \tag{A.26}$$

where c is a constant. The idea is that $q\bar{q}$ "production" should feel the gluon fields in a disc of area $|P_j|$ and thickness a.

We can now insert the ansatz (A.25) in (A.13). The resulting expression for $W(C)$ is of the form:

$$\begin{aligned} W(C) \cong \frac{1}{3}\mathrm{Tr}\langle V_{0,N_1}[\langle V(P_{N_1})\rangle \\ -w(P_{N_1})q(Y_{N_1})\bar{q}(Y_{N_1})]V_{N_1,N_1-1} \\ \ldots V_{2,1}[\langle V(P_1)\rangle - w(P_1)q(Y_1)\bar{q}(Y_1)] \cdot V_{1,0}\rangle. \end{aligned} \tag{A.27}$$

If we mutliply out these brackets we get terms where we have again the Wegner-Wilson loop along C_1 (Fig. A.3) but then also terms where quarks and antiquarks at different points Y_k, Y_l are connected. For two neighbouring points, for instance, $k = l+1, l$ this would read:

$$\ldots q(Y_{l+1})[-w(l)\bar{q}(Y_{l+1})V_{l+1,l}q(Y_l)]\bar{q}(Y_l)\ldots$$

The importance of these terms will increase with increasing $w(P_j)$. Starting from (A.27) we can now inscribe plaquettes into the rectangle S_1 and transport everything, including the quark variables $q(Y_k)\bar{q}(Y_k)(k = 1, \ldots, N_1)$ to a curve C_2 etc.

To get an orientation we will now make a drastic approximation: For a loop with $R \ll R_c$, where R_c is some critical value to be determined, we neglect the dynamical quarks, saying that the $w(P_j)$ factors will still be too small. The result for $W(C)$ and $V(R)$ in the range $a \lesssim R \ll R_c$ will then be as in (A.20):

$$\begin{aligned} &W(C) = \exp(-RT\sigma), \\ &V(R) = \sigma R, \\ &(a \lesssim R \ll R_c), \end{aligned} \tag{A.28}$$

where we set $\sigma_{\mathrm{true}} \cong \sigma$.

In the other limiting case, $R \gg R_c$, we can expect the $q\bar{q}$ terms in (A.27) to dominate. Indeed, let us start with the loop C with sides R and T of Fig. A.3 and inscribe first plaquettes $P_1, \ldots, P_{N_1}$ of size a'^2. With the replacement (A.25) this gives for $W(C)$ the expression (A.27). Now we inscribe plaquettes of size a'^2 in the loop C_1 and parallel transport everything to a smaller loop C_2. The new element is that we will now have to parallel-transport also the quark variables, but this poses no problem. We will again obtain an expression like the one shown in (A.27) but now for the curve C_2 and with increased weight factors w in front of the $q\bar{q}$ terms. We assume that we continue this procedure. Finally, we will obtain an expression like (A.27) but with the terms involving the quark variables at the reference points Y_j dominating the expression. Thus we set:

$$W(C) \cong \frac{1}{3}\mathrm{Tr}\langle[-w(N)q(N)\bar{q}(N)]V_{N,N-1}\ldots[-w(1)q(1)\bar{q}(1)]V_{1,N}\rangle, \tag{A.29}$$

where the points $1, ..., N$ are on a curve C' of sides R', T'. This summing up of quark contributions from various plaquettes of size a'^2 should be reasonable as long as the quarks are inside an area corresponding to their own correlation length, a_χ, the chiral correlation length. This can be estimated from the behaviour of the non-local $\bar{q}q$ condensate [86-88] for which one typically makes an ansatz of the form ($q = u, d$):

$$\langle \bar{q}(X_2) V_{2,1}\, q(X_1) \rangle = \langle \bar{q}q \rangle e^{-|X_2 - X_1|^2/a_\chi^2}. \tag{A.30}$$

Here $\langle \bar{q}q \rangle$ is the local quark condensate for which we set, neglecting isospin breaking (cf. [39]):

$$\langle \bar{q}q \rangle = \frac{1}{2} \langle \bar{u}u + \bar{d}d \rangle = -(0.23 \text{ GeV})^3. \tag{A.31}$$

For the correlation length a_χ one estimates (cf. [88]):

$$a_\chi \cong 1 \text{ fm}. \tag{A.32}$$

It will be reasonable to choose the reference points $1, ..., N$ on C' also such that the distance between neighbouring points is a_χ. Then

$$\begin{aligned} R' &= R - 2a_\chi, \\ T' &= T - 2a_\chi, \\ N &= \frac{2T' + 2R'}{a_\chi} = \frac{2(T+R)}{a_\chi} - 8. \end{aligned} \tag{A.33}$$

We will now estimate $W(C)$ from (A.29) as product of the expectation values of the nonlocal $\bar{q}q$ condensate. We set

$$w(1) = w(2) ... = w(N) = c \cdot a a_\chi^2, \tag{A.34}$$

where c should be a numerical constant of order 1. Furthermore we have from (A.30)

$$\begin{aligned} &\langle \bar{q}_{A,f,\alpha}(j) (V_{j,j-1})_{AB}\, q_{B,f',\alpha'}(j-1) \rangle \\ &= \frac{1}{4} \delta_{f,f'} \delta_{\alpha,\alpha'} \langle \bar{q}q \rangle \exp \left[-\frac{|X_j - X_{j-1}|^2}{a_\chi^2} \right] \\ &= \frac{1}{4e} \delta_{f,f'} \delta_{\alpha,\alpha'} \langle \bar{q}q \rangle \end{aligned} \tag{A.35}$$

for $|X_j - X_{j-1}| = a_\chi$. This gives:

$$\begin{aligned} W(C) \cong &-\frac{1}{3} \cdot \{ [-w(N) \langle \bar{q}_{f_1,\alpha_1}(1) V_{1,N}\, q_{f_N,\alpha_N}(N) \rangle] \\ &\quad ... [-w(1) \langle \bar{q}_{f_2,\alpha_2}(2) V_{2,1}\, q_{f_1,\alpha_1}(1) \rangle] \} \\ &= -\frac{8}{3} \left[-\frac{w(1) \langle \bar{q}q \rangle}{4e} \right]^N \\ &= -\frac{8}{3} \left[-\frac{w(1) \langle \bar{q}q \rangle}{4e} \right]^{[\frac{2}{a_\chi}(T+R) - 8]} . \end{aligned} \tag{A.36}$$

For the potential this leads to

$$V(R) = -\lim_{T\to\infty}\frac{1}{T}\ln W(C) = \frac{2}{a_\chi}\ln\frac{4e}{[-w(1)\langle\bar{q}q\rangle]} \equiv V_\infty. \tag{A.37}$$

Thus we find indeed a constant potential for $R \to \infty$.

For intermediate values of R the potential should change smoothly from the linear rise to the constant behaviour. As a crude approximation we will assume the potential to be continuous but turn abruptly from one to the other behaviour at a critical length $R = R_c$ which we define as

$$R_c = V_\infty/\sigma. \tag{A.38}$$

Our simple ansatz for $V(R)$ is then as follows (cf. Fig. A.4):

$$V(R) = \begin{cases} \sigma R & \text{for } R \le R_c, \\ V_\infty = \sigma R_c & \text{for } R \ge R_c. \end{cases} \tag{A.39}$$

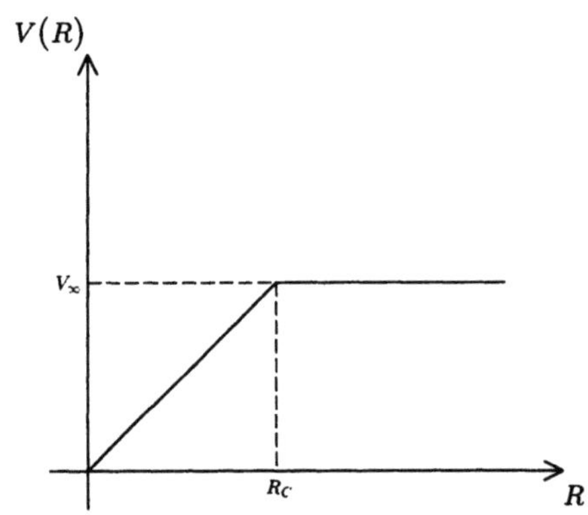

Fig. A.4. The potential $V(R)$ as defined in (A.39) with R_c the "string breaking" radius.

How does the numerics work out taking (A.37) and (A.38) seriously? We can estimate the constant V_∞ crudely from the mass difference of two B-mesons and the Υ resonance

$$V_\infty \cong 2m_B - m_\Upsilon \cong 1.1 \text{ GeV}. \tag{A.40}$$

With σ from (2.85) we find for R_c (A.38):

$$R_c \cong 1,2 \text{ fm}. \tag{A.41}$$

A much more elaborate estimate of the string-breaking distance gives [89] $R_c \leq 1{,}6$ to $2{,}1$fm. The lattice calculations of [90] suggest an even larger value for R_c. From (A.40) and (A.37), (A.32) we get for $w(1)$ and c:

$$
\begin{aligned}
w(1) &= -\frac{4\exp[1-\frac{1}{2}a_\chi V_\infty]}{\langle \bar{q}q\rangle} = 0.42 \text{ fm}^3, \\
c &= \frac{w(1)}{a_\chi^2 a} = 1.2.
\end{aligned}
\tag{A.42}
$$

Thus the probability factor $w(1)$ comes out as we estimated it on geometrical grounds in (A.26).

With these remarks we close this appendix. Of course, much more work is needed in order to decide if our simple ansatz for incorporating dynmical fermions in the stochastic vacuum model is viable or not.

Appendix B: The Scattering of Gluons

In this appendix we will discuss the contribution of gluons to the scattering amplitude of a general parton reaction, generalizing (3.50):

$$
G(1)+G(2)+...+q(k)+\bar{q}(k')+... \rightarrow G(3)+G(4)+...+q(l)+\bar{q}(l')+.... \tag{B.1}
$$

Here, as in all of Sect. 3, we use the convention that partons with odd (even) number are moving fast in positive (negative) x^3-direction. Consider thus the transition of a right-moving gluon:

$$
G(1) \equiv G(P_1,\epsilon_1,a_1) \rightarrow G(3) \equiv G(P_3,\epsilon_3,a_3), \tag{B.2}
$$

where $a_{1,3}$ ($1 \leq a_{1,3} \leq 8$) are the colour indices, P_1, P_3 are the momenta with $P_{1+}, P_{3+} \rightarrow \infty$, and $\boldsymbol{\epsilon}_{1,3}$ are the polarization vectors which satisfy:

$$
\begin{aligned}
&\mathbf{P}_1\boldsymbol{\epsilon}_1 = 0, \\
&\mathbf{P}_3\boldsymbol{\epsilon}_3 = 0, \\
&|\boldsymbol{\epsilon}_1| = |\boldsymbol{\epsilon}_3| = 1.
\end{aligned}
\tag{B.3}
$$

In the high energy limit the vectors $\boldsymbol{\epsilon}_{1,3}$ are transverse with corrections of order $|\mathbf{p}_{T1,3}|/P_{+1,3}$ which we can neglect. We will argue that such a gluon in a high energy soft reaction is equivalent to a quark-antiquark pair with the same quantum numbers in the limit of the q and $\bar{q}$ being so close to each other in position space that their separation cannot be resolved in the collision. From the contribution of such a $q\bar{q}$ pair to the scattering amplitude we will get in the above limit the gluon contribution.

We start by constructing wave packets similar to the mesonic wave packets of (3.66)

$$
\begin{aligned}
|q\bar{q}; P_j, \epsilon_j, a_j\rangle = \int d^2p_T \int_0^1 d\zeta \frac{1}{(2\pi)^{3/2}} h^j(\zeta, |\mathbf{p}_T|) \cdot \\
\frac{1}{2}(\lambda_{a_j})_{A_j A'_j} (\boldsymbol{\epsilon}_j \cdot \boldsymbol{\sigma}\epsilon)_{s_j s'_j} |q(p_j, s_j, A_j)\bar{q}(P_j - p_j, s'_j, A'_j)\rangle \\
(j = 1, 2, 3, 4),
\end{aligned}
\tag{B.4}
$$

where p_j is as in (3.67), (3.68) and

$$
\epsilon = \begin{pmatrix} 0 & 1 \\ -1 & 0 \end{pmatrix}. \tag{B.5}
$$

We require the functions h^j to satisfy:

$$
\begin{aligned}
(h^j(\zeta, |\mathbf{p}_T|))^* &= h^j(\zeta, |\mathbf{p}_T|), \\
h^j(\zeta, |\mathbf{p}_T|) &= h^j(1 - \zeta, |\mathbf{p}_T|), \\
\int d^2p_T \int_0^1 d\zeta \; 2\zeta(1-\zeta)|h^j(\zeta, |\mathbf{p}_T|)|^2 &= 1.
\end{aligned}
\tag{B.6}
$$

This gives us the normalization of the states (B.4) as

$$
\begin{aligned}
&\langle q\bar{q}; P'_j, \epsilon'_j, a'_j | q\bar{q}, P_j, \epsilon_j, a_j\rangle \\
&= \boldsymbol{\epsilon}'^{*}_j \cdot \boldsymbol{\epsilon}_j \delta_{a'_j, a_j} (2\pi)^3 2P^0_j \delta^3(\mathbf{P}'_j - \mathbf{P}_j).
\end{aligned}
\tag{B.7}
$$

The $q\bar{q}$ states (B.4) have the same transformation properties under a parity (P) charge conjugation (C) and time reversal (T) transformation as a gluon state:

$$
U(\mathrm{P})|q\bar{q}; P^0, \mathbf{P}, \epsilon, a\rangle = -|q\bar{q}; P^0, -\mathbf{P}, \epsilon, a\rangle, \tag{B.8}
$$

$$
U(\mathrm{C})|q\bar{q}; P^0, \mathbf{P}, \epsilon, a\rangle = -|q\bar{q}; P^0, \mathbf{P}, \epsilon, b\rangle \cdot \frac{1}{2}\mathrm{Tr}(\lambda_b \lambda_a^T), \tag{B.9}
$$

$$
V(\mathrm{T})|q\bar{q}; P^0, \mathbf{P}, \epsilon, a\rangle = -|q\bar{q}; P^0, -\mathbf{P}, \epsilon^*, b\rangle \frac{1}{2}\mathrm{Tr}(\lambda_b \lambda_a^T). \tag{B.10}
$$

Here $U(\mathrm{P}), U(\mathrm{C})$ are the unitary operators, $V(\mathrm{T})$ is the antiunitary operator representing the P, C and T transformations, respectively, in the Fock space of parton states.

As in (3.71) we define the wave functions in transverse position space and longitudinal momentum fraction:

$$
\varphi^j(\zeta, \mathbf{y}_T) := \sqrt{2\zeta(1-\zeta)} \frac{1}{2\pi} \int d^2p_T \exp(i\mathbf{p}_T \cdot \mathbf{y}_T) h^j(\zeta, |\mathbf{p}_T|). \tag{B.11}
$$

Here we also define for the right (left) movers the wave functions in $y_-, \mathbf{y}_T$ space ($y_+, \mathbf{y}_T$ space) as:

$$\tilde{\varphi}^j(y_-, \mathbf{y}_T) := \frac{1}{\sqrt{4\pi}}\sqrt{P_{j+}}\int_0^1 \mathrm{d}\zeta\, \varphi^j(\zeta, \mathbf{y}_T)\exp[-\frac{i}{2}P_{j+}(\zeta - \frac{1}{2})y_-],$$
$$(j \text{ odd}), \tag{B.12}$$

$$\tilde{\varphi}^j(y_+, \mathbf{y}_T) := \frac{1}{\sqrt{4\pi}}\sqrt{P_{j-}}\int_0^1 \mathrm{d}\zeta\, \varphi^j(\zeta, \mathbf{y}_T)\exp[-\frac{i}{2}P_{j-}(\zeta - \frac{1}{2})y_+],$$
$$(j \text{ even}). \tag{B.13}$$

The normalization condition (B.6) implies:

$$\begin{aligned} &\int \mathrm{d}^2 y_T \int_0^1 \mathrm{d}\zeta |\varphi^j(\zeta, \mathbf{y}_T)|^2 = 1, \\ &\int \mathrm{d}y_- \int \mathrm{d}^2 y_T |\tilde{\varphi}^j(y_-, \mathbf{y}_T)|^2 = 1 \qquad (j \text{ odd}). \\ &\int \mathrm{d}y_+ \int \mathrm{d}^2 y_T |\tilde{\varphi}^j(y_+, \mathbf{y}_T)|^2 = 1 \qquad (j \text{ even}). \end{aligned} \tag{B.14}$$

To realize the condition that the $q\bar{q}$ pair acts like a gluon, we require for right (left) movers that they have similar longitudinal momenta and that their wave function in the relative q-$\bar{q}$ coordinates, $y_-, \mathbf{y}_T(y_+, \mathbf{y}_T)$ is nearly a δ function. To be concrete, we require:

$$\varphi^j(\zeta, \mathbf{y}_T) \neq 0 \quad \text{only for} \quad |\zeta - \frac{1}{2}| \leq \xi_0, \tag{B.15}$$

where

$$0 < \xi_0 \ll \frac{1}{2}, \tag{B.16}$$

and for j odd:

$$\tilde{\varphi}^j(y_-, \mathbf{y}_T) \neq 0 \quad \text{only if} \quad |y_-| \lesssim \frac{1}{P_+\xi_0} \quad \text{and} \quad |\mathbf{y}_T| \ll a. \tag{B.17}$$

For right movers we replace plus by minus signs in (B.17). Any $q\bar{q}$ wave packets with these properties should then look identical to a gluon for an observer in the femto universe with regard to "soft" scatterings.

Now we replace $G(1)$ and $G(3)$ in reaction (B.1) by the $q\bar{q}$ wave packets (B.4). According to the rules derived in Sect. 3.5 the scattering of the $q\bar{q}$ system

$$q(1)\bar{q}(1') \to q(3)\bar{q}(3') \tag{B.18}$$

gives the following factor in the S-matrix (cf. (3.58) (3.59)):

$$\mathcal{S}_{q+}(3,1)\mathcal{S}_{\bar{q}+}(3',1')$$

which still has to be integrated over the wave functions (B.4). In this way we obtain for the $q\bar{q}$-pair:

$$\begin{aligned}\mathcal{S}_{q\bar{q}}(3,1) = \int \mathrm{d}^2p'_T \int_0^1 \mathrm{d}\zeta' \int \mathrm{d}^2p_T \int_0^1 \mathrm{d}\zeta \\ \frac{1}{(2\pi)^3} h^3(\zeta', |\mathbf{p}'_T|) h^1(\zeta, |\mathbf{p}_T|) \\ \frac{1}{2}(\lambda_{a_3})_{A'_3,A_3} (\epsilon_3{}^* \cdot \epsilon^T \sigma)_{s'_3,s_3} \\ \frac{1}{2}(\lambda_{a_1})_{A_1,A'_1} (\epsilon_1 \cdot \sigma\epsilon)_{s_1,s'_1} \\ \mathcal{S}_{q+}(3,1)\ \mathcal{S}_{\bar{q}+}(3',1'),\end{aligned} \tag{B.19}$$

$$\begin{aligned}\mathcal{S}_{q\bar{q}}(3,1) = \int \mathrm{d}z_- \int \mathrm{d}y_- \int \mathrm{d}^2 z_T \int \mathrm{d}^2 y_T \sqrt{P_{3+}P_{1+}} \epsilon_3^* \cdot \epsilon_1 \\ \exp[\frac{i}{2}(P_{3+} - P_{1+}) z_- - i(\mathbf{P}_3 - \mathbf{P}_1)_T \mathbf{z}_T] \\ \tilde{\varphi}^{3*}(y_-, \mathbf{y}_T) \tilde{\varphi}^1(y_-, \mathbf{y}_T) \\ \frac{1}{2}\mathrm{Tr}[\lambda_{a_3} V_-(\infty, z_- + \frac{1}{2}y_-, \mathbf{z}_T + \frac{1}{2}\mathbf{y}_T) \\ \lambda_{a_1} V_-^\dagger(\infty, z_- - \frac{1}{2}y_-, \mathbf{z}_T - \frac{1}{2}\mathbf{y}_T)].\end{aligned} \tag{B.20}$$

With our assumptions (B.17) $\tilde{\varphi}^{3*}(y_-, \mathbf{y}_T)\tilde{\varphi}^1(y_-, \mathbf{y}_T)$ acts like a δ-function at $y_- = 0,\ \mathbf{y}_T = 0$ and we get:

$$\begin{aligned}\mathcal{S}_{q\bar{q}}(3,1) = \sqrt{P_{3+}P_{1+}}\ \epsilon_3^* \cdot \epsilon_1 \\ \int \mathrm{d}y_- \int \mathrm{d}^2 y_T\ \tilde{\varphi}^{3*}(y_-, \mathbf{y}_T) \tilde{\varphi}^1(y_-, \mathbf{y}_T) \\ \int \mathrm{d}z_- \int \mathrm{d}^2 z_T\ \exp[\frac{i}{2}(P_{3+} - P_{1+}) z_- - i(\mathbf{P}_3 - \mathbf{P}_1)_T \cdot \mathbf{z}_T] \\ \frac{1}{2}\mathrm{Tr}[\lambda_{a_3} V_-(\infty, z_-, \mathbf{z}_T) \lambda_{a_1} V_-^\dagger(\infty, z_-, \mathbf{z}_T)].\end{aligned} \tag{B.21}$$

An easy exercise shows that

$$\frac{1}{2}\mathrm{Tr}[\lambda_{a_3} V_-(\infty, z_-, \mathbf{z}_T) \lambda_{a_1} V_-^\dagger(\infty, z_-, \mathbf{z}_T)] = \mathcal{V}_-(\infty, z_-, \mathbf{z}_T)_{a_3,a_1}, \tag{B.22}$$

where $\mathcal{V}_-$ is the connector analogous to (3.35) but for the adjoint representation (cf. (2.22)):

$$\begin{aligned}\mathcal{V}_-(\infty, z_-, \mathbf{z}_T) = \mathrm{P}\{\exp[-\frac{i}{2} g \int_{-\infty}^{\infty} \mathrm{d}z_+ G_-^a(z_+, z_-, \mathbf{z}_T) T_a]\}, \\ (T_a)_{bc} = \frac{1}{i} f_{abc}.\end{aligned} \tag{B.23}$$

The f_{abc} are the structure constants of $SU(3)_c$. Inserting (B.22) in (B.21) gives

$$\mathcal{S}_{q\bar{q}}(3,1) = \int \mathrm{d}y_- \int \mathrm{d}^2 y_T \; \tilde{\varphi}^{3*}(y_-, \mathbf{y}_T)\tilde{\varphi}^1(y_-, \mathbf{y}_T)\epsilon^*_{3j_3}\epsilon_{1j_1}\mathcal{S}_{G+}(3,1), \quad \text{(B.24)}$$

where

$$\begin{aligned} \mathcal{S}_{G+}(3,1) &= \sqrt{P_{3+}P_{1+}}\;\delta_{j_3,j_1} \int \mathrm{d}z_- \int \mathrm{d}^2 z_T \\ &\exp\big[\frac{i}{2}(P_{3+} - P_{1+})z_- - i(\mathbf{P}_{3T} - \mathbf{P}_{1T})\cdot \mathbf{z}_T\big] \\ &\mathcal{V}_-(\infty, z_-, \mathbf{z}_T)_{a_3,a_1}. \end{aligned} \quad \text{(B.25)}$$

In a general scattering reaction (B.1) the factor $\mathcal{S}_{q\bar{q}}(3,1)$ (B.24) has to be inserted with other factors $\mathcal{S}_q, \mathcal{S}_{\bar{q}}, \ldots$ and then integrated over all gluon potentials as explained in Sect. 3.5. We note that $\mathcal{S}_{q\bar{q}}(3,1)$ in (B.24) factorizes into $\mathcal{S}_{G+}(3,1)$ (B.25) times the overlap of the internal wave functions of the incoming and outgoing $q\bar{q}$ pairs:

$$\int \mathrm{d}y_- \int \mathrm{d}^2 y_T \; \tilde{\varphi}^{3*}(y_-, \mathbf{y}_T)\tilde{\varphi}^1(y_-, \mathbf{y}_T). \quad \text{(B.26)}$$

This means that the $q\bar{q}$ pair will come out with some distribution in total momentum P_3 and polarization vector $\epsilon_3 = \epsilon_1$ but always with internal wave function $\tilde{\varphi}^1(y_-, \mathbf{y}_T)$. With the conditions (B.15)-(B.17) $\tilde{\varphi}^1(y_-, \mathbf{y}_T)$ leads to a "permissible" internal wave function for a $q\bar{q}$ pair of momentum P_3, to be regarded as a gluon of momentum P_3 by our observer in the femto universe. Indeed we have from (B.12):

$$\begin{aligned} \tilde{\varphi}^1(y_-, \mathbf{y}_T) &= \sqrt{\frac{P_{1+}}{4\pi}} \int_0^1 \mathrm{d}\zeta \; \varphi^1(\zeta, \mathbf{y}_T) \exp[-\frac{i}{2}P_{1+}(\zeta - \frac{1}{2})y_-] \\ &= \sqrt{\frac{P_{1+}}{4\pi}} \int_{-1/2}^{1/2} \mathrm{d}\xi \; \varphi^1(\frac{1}{2} + \xi, \mathbf{y}_T) \exp[-\frac{i}{2}P_{1+}\xi \cdot y_-] \\ &= \sqrt{\frac{P_{3+}}{4\pi}} \int_{\xi_-}^{\xi_+} \mathrm{d}\xi' \; \varphi'^3(\frac{1}{2} + \xi', \mathbf{y}_T) \exp[-\frac{i}{2}P_{3+}\xi' \cdot y_-] \end{aligned} \quad \text{(B.27)}$$

Here we define the internal wave function of the outgoing $q\bar{q}$ pair as

$$\varphi'^3\left(\frac{1}{2} + \xi', \mathbf{y}_T\right) = \sqrt{\frac{P_{3+}}{P_{1+}}}\varphi^1\left(\frac{1}{2} + \frac{P_{3+}}{P_{1+}}\xi', \mathbf{y}_T\right) \quad \text{(B.28)}$$

and

$$\xi_\pm = \pm\frac{P_{1+}}{P_{3+}} \cdot \frac{1}{2}. \quad \text{(B.29)}$$

The condition (B.15) for φ^1 guarantees a similar condition for φ'^3 if P_{3+}/P_{1+} is of order 1, as we will always assume. Thus the integration limits $\xi_\pm$ in

(B.27) can be replaced by $\pm\frac{1}{2}$ and also the normalization conditions (B.14) can easily be checked for φ'^3.

To summarize: In this appendix we have shown that suitable $q\bar{q}$ pairs which are indistinguishable from gluons for our observer in the femto universe scatter as entities in a soft reaction. Their internal wave function in momentum space is modified, but in a way not observable in a soft reaction. Thus we can consider $\mathcal{S}_{G+}(3,1)$ in (B.25) as the scattering contribution of a right-moving gluon in the transition (B.2). We quoted this result already in (3.64) in Sect. 3.5. For left moving gluons we just have to exchange everywhere + and − components.

Appendix C: The Scattering of Baryons

In this appendix we discuss high energy soft reactions, in particular elastic reactions involving baryons and antibaryons. We represent baryons by qqq wave packets:

$$|B_j(P_j)\rangle = \frac{1}{6\cdot(2\pi)^3}\int \mathrm{d}\mu\; h^j(f^i, s^i, \zeta^i, \mathbf{p}_T^i)\cdot$$
$$\epsilon_{A^1A^2A^3}|q(p_j^1, f^1, s^1, A^1), q(p_j^2, f^2, s^2, A^2), q(p_j^3, f^3, s^3, A^3)\rangle. \quad \text{(C.1)}$$

Here f^i, s^i, A^i $(i = 1,2,3)$ are the flavour, spin, and colour indices of the quarks and ϵ_{ABC} $(\epsilon_{123} = 1)$ is the totally antisymmetric tensor. For the momenta p_j^i we set for right-moving baryons (j odd, $P_{j+} \to \infty$):

$$\begin{aligned} p_{j+}^i &= \zeta^i P_{j+}, \\ \mathbf{p}_{jT}^i &= \frac{1}{3}\mathbf{P}_{jT} + \mathbf{p}_T^i, \\ (i &= 1,2,3) \end{aligned} \quad \text{(C.2)}$$

and for left-moving baryons (j even, $P_{j-} \to \infty$):

$$\begin{aligned} p_{j-}^i &= \zeta^i P_{j-}, \\ \mathbf{p}_{jT}^i &= \frac{1}{3}\mathbf{P}_{jT} + \mathbf{p}_T^i, \\ (i &= 1,2,3). \end{aligned} \quad \text{(C.3)}$$

The integral with the measure $\mathrm{d}\mu$ stands for

$$\int \mathrm{d}\mu \equiv \int\!\!\int\!\!\int \prod_{i=1}^{3} \mathrm{d}^2 p_T^i \,\delta^2\Big(\sum_{i=1}^{3} \mathbf{p}_T^i\Big)\cdot$$
$$\int_0^1\!\int_0^1\!\int_0^1 \prod_{i=1}^{3} \mathrm{d}\zeta^i \,\delta\Big(1-\sum_{i=1}^{3}\zeta^i\Big). \quad \text{(C.4)}$$

The flavour and spin of the baryon states (C.1) is, of course, determined by the functions h^j which must be totally symmetric under simultaneous exchange of the arguments $(f^i, s^i, \zeta^i, \mathbf{p}_T^i)$ for $i = 1, 2, 3$. In the following we will collectively set $(f, s, \zeta, \mathbf{p}_T) \equiv \alpha$ and $q(\alpha, A) \equiv q(p_j, f, s, A)$. We have then

$$h^j(\alpha, \beta, \gamma) = h^j(\beta, \alpha\gamma) = h^j(\alpha, \gamma, \beta). \tag{C.5}$$

$$|B_j(P_j)\rangle = \frac{1}{6.(2\pi)^3} \int \mathrm{d}\mu \; h^j(\alpha, \beta, \gamma)\epsilon_{ABC}|q(\alpha, A)q(\beta, B)q(\gamma, C)\rangle. \tag{C.6}$$

The normalization condition

$$\langle B_j(P_j')|B_j(P_j)\rangle = (2\pi)^3 2P_j^0 \delta^3(\mathbf{P}_j' - \mathbf{P}_j) \tag{C.7}$$

requires h^j to satisfy:

$$\int \mathrm{d}\mu \; 4\zeta^1\zeta^2\zeta^3 \; h^{j*}(\alpha, \beta, \gamma)h^j(\alpha, \beta, \gamma) = 1$$
$$\text{(no summation over} j). \tag{C.8}$$

We define the wave functions φ^j in transverse position and longitudinal momentum space and the transition profile functions w_{kj}^B for $B_j \to B_k$ as:

$$\varphi^j(f^i, s^i, \zeta^i, \mathbf{x}_T^i) := \frac{1}{(2\pi)^2} \int \prod_{i=1}^{3} \mathrm{d}^2 p_T^i \cdot \delta^2(\sum_{i=1}^{3} \mathbf{p}_T^i) \cdot$$
$$6 \cdot (\zeta^1\zeta^2\zeta^3)^{1/2} \exp(i \sum_{i=1}^{3} \mathbf{p}_T^i \cdot \mathbf{x}_T^i) h^j(\alpha^1, \alpha^2, \alpha^3), \tag{C.9}$$

$$w_{k,j}^B(\mathbf{x}_T^1, \mathbf{x}_T^2, \mathbf{x}_T^3) := \int_0^1 \int_0^1 \int_0^1 \prod_{i=1}^{3} \mathrm{d}\zeta^i \; \delta(1 - \sum_{i=1}^{3} \zeta^i)$$
$$\sum_{f^i, s^i} \varphi^{3*}(f^1, s^1, \zeta^1, \mathbf{x}_T^1; f^2, s^2, \zeta^2, \mathbf{x}_T^2; f^3, s^3, \zeta^3, \mathbf{x}_T^3)$$
$$\varphi^1(f^1, s^1, \zeta^1, \mathbf{x}_T^1; f^2, s^2, \zeta^2, \mathbf{x}_T^2; f^3, s^3, \zeta^3, \mathbf{x}_T^3). \tag{C.10}$$

The symmetry relations for h (C.5) and the normalization condition (C.8) imply:

$$w_{k,j}^B(\mathbf{x}_T^1, \mathbf{x}_T^2, \mathbf{x}_T^3) = w_{k,j}^B(\mathbf{x}_T^2, \mathbf{x}_T^1, \mathbf{x}_T^3) = w_{k,j}^B(\mathbf{x}_T^1, \mathbf{x}_T^3, \mathbf{x}_T^2), \tag{C.11}$$

$$\int \prod_{i=1}^{3} \mathrm{d}^2 x_T^i \; \delta^2(\mathbf{x}_T^1 + \mathbf{x}_T^2 + \mathbf{x}_T^3) w_{jj}^B(\mathbf{x}_T^1, \mathbf{x}_T^2, \mathbf{x}_T^3) = 1$$
$$\text{(no summation over } j). \tag{C.12}$$

As a concrete scattering reaction let us consider meson-baryon scattering:

$$B_1(P_1) + M_2(P_2) \to B_3(P_3) + M_4(P_4). \tag{C.13}$$

From the rules of Sect. 3.5 we get

$$\begin{aligned}
&S_{fi} \equiv \langle B_3(P_3), M_4(P_4)|S|B_1(P_1)M_2(P_2)\rangle = \\
&\frac{1}{6(2\pi)^6}\int \mathrm{d}\mu'\; h^{3*}(\alpha',\beta',\gamma')\epsilon_{A'B'C'} \int \mathrm{d}\mu\; h^1(\alpha,\beta,\gamma)\epsilon_{ABC} \\
&\frac{1}{3(2\pi)^3}\int \mathrm{d}^2p'_T \int_0^1 \mathrm{d}\zeta' \int \mathrm{d}^2p_T \int_0^1 \mathrm{d}\zeta\; h^{4*}_{s_4s'_4}(\zeta',\mathbf{p}'_T)h^2_{s_2s'_2}(\zeta,\mathbf{p}_T) \\
&\langle \mathcal{S}_{q+}(\alpha',\alpha)_{A'A}\mathcal{S}_{q+}(\beta',\beta)_{B'B}\mathcal{S}_{q+}(\gamma',\gamma)_{C'C} \\
&\quad \delta_{A'_2A_2}\delta_{A'_4A_4}\mathcal{S}_{q-}(4,2)_{A_4A_2}\mathcal{S}_{\bar{q}-}(4',2')_{A'_4A'_2}\rangle_G .
\end{aligned} \tag{C.14}$$

After some straightforward algebra we get:

$$\begin{aligned}
S_{fi} &= \delta_{fi} + i(2\pi)^4\delta(P_3 + P_4 - P_1 - P_2)\mathcal{T}_{fi}, \\
\mathcal{T}_{fi} &= -\frac{i}{(2\pi)^6}\int \mathrm{d}\mu'\; h^{3*}(\alpha',\beta',\gamma') \\
&\int \mathrm{d}\mu\; h^1(\alpha,\beta,\gamma)\prod_{i=1}^{3}(\delta_{f'^i,f^i}\delta_{s'^i,s^i})(P_{1+})^3\left[\prod_{i=1}^{3}\zeta'^i\zeta^i\right]^{1/2} \\
&\frac{1}{(2\pi)^3}\int \mathrm{d}^2p'_T \int_0^1 \mathrm{d}\zeta' \int \mathrm{d}^2p_T \int_0^1 \mathrm{d}\zeta\; h^{4*}_{s,r}(\zeta',\mathbf{p}'_T)h^2_{s,r}(\zeta,\mathbf{p}_T) \\
&2(P_{2-})^2\left[\zeta'(1-\zeta')\zeta(1-\zeta)\right]^{1/2} \\
&3^3\cdot\int \mathrm{d}^2b_T \exp(i\mathbf{q}_T\cdot\mathbf{b}_T)\int\prod_{i=1}^{3}(\mathrm{d}x_-^i\,\mathrm{d}^2x_T^i)\delta\left(\sum_{i=1}^{3}x_-^i\right)\;\delta^2\left(\sum_{i=1}^{3}\mathbf{x}_T^i\right) \\
&\int \mathrm{d}y_+\mathrm{d}^2y_T \exp\Big\{i\sum_{i=1}^{3}\left[\frac{1}{2}P_{1+}(\zeta'^i-\zeta^i)x_-^i - (\mathbf{p}'^i_T-\mathbf{p}^i_T)\cdot\mathbf{x}_T^i\right] \\
&+i\frac{1}{2}P_{2-}(\zeta'-\zeta)y_+ - i(\mathbf{p}'_T-\mathbf{p}_T)\cdot\mathbf{y}_T\Big\} \\
&\langle\{V_-(\infty,x_-^1,\frac{1}{2}\mathbf{b}_T+\mathbf{x}_T^1)_{A'A}V_-(\infty,x_-^2,\frac{1}{2}\mathbf{b}_T+\mathbf{x}_T^2)_{B'B} \\
&\quad V_-(\infty,x_-^3,\frac{1}{2}\mathbf{b}_T+\mathbf{x}_T^3)_{C'C}\;\frac{1}{6}\varepsilon_{A'B'C'}\;\varepsilon_{ABC} \\
&\quad \frac{1}{3}\mathrm{Tr}[V_+(y_+,\infty,-\frac{1}{2}\mathbf{b}_T+\frac{1}{2}\mathbf{y}_T)V_+^\dagger(0,\infty,-\frac{1}{2}\mathbf{b}_T-\frac{1}{2}\mathbf{y}_T)] - 1\}\rangle_G,
\end{aligned} \tag{C.15}$$

where

$$\mathbf{q}_T = (\mathbf{P}_1 - \mathbf{P}_3)_T. \tag{C.16}$$

Now we make the transformation of variables

$$x_-^i \to \frac{2}{P_{1+}} x_-^i,$$
$$y_+ \to \frac{2}{P_{2-}} y_+ \tag{C.17}$$

and use $P_{1+} \to \infty$, $P_{2-} \to \infty$. With the same arguments which led us from (3.77) to (3.79) we get

$$\begin{aligned}\mathcal{T}_{fi} = &-2is \int \mathrm{d}^2 b_T \exp(i\mathbf{q}_T \cdot \mathbf{b}_T) \int \prod_{i=1}^{3} \mathrm{d}^2 x_T^i \, \delta^2(\mathbf{x}_T^1 + \mathbf{x}_T^2 + \mathbf{x}_T^3) \\ &w_{3,1}^B(\mathbf{x}_T^1, \mathbf{x}_T^2, \mathbf{x}_T^3) \int \mathrm{d}^2 y_T w_{4,2}^M(\mathbf{y}_T) \\ &\langle \{ V_-(\infty, 0, \tfrac{1}{2}\mathbf{b}_T + \mathbf{x}_T^1)_{A'A} V_-(\infty, 0, \tfrac{1}{2}\mathbf{b}_T + \mathbf{x}_T^2)_{B'B} \\ &V_-(\infty, 0, \tfrac{1}{2}\mathbf{b}_T + \mathbf{x}_T^3)_{C'C} \frac{1}{6} \epsilon_{A'B'C'} \epsilon_{ABC} \\ &\frac{1}{3} \mathrm{Tr}\big[V_+(0, \infty, -\tfrac{1}{2}\mathbf{b}_T + \tfrac{1}{2}\mathbf{y}_T) V_+^\dagger(0, \infty, -\tfrac{1}{2}\mathbf{b}_T - \tfrac{1}{2}\mathbf{y}_T\big] - 1 \} \rangle_G .\end{aligned} \tag{C.18}$$

Here $w_{4,2}^M(y_T)$ is the transition profile function for the mesons as defined in (3.72) and $w_{3,1}^B$ is the corresponding function for the baryons (cf. (C.10)).

In the next step we follow [59], [45] and use relations which are valid for any 3×3 matrix: $H = (H_{AB})$:

$$H_{A'A''} H_{B'B''} H_{C'C''} \cdot \epsilon_{A''B''C''} = \det H \cdot \epsilon_{A'B'C'}, \tag{C.19}$$

$$\begin{aligned}&\det H \cdot \epsilon_{A'B'C'} \epsilon_{ABC} = \\ &H_{A'A} H_{B'B} H_{C'C} + H_{A'B} H_{B'C} H_{C'A} + H_{A'C} H_{B'A} H_{C'B} \\ &- H_{A'B} H_{B'A} H_{C'C} - H_{A'A} H_{B'C} H_{C'B} - H_{A'C} H_{B'B} H_{C'A} .\end{aligned} \tag{C.20}$$

We take H equal to the antiquark line integral at the central point of the baryon in transverse space:

$$H = V_-^*(\infty, 0, \frac{1}{2}\mathbf{b}_T). \tag{C.21}$$

As a $SU(3)$-connector V_-^* satisfies

$$\det V_-^*(\infty, 0, \frac{1}{2}\mathbf{b}_T) = 1. \tag{C.22}$$

This is easy to prove. From (C.19) we see that $\det V_-^*$ is the connector in the singlet part of the product of three $SU(3)$ antiquark representations: $\bar{3} \times \bar{3} \times \bar{3}$.

But for the singlet representation the connector equals 1 since we have to set $T_a = 0$ in (2.22).

In the following we will use as shorthand notation

$$\begin{aligned}
&V(i) \equiv V_-(\infty, 0, \frac{1}{2}\mathbf{b}_T + \mathbf{x}_T^i), \\
&(i = 1, 2, 3), \\
&V(0) \equiv V_-(\infty, 0, \frac{1}{2}\mathbf{b}_T). \qquad \text{(C.23)}
\end{aligned}$$

With this we get for the qqq-contribution to the integrand in the functional integral in (C.18) using (C.20) to (C.22):

$$\begin{aligned}
&\mathcal{W}_+^B(\frac{1}{2}\mathbf{b}_T, \mathbf{x}_T^1, \mathbf{x}_T^2, \mathbf{x}_T^3) := \frac{1}{6} V(1)_{A'A} V(2)_{B'B} V(3)_{C'C}\ \epsilon_{A'B'C'}\epsilon_{ABC} \\
&= \frac{1}{6}\{\mathrm{Tr}[V(1)V^\dagger(0)] \cdot \mathrm{Tr}[V(2)V^\dagger(0)] \cdot \mathrm{Tr}[V(3)V^\dagger(0)] \\
&\quad + \mathrm{Tr}[V(1)V^\dagger(0)V(2)V^\dagger(0)V(3)V^\dagger(0)] \\
&\quad + \mathrm{Tr}[V(1)V^\dagger(0)V(3)V^\dagger(0)V(2)V^\dagger(0)] \\
&\quad - \mathrm{Tr}[V(1)V^\dagger(0)V(2)V^\dagger(0)] \cdot \mathrm{Tr}[V(3)V^\dagger(0)] \\
&\quad - \mathrm{Tr}[V(2)V^\dagger(0)V(3)V^\dagger(0)] \cdot \mathrm{Tr}[V(1)V^\dagger(0)] \\
&\quad - \mathrm{Tr}[V(1)V^\dagger(0)V(3)V^\dagger(0)] \cdot \mathrm{Tr}[V(2)V^\dagger(0)]\}\,. \qquad \text{(C.24)}
\end{aligned}$$

As for the meson case (cf. (3.82)) we will now add suitable connectors at $\pm\infty$. In this way all the traces in (C.24) become closed light-like Wegner-Wilson loops. To give an example: the term

$$\mathrm{Tr}[V(1)V^\dagger(0)V(2)V^\dagger(0)]$$

should then be read as the loop in the hyperplane $x_- = 0$ in the limit $T \to \infty$ which connects the following points $(x_+, x_-, \mathbf{x}_T)$:

$$\begin{aligned}
&(T, 0, \frac{1}{2}\mathbf{b}_T), \\
&(-T, 0, \frac{1}{2}\mathbf{b}_T), \\
&(-T, 0, \frac{1}{2}\mathbf{b}_T + \mathbf{x}_T^2), \\
&(T, 0, \frac{1}{2}\mathbf{b}_T + \mathbf{x}_T^2), \\
&(T, 0, \frac{1}{2}\mathbf{b}_T), \\
&(-T, 0, \frac{1}{2}\mathbf{b}_T), \\
&(-T, 0, \frac{1}{2}\mathbf{b}_T + \mathbf{x}_T^1),
\end{aligned}$$

$$(T, 0, \frac{1}{2}\mathbf{b}_T + \mathbf{x}_T^1),$$
$$(T, 0, \frac{1}{2}\mathbf{b}_T) \tag{C.25}$$

on straight lines in the order indicated.

Inserting (C.24) in (C.18) and denoting the mesonic Wegner-Wilson loop as defined in (3.83) by $\mathcal{W}_-^M$, we get finally for the T-matrix element of baryon-meson scattering:

$$\begin{aligned}
\mathcal{T}_{fi} &= -2is \int \mathrm{d}^2 b_T \exp(i\mathbf{q}_T \cdot \mathbf{b}_T) \\
&\int \prod_{i=1}^{3} \mathrm{d}^2 x_T^i \, \delta^2(\mathbf{x}_T^1 + \mathbf{x}_T^2 + \mathbf{x}_T^3) w_{3,1}^B(\mathbf{x}_T^1, \mathbf{x}_T^2, \mathbf{x}_T^3) \\
&\int \mathrm{d}^2 y_T \; w_{4,2}^M(\mathbf{y}_T) \\
&\left\langle \mathcal{W}_+^B \left(\frac{1}{2}\mathbf{b}_T, \mathbf{x}_T^1, \mathbf{x}_T^2, \mathbf{x}_T^3\right) \mathcal{W}_-^M \left(-\frac{1}{2}\mathbf{b}_T, \mathbf{y}_T\right) - 1 \right\rangle_G .
\end{aligned} \tag{C.26}$$

This formula is the starting point for the evaluation of the baryon-meson elastic scattering amplitude: One can now apply the Minkowskian version of the SVM to calculate the functional integral $\langle \quad \rangle_G$ in an appropriate way. Then one has to fold the result with the profile functions of the mesonic and baryonic transitions. At the present state one has to make a suitable ansatz for these profile functions.

For the case of right-moving antibaryons in an elastic reaction we just have to substitute the loop factor $\mathcal{W}_+^B$ by $\mathcal{W}_+^{\bar{B}}$ which is obtained by replacing the quark connectors V_- by the antiquark connectors V_-^* and vice versa in (C.23), (C.24). In an equivalent way we can get $W_+^{\bar{B}}$ from $\mathcal{W}_+^B$ by reversing the directional arrows on all Wegner-Wilson loops obtained in the way discussed above from (C.24). For left-moving baryons and antibaryons we have to exchange $+$ and $-$ components.

For the further treatment of scattering amplitudes involving mesons, baryons and antibaryons, for many results and a comparison with experiments we refer to [59], [45], [62].

References

[1] H. Fritzsch, , M. Gell-Mann, H. Leutwyler: Phys. Lett. **B47**, 365 (1973); M. Gell-Mann: Acta Physica Austriaca, Suppl. 9, 733 (1972)

[2] G. 't Hooft: remarks at a conference in Marseille (1972); H. D. Politzer: Phys. Rev. Lett. **30**, 1346 (1973); D. Gross, F. Wilczek: Phys. Rev. Lett. **30**, 1343 (1973)

[3] K. G. Wilson: Phys. Rev. **D10**, 2445 (1974)

[4] P. Lepage: These proceedings

[5] G. Altarelli: Phys. Rep. **81**, 1 (1982)

[6] J. L. Cardy, G. A. Winbow: Phys. Lett. **B52**, 95 (1974); C. E. DeTar, S. D. Ellis, P. V. Landshoff: Nucl. Phys. **B87**, 176 (1975)

[7] H. D. Politzer: Nucl. Phys. **B129**, 301 (1977); C. T. Sachrajda: Phys. Lett. **B73**, 185 (1978); D. Amati, R. Petronzio, G. Veneziano: Nucl. Phys. **B140**, 54 (1978); **B146**, 29 (1978); R. K. Ellis, H. Georgi, M. Machacek, H. D. Politzer, G. G. Ross: Nucl. Phys. **B152**, 285 (1979); S. B. Libby, G. Sterman: Phys. Rev. **D18**, 3252 (1978); S. Gupta, A. H. Mueller: Phys. Rev. **D20**, 118 (1979)

[8] J. C. Collins, D. E. Soper, G. Sterman: Nucl. Phys. **B261**, 104 (1985); **B308**, 833 (1988); G. Sterman: Phys. Lett. **B179**, 281 (1986); Nucl. Phys. **B281**, 310 (1987); J. C. Collins, D. E. Soper: Ann. Rev. Nucl. Part. Sci. **37**, 383 (1987); T. Matsuura, W. L. van Neerven: Z. Phys. **C38**, 623 (1988); J. C. Collins, D. E. Soper, G. Sterman: "Factorization of Hard Processes in QCD", in "Perturbative QCD", A. H. Mueller, ed., World Scientific, Singapore 1990

[9] G. T. Bodwin, Phys. Rev. **D31**, 2616 (1985); G. T. Bodwin, S. J. Brodsky, G. P. Lepage: Phys. Rev. **D39**, 3287 (1989)

[10] A. H. Mueller, Les Houches Lectures 1991

[11] G. Sterman; these Proceedings

[12] J. Ellis, M. K. Gaillard, W. J. Zakrzewski: Phys. Lett. **B81**, 224 (1979)

[13] R. Doria, J. Frenkel, J. C. Taylor: Nucl. Phys. **B168**, 93 (1980)

[14] O. Nachtmann, A. Reiter: Z. Phys. **C24**, 283 (1984)

[15] T. T. Wu, C. N. Yang: Phys. Rev. **B137**, 708 (1965); T. T. Chou, C. N. Yang: Phys. Rev. **B170**, 1591 (1968); Phys. Rev. **D19**, 3268 (1979); Phys. Lett. **B128**, 457 (1983); Phys. Lett. **B244**, 113 (1990); J. Dias de Deus, P. Kroll: Phys. Lett. **B60**, 375 (1976); Nuovo Cimento **A37**, 67 (1977); P. Kroll: Z. Phys. **C15**, 67 (1982); J. Hüfner, B. Povh: Phys. Rev. **D46**, 990 (1992)

[16] C. Bourrely, J. Soffer, T. T. Wu: Nucl. Phys. **B247**, 15 (1984); Phys. Rev. Lett. **54**, 757 (1985); Phys. Lett. **B196**, 237 (1987); Z. Phys. **C37**, 369 (1988); R. J. Glauber, J. Velasco: Phys. Lett. **B147**, 380 (1984); R. Henzi, P. Valin: Phys. Lett. **B149**, 239 (1984)

[17] E. M. Levin, L. L. Frankfurt: JETP Lett. **2**, 65 (1965); H. J. Lipkin, F. Scheck: Phys. Rev. Lett. **16**, 71 (1966); H. J. Lipkin: Phys. Rev. Lett. **16**, 1015 (1966);

J. J. J. Kokkedee, L. Van Hove: Nuovo Cimento **A42**, 711 (1966);
V. V. Anisovich et al.: "Quark Model and High Energy Collisions", World Scientific, Singapore 1985

[18] P. D. B. Collins: "An Introduction to Regge Theory", Cambridge University Press, Cambridge, U.K. 1977;
L. Caneschi, ed.: *Regge Theory of low p_T hadronic interaction*, North Holland, Amsterdam 1989

[19] G. 't Hooft: Nucl. Phys. **B72**, 461 (1974);
G. Veneziano: Nucl. Phys. **B74**, 365 (1974); Phys. Lett. **B52**, 220 (1974); Nucl. Phys. **B117**, 519 (1976);
M. Ciafaloni, G. Marchesini, G. Veneziano: Nucl. Phys. **B98**, 472, 493 (1975);
A. Capella, U. Sukhatme, Chung-I Tan, J. Tran Thanh Van: Phys. Lett. **B81**, 68 (1979);
A. Capella, U. Sukhatme, J. Tran Thanh Van: Z. Phys. **C3**, 329 (1980);
B. Andersson, G. Gustafson, G. Ingelman, T. Sjöstrand: Phys. Rep. **C97**, 31 (1983);
X. Artru: Phys. Rep. **C97**, 147 (1983);
B. Andersson, G. Gustafson, B. Nilsson-Almqvist: Nucl. Phys. **B281**, 289 (1987);
K. Werner: Phys. Rep. **C232**, 87 (1993)

[20] R. C. Hwa: Phys. Rev. **D22**, 759, 1593 (1980);
R. C. Hwa, M. Sajjad Zahir: Phys. Rev. **D23**, 2539 (1981);
R. C. Hwa: "Central production and small angle elastic scattering in the valon model", Proc. 12th Int. Symp. Multiparticle Dynamics, Notre Dame, 1981 (W. D. Shephard, V. P. Kenny, eds.) World Scientific, Singapore 1982

[21] E. A. Kuraev, L. N. Lipatov, V. S. Fadin: Sov. Phys. J.E.T.P. **44**, 443 (1976), **45**, 199 (1977);
L. N. Lipatov: Sov. J. Nucl. Phys. **23**, 338 (1976);
Ya. Ya. Balitskii, L. N. Lipatov: Sov. J. Nucl. Phys. **28**, 822 (1978);
L. N. Lipatov: Sov. Phys. J.E.T.P. **63**, 904 (1986);
L. N. Lipatov: "Pomeron in Quantum Chromodynamics", in "Perturbative Quantum Chromodynamics" (A. H. Mueller, Ed.), World Scientific, Singapore, 1989;
A. R. White: Int. J. Mod. Phys. **A6**, 1859 (1990);
J. Bartels: Z. Phys. **C60**, 471 (1993)

[22] F. E. Low: Phys. Rev. **D12**, 163 (1975);
S. Nussinov: Phys. Rev. Lett. **34**, 1286 (1975);
J. F. Gunion, D. E. Soper: Phys. Rev. **D15**, 2617 (1977)

[23] A. Donnachie, P. V. Landshoff: Nucl. Phys. **B244**, 322 (1984); Nucl. Phys. **B267**, 690 (1986); Phys. Lett. **B185**, 403 (1987)

[24] W. Heisenberg: Z. Phys. **133**, 65 (1952)

[25] H. Cheng, T. T. Wu: "Expanding Protons", MIT Press, Cambridge, Mass. 1987, and references cited therein.

[26] P. V. Landshoff, O. Nachtmann: Z. Phys. **C35**, 405 (1987)

[27] G. K. Savvidy: Phys. Lett. **71B**, 133 (1977)

[28] A. I. Vainshtein, V. I. Zakharov, M. A. Shifman: JETP Lett. **27**, 55 (1978)

[29] N. K. Nielsen, P. Olesen: Nucl. Phys. **B144**, 376 (1978)

[30] M. A. Shifman, A. I. Vainshtein, V. I. Zakharov: Nucl. Phys. **B147**, 385, 448, 519 (1979)

[31] G. 't Hooft: Cargèse Lectures, 1979, ed. G. 't Hooft et al., Plenum, New York, London 1980;
G. 't Hooft: Acta Phys. Austriaca, Suppl. **22**, 531 (1980)
[32] G. Mack: Acta Phys. Austriaca, Suppl. **22**, 509 (1980)
[33] J. Ambjørn, P. Olesen: Nucl. Phys. **B170**, 60, 265 (1980)
[34] H. Leutwyler: Phys. Lett. **96B**, 154 (1980)
[35] T. H. Hansson, K. Johnson, C. Peterson: Phys. Rev. **D26**, 2069 (1982)
[36] E. V. Shuryak: Phys. Rep. **C115**, 151 (1984), and references cited therein
[37] H. M. Fried, B. Müller (eds.): "QCD Vacuum Structure", World Scientific, Singapore 1993
[38] J. C. Maxwell: Philos. Mag. **21**, 281 (1861), reproduced in: *The Scientific Papers of James Clerk Maxwell*, Vol. 1, p. 488; W. D. Niven, ed., Cambridge Univ. Press 1890
[39] H. G. Dosch: "Nonperturbative Methods in Quantum Chromodynamics", Progr. in Part. and Nucl. Phys. **33**, 121 (1994)
[40] H. G. Dosch: Phys. Lett. **B190**, 177 (1987);
H. G. Dosch, Yu. A. Simonov: Phys. Lett. **B205**, 339 (1988);
Yu. A. Simonov: Nucl. Phys. **B307**, 512 (1988)
[41] J. Schwinger: Phys. Rev. **82**, 664 (1951)
[42] I. Ya. Aref'eva: Theor. Mat. Phys. **43**, 353 (1980);
N. E. Bralić: Phys. Rev. **D22**, 3090 (1980);
P. M. Fishbane, S. Gasioroviwcz, P. Kaus: Phys. Rev. **D24**, 2324 (1981);
L. Diosi: Phys. Rev. **D27**, 2552 (1983);
Yu. A. Simonov: Sov. J. Nucl. Phys. **48**, 878 (1988)
[43] N. G. Van Kampen: Physica **74**, 215, 239 (1974); Phys. Rep. **C24**, 172 (1976)
[44] A. DiGiacomo, H. Panagopoulos: Phys. Lett. **B285**, 133 (1992)
[45] H. G. Dosch, E. Ferreira, A. Krämer: Phys. Rev. **D50**, 1992 (1994)
[46] V. A. Novikow, M. A. Shifman, A. I. Vainshtein, V. I. Zakharov: Nucl. Phys. **B191**, 301 (1981)
[47] F. Borkowski et al., Nucl. Phys. **A222**, 269 (1974)
[48] F. Borkowski et al.: Nucl. Phys. **B93**, 461 (1975)
[49] S. R. Amendolia et al.: Nucl. Phys. **B277**, 168 (1986)
[50] B. Povh, J. Hüfner, Phys. Rev. Lett. **58**, 1612 (1987)
[51] E. Eichten, K. Gottfried, T. Kinoshita, K. D. Lane, T. M. Yan: Phys. Rev. **D21**, 203 (1980)
[52] M. Rueter, H. G. Dosch: Z. Phys. **C66**, 245 (1995)
[53] O. Nachtmann: Ann. Phys. **209**, 436 (1991)
[54] B. Andersson, G. Gustafson, G. Ingelman, T. Sjöstrand: Phys. Rep. **C97**, 33 (1983)
[55] W. Buchmüller, A. Hebecker: "Semiclassical approach to structure functions at small x", report DESY 95-208 (1995)
[56] H. Verlinde, E. Verlinde: "QCD at high energies and two-dimensional field theory" report PUPT-1319, IASSNS-HEP-92/30 (1993)
[57] L. Van Hove: Phys. Lett. **24B**, 183 (1967)
[58] L. Lukaszuk, B. Nicolescu: Nuovo Cimento Lett. **8**, 405 (1973);
D. Bernard, P. Gauron, B. Nicolescu: Phys. Lett. **B199**, 125 (1987);
E. Leader: Phys. Lett. **B253**, 457 (1991)

[59] A. Krämer, H. G. Dosch: Phys. Lett. **B252**, 669 (1990), **B272**, 114 (1991); H. G. Dosch, E. Ferreira, A. Krämer: Phys. Lett. **B289**, 153 (1992)
[60] E. Berger, "Nichtstörungstheoretische Methoden zur Beschreibung von Reaktionen mit kleinem Impulstransfer in einem abelschen Modell", diploma thesis, Univ. of Heidelberg (1996), unpublished
[61] G. B. West: Phys. Lett. **B115**, 468 (1982);
C. D. Roberts, A. G. Williams: Progr. Part. and Nucl. Phys. **33**, 477 (1994), and references cited therein;
K. Büttner, M. R. Pennington: "Infrared behaviour of the gluon propagator: Confining or confined?", Durham Univ. report DTP-95/32 (1995)
[62] M. Rueter, H. G. Dosch: "Nucleon Structure and High Energy Scattering", Univ. of Heidelberg report HD-THEP-96-04, hep-ph/9603214 (1996)
[63] R. P. Feynman: Phys. Rev. Lett. **23**, 1415 (1969); "Photon-Hadron Interactions", W. A. Benjamin, Reading, Mass. 1972
[64] I. M. Ternov, Yu. M. Loskutov, L. I. Korovina: Zh. Eskp. Teor. Fiz. **41**, 1294 (1961) (Sov. Phys. - JETP **14**, 921 (1962));
A. A. Sokolov, I. M. Ternov: Dokl. Akad. Nauk SSSR **153**, 1052 (1963) (Sov. Phys.-Dokl. **8**, 1203 (1964))
[65] J. D. Jackson: Rev. Mod. Phys. **48**, 417 (1976)
[66] G. W. Botz, P. Haberl, O. Nachtmann: Z. Phys. **C67**, 143 (1995)
[67] F. E. Low: Phys. Rev. **110**, 974 (1958)
[68] P. V. Chliapnikov et al.: Phys. Lett. **141B**, 276 (1984)
[69] F. Botterweck et al.: Z. Phys. **C51**, 541 (1991)
[70] S. Abatzis et al.: Nucl. Phys. **A525**, 487c (1991)
[71] S. Banerjee et al. (SOPHIE/WA83 coll.): Phys. Lett. **305B**, 182 (1993)
[72] J. Antos et al.: Z. Phys. **C59**, 547 (1993)
[73] E. Fermi: Z. Phys. **29**, 315 (1924);
C. F. v. Weizsäcker: Z. Phys. **88**, 612 (1934);
E. J. Williams: Kgl. Danske Videnskab. Selskab. Mat.-Fiz. Medd. **13**, No 4 (1935)
[74] V. M. Budnev et al.: Phys. Rep. **C15**, 181 (1975)
[75] S. Platchkov et al.: Nucl. Phys. **A510**, 740 (1990)
[76] M. Meyerhoff et al.: "First measurement of the electric form factor of the neutron in the exclusive quasielastic scattering of polarized electrons from polarized ^3He", Univ. Mainz Report (1994)
[77] H. Leeb, C. Teichtmeister: Phys. Rev. **C48**, 1719 (1993)
[78] P. E. Bosted et al., Phys. Rev. Lett. **68**, 3841 (1992)
[79] R. D. Carlitz, S. D. Ellis, R. Savit: Phys. Lett. **68B**, 443 (1977);
N. Isgur, G. Karl, R. Koniuk: Phys. Rev. Lett. **41**, 1269 (1978);
D. Gromes: "Ordinary Hadrons", in Proc. Yukon Adv. Study Inst.: *The Quark Structure of Matter*, eds. N. Isgur, G. Karl, P. J. O'Donnell (World Scientific, Singapore 1985)
[80] A. Brandenburg, E. Mirkes, O. Nachtmann: Z. Phys. **C60**, 697 (1993)
[81] S. Falciano et al., (NA10 coll.): Z. Phys. **C31**, 513 (1986);
M. Guanziroli et al., (NA10 coll.): Z. Phys. **C37**, 545 (1988)
[82] W. Buchmüller, A. Hebecker: Phys. Lett. **B355**, 573 (1995);
A. Edin, G. Ingelman, J. Rathsman: Phys. Lett. **B366**, 371 (1996)
[83] A. Efremov, D. Kharzeev: Phys. Lett. **B366**, 311 (1996)

[84] W. Bonivento et al. (DELPHI Coll.): "A measurement of quark spin correlations in hadronic Z^0 decays." Contribution to the EPS-HEP 95 conference, Brussels, August 1995
[85] H. B. Nielsen, A. Patkos: Nucl. Phys. **B195**, 137 (1982)
[86] S. V. Mikhailov, A. V. Radyushkin: Sov. J. Nucl. Phys. **49**, 494 (1989)
[87] E. V. Shuryak: Nucl. Phys. **B328**, 85 (1989)
[88] P. Ball, V. M. Braun, H. G. Dosch: Phys. Rev. **D44**, 3567 (1991)
[89] C. Alexandrou, S. Güsken, F. Jegerlehner, K. Schilling, R. Sommer: Nucl. Phys. **B414**, 815 (1994);
R. Sommer: report DESY 94-011 (1994), unpublished
[90] U. M. Heller, K. M. Bitar, R. G. Edwards, A. D. Kennedy: Phys. Lett. **B335**, 71 (1994)

Perturbation Theory in a Nonperturbative QCD Background

Yuri A. Simonov

Institute of Theoretical and Experimental Physics,
117259 Moscow, B. Cheremushkinskaya 25, Russia

1 Introduction. The Double–Faced Nature of QCD

QCD as a science started with asymptotic freedom (AF) [1-3] which stated that the coupling constant at the scale Q decreases as

$$\alpha_s(Q) = \frac{g^2(Q)}{4\pi} = \frac{4\pi}{b_0 \ln \frac{Q^2}{\Lambda^2_{\mathrm{QCD}}}}, \quad b_0 = \frac{11}{3}N_c - \frac{2}{3}n_f, \tag{1.1}$$

where Λ_{QCD} is the only parameter of the theory. The main development for many years has been connected to the perturbative treatment, similar to QED. It is no wonder therefore that when one speaks of QCD one usually means perturbative QCD (PQCD). There is a huge amount of information on PQCD, both purely theoretical and relating to experiment, many books and reviews have been published (see e.g. [4-8] and lectures at this school [9,10]). The overall agreement between PQCD and experiment looks impressive (with few exceptions to be discussed below). This agreement is also surprising, since PQCD describes perturbative processes between quarks and gluons, which are necessarily confined inside hadrons, and this latter interaction is totally outside the realm of PQCD.

So here appears another face of QCD – the nonperturbative one. Nonperturbative QCD (NPQCD) is only doing its first steps as a science, although nonperturbative effects in QCD have been well known for a long time. The necessary existence of nonperturbative dynamics can be easily recognized on dimensional grounds. Indeed, any quantity of the dimension of mass, m, can be expressed in PQCD through the only dimensional parameter of the theory – Λ_{QCD} (we consider massless quarks to simplify the matter). Using variants of (1.1) for the bare coupling $g(\Lambda)$ corresponding to the cutoff momentum Λ, one can write for m

$$m = \mathrm{const}\,\Lambda_{\mathrm{QCD}} = \mathrm{const}\,\Lambda \exp(-\frac{8\pi^2}{bg^2(\Lambda)}). \tag{1.2}$$

One can immediately see in (1.2) that $m(g)$ cannot be expanded in a power series in g^n, hence the perturbative treatment of dimensional quantities is impossible.

Examples of nonperturbative quantities are numerous. We start with the string tension σ, which defines the linear confining potential $V(r) = \sigma r$, or the area law of the Wilson loop [11], which was measured on the lattice by many groups, see e.g. [12]. Indeed the dimension $[\sigma] = [m^2]$ and the scaling law (1.2) (squared) was checked on the lattice [13], where the cut-off Λ is associated with the lattice unit a, and $g(\Lambda)$ is the bare lattice coupling g_0, namely

$$\sigma a^2 \sim \exp(-\frac{16\pi^2}{bg_0^2})(\frac{g_0^2 b}{16\pi^2})^{-\gamma_1/b}, \tag{1.3}$$

where we also keep the 2-loop correction,

$$\gamma_1 = (\frac{1}{16\pi^2})^2(\frac{34N_c^2}{3} - \frac{10N_c}{3}n_f - \frac{n_f}{N_c}(N_c^2 - 1)).$$

Lattice data [12,13] usually exploit σ as an external scale in their calculations, so that all masses can be expressed through σ. From phenomenology (Regge slope $\alpha^1 = 1/2\pi\sigma$) $\sigma \simeq 0.2\text{GeV}^2$ and we also shall use this value as it used in most lattice calculations.

At this point an interesting question arises: in PQCD there is only one parameter Λ_{QCD}, in NPQCD one has instead the string tension σ (or any other NPQCD parameter, see below). In a full theory there should be a possibility to use only one of these two parameters, and to predict their ratio. This is indeed so in lattice QCD, and this ratio was determined in different ways – we shall come to this question in Sect. 4 of the lectures.

Another interesting question is how the NPQCD mass parameters can be settled down to some fixed values and how nonperturbative dynamics works – this we shall discuss in the next section, using the instanton gas as an example.

Other examples of NPQCD quantities: the quark condensate $< \bar{q}q >$ of dimension $[m^3]$, topological susceptibility χ of dimension $[m^4]$, and finally the infinite set of gluonic field correctors:

$$g^n < F_{\mu_1\nu_1}(x_1)\Phi(x_1x_2)F_{\mu_2\nu_2}(x_2)(x_2)...F_{\mu_n\nu_n}(x_n)\Phi(x_{n_1}x_1) >\equiv \Delta_n(x_1...x_n), \tag{1.4}$$

where

$$\Phi(x,y) = P\exp\, ig\int_y^x A_\mu(z)dz_\mu$$

is the parallel transporter. When $x_1 = x_2 = ...x_n$, all parallel transporters disappear and one has a gluonic condensate of dimension $[m^{4n}]$, which signals the presence of nonperturbative effects.

The lowest nonzero condensate is for $n = 2$; it was first introduced in [14] in the framework of the operator product expansion (OPE) and connected to the QCD sum rules [14], which we shall discuss in Sect. 10. Of special importance is the $n = 2$ condensate for the scale anomaly [15]; it enters the nonperturbative energy shift of the vacuum [16]

$$\varepsilon_{\text{nonpert}} = \frac{\beta(\alpha_s)}{16\alpha_s} < F^a_{\mu\nu}(0)F^a_{\mu\nu}(0) > , \tag{1.5}$$

where $\beta(\alpha_s)$ is the Gell-Mann-Low β–function of QCD, with the known first three coefficients [17] (for the MS-type scheme, $N_c = 3$) (note an extra factor 2π in our definition of β_n as compared to [17])

$$\frac{1}{2\pi}\beta_{\text{MS}}(\alpha) = \frac{1}{\pi}\mu^2\frac{\partial\alpha_{\text{MS}}}{\partial\mu^2} = -\sum_{n\geq 0}\beta_n(\frac{\alpha_{\text{MS}}}{\pi})^{n+2} , \tag{1.6}$$

$$\beta_0 = \frac{1}{4}(11 - \frac{2}{3}n_f) \approx 2,75 - 0.1667n_f ,$$

$$\beta_1 = 102 - \frac{38}{3}n_f \approx 6.375 - 0.7917n_f ,$$

$$\beta_2 = (\frac{2857}{2} - \frac{5033}{18}n_f + \frac{325}{54}n_f^2)\frac{1}{64} \approx 22.32 - 4.369n_f - 10.094n_f^2 .$$

In the lowest order one has

$$\varepsilon_{\text{nonpert}} \approx -\frac{\alpha_s}{32\pi} < F^a_{\mu\nu}(0)F^a_{\mu\nu}(0) > \tag{1.7}$$

and therefore the nonperturbative gluonic condensate (positive in the Euclidean vacuum) shifts the vacuum energy downwards, making it advantageous to have the vacuum nonperturbative.

This is striking: the asymptotic freedom is responsible for the negative sign in (1.6) and (1.7) and hence it is responsible for the nonperturbative QCD vacuum too! (A word of caution: from a rigorous point of view a nonzero value of $< F^a_{\mu\nu}(0)F^a_{\mu\nu}(0) >$ is not proved by theory, but rather preferred phenomenologically [14], also the effective α_s in $\beta(\alpha_s)$ of (1.5) can be large in principle, and the use of expansion (1.6) not possible).

From some fundamental point of view, (1.7) tells us that it is the gluonic condensate $< F^2 >$, which sets the scale of the strong interaction (and of our world), therefore in consistent nonperturbative calculations one could use this scale as the most fundamental one and express all other dimensional quantities through it, e.g., $\Lambda_{\text{QCD}}, \sigma, < \bar{q}q >$ etc.; we shall come back to this point in the next sections.

Let us turn now to the nonlocal field correlator (1.4); the simplest one for $n = 2$ is bilocal and can be expressed in terms of two Lorentz–invariant functions, D and $D_1(z)$ [18]

$$g^2\text{tr} < F_{\mu\nu}(x_1)\Phi(x_1x_2)F_{\lambda\sigma}(x_2)\Phi(x_2,x_1) >= N_c[(\delta_{\mu\lambda}\delta_{\nu\sigma} - \delta_{\mu\sigma}\delta_{\nu\lambda})D(u)+$$

$$+\frac{1}{2}(\frac{\partial}{\partial x_{1\mu}}u_\lambda \cdot \delta_{\nu\sigma} + \frac{\partial}{\partial x_{1\lambda}}u_\mu\delta_{\nu\sigma} + \text{perm})D_1(u)] , \tag{1.8}$$

where $u \equiv x_1 - x_2$.

It is clear that $D(u)$ and $D_1(u)$ contain two types of masses (lengths). The condensate $< F^2(0) >$ is expressed through $D(0) + D_1(0)$, and it was discussed above. But the dependence on u implies another length (mass), typically one can expect a nonperturbative dependence like

$$D(u) = D(0) \exp(-|u|/T_{\mathrm{g}}) , \tag{1.9}$$

$$D_1(u) = D_1(0) \exp(-|u|/T'_{\mathrm{g}}) , \tag{1.10}$$

in addition to the perturbative contribution which appears in $D_1(u)$, to lowest order [19]

$$D_1^{\mathrm{pert}}(u) = \frac{16\alpha_{\mathrm{s}}(u)}{3\pi u^4} . \tag{1.11}$$

The notion of T_{g}, which we shall call the gluonic correlation length of the vacuum, was first introduced in [20] in the treatment of heavy quarkonia.

Actually it has much more universal meaning and the value of T_{g} is crucial to our understanding of all nonperturbative dynamics [21].

Namely, as we shall see below, the QCD string is formed at distances of the order of T_{g} [18], and the width of the string between quarks is defined by T_{g} [22].

As we shall discuss later in Sect. 10, the applicability of OPE and QCD sum rules is limited to distances less than T_{g}, and its actual value is of great importance. As was shown in [21,23] the applicability of QCD sum rules corresponds to the limit $T_{\mathrm{g}} \to \infty$, while the limit $T_{\mathrm{g}} \to 0$ yields the local potential picture of quark and gluon interaction, e.g. the potential picture of heavy quarkonia. As is well known [24] the potential model is much more predictive for charmonium and bottomonium (and even for light mesons), than the QCD sum rules, which may imply that the limit $T_{\mathrm{g}} \to 0$ is close to reality (i.e. $T_{\mathrm{g}} \ll T_{\mathrm{q}}$, where T_{q} is a characteristic period of quark motion, which is of the order of the hadron size R for light quarks [20], [21]).

Another argument in favour of small T_{g} comes from the lattice calculations of the static quark–antiquark potential $V(r)$ [12], which is well fitted by the linear plus Coulomb terms, and a linear term is visible already for $r \geq 0.25$fm. From the general expression for $V(r)$ [23]

$$V(r) = -C_2(j)\frac{\alpha_{\mathrm{s}}(r)}{r} + 2r\int_0^r d\lambda \int_0^\infty d\nu D(\lambda,\nu) + \int_0^r \lambda d\lambda \int_0^\infty d\nu[-2D(\lambda,\nu) + D_1(\lambda,\nu)] \tag{1.12}$$

the asymptotic behaviour at $r \gg T_{\mathrm{g}}$ is

$$V(r) = \sigma r + C - \frac{C_2(j)\alpha_{\mathrm{s}}(r)}{r} , \tag{1.13}$$

where the string tension and the constant C are expressed through D and D_1, e.g

$$\sigma = \frac{1}{2}\int_{-\infty}^{\infty}\int d^2 u D(u) = 2\int_0^{\infty} d\lambda \int_0^{\infty} D(\lambda,\nu)\,. \tag{1.14}$$

The lattice measurements [12] demonstrate that the asymptotic regime (1.13) is valid in all measured regions $0 \lesssim r \lesssim 1.5$fm, which means that T_g is small – of the order of lattice unit a (typically $a = 0.1 \div 0.2$fm).

We have several sources of direct information about T_g. In direct lattice calculations [25] $D(x)$ and $D_1(x)$ have been measured by the cooling method and were found in the interval $x \geq 0.1$fm to follow the dependence (1.9), (1.10) with excellent accuracy; it was also found that $D_1(0) \approx \frac{1}{3}D(0)$, moreover [25]

$$T_g \approx 0.2\text{fm}\,. \tag{1.15}$$

A somewhat larger value ($T_g \approx 0.35$fm) has been found in [26], but in a less direct way.

To understand nonperturbative dynamics, it is always useful to model it with some explicit NP configurations. Take, e.g., the instanton gas model [27] which has proved to be phenomenologically reasonable for at least some correlators [28].

In the case of a dilute instanton gas the field correlators can be computed explicitly and all correlators (1.4) have a power–like behaviour $(x^2 + T_g^2)^{-n}$ with a characteristic length T_g equal to the instanton radius ρ.

The latter is known in the instanton gas model [27], $\rho \approx 0.3$fm, and one obtains the result that the correlation length of the the vacuum $T_g \approx 0.3$fm is again close to the previous estimates. Thus we shall always have in mind the new fundamental length of the QCD vacuum – $T_g \approx 0.2 \div 0.3$fm, and sometimes we will take the limit $T_g \to 0$ while the string tension σ is kept fixed.

To summarize the introduction, we have shown
i) the necessity of a nonperturbative component in the vacuum fields in addition to the usual perturbative one, which cannot be obtained from the latter,
ii) nonperturbative dynamics, as reflected by nonperturbative field correlators, contains two types of nonperturbative parameters: condensates and the correlation length of the vacuum T_g.

There are two types of possible dynamics:
a) $T_g \ll R_h \sim (\Lambda_{QCD})^{-1}$
b) $T_g \gg R_h$, where R_h is the typical hadronic radius.

In most cases type a) is operating.
iii) In the full dynamics only one input is necessary – only one mass parameter. Therefore the NP parameters should be expressed through Λ_{QCD}, or Λ_{QCD} and all NP parameters through one of them – say, the condensate $< F^2 >$.

2 Combining Perturbative and Nonperturbative Methods. A New Background Field Method

We have seen in the previous section that NP effects should emerge from the theory on the same grounds as the perturbative ones and they should somehow be built into the total gluonic field A_μ. At the basis of the whole theory of course lies the standard QCD Lagrangian , or the QCD action which we write in Euclidean space-time.

$$S_{\rm E}^{\rm QCD} = \int d^4x \{ \frac{1}{4}(F^a_{\mu\nu})(F^a_{\mu\nu}) + \bar{\psi}^{\rm E}(-i\hat{D} - im)\psi^{\rm E} \} , \tag{2.1}$$

where

$$F_{\mu\nu} = F^a_{\mu\nu} t^a = \partial_\mu A_\nu - \partial_\nu A_\mu - ig[A_\mu, A_\nu] ,$$

$$\{\gamma^{\rm E}_\mu, \gamma^{\rm E}_\nu\} = 2\delta_{\mu\nu}, \; D_\mu = \partial_\mu - igA_\mu, \;\; {\rm tr} t^a t^b = \frac{1}{2}\delta^{ab}, \; a, b = 1, ...N_{\rm c} .$$

The transition to Minkowski space–time is

$$x_0 = -ix_4, \;\; A^{\rm M}_0 = iA_4, \;\; A^{\rm M}_n = -A_n, \tag{2.2}$$

$$\psi^{\rm M} = \psi^{\rm E}, \;\; \bar{\psi}^{\rm M} = -i\bar{\psi}^{\rm E} .$$

Let us now discuss how perturbative and nonperturbative field components might enter into physical processes. We start with the general definition of nonperturbative fields, which we shall denote by B_μ, and perturbative fluctuations, denoted by a_μ, so that the total gluonic field A_μ is

$$A_\mu(x) = B_\mu(x) + a_\mu(x) . \tag{2.3}$$

In addition to the definition (2.3), one should prescribe some principle of how to separate A_μ into the perturbative and nonperturbative parts.

This is necessary, since otherwise in the general expression for the partition function (generating functional)

$$Z(J) = \frac{1}{N} \int e^{-S_{\rm E}(A) + \int J_\mu(x) A_\mu(x) d^4x} DAD\psi D\bar{\psi} \tag{2.4}$$

a double-counting problem appears (if, e.g., $DA \to DB\, Da$). One can avoid double counting and the arbitrariness of any separation principle by using the so–called 'tHooft identity[1] which leads, after insertion of (2.3) into (2.4), to

$$Z(J) = \frac{1}{N'} \int DB\eta(B) e^{+\int JBd^4x} \int DaD\psi D\bar{\psi} e^{-S_{\rm E}(B+a) + \int Jad^4x} , \tag{2.5}$$

[1] Private communication to the author, December 1993

where

$$N' = N \int DB\eta(B).$$

Here $\eta(B)$ is an arbitrary weight of integration. The proof of (2.5) is trivial, when one notices that the shift in the integration variable, $DaD(B+a)$, makes the last intergral in (2.5) independent of B, and the B integration then produces a factor which cancels with the last factor in N'. This happens for any weight factor $\eta(B)$, therefore the final answer for, e.g., the vacuum energy or a Green's function, does not depend on $\eta(B)$.

Suppose now that one is expanding (2.5) in a power series in g, i.e., considering a_μ as a small fluctuation around the background B_μ. Then the finite sum of any N first terms of the integral (we omit for simplicity quark degrees of freedom) over a_μ

$$I(B,J) \equiv \int Da_\mu e^{-S_{\mathrm{E}}(B+a)+\int Jad^4x} = \tag{2.6}$$

$$= \sum_{n=0}^{\infty} g^n I_n(B,J)$$

depends on B_μ, i.e. $I_n(B,J)$ depends on B_μ, while the total sum (2.6) (if it exists) is independent B_μ. Then the final result coincides with the original form (2.4).

The idea of this new background method is based on the assumption (which we shall partly justify) that the background perturbation series (2.6) converges better than the free perturbation series, corresponding to $B_\mu = 0$.

In the latter case there are two main defects, which are interconnected:
a) the existence of the Landau pole in the running coupling constant (1.1) at $Q^2 = \Lambda^2_{\mathrm{QCD}}$, where α_s diverges and the perturbation theory has no sense at all;
b) the existence of the so-called IR renormalons [29] which create singularities in the Borel plane for the Borel transformed perturbation series, and as a result, the sum of the perturbation series cannot be given a rigorous meaning (in the Borel sense).

As we shall show in the next sections, both difficulties a) and b) disappear when one is using the background perturbation series (2.6) with the realistic background corresponding to the confining vacuum.

At this point one must say some words about the choice of the weight function $\eta(B)$, which practically defines the result, when only few terms of the perturbation theory are taken into account. Here we only suggest some procedure which looks reasonable in practice and was successfully used in applications [30-33], leaving a more rigourous and refined theory for the future.

Let us fix $\eta(B)$ by the requirement that all gluonic field correlators (1.4) for $n = 2, 3, \ldots\infty$ are equal to some known functions, i.e.,

$$\Delta_n(x_1, \ldots x_n) = \Delta_n^{(0)}(x_1, \ldots x_n), \tag{2.7}$$

where $\Delta_n^{(0)}(x_1,...x_n)$ are an input of the theory. Thus $\eta(B)$ may be taken as a product of (infinite–dimensional) δ –functionals:

$$\eta(B) = \text{const} \prod_{n=2}^{\infty} \delta(\Delta_n(x_1,...x_n) - \Delta_n^{(0)}(x_1...x_n)) \,. \tag{2.8}$$

With this choice of $\eta(B)$ one obtains $Z(J)$ as a generating functional of a stochastic ensemble with some fixed nonperturbative correlators $\Delta_n^{(0)}(x_1..x_n)$.

As an example, one can fix the lowest correlator $\Delta_2^{(0)}(x_1,x_2)$ (which is equivalent to fixing the functions $D(x)$ and $D_1(x)$ in (1.8)), and express all higher correlators (1.4) through the lowest, $\Delta_2^{(0)}$, by vacuum insertion (all correlators with odd n vanish in this case). In this way one obtains the Gaussian stochastic ensemble, where the nonperturbative background is Gaussian (see [18,21] and the lectures by Nachtmann at this School [33], where the Gaussian background is discussed in more detail), while the series (2.6) represents all quantum fluctuations around the background.

Till now all the strategy here and in the papers [30-33] was to use $\Delta_2^{(0)}$ as an input, taken from the lattice data, like that in [25], or fitting it to the experiment.

One may ask the question, whether it is possible to find equations for $\Delta_n^{(0)}$ (or more generally for $\eta(B)$), in which case our theory would be complete. We shall describe below these equations in general terms, leaving a more detailed discussion for future publications, since no concrete analysis is available so far.

To this end we put $J = 0$ in (2.5) and generalize (2.5) to nonzero temperature $T \geq 0$, which means that the integration over x_4 is in the interval $[0, 1/T]$, and periodic boundary conditions are imposed on B_μ and a_μ (see [31] for more details). Then one can define the free energy of the vacuum in a given background $F(B,T)$ and the free energy of vacuum $F(\Delta,T)$ with a fixed set of $\Delta_n^{(0)}$ (one can use any other complete gauge–invariant set instead of $\Delta_n^{(0)}$) as follows

$$\int Da e^{S_{\mathrm{E}}(B+a)} \equiv e^{-F(B,T)} \,, \tag{2.9}$$

$$\frac{\int DB\eta(B)e^{-F(B,T)}}{\int DB\eta(B)} \equiv e^{-F(\Delta,T)} \,. \tag{2.10}$$

Let us now use the thermodynamical property of the vacuum, namely the corollary of the second law of thermodynamics, stating that for a given temperature the free energy corresponds to the phase, which ensures its minimal possible value. Feynman [34] has formulated a variational principle using this minimal property of the free energy, and this principle was used in the last paper of [27] for the instanton gas model.

To apply this principle to $F(\Delta,T)$ we first fix the only QCD parameter which defines the scale λ of the theory, it can be the gluonic condensate

$< F^2 >$ or $\Lambda_{\rm QCD}$, and consider all other paremeters in $\Delta_n^{(0)}$ as free and to be determined from the minimum of $F(\Delta, T)$, i.e., we have the equations

$$\frac{\delta F(\Delta, T)}{\delta \Delta_n^{(0)}(x_1, x ... x_n)} |_{\lambda=\text{fixed}} = 0 \ . \tag{2.11}$$

No analysis of these equations has been done so far. The only and simplest use was done in [31], where only one parameter out of the set $\{\Delta_n^{(0)}\}$ was taken – the magnetic part of the gluonic condensate $< B_i^2(0) >$ – and it was shown that in the deconfined phase it is advantageous to have nonzero $< B_i^2(0) >$, which differs only slightly from its zero temperature value. Recently this structure of the QCD vacuum at $T > T_c$ was confirmed by lattice data [25].

3 Two Examples: Connecting Perturbative and Nonperturbative Parameters

In the previous section we have noted that gluodynamics as a theory is fixed by one external mass parameter which can be taken either nonperturbatively, like the gluonic condensate $< F^2(0) >$ or the string tension, or perturbatively like $\Lambda_{\rm QCD}$, and the theory should provide an internal connection between these two types of parameters.

3.1 First Example: Instanton Gas Model

It is instructive to illustrate this connection using the instanton gas model as an example [27]. To this end we shall use the partition function of the dilute instanton gas with the one-loop and two-loop renormalization. For a single instanton the partition function is [35]

$$Z(\text{one instanton}) = \int d^4 R d\rho d\Omega d(\rho) \, , \tag{3.1}$$

where R, ρ, Ω are collective coordinates of the instanton center, radius and color orientation, respectively, and the one loop renormalized instanton density is

$$d_{1 \text{ loop}}(\rho) = \frac{1}{\rho^5} (\beta(M))^{2N_c} e^{-\beta(\rho)}, \ \ \beta(\rho) = b \ln \frac{1}{C_{N_c}^{1/b} \Lambda \rho} \, , \tag{3.2}$$

where

$$b = \frac{11}{3} N_c, \ \ C_{N_c} = \frac{4.6 \exp(-1.68 N_c)}{\pi^2 (N_c - 1)! (N_c - 2)!} \, .$$

The Pauli–Villars scheme was used, while M is the cut–off mass of the scheme.

The factor $\beta(M)$ in front of the exponential is renormalized in the 2 -loop approximation, which yields for $d(\rho)$ [35,27]

$$d_{2\ \mathrm{loop}}(\rho) = \frac{C_{N_c}}{\rho^5}(\beta_I(\rho))^{2N_c} \exp\{-\beta' I(\rho) + n\frac{\ln \beta_I(\rho)}{\beta_I(\rho)}\}, \tag{3.3}$$

where we have denoted

$$\beta'_I(\rho) = \beta_I(\rho) + \frac{b'}{2b}\ln\frac{2\beta_I(\rho)}{b}, \quad \beta_I(\rho) = b\ln\frac{1}{\Lambda\rho}, \tag{3.4}$$

$$n = (2N_c - \frac{b'}{2b})\frac{b'}{2b}, \quad b' = \frac{34}{3}N_c^2.$$

For a single instanton the density (3.3) depends on $\rho\Lambda$ in the leading term as $d(\rho) \sim \frac{(\rho\Lambda)^b}{\rho^5} \sim \rho^6$ (for $N_c = 3$) which signals the familiar IR divergence [35]. In the instanton gas this divergence is believed to be removed by the interinstanton repulsion [27]. There are arguments [36] that actually instantons (I) and antiinstantons ($\bar{I}$) prefer to annihilate in the $I\bar{I}$ gas, but for our purpose we adopt the reasoning of [27] that in the $I\bar{I}$ gas the repulsion dominates which stabilizes the system and yields the equilibrium values of the average size $\bar{\rho}$ and average distance between close neighbors $\bar{R}$ (or average density $\frac{\bar{N}}{V} = \bar{R}^{-4}$).

To this end we write down the partition function of the $I\bar{I}$ gas, normalized to its perturbative value (see last paper of [27])

$$\frac{Z}{Z_{\mathrm{pert}}} = \frac{1}{N_+!N_-!}\prod_{i=1}^{N_++N_-} d^4R^{(i)}d\rho^{(i)}d\Omega^{(i)}d(\rho^{(i)})e^{-\beta(\bar{\rho})u_{\mathrm{int}}}, \tag{3.5}$$

where u_{int} is the classical interaction between (anti)instantons, which can be approximated by the sum of pair interactions. Its averaged value is repulsive and is given by [27]

$$\int d\Omega^{(1)}d\Omega^{(2)}d^4R^{(1)}d^4R^{(2)}u_{\mathrm{int}}(R^{(1)}-R^{(2)},\rho^{(1)},\rho^{(2)},\Omega^{(12)}) = V\gamma^2(\rho^{(1)}\rho^{(2)})^2, \tag{3.6}$$

where

$$\gamma^2 = \frac{27\pi^2 N_c}{4(N_c^2-1)}.$$

Now we can use the principle of minimality of the free energy F, or equivalently, the principle of maximality of the partition function $Z \sim \exp(-F)$.

To this end we can use the number of instantons N (in the fixed 4–volume V), and the instanton size ρ as variational parameters in the spirit of the relation (2.11). This allows us to express the average density $\frac{\bar{N}}{V}$ (or the gluonic condensate) through Λ_{QCD} [27].

In the one–loop approximation

$$\frac{< F^2 >}{32\pi^2} = \frac{\bar{N}}{V} = \frac{1}{R^4} = \Lambda^4[\Gamma(\nu)C_{N_c}\tilde{\beta}^{2N_c}(\beta\gamma^2\nu)^{-\frac{\nu}{2}}]^{\frac{2}{\nu+2}}\,, \tag{3.7}$$

where $\tilde{\beta}$ is still defined at bare mass, and $\beta = \beta(\bar{\rho})$, $\nu = \frac{b-4}{2}$. To improve it, let us exploit the 2–loop expression (3.3) which yields

$$\frac{< F^2_{\mu\nu}(0) >}{32\pi^2} = \frac{\bar{N}}{V} = \frac{1}{(\bar{\rho}^2)^2}\{\Gamma(\nu)C_{N_c}(\beta\gamma^2\nu)^{-\frac{\nu}{2}}\beta_I^{2N_c}\exp[-\beta_I' + n\frac{\ln\beta_I}{\beta_I}]\}^{\frac{2}{\nu+2}}\,, \tag{3.8}$$

where $\beta = \frac{b}{2}\ln(C_{N_c}^{2/b}\Lambda^2\bar{\rho}^2)^{-1}$, and β_I and β_I' are defined for the equilibrium values of $\rho = \bar{\rho}$. The latter is obtained from the extremum of Z to be

$$(\bar{\rho}^2)^2 = \frac{\nu\bar{R}^4}{\beta\gamma^2}. \tag{3.9}$$

Finally one obtains for the SU(2) group and Pauli-Villars scheme

$$\frac{< F^2(0) >}{32\pi^2} = \frac{1}{\bar{R}^4} = (\frac{0.7}{\Lambda_{\mathrm{PV}}})^4;\;\; \bar{\rho}^2 = (3\Lambda_{\mathrm{PV}})^{-2}\,. \tag{3.10}$$

Thus the gluonic condensate (purely nonperturbatively) is expressed through the only perturbative parameter Λ_{PV} as planned.

Let us reflect for the moment on the reason why we have been able to connect perturbative and nonperturbative constants in our instanton model. The reason is that quantum fluctuations can be reliably calculated around small objects. Also the instanton size $\rho \approx 1\mathrm{GeV}^{-1}$ was considered above and in [27] to be small enough to use one- and two- loop expressions. The latter establish the scale Λ_{QCD}, and especially through the terms βu_{int} (which stabilizes the distribution in ρ) and β_I, β_I' this scale finally enters into $< F^2(0) >$ and other nonperturbative quantities. There are arguments [27] that the loop expansion is here actually in powers of $\frac{1}{\beta_I}$ which makes it reasonably convergent even for $\bar{\rho} = (600\mathrm{MeV})^{-1}$ – the resulting equilibrium radius of SU(3).

This example serves a methodic purpose rather than a physical one since the instanton model of the QCD vacuum lacks confinement, and, in addition, may be unstable as to annihilation of I and $\bar{I}$ [36]. However, the quasiclassical method described above can work for another application – the dyonic model of the QCD vacuum [37] which only recently was started and which seemingly contains confining properties. In this case one can account for quantum fluctuations around dyons and connect Λ_{QCD} from renormalized $\alpha_s(\rho)$ to $< F^2(0) >$, and, in addition, to the string tension σ. How this is done on the lattice we discuss in the next example.

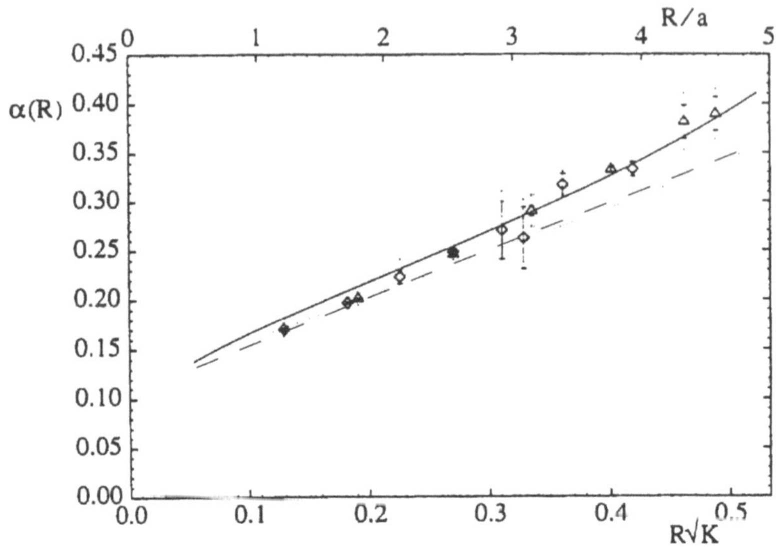

Fig. 1. The effective running coupling constant $\alpha(R)$ obtained from the force between static quarks at separation R. The scale is set by the string tension K. Data at $\beta = 6.5$ are represented by diamonds, and at $\beta = 6.2$ by triangles. The dotted error bars represent an estimate of the systematic error due to lattice artefact correction as described in the text. The curves are the two–loop perturbative expression with $\alpha(6.5)\Lambda_R = 0.060$ (dashed) and 0.070 (continuous) (from Ref. [38]).

3.2 Another Example: Connection of Λ_{QCD} and σ on the Lattice

We consider here as an example the UKQCD paper [38] where the static $q\bar{q}$ potential $V(R)$ was computed in quenched gluodynamics on a 36^4 lattice. At $\beta = \frac{6}{g_0^2} = 6.5$ the potential $V(R)$ was fitted at large R as

$$V(R) = C - \frac{E}{R} + \sigma R\,, \tag{3.11}$$

while at small R the expected behaviour is the two–loop expression (which fixes the renormalization "scheme")

$$\frac{dV}{dR} = \frac{4}{3}\frac{\alpha(R)}{R^2}\,, \tag{3.12}$$

with

$$\alpha(R) = \{4\pi[b_0 \ln \frac{1}{(R\Lambda_R)^2} + \frac{b_1}{b_0} \ln\ln \frac{1}{(R\Lambda_R)^2}]\}^{-1} \tag{3.13}$$

and $b_0 = 11/16\pi^2$, $b_1 = 102b_0^2/121$. Moreover there is the connection [39] $\Lambda_R = 1.048\Lambda_{\overline{\rm MS}}$, and the usual lattice $\Lambda_{\rm L}$ is $\Lambda_{\rm L} = 0.03471\Lambda_{\overline{\rm MS}}$.

The scale is introduced by putting $\sigma = (440\text{MeV})^2$, and the weak scaling of the data was checked by measuring the ratio of physical quantities as functions of β and demonstrating its independence of β.

Finally the measurement of $\alpha(R)$ at small R allows to compare data to the two–loop expansion (3.13) and find the resulting Λ_R. This comparison is done in Fig. 1, taken from [38] where R is scaled in units of $R\sqrt{\sigma}$.

The resulting value of $\Lambda_{\overline{\rm MS}}$ obtained via (3.13) and Fig. 1 is equal to $\Lambda_{\overline{\rm MS}} = 256\text{MeV}$. This value is obtained in the region of distances $R = 1 - 0.3$ GeV^{-1}, which are not so small to completely neglect nonperturbative effects there, and one should take them into account in a more precise treatment.

These two examples have served to show how perturbative and nonperturbative physics combine to depend on only one mass scale – this is demonstrated in the analytic model in the first example and by the numerical calculations on the lattice in the second.

4 Perturbative *vs* Nonperturbative Methods: Two Types of Expansions and the Background Lagrangian

As was discussed in Sect. 2, we combine the perturbative field a_μ and the NP degrees of freedom B_μ in one gluonic field A_μ, namely

$$A_\mu = B_\mu + a_\mu\,. \tag{4.1}$$

The gauge transformation of A_μ is

$$A_\mu \to A'_\mu = U^+(A_\mu(x) + \frac{i}{g}\partial_\mu)U\,. \tag{4.2}$$

At this point one must distinguish two different physical situations, which require different types of expansions. Consider first systems of small size, e.g., heavy quarkonia which are mostly governed by the color Coulomb interaction and have a radius of the order $(M\alpha_s(M))^{-1}$, where M is the quark mass. For the ground state bottomonium this radius is around 0.2 fm and for charmonium 0.4 fm.

In this case we have the first type of expansion: at the zeroth order all gluon exhanges are taken into account (practically Coulomb exchanges and a few first radiative corrections), while in the first order one treats the nonperturbative contribution as a correction.

This expansion is considered in detail in the next section.

The second type of expansion takes into account the NP interaction fully through NP vacuum correlators already in the zeroth order – this is the NP background – and in the next orders the usual background perturbation theory [40-42] is developed with appropriate modifications, described in Sect. 2.

In the rest of this section we shall discuss the latter formalism with some modifications due to the independent integral over the background field, as in 'tHooft's identity (2.5).

It is convenient to prescribe to B_μ, a_μ the following gauge transformations

$$a_\mu \to U^+ a_\mu U\,, \tag{4.3}$$

$$B_\mu \to U^(B_\mu + \frac{i}{g}\partial_\mu)U\,, \tag{4.4}$$

and to impose on a_μ the background gauge condition [40-42]

$$D_\mu a_\mu = \partial_\mu a^a_\mu + g f^{abc} B^b_\mu a^c_\mu = 0\,. \tag{4.5}$$

In this case ghost fields have to be introduced and one can write the resulting partition function as

$$Z = \frac{1}{N'} \int DB \eta(B) e^{\int J_\mu B_\mu d^4x} Z(J,B)\,, \tag{4.6}$$

where

$$Z(J) = \int Da\ \det(\frac{\delta G^a}{\delta w^b}) \exp \int d^4x [L(a) - \frac{1}{2}(G^a)^2 + J^a_\mu a^a_\mu]\,, \tag{4.7}$$

where $L(a) = L_0 + L_1(a) + L_2(a) + L_{\text{int}}(a)$,

$$\begin{aligned} L_2(a) &= +\frac{1}{2} a_\nu (\hat{D}^2_\lambda \delta_{\mu\nu} - \hat{D}_\mu \hat{D}_\nu + ig\hat{F}_{\mu\nu}) a_\mu = \\ &= \frac{1}{2} a^c_\nu [D^{ca}_\lambda D^{ad}_\lambda \delta_{\mu\nu} - D^{ca}_\mu D^{ad}_\nu - g\ f^{cad} F^a_{\mu\nu}] a^d_\mu\ , \end{aligned} \tag{4.8}$$

$$\begin{aligned} D^{ca}_\lambda &= \partial_\lambda \cdot \delta_{ca} + g\ f^{cba} B^b_\lambda \equiv \hat{D}_\lambda\,, \\ L_0 &= -\frac{1}{4}(F^a_{\mu\nu}(B))^2\ , \quad L_1 = a^c_\nu D^{ca}_\mu (B) F^a_{\mu\nu}\,, \\ L_{\text{int}} &= -\frac{1}{2}(D_\mu(B) a_\nu - D_\nu(B) a_\mu)^a g\ f^{abc} a^b_\mu a^c_\nu - \frac{1}{4} g^2 f^{abc} a^b_\mu a^c_\nu f^{aef} a^e_\mu a^f_\nu\,. \end{aligned} \tag{4.9}$$

The background gauge condition is

$$G^a = \partial_\mu a^a_\mu + g f^{abc} B^b_\mu a^c_\mu = (D_\mu a_\mu)^a \tag{4.10}$$

and the ghost vertex [14] is obtained from $\frac{\delta G^a}{\delta w^b} = (D_\mu(B) D_\mu(B+a))_{ab}$ to be

$$L_{\text{ghost}} = -\theta^+_a (D_\mu(B) D_\mu(B+a))_{ab} \theta_b\,. \tag{4.11}$$

The linear part of the Lagrangian L_1 disappears if B_μ satisfies the classical equations of motion. We do not impose this condition on B_μ, and we show below and in [30] that L_1 gives no contribution to the effect we are primarily interested in – the modification of the AF logarithm.

We now can identify the propagator of a_μ from the quadratic terms in the Lagrangian $L_2(a) - \frac{1}{2\xi}(G^a)^2$

$$G^{ab}_{\nu\mu} = [\hat{D}^2_\lambda \delta_{\mu\nu} - \hat{D}_\mu \hat{D}_\nu + ig\hat{F}_{\mu\nu} + \frac{1}{\xi}\hat{D}_\nu \hat{D}_\mu]^{-1}_{ab} \,. \tag{4.12}$$

It will be convenient sometimes to choose $\xi = 1$ and end up with the well-known form of the propagator in – what one would call – the background Feynman gauge

$$G^{ab}_{\nu\mu} = (\hat{D}^2_\lambda \cdot \delta_{\mu\nu} - 2ig\hat{F}_{\mu\nu})^{-1} \,. \tag{4.13}$$

Integration over ghost and gluon degrees of freedom in (4.7) yields

$$Z(J,B) = \text{const}(\det W(B))^{-1/2}_{\text{reg}}[\det(-D_\mu(B)D_\mu(B+a)]_{a=\frac{\delta}{\delta J}} \times \tag{4.14}$$

$$\times\{1 + \sum_{l=1}^{\infty} \frac{S_{\text{int}}}{l!}(a = \frac{\delta}{\delta J})\} \exp(-\frac{1}{2}JW^{-1}J) \left|_{J_\nu = D_\mu(B)F_{\mu\nu}(B)}\right. ,$$

where $W = G^{-1}$, and G is defined in (4.13).

Let us mention an important property of the background Lagrangian (4.7): under gauge tranformations the fields a_μ, B_μ transform as in (4.3)-(4.4), and all terms of (4.7), including the gauge fixing one $\frac{1}{2}(G^a)^2$, are gauge invariant. That was actually one of the aims put forward by G. 'tHooft in [41], and it has important consequences too:
(i) any amplitude in a perturbative expansion in ga_μ of (4.7) and (4.14), corresponding to a generalized Feynman diagram, is separately gauge invariant (for colorless initial and final states of course).
(ii) Due to the gauge invariance of all terms, the renormalization is specifically simple in the background field formalism [42], since the counterterms enter only in gauge–invariant combinations, e.g., $F^2_{\mu\nu}$, and hence the Z–factors Z_g and Z_A are connected. We shall exploit this fact in Sects. 6 and 7.

Let us now turn to the term L_1 in (4.9) $L_1 = 2\text{tr}(a_\nu D_\mu F_{\mu\nu})$, which is usually missing in the standard background field formalism [40-42], since one assumes there that the background B_μ is a classical solution,

$$D_\mu(B)F_{\mu\nu}(B) = 0. \tag{4.15}$$

Here we do not impose the condition (4.15) and consider any background, classical or a purely quantum fluctuation. Let us estimate the influence of the vertex L_1. In general it leads to a shift of the current J_μ in the expression for the perturbative series (4.14). Physically it means that at each point the background B_μ can generate a perturbative gluon via the vertex $(a_\mu D_\nu F_{\nu\mu})$, and this vertex is proportional to the degree of "nonclassicality" of B_μ. For the quasiclassical vacuum, like the instanton model, the average value $< (D_\mu F_{\mu\nu})^2 >$ over the instanton ensemble is less than $0(\rho^4/R^4)$ and is small (at most of the order of few percent), while the average of $< D_\mu F_{\mu\nu} >$ vanishes in the symmetric vacuum. All this is true, provided the instanton gas stabilizes at small density.

Let us estimate the effect of L_1 in the general quantum case. To this end we calculate the contribution of L_1 to the gluon propagator as in [30].

If one denotes by $<>_a$ the integral Da_μ with the weight $L(a)$ as in (4.7), we obtain

$$\Delta^B_{\mu_1\mu_2}(x_1,x_2) \equiv < a_{\mu_1}(x_1)a_{\mu_2}(x_2) >_a = \int d^4y_1 d^4y_2 G_{\mu_1\nu_1}(x_1,y_1) D_\rho F_{\rho\nu_1}(y_1) D_\lambda F_{\lambda\nu_2}(y_2) G_{\nu_2\mu_2}(y_2,x_2) \,. \quad (4.16)$$

The gluon Green's function $G_{\mu\nu}$ is given in (4.13) and depends on the background field B_μ, as well as D_μ and $F_{\rho\lambda}$. To get a simple estimate of Δ we replace $G_{\mu\nu}$ by the free Green's function $G^{(0)}_{\mu\nu}$ and take into account that

$$< D_\rho F_{\rho\nu_1}(y_1) D_\lambda F_{\lambda\nu_2}(y_2) >_B \Rightarrow \frac{\partial}{\partial y_{1\rho}} \frac{\partial}{\partial y_{2\lambda}} < F_{\rho\nu_1}(y_1) F_{\lambda\nu_2}(y_2) > . \quad (4.17)$$

For the latter we use the representation [13] in terms of two independent Lorentz structures, $D(y_1 - y_2)$ and $D_1(y_1 - y_2)$. The contribution of D in the momentum space is (that of D_1 is of similar character)

$$\Delta^B_{\mu_1\mu_2}(k) \sim \frac{(k^2\delta_{\mu_1\mu_2} - k_{\mu_1}k_{\mu_2})}{k^4} D(k) \,, \quad (4.18)$$

where

$$D(k) = \int d^4y e^{iky} D(y) \,. \quad (4.19)$$

Insertion in (4.19) of the exponential fall-off for $D(y)$ found in lattice calculations [25] yields

$$D(k) = \frac{< F^2(0) >}{(N_c^2 - 1)} \cdot \frac{\pi^2 \mu}{(\mu^2 + k^2)^{5/2}} \,. \quad (4.20)$$

With $\mu \approx 1\text{GeV}$, $D(k) \approx 0.12$. Thus $\Delta^B(k)$ is a soft correction, quickly decreasing with k as k^{-5} to the perturbative gluon propagator.

5 Nonperturbative Effects as a Correction: An Example of Heavy Quarkonia

In this section we discuss in detail the first type of expansions mentioned in the previous section: when the NP contribution is considered as a (small) correction to a basically perturbative result. As an example let us consider the spectrum of heavy quarkonia, and calculate the NP shift of the Coulombic levels of heavy quark–antiquark system, $q\bar{q}$, following mostly the recent paper [43]. When the quark mass m is large, the spatial and temporal extensions of the n-th bound state are

$$\bar{r}_n \simeq \frac{n}{m\alpha_s}, \qquad \bar{t}_n \simeq \frac{n^2}{m\alpha_s^2}, \tag{5.1}$$

and for low $n \sim 1$ these may be small enough to disregard the NP interaction in the first approximation. So for the spin–averaged spectrum we can write,

$$M(n,l) = 2m\{1 - \frac{C_F\alpha_s^2}{8n^2} + 0(\alpha_s^3) + \Delta_{NP}\}, \tag{5.2}$$

where the term $0(\alpha_s^3)$ was calculated in [44], and δ_{NP} is the expected non-perturbative correction, which should be small for states of small spatial extension.

This conclusion follows from lattice (and phenomenological) representations of the static $q\bar{q}$ potential, see e.g. [38] and [24],

$$V(r) = -\frac{4\alpha_s(r)}{3r} + \sigma r + \text{const}. \tag{5.3}$$

From $\sigma = 0.2\text{GeV}^2$ and $\alpha_s(r) \sim 0.3$ at $r \approx 0.2$fm (see [38] and Fig. 1) one may deduce that the first term on the l.h.s. of (5.3) – perturbatively – matches the second, the NP term at $r \approx 0.3$fm. Hence the hint is that the states of radius $r \ll 0.3$fm are mainly governed by (color) Coulomb dynamics, while those with $r \gg 0.3$fm are mostly NP states. So we expect, e.g., the $n = 1$ bottomonium state to be largerly Coulombic.

At the end of this section we shall recapitulate the result of a more detailed analysis, which we start now, and compare it with our initial estimates. To proceed we turn to the method of [18] and [45].

We present the general path–integral formalism for the $q\bar{q}$ system interacting via perturbative gluon exchanges and nonperturbative correlators.

We start with the quark Green's function in the form of the proper–time and path integral (the Feynman–Schwinger representation [45, 23])

$$S(x,y) = i(m - \hat{D}) \int_0^\infty ds Dz e^{-K} \Upsilon(x,y), \tag{5.4}$$

where Υ contains spin insertions into the parallel transporter

$$\Upsilon(x,y) = P_A P_F \exp[ig\int_y^x A_\mu dz_\mu + \int_0^s d\tau \Sigma F(z(\tau))], \qquad (5.5)$$

$$K = m^2 s + \frac{1}{4}\int_0^s \dot{z}_\mu^2 d\tau,$$

and double ordering in A_μ and $F_{\mu\nu}$ is implied by the operators P_A, P_F.

We have also introduced the 4×4 matrix in Dirac indices

$$\Sigma F \equiv \sigma_i \begin{pmatrix} \mathbf{B}_i & \mathbf{E}_i \\ \mathbf{E}_i & \mathbf{B}_i \end{pmatrix}. \qquad (5.6)$$

Neglecting spins one has instead of (5.5)

$$\Upsilon(x,y) \to \Phi(x,y) \equiv P_A \exp\, ig\int_y^x A_\mu dz_\mu . \qquad (5.7)$$

In terms of q$\bar{\text{q}}$-Green's functions (5.4) and initial and final state matrices Γ_i, Γ_f (such that $\bar{q}(x)\Gamma_f\Phi(x,\bar{x})q(\bar{x})$ is the final q$\bar{\text{q}}$ state) the total relativistic gauge–invariant q$\bar{\text{q}}$ Green's function in the quenched approximation is

$$G(x,\bar{x},y\bar{y}) =< \mathrm{tr}(\Gamma_f S_1(x,y)\Gamma_i\Phi(y,\bar{y})S_2(\bar{y},\bar{x})\Phi(\bar{x},x))| > \qquad (5.8)$$
$$- < \mathrm{tr}(\Gamma_f S_1(x,\bar{x})\Phi(\bar{x},x))tr(\Gamma_i S_2(\bar{y},y)\Phi(y,\bar{y})) > .$$

The angular brackets in (5.8) imply averaging over the gluonic field A_μ.

Since we are interested primarily in heavy quarkonia, it is reasonable to do a systematic nonrelativistic approximation. To this end we introduce, as in [23], the real evolution parameter t (time) instead of the proper time τ in K, ($\bar{\tau}$ in $\bar{K}$) and the dynamical mass parameters $\mu, \bar{\mu}$

$$\frac{dt}{d\tau} = 2\mu_1,\ \frac{dt}{d\bar{\tau}} = 2\mu_2;\ \int_0^s \dot{z}_\mu^2(\tau)d\tau = 2\mu_1\int_0^T dt(\frac{dz_\mu(t)}{dt})^2 . \qquad (5.9)$$

Here we have denoted T by

$$T \equiv \frac{1}{2}(x_4+\bar{x}_4). \qquad (5.10)$$

The nonrelativistic approximation is obtained, when one writes for $z_4(t), \bar{z}_4(t)$

$$z_4(t) = t + \zeta(t),\ \ 2\mu_1 = \frac{T}{s_1}, \qquad (5.11)$$

$$\bar{z}_4(t) = t + \bar{\zeta}(t),\ \ 2\mu_2 = \frac{T}{s_2},$$

and expands in fluctuations $\zeta, \bar{\zeta}$, which are $0(\frac{1}{\sqrt{m}})$. Note that the integration over ds_1ds_2 goes over into $d\mu_1 d\mu_2$. Physically the condition (5.11) means that we neglect trajectories with backtracking of $z_4, \bar{z}_4$, i.e., neglect q$\bar{\text{q}}$ pair creation. One can convince oneself that insertion of (5.11) into $K, \bar{K}$ allows one to determine μ_1, μ_2 from the extremum in $K, \bar{K}$ to be

$$\mu_1 = m_1 + 0(1/m_1), \quad \mu_2 = m_2 + 0(1/m_2), \tag{5.12}$$

and one can further make a systematic expansion in powers $1/m_i$ (for details see [23]). At least to lowest orders in $1/m_i$ this procedure is equivalent to the standard (gauge–noninvariant) nonrelativistic expansion [59].

Let us keep the leading term of this expansion

$$G(x\bar{x}, y\bar{y}) = 4m_1 m_2 e^{-(m_1+m_2)T} \int D^3 z D^3 \bar{z} e^{-K_1-K_2} < W(C) >, \tag{5.13}$$

where $K_1 = \frac{m_1}{2}\int_0^T \dot{z}_i^2(t)dt$, $K_2 = \frac{m_2}{2}\int_0^T \dot{\bar{z}}_i^2(t)dt$ and $< W(C) >$ is the Wilson loop operator with closed contour C comprising q and q̄ paths, and initial and final state parallel transporters $\Phi(x,\bar{x})$ and $\Phi(y,\bar{y})$.

The representation (5.13) will be our main object of study in this section.

For a heavy qq̄ system the perturbative interaction contains an expansion in powers of $\frac{\alpha_s}{v}$ (v – velocity in the c.m. system) which should be kept completely, and a nonperturbative interaction, of which we keep here only the lowest term.

We represent the total gluonic field A_μ as

$$A_\mu = B_\mu + a_\mu , \tag{5.14}$$

where B_μ is the NP background, while a_μ is a perturbative fluctuation. The principle of separation in (5.14) is immaterial for our purposes, the only requirement is that the NP gluonic condensate (and correlator) consists of the field B_μ (for further discussion of this point see (2.5) and the text there).

It is convenient to split the gauge transformation as

$$B_\mu \to V^+ B_\mu V, \qquad a_\mu \to V^+(a_\mu + \frac{i}{g}\partial_\mu)V , \tag{5.15}$$

so that the parallel transporter

$$\Phi(a;x,y) \equiv P\exp\, ig\int_y^x a_\mu dz_\mu \tag{5.16}$$

transforms as

$$\Phi(a;x,y) \to V^+(x)\Phi(a;x,y)V(y) . \tag{5.17}$$

The Wilson loop average in (5.13) can be written using (5.14) as

$$< W(C) > \equiv < \mathrm{tr}P\exp\left(ig\int_C A_\mu dz_\mu\right) > = < \mathrm{tr}P\exp\left(ig\int_C a_\mu dz_\mu\right) > + \tag{5.18}$$
$$+\frac{(ig)^2}{2!}\int_C dz_\mu \int_C dz'_\mu < \mathrm{tr}P\Phi(a;z,z')B_\nu(z')\Phi(a;z',z)B_\mu(z) > + ...$$
$$= W_0 + W_2 + ...$$

where we have omitted the term linear in B_μ since it vanishes when averaged over the field B_μ. The dots imply terms of higher powers in B_μ.

The contour and points z, z' are shown in Fig. 2.

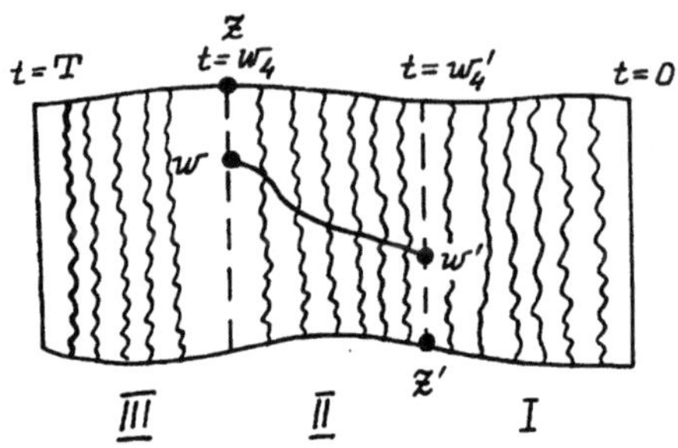

Fig. 2. The Wilson loop for heavy quark and antiquark trajectories from the initial time $t = 0$ to the final time $t = T$. The nonperturbative background contributes at $t = w_4$ and $t = w_4'$.

Let us discuss the first term on the r.h.s. of (5.18). It is the Wilson loop average of the usual perturbative fields, discussed extensively in [60]. One can use the cluster expansion for W_0 to obtain:

$$W_0 = Z \exp(\varphi_2 + \varphi_4 + \varphi_6 + \ldots), \tag{5.19}$$

$$\varphi_2 \equiv -\frac{g^2}{8\pi^2} \int\int \frac{dz_\mu dz'_\mu}{(z-z')^2} C_2, \quad C_2 = \frac{N_c^2 - 1}{2N_c}, \tag{5.20}$$

where in the integral φ_2 regularization is implied to be absorbed into the Z factor. Note that φ_2 contains all ladder–type exchanges, as shown in Fig. 2, and in addition also "Abelian–crossed" diagrams — those where the times of vertices cannot be ordered, while the color generators t_i^a are always kept in the same order, as in the ladder diagrams of Fig. 2. Therefore all crossed diagrams (minus "Abelian–crossed") are contained in φ_4 and contribute of $0(1/N_c)$ as compared to ladder ones (cf. the discussion in [60]). In addition φ_4 contains "Mercedes–Benz diagrams", shown in Fig. 3, again repeated infinitely many times. It is remarkable that each term φ_2, φ_4 etc. in (5.19) sums up an infinite series of diagrams.

In particular $\exp \varphi_2$ contains all terms with powers of $\frac{\alpha_s}{v}$, as we shall see below.

For heavy (and slow) quarks one can write:

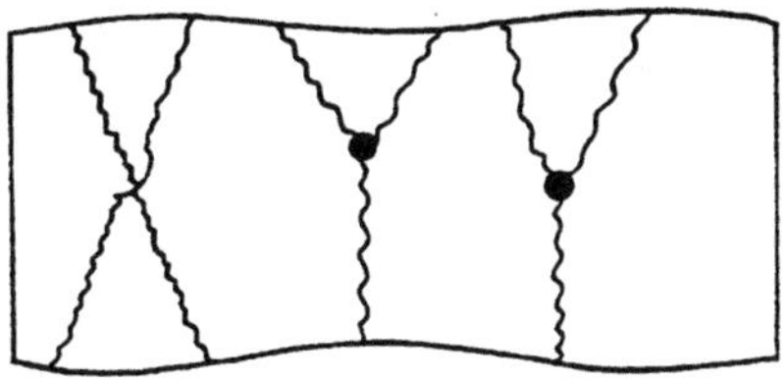

Fig. 3. The Wilson loop for heavy quark and antiquark trajectories from the initial time $t = 0$ to the final time $t = T$. The nonperturbative background contributes at $t = w_4$ and $t = w_4'$.

$$\varphi_2 = \frac{g^2}{4\pi^2}\int_0^T\int_0^T \frac{dtdt' C_2(1+\dot{\mathbf{z}}_i\dot{\mathbf{z}}_i')}{\mathbf{r}^2+(t-t')^2} =$$
$$= \int_0^T \frac{C_2\alpha_s}{|\mathbf{r}|}(1+0(v^2/c^2)dt = \varphi_2^{(0)} + 0(v^2/c^2)\,, \qquad (5.21)$$

where $\mathbf{r} = \mathbf{z}-\mathbf{z}'$. In this way one obtains a singlet one–gluon–exchange (OGE) potential, the effective time difference being $\Delta t = |t-t'| \sim |\mathbf{r}|$.

In addition one also obtains radiative corrections due to transverse gluon exchange, which have been computed earlier in a different way [61].

For the q$\bar{\text{q}}$ mass, φ_2 yields a correction $0(\alpha_s^2)$ (Coulombic energy), in the wave–function one should keep the Coulomb potential to all orders because of its singular character. In fact we shall not expand $\exp\varphi_2^{(0)}$, while expanding other radiative corrections.

We now turn to W_2 and draw the leading (in N_c) set of diagrams which one obtains when one connects points with the same time coordinates on the q and $\bar{\text{q}}$ lines by propagators (Coulombic or instantaneous gluon exchanges). One has

$$< a_\mu^a(x)a_\nu^b(y) >= \frac{\delta_{ab}\delta_{\mu\nu}}{4\pi^2(x-y)^2}. \qquad (5.22)$$

Here we have chosen for simplicity the Feynman gauge, since W_0 and W_2 are gauge invariant.

Expanding W_2 of (5.18) in powers of a_μ one has, due to (5.22), a typical term of the form

$$\text{tr}(t^{b_k}t^{b_{k-1}}...t^{b_1}t^{b_1}t^{b_2}...t^{b_k}t^{a_1}...t^{a_n}t^a B_\mu^a t^{a_n}...t^{a_1}B_\mu^b t^b...) \qquad (5.23)$$
$$\to (C_2)^k tr(t^{a_1}...t^{a_n}t^a B_\mu^a t^{a_n}...t^{a_1}B_\mu^b t^b...)\,.$$

Now due to the equality

$$t^c t^a t^c = -\frac{1}{2N_c}t^a, \qquad (5.24)$$

one obtains, for all exchanges in the time interval between times of $B_\mu(z)$ and $B_\nu(z')$, a factor $(-\frac{1}{2N_c})$ instead of the factor C_2 for all other exchanges. As a result W_2 can be written as

$$W_2 = \frac{(ig)^2}{2} \int_C dz_\mu \int_C dz'_\nu < \mathrm{tr} B_\nu B_\mu > e^{\int_{t_2}^T dt V_C^0 + \int_0^{t_1} dt V_C^0 + \int_{t_1}^{t_2} V_C^8 dt}, \quad (5.25)$$

where

$$V_C^0 = C_2 \frac{\alpha_s}{r}, \quad V_C^8 = -\frac{1}{2N_c}\frac{\alpha_s}{r}. \quad (5.26)$$

One can easily identify $-V_C^0$ and $-V_C^8$ as a singlet and octet $q\bar{q}$ potential, as considered in [46]. And for $< \mathrm{tr} B_\nu B_\mu >$ one can use the modified Fock–Schwinger gauge [47] to obtain:

$$B_\mu(z) = \int_{x_0}^{z} dw_\rho \alpha(w) F_{\rho\mu}(w) \quad (5.27)$$

and finally:

$$\int dz_\mu dz'_\nu < B_\mu(z) B_\nu(z') > = \int d\sigma_{\mu\rho} d\sigma'_{\nu\lambda} < F_{\mu\rho}(w) F_{\nu\lambda}(w') >, \quad (5.28)$$

where we have introduced a surface element

$$d\sigma_{\mu\rho} = dz_\mu dw_\rho \alpha(w). \quad (5.29)$$

To make (5.28) fully gauge–invariant one can introduce in the integral in (5.27) factors $\Phi(x_0, w)$ identically equal to unity in the Fock–Schwinger gauge. So finally one obtains

$$< \mathrm{tr} B_\nu B_\mu > = \int d\sigma_{\mu\rho} d\sigma_{\nu\lambda} \quad (5.30)$$
$$\times \mathrm{tr}\{\Phi(x_0, w) F_{\mu\rho}(w) \Phi(w, x_0) \Phi(x_0, w') F_{\nu\lambda}(w') \Phi(w', x_0)\}.$$

One can see that (5.30) is fully gauge invariant.

In the next section we evaluate W_2 explicitly.

We derive in this section SI corrections to energy levels and wave functions due to NP field correlators. As shown in Fig. 2 we divide the total time interval T into three parts

$$\begin{aligned} &(1) \quad 0 \le t \le w'_4, \\ &(2) \quad w'_4 \le t \le w_4, \\ &(3) \quad w_4 \le t \le T, \end{aligned}$$

where t is the c.m. time. In $K + \bar{K}$ one can separate c.m. and relative coordinates

$$R_i = \frac{m_1 z_i + m_2 \bar{z}_i}{m_1 + m_2}, \quad r_i(t) = z_i(t) - \bar{z}_i(t), \quad (5.31)$$

so that

$$K+\bar{K}=\int_0^T \frac{M\dot{R}^2}{2}dt+\frac{\tilde{m}}{2}\int_0^T \dot{r}^2(t)dt, \tag{5.32}$$

with $\tilde{m}=\frac{m_1m_2}{m_1+m_2}$, $M=m_1+m_2$.

Separating out the trivial c.m. motion, in parts (1) and (3) one has the path integrals, representing actually the singlet Coulomb Green's function

$$G_C^{(1)}(r(t_1),r(t_2);t_1-t_2)=\int D\mathbf{r}(t)e^{-\frac{\tilde{m}}{2}\int_{t_2}^{t_1}\dot{\mathbf{r}}^2 dt+C_2\alpha_s\int_{t_2}^{t_1}\frac{dt}{|r(t)|}}. \tag{5.33}$$

In part (2), on the other hand, one has an octet Coulomb Green's function

$$G_C^{(8)}(r(w_4),r(w_4');w_4-w_4')=\int D\mathbf{r}(t)e^{-\frac{\tilde{m}}{2}\int_{w_4'}^{w_4}dt\dot{\mathbf{r}}^2-\frac{C_2\alpha_s}{2N_c}\int_{w_4'}^{w_4}\frac{dt}{|\mathbf{r}(t)|}}. \tag{5.34}$$

Insertion of (5.33) and (5.34) into W_2 (Eq. (5.25)), yields for the correction to the total q$\bar{\text{q}}$-Green's function $G=G^{(0)}+\Delta G$,

$$\begin{aligned}\Delta G=-\frac{g^2}{2}G_C^{(1)}(r(T),r(w_4);T-w_4)d^3r(w_4)\times G_C^{(8)}(r(w_4),r(w_4'),w_4-w_4')\times \\ \times\int d\sigma_{\mu\nu}(w)\int d\sigma_{\mu'\nu'}(w')<F_{\mu\nu}(w)F_{\mu'\nu'}(w')>d^3r(w_4')\times \\ \times G_C^{(1)}(r(w_4'),r(0);w_4').\end{aligned} \tag{5.35}$$

It is convenient to attach $d\sigma_{\mu\nu}$ to the minimal surface inside the contour C, formed by connecting the points $z(t)$ and $\bar{z}(t)$, which is a straight line:

$$w_\mu(t,\beta)=z_\mu(t)\beta+\bar{z}_\mu(t)(1-\beta)=R_\mu+r_\mu(\beta-\frac{m_1}{m_1+m_2}), \tag{5.36}$$

$$w_4=z_4=\bar{z}_4=t.$$

The surface element can be written as

$$d\sigma_{\mu\nu}(w)=(w_\mu'\dot{w}_\nu-\dot{w}_\mu w_\nu')dz_0d\beta\equiv a_{\mu\nu}dz_0d\beta, \tag{5.37}$$

with $w_\mu'=\frac{\partial w_\mu}{\partial\beta}=r_\mu; \dot{w}_\mu=\frac{\partial w_\mu}{\partial t}$. We also have in the c.m. system

$$a_{i4}=r_i,\quad a_{ij}=e_{ijk}L_k\frac{1}{i\tilde{m}}(\beta-\frac{m_1}{m_1+m_2}) \tag{5.38}$$

and the Minkowskian angular momentum is

$$L_i=e_{ikl}\,r_k\cdot\frac{1}{i}\frac{\partial}{\partial r_l}. \tag{5.39}$$

In a nonrelativistic approximation we expand in powers of $\frac{1}{m_1},\frac{1}{m_2}$, hence a_{ij} can be neglected in lowest order, and one is left with only a_{i4}, i.e., in

(5.35) only electric field correlators should be kept, which have the following representation in terms of the two Lorentz invariants D and D_1 [18]

$$g^2 \mathrm{tr} < E_i(w)E_k(w') >= \frac{1}{12}[\delta_{ik}(D(w-w')+D_1(w-w')+h_4^2\frac{\partial D_1}{\partial h^2})+h_ih_k\frac{\partial D_1}{\partial h^2}], \tag{5.40}$$

where $h_i = w_i - w_i'$, and D and D_1 are normalized as

$$D(0) + D_1(0) = g^2 < \mathrm{tr}F^2_{\mu\nu}(0) >= \frac{1}{2}4\pi^2 G_2 . \tag{5.41}$$

G_2 is the standard definition of the gluonic condensate [14]

$$G_2(\mathrm{stand}) = \frac{\alpha_s}{\pi} < F^a_{\mu\nu}F^a_{\mu\nu} >= 0.012\mathrm{GeV}^4 . \tag{5.42}$$

Insertion of (5.40) into (5.35), neglecting in (5.40) the terms $h_ih_k \sim 0(\frac{1}{\tilde{m}^2})$, yields

$$\Delta G = -\frac{1}{24}G_C^{(1)}(r(T),r)d^3rG_C^{(8)}(r,r')d^3r'\times \tag{5.43}$$
$$\times r_i d\beta dw_4 r_i' d\beta' dw_4' \Delta(w-w')G_C^{(1)}(r',r(0)) ,$$

where we have denoted

$$\Delta(w-w') = D(w-w') + D_1(w-w') + h_4^2\frac{\partial D_1}{\partial h^2} . \tag{5.44}$$

Using the spectral decomposition for G_C

$$G_C^{(1,8)}(r,r',t) =< r|e^{-H_C^{(1,8)}t}|r' >= \sum_n \psi_n^{(1,8)}(r)e^{-E_n^{(1,8)}t}\psi_n^{(1,8)+}(r') \tag{5.45}$$

one can rewrite (5.35) for the matrix element of ΔG between singlet Coulomb wave functions

$$\begin{aligned} < n|\Delta G|n > &= -\frac{e^{-E_n^{(1)}T}}{24}T\int\frac{dp_4 d\mathbf{p}}{(2\pi)^4}\tilde{\Delta}(p)d\beta d\beta' \\ &\sum_{k=0,1,\dots} < n|r_i e^{i\mathbf{p}(\beta-\frac{m_1}{m_1+m_2})\mathbf{r}}|k > \times \\ &\times\frac{< k|r_i' e^{-i\mathbf{p}(\beta-\frac{m_1}{m_1+m_2})\mathbf{r}'}|n >}{E_k^{(8)}-E_n-ip_4} . \end{aligned} \tag{5.46}$$

Here the set of states $|k >$ with eigenvalues $E_k^{(8)}$ refer to the octet Hamiltonian, which can read off from (5.34)

$$H^{(8)} = \frac{\mathbf{p}^2}{2\tilde{m}} + \frac{C_2\alpha_s}{2N_c|\mathbf{r}|} , \tag{5.47}$$

and $\tilde{\Delta}(p)$ is the Fourier transform of (5.44).

The correlator $\Delta(x)$ depends on x as $\Delta(x) = f(\frac{|x|}{T_g})$, and decays exponentially at large $|x|$ [25]. For what follows it is crucial to compare two parameters: T_g and the Coulombic size of the n-th state of the $q\bar{q}$ system; $R_n = \frac{n}{\tilde{m}c_2\alpha_s}$

In the WL case [46] it is assumed explicitly or implicitly that

$$\text{Case (i)} \qquad T_g \gg R_n\,.$$

In the opposite case

$$\text{Case (ii)} \qquad T_g \ll R_n$$

a completely different dynamics occurs, as we shall see.

The approximation used in [48,49] corresponds to the replacement

$$\bar{\Delta}(p) = \frac{2\mu(2\pi)^3\delta(p)}{P_4^2+\mu^2}D(0)\,. \tag{5.48}$$

In this case one obtains from (5.46)

$$< n|\Delta G|n> = -\frac{e^{-E_n^{(1)}T}}{24}T2\pi^2 G_2 \sum_{k=0,1,\dots} \frac{< n|r_i|k>< k|r_i'|n>}{E_k^{(8)} - E_n + \mu}\,. \tag{5.49}$$

Writing $G_0 + \Delta G = \text{const}\, e^{-(E_n^{(1)}+\Delta E_n)T} \approx \text{const}\; e^{-E_n^{(1)}T}(1 - \Delta E_n T)$ one finally obtains for ΔE_n:

$$\Delta E_n = \frac{\pi^2 G_2}{18}\frac{< n|r_i|k>< k|r_i|n>}{E_k^{(8)} - E_n + \mu}\,. \tag{5.50}$$

From a physical point of view it is better to start from the general answer (5.46), and to realize that $(-ip_4)$ in the denominator there is of the order of μ (it is effectively replaced by μ in the integration in (5.49)); when $\mu \sim 1\text{GeV}$ it is much larger than $E_k^{(8)} \sim m_q\alpha^2$.

Neglecting $E_k^{(8)}$ in (5.46) one readily obtains

$$\Delta E_n = \frac{\pi^2 G_2}{18}\frac{< n|\mathbf{r}^2|n>}{|E_n| + \mu}\,. \tag{5.51}$$

This result can be called a small correlation length limit, since it is obtained when $\mu \to \infty$, or $T_g \to 0$.

Let us now perform a phenomenological analysis of our equations, applying (5.2) with Δ_{NP} given by (5.50) to bottomonium and charmonium (for more detail see [43,44]).

Following [44] one should choose the normalization point μ_0 for $\alpha_s(\mu_0)$ depending on the state. We choose $\mu_0 = 1.5\text{GeV}$ for $n = 1$, $b\bar{b}$ and $\mu_0 = 0.95\text{GeV}$ for $n = 2$, $b\bar{b}$ and all $c\bar{c}$ states. The free renormalized $\alpha_s(\mu_0) = 0.27$ in the first case, and 0.35 in the second.

To get rid of quark mass effects to lowest order, we consider the difference of levels (actually of spin averaged levels for a given (n, l)). This is shown in Table 1. The left column displays the result of the Leutwyler–Voloshin (LV) approximation ($T_g \to \infty$, $\mu \to 0$). One can compute from (5.2) and (5.51) that the nonperturbative effects are only small ($\Delta_{NP} \approx 0.1|E_{coul}|$) for the ground state of $b\bar{b}$. The LV approximation is violated in all other cases, and one should take into account $\mu \neq 0$.

Table 1. Predicted splittings and experiment

splitting	$\mu = 0$	[50]	$\mu = 0.40$	exp.
$2S - 1S(b\bar{b})$	479	554	522	558 MeV
$2S - 2P(b\bar{b})$	181	112	147	123 MeV
$3S - 2S(b\bar{b})$	4570	342	614	332 MeV
$2S - 1S(c\bar{c})$	9733	582	1626	670 MeV

To estimate μ needed for our input, we use the expression for the string tension in terms of $D(x)$, (1.14), fix $\sigma = 0.2\text{GeV}^2$ and the gluonic condensate in (5.41) with $D_1 = 0$ at the standard value (5.42). This yields $\mu = 0.35 - 0.4$GeV The predicted energy splittings are listed in the last but one column in Table 1. One can see that the agreement with experiment is not good, especially for higher $b\bar{b}$ states and for $c\bar{c}$.

To understand what is bad with our treatment of the NP interaction, one can do another analysis, taking the limit $T_g \to 0$, or $\mu \to \infty$. Formally the correction Δ_{NP} in this limit vanishes as $1/\mu$, and one might think that the NP effects are unimportant in this limit. However, this conclusion is incorrect, and actually the NP effects in this limit become very important, if one takes into account that the limit $T_g \to 0$ ($\mu \to \infty$) should be taken at a constant value of the string tension σ, which is the main NP characteristics in the $T_g = 0$ limit.

Indeed, one can rewrite Δ_{NP} as [43]

$$\Delta_{NP} = \frac{\pi \varepsilon_{nl}(\mu) n^6 < \alpha_s G^2 >}{2(m C_F \tilde{\alpha}_s)^4}, \tag{5.52}$$

where

$$\varepsilon_{nl}(\mu) \simeq \frac{\varepsilon_{nl}(\text{coul})}{1 + \rho_{nl}\eta_n}, \tag{5.53}$$

and $\rho_{10} = 0.62; \rho_{20} = 0.76, \rho_{21} = 0.70, \rho_{30} = 0.9$

$$\eta_n \equiv (\frac{2n}{C_F \tilde{\alpha}_s})^2 \frac{\mu}{m}. \tag{5.54}$$

Now taking into account that $\sigma \sim < \alpha_s G^2 > T_g^2$, and $\mu = 1/T_g$ one has an estimate

$$\Delta_{NP}(T_g) \sim \frac{\text{const}\sigma}{T_g}, \quad T_g \to 0. \tag{5.55}$$

Hence the NP effects become very important in this limit and should not be taken as a perturbation.

This is done in the framework of the local potential picture in [50], where the NP potential is expressed via correlators $D(x)$, $D_1(x)$ as in (1.12). Results of the calculations of [50] yield a very consistent picture both for levels and wave functions of bottomonium and quarkonium. To compare with LV and our results here, the results of [50] are listed in the center column of Table 1, demonstrating much better agreement with experiment than the results for $\mu = 0$ (LV approximation) or $\mu = 0.40$.

Note, that in [50] the NP interaction was not treated as a perturbation, but exactly via the NP part of the potential, which explains its better applicability.

Thus in reality, because of the small $T_g \approx 0.2$fm, the potential picture is more adequate for quarkonia than the VL formalism or QCD sum rules, uncluding even the bottomonium case.

6 Perturbation Series in the NP Background. Definition of the Running Coupling Through the Static Potential

As was discussed in the introduction and in Sect. 2, the real QCD vacuum contains both perturbative and NP configurations, and the latter as we shall see below, play a very important role in the properties of the perturbation series – both in its convergence (to be discussed in Sect. 9) and in the definition of lowest order terms. Therefore the free perturbation series usually discussed has to do with reality only in some limiting cases, as will be discussed in this and subsequent sections. It is important that the NP background enters already at the stage of the definition of the (renormalized) coupling constant. In the usual language of perturbative QCD, what one should do with the running α_s, taking into account higher orders, is to fix the scheme and set the scale [51]. The NP background puts some new meaning into the notion of the scheme fixing. In the case of the free vacuum one can use different schemes of regularizations (e.g. MS or $\overline{\text{MS}}$ schemes) and renormalize the coupling g^2, fixing the value of the vertex at some Euclidean momentum. For example, one may require that the quark (or gluon) propagator has the form of the free propagator at some momentum p, $p^2 = -\mu^2$ [5].

In the case of the NP vacuum this kind of definition is not possible anymore, and one should instead introduce a new NP notion of the regularized coupling. We consider here three examples of such a definition – two in this section and the third one in the next section.

i) Definition of the running coupling through the interaction energy of heavy quarks (HQ)

The interaction energy of a heavy quark and antiquark $E(R)$ at a distance R between them can be written as

$$E(R) = E_{\rm NP}(R) - \frac{4}{3}\frac{\bar{\alpha}_s(R)}{R}, \tag{6.1}$$

where $E_{\rm NP}$ is the purely NP part of the energy, while $\bar{\alpha}_s(R)$ is a renormalized running coupling at distance R

$$\bar{\alpha}_s(R) = \frac{\bar{g}^2(R)}{4\pi} = \frac{g_0^2}{4\pi}(1 + \frac{g_0^2}{4\pi} f(R) + ... \tag{6.2}$$

and g_0 is the bare coupling. $f(R)$ is to be calculated from the 1-loop diagrams in the background NP field.

In the case of the free vacuum $\bar{\alpha}_s^{(0)}(R)$ has been computed to one loop in [16] through the bare coupling

$$g^2(R) = g_0^2 + g_0^4 \frac{11}{48\pi^2} C_2 \ln(\frac{R}{\delta})^2, \tag{6.3}$$

where $\delta \sim 1/M$ is the UV cut-off.

Renormalizing via the $\overline{\rm MS}$ scheme with the normalization mass μ, one gets [62]

$$\alpha_{\overline{\rm MS}}(\mu, R) = \alpha_{\overline{\rm MS}}\{1 + \frac{\alpha_s}{\pi}[\frac{b_0}{2}(\ln \mu R + \gamma_{\rm E}) + \frac{5}{12}, b_0 - \frac{2}{3}C_A]\} \tag{6.4}$$

with $b_0 = \frac{11}{3}N_c - \frac{2}{3}n_f$, $\gamma_{\rm E} = 0.5777...$

The main feature of (6.3) and (6.4) is the logarithmic growth of $\alpha_s(R)$ at large distances, and this feature produces both the Landau ghost pole [63] and the IR renormalon problem [29]. In the next sections we shall demonstrate that the strong confining NP background eliminates the logarithmic growth of $\alpha_s(R)$, and in this way solves both the ghost pole and the IR renormalon problem.

To calculate $\bar{\alpha}_s(R)$ in the NP background we shall use the gauge-invariant definition

$$E(R) = - \lim_{T\to\infty} \{\frac{1}{T} \ln < W_C(B + a) >^{\rm reg}_{B,a}\}, \tag{6.5}$$

where the contour C of the Wilson loop $< W_C >$ is a rectangle $R \times T$ shown in Fig. 4.

The averaging over fields B_μ, a_μ and the regularization assumed in (6.5) will be specified in the next section. Note that $E(R)$ in (6.5) contains both PTH and NP contributions as in (6.1). We again stress at this point that in our approach there is no problem of separation of a given $E(R)$ into $E_{\rm NP}$ and a PTH piece (which is, of course, not well defined). Instead we assume, as in

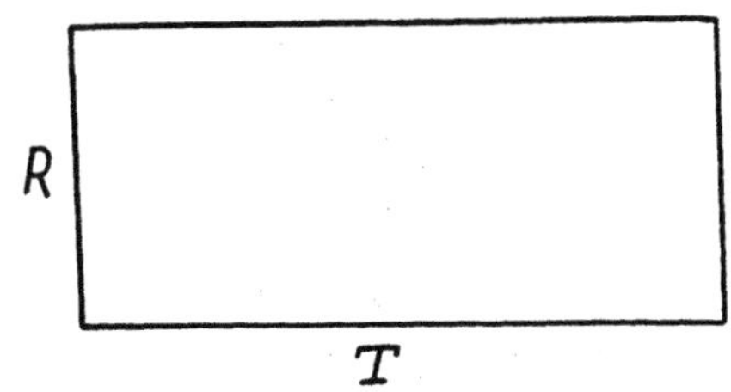

Fig. 4. The rectangular Wilson loop of the size $R \times T$ used to define the static quark energy $E(R)$.

(2.5), that an ensemble of fields $\{B_\mu\}$ is given to us; it is characterized by correlators

$$g^n < F_{\mu_1\nu_1}(B,x_1)...F_{\mu_n\nu_n}(B,x_2) >_B \tag{6.6}$$

and in particular by the string tension σ, which can be expressed through the correlators. Our task is to construct the PTH in this given background and to express in particular $\bar{\alpha}_s(R)$ through the nonperturbative parameters.

To perform renormalization in the background field formalism, one may use its gauge invariance [41,42]. One can visualize that all terms in (4.8) are gauge invariant with respect to gauge transformations "rotating" both B_μ and a_μ. It is convenient to split gauge transformations as follows [42]:

$$\begin{aligned} B_\mu &\to U^+(B_\mu + \frac{i}{g}\partial_\mu)U \ , \\ a_\mu &\to U^+ a_\mu U \ , \quad B_\mu = B_\mu^a t^a . \end{aligned} \tag{6.7}$$

If we consider, following [42], only amplitudes with no external a_μ lines (and we need only those), then renormalization of a_μ and θ lines is inessential (the corresponding Z factors compensate each other), while g, B, ξ renormalize as

$$g_0 = Z_g g \ , \ B_0 = Z_B^{1/2} B, \ \ \xi_0 = Z'\xi \, . \tag{6.8}$$

Now because of the explicit gauge invariance, Z_g and Z_B are connected [42]

$$Z_g^{-1} \ = \ Z_B^{1/2} \, . \tag{6.9}$$

As a consequence, the quantities gB_μ and $gF_{\mu\nu}(B)$ are RG invariant, and therefore all characteristics of NP fields, like correlators (6.6) or the string tension are RG invariant and should be considered on the same footing as the external momenta in the amplitudes.

ii) Definition through the B-field two-point function

This is equivalent to the method used in [42], where the two 1-loop diagrams of Fig. 5 and Fig. 6 with two external B lines have been used. To perform the averaging over fields $\{B_\mu\}$ we multiply each diagram in the coordinate space with the parallel transporter in the adjoint representation,

$$\hat{\Phi}(x,y;B) = P\exp\, ig \int_y^x \hat{B}_\mu dz_\mu \tag{6.10}$$

and consider both gluon and ghost propagating in the field B_μ with propagators $G^{ab}_{\mu\nu}(x,y,B)$ and $G^{ab}(xy;B)$, respectively. The product

$$H(x,y) \equiv < \hat{\Phi}^{ef}(x,y,B)\Gamma^{eac}_\mu G^{ab}_{\mu\nu}(x,y,B)G^{cd}(x,y,B)\Gamma^{fbd}_\nu >_B \tag{6.11}$$

depicted in Fig. 7, yields the one-loop correction to the renormalized charge, which, in the absence of the field B_μ reduces to the second term on the r.h.s. of (6.3). Using the results of the next section one can again prove that a dependence like $\ln|x-y|$ disappears in (6.11) when $|x-y| \gg m^{-1}, m > 1\text{GeV}$. This will be clear from the discussion in the next two subsections.

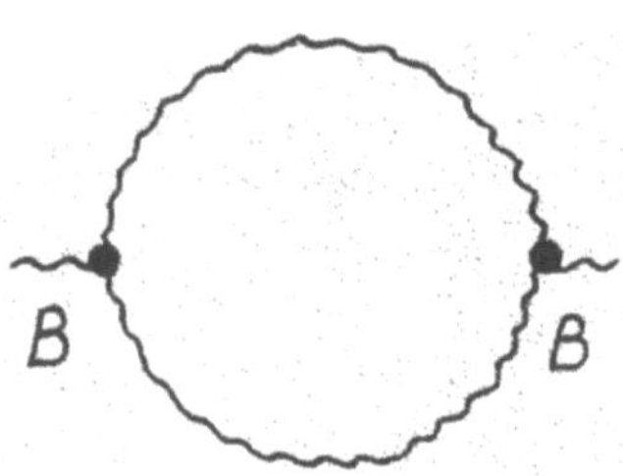

Fig. 5. Gluon loop with external lines corresponding to the background field B.

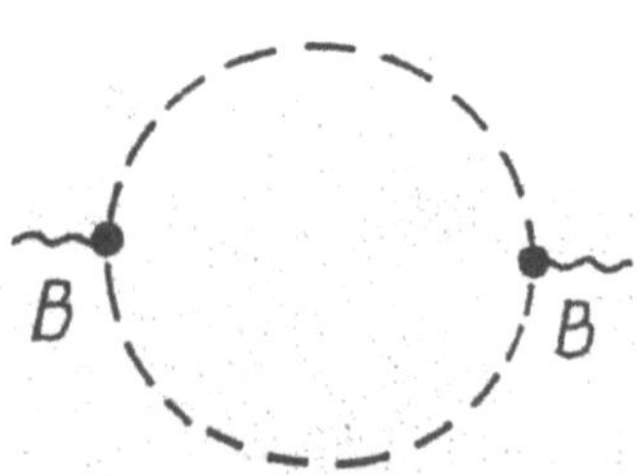

Fig. 6. The same as in Fig. 5, but for the ghost loop.

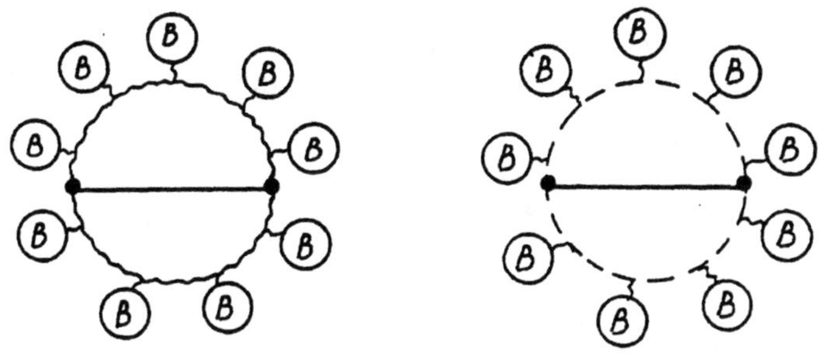

Fig. 7. The nonperturbative generalization of the diagrams of Figs. 5 and 6 with full account of the background in the propagators of the gluon (a) and the ghost (b).

In a similar way one calculates two- and more loop corrections, using the same two-point diagrams as in [42].

iii) Calculation of $\alpha_s(R)$ through the Wilson loop formalism

In this section we shall follow the way outlined in the first definition of $\alpha_s(R)$, i.e., we shall use $E(R)$ given by (6.5). To this end we need the PTH expansion of the Wilson loop $< W(B+a) >$, i.e., the expansion in powers of the gluon operator a_μ. We start with the Wilson contour C of an arbitrary configuration and divide the contour into N pieces $\Delta x_\mu(n)$, writing identically

$$\begin{aligned} W\ (B+a) &\equiv \frac{1}{N_c}\mathrm{tr}P\exp ig\int_C (B_\mu + a_\mu)dx_\mu \\ &= \frac{1}{N_c}\mathrm{tr}P\prod_{n=1}^{N}[1+ig(B_\mu(x(n))+a_\mu(x(n)))\Delta x_\mu(n)] \qquad (6.12) \\ &= \frac{1}{N_c}\mathrm{tr}P\left\{\prod_{n=1}^{N}(1+igB_\mu\Delta x_\mu)+\prod_{n=1}^{k-1}(1+igB_\mu\Delta x_\mu)iga_\mu(k)\Delta x_\mu(k)\ \times \right. \\ &\left. \times \prod_{n=k+1}^{N}(1+igB_\mu\Delta x_\mu(n))+\ldots\right\}. \end{aligned}$$

One can represent the sum (6.12) as a sum of Wilson loops $W^{(n)}(B)$ with n insertions of the field $iga_\mu(k)$ on the contour C:

$$W(B+a) = W(B) + \sum_{n=1}^{\infty}(ig)^n W^{(n)}(B;X(1)\ldots X(n))dx_{\mu_1}(1)\ldots dx_{\mu_n}(n). \qquad (6.13)$$

Now we must integrate over the fields Da_μ with the weight given in (4.6-4.7). In lowest order $< W(B) >_a = W(B)$ for the properly normalized weight, and in order $O(g^2)$ one has a one-"gluon" exchange diagram shown in Fig. 8

$$-g^2 < W^{(2)} >_a dxdy = \tag{6.14}$$
$$- g^2 \int \Phi^{\alpha\beta}(x,y;B) G^{\delta\alpha;\beta\gamma}_{\mu\nu}(x,y;B) \Phi^{\gamma\delta}(y,x,B) dx_\mu dy_\nu ,$$

where $\Phi^{\alpha\beta}$ are the parallel transporters (6.10) with fundamental indices written explicitly, $\alpha, \beta = 1, ... N_c$,

$$G^{\delta\alpha,\beta\gamma}_{\mu\nu} = t_a^{\delta\alpha} t_b^{\beta\gamma} G^{ab}_{\mu\nu} \ , \tag{6.15}$$

and $G^{ab}_{\mu\nu}$ is the gluon propagator in the NP background field B_μ given in (4.12) and (4.13).

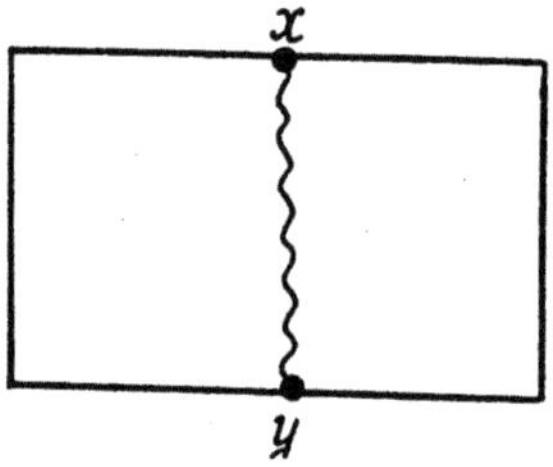

Fig. 8. The one–gluon exchange in the Wilson loop between the points x and y.

It is appropriate at this point to specify the notion of the gluon in the background and to distinguish it from the free gluon, interacting only perturbatively. We propose to call it the "hybrid-gluon", or h-gluon, since, as we shall see, in most cases it will acquire a mass and behave as a hybrid.

In next order, $0(g^4)$, one obtains the five diagrams shown in Figs. 9-13.

The next step is to average over the ensemble of NP fields $\{B_\mu\}$. It is not necessary to specify the ensemble and prescribe the measure; we can use the cluster expansion to obtain for the simplest term [18]:

$$< W(B) >_B = \exp \sum_{n=1}^{\infty} \frac{(ig)^n}{n!} \int d\sigma(1) ... d\sigma(n) < F(1) ... F(n) > . \tag{6.16}$$

The exact gauge invariant form of the correlator $< F(1) ... F(n) >$ is discussed in [18]. The lowest order term in (6.16), e.g., has two independent structures: the Kronecker one and the full derivative term [18]

$$g^2 < F_{\mu\nu}(u) \Phi(u,v) F_{\rho\sigma}(v) \Phi(v,u) > = (\delta_{\mu\rho}\delta_{\nu\sigma} - \delta_{\mu\sigma}\delta_{\nu\rho}) D(u-v) + \partial(D_1) . \tag{6.17}$$

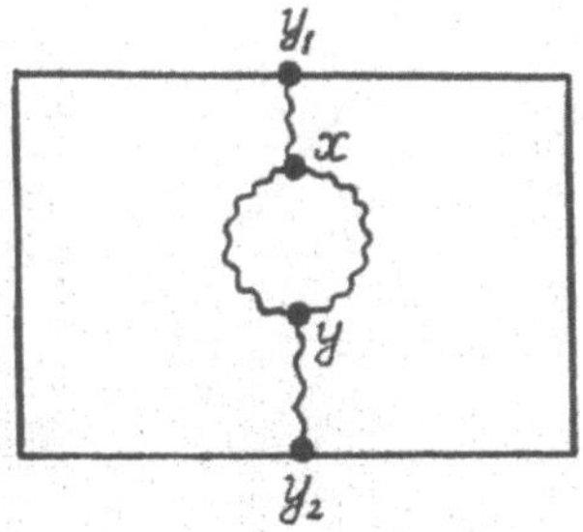

Fig. 9. The gluon–loop diagram for the gluon exchange inside the Wilson loop.

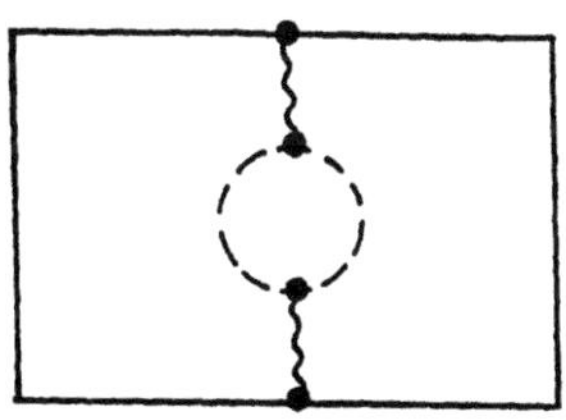

Fig. 10. The same as in Fig. 9 for the ghost–loop diagram.

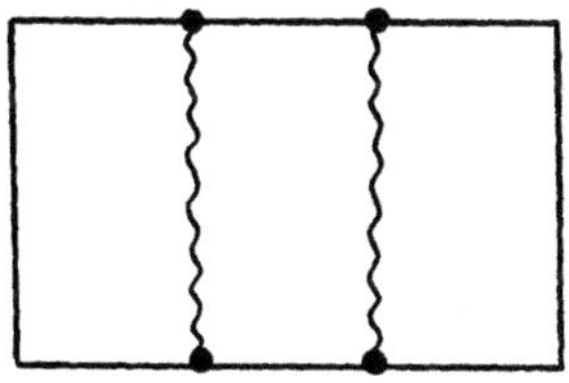

Fig. 11. Two–gluon exchange diagram from the iteration of one–gluon–exchange.

It is D and the Kronecker structures of higher correlators that bring about the area law of the Wilson loop [18]:

$$< W(B) >_B \cong \exp(-\sigma S_{\min}) \quad , \quad S \quad \text{large}, \tag{6.18}$$

where

$$\sigma = \frac{1}{2} \int \int_{-\infty}^{\infty} d^2 u D(|u|) + \int < FFFF > + \dots . \tag{6.19}$$

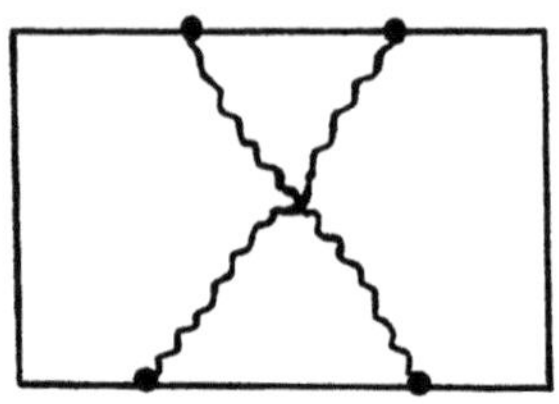

Fig. 12. Two–gluon–exchange crossed diagram, containing both the iteration of one–gluon exchange (abelian part) and the nonabelian part.

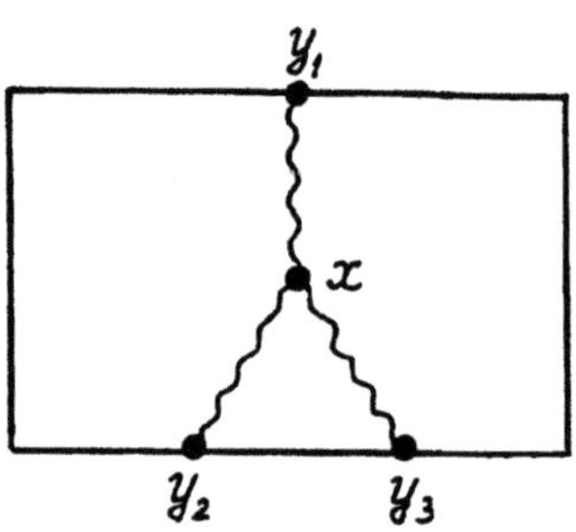

Fig. 13. Gluon exchange with the three–gluon vertex in the Wilson loop.

Thus any finite number of terms in the sum (6.16) or the whole sum, if it converges, yields the area law for the Wilson loop. In what follows we shall assume that the area law with $\sigma \cong \sigma_{\text{phen}}$ holds for all Wilson loops provided their size (both width and length) is large [18], i.e., for the rectangle $R \times T$ in Fig. 4,

$$R, T \gg T_g\,, \tag{6.20}$$

where T_g is the gluonic correlation length [20], defined by the range of correlators $D(u)$ etc.

We stress at this point that the assumption (6.18) is very natural from the point of view of the cluster expansion [18], and moreover it is supported by the lattice calculations [13].

With the help of (6.16) one easily obtains the first term in (6.1); using [23] one has

$$E_{\rm NP}(R) = 2R\int_0^R d\lambda \int_0^\infty d\nu D(\lambda,\nu) + \int_0^R \lambda d\lambda \int_0^\infty d\nu[-2D(\lambda,\nu)+D_1(\lambda,\nu)] \tag{6.21}$$

$$+\text{higher correlators.}$$

We note that at small R, $R < T_g$, [47,23]

$$E_{\rm NP}(R) \sim CR^2 + 0(R^4)\,, \tag{6.22}$$

where $0(R^4)$ comes from higher order correlators, while at large R [23]

$$E_{\rm NP}(R) = \sigma R - C_0\,. \tag{6.23}$$

We refer the reader to [21] for a detailed discussion.

Now we turn to the h-gluon exchange diagrams of Fig. 8-13. First we must regularize them, following the method of [60]. It was shown there that all divergencies of smooth Wilson loops are regularized through the introduction of some minimal distance δ on the contour C or by dimensional regularization and renormalization by introduction of a common Z factor (heavy mass renormalization) and charge renormalization, generated by the divergencies in the diagrams of Figs. 9,10,12,13.

We shall follow the same procedure as in [60], but we are interested not only in the divergent parts of diagrams as in [60], but also in the large R dependence with and without background field B_μ.

We start with the simplest h-gluon exchange diagram, Fig. 8, given by (6.14). The first step is to use for $G_{\mu\nu}$ the Feynman-Schwinger representation (FSR) [23,45]

$$G^{ab}_{\mu\nu}(x,y;B) = \int_0^\infty ds e^{-K}(Dz)_{xy}\hat{\Phi}_F(x,y;B)^{ab}_{\mu\nu}\,, \tag{6.24}$$

where

$$K = \frac{1}{4}\int_0^S \dot{z}^2_\mu(\lambda)d(\lambda)\,,$$

$$\Phi_F(x,y,B) = P_F P \exp\left(ig\int_y^x \hat{B}_\mu dz_\mu\right)\exp\left(2\,ig\int_0^S \hat{F}_{\lambda\rho}(z(\lambda))d\lambda\right), \tag{6.25}$$

and Φ_F differs from the usual parallel transporter by the ordered (due to P_F) insertions of the gluon color magnetic moment interaction $2g\hat{F}_{\lambda\rho}$. The use of the FSR is the central point of the method since it enables one to write explicitly the dependence on B_μ - it enters via Φ_F. As a next step we must average (6.14) over the ensemble $\{B_\mu\}$, i.e., compute an average of an expression of the form

$$< W^{(2)} >_B \sim < \Phi t\hat{\Phi}_F t\Phi >_B\,, \tag{6.26}$$

where $\hat{\Phi}_F$ is in the adjoint representation, while Φ is in the fundamental one. To understand better the structure of (6.26) let us take the limit $N_c \to \infty$ and represent the adjoint line in $\hat{\Phi}_F$ by the double fundamental line, as suggested by 'tHooft [64]. We obtain two adjacent closed contours C_1 and C_2 as shown in Fig. 14. When both contours are large, i.e., for $R, T \gg T_g$, and $N_c \to \infty$ one has

$$< W^{(2)} >_B \sim < W_{C_1}(B) >_B < W_{C_2} >_B + 0(\frac{1}{N_c^2}) \sim \exp(-\sigma(S_1 + S_2)), \tag{6.27}$$

where S_1, S_2 are minimal areas inside the contours C_1, C_2.

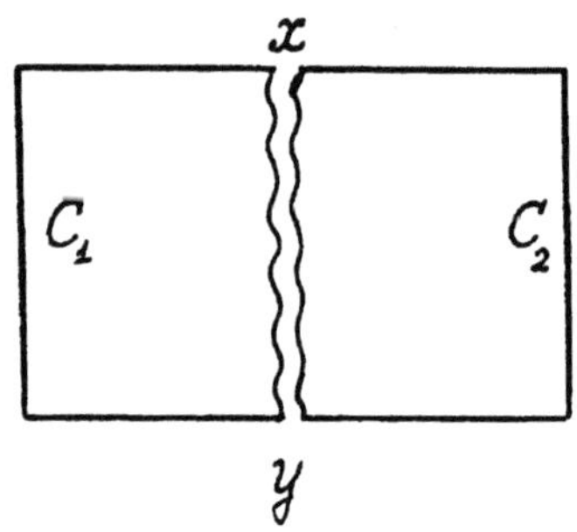

Fig. 14. At large N_c the one–gluon–exchange diagram of Fig. 8 goes over into the product of two Wilson loops.

Combining now (6.14),(6.24) and (6.27) we obtain

$$-g^2 < W^{(2)} >_B = -g^2 \int_0^\infty ds (Dz)_{xy} dx_\mu dy_\nu \exp(-K - \sigma(S_1 + S_2))\delta_{\mu\nu} \,. \tag{6.28}$$

(Note that we have omitted for simplicity the $\hat{F}_{\lambda\rho}$ term in (6.25), which brings about spin-orbit corrections.)

Equation (6.28) describes the propagation of a scalar particle interacting with the contour via two strings, the world surface of which is given by S_1, S_2.

We now show that for our choice of the contour in Figs. 8,14, with the horizontal size T much larger than the vertical size R, the resulting interaction tends to zero. Indeed for a given point $z(\lambda)$ on the line of the propagator, see Figs. 8,14, the interaction does not depend on $z_4(\lambda)$ and $z_1(\lambda)$ which are in the plane of the contour $R \times T$, but rather depends only on the transverse coordinates z_2, z_3. The interaction at the "time" λ can be written as

$$V \equiv \sigma(\sqrt{z_2^2 + z_3^2 + z_4^2} + \sqrt{z_2^2 + z_3^2 + (T - z_4)^2}) - \sigma T \,, \tag{6.29}$$

where we have normalized to the case with no h-gluon exchange. It is easy to see that for a generic case $T - z_4, z_4 \gg z_2, z_3$ (according to (6.51) we must finally take the limit $T \to \infty$) one has

$$V \approx \frac{\sigma}{2}\left(\frac{z_2^2 + z_3^2}{|z_4|} + \frac{z_2^2 + z_3^2}{|T - z_4|}\right) \to 0 . \tag{6.30}$$

Thus we come back to the case of a free gluon exchange and obtain from (6.28)

$$-g^2 \frac{< W^{(2)} >_B}{< W >_B} = -\frac{g^2 C_2(f)}{4\pi^2} \int\int \frac{dx_4 dy_4}{(x_4 - y_4)^2 + R^2} = \frac{g^2 C_2(f)}{4\pi R} . \tag{6.31}$$

In a similar way we treat the diagrams with more gluon exchanges, like in Figs. 11,15 and exponentiate the result (6.31), obtaining the lowest order term $(-\frac{4}{3}\frac{\bar{\alpha}_s(R)}{R})$ in (6.8). One should mention that in multiple gluon exchanges, like Fig. 15, there appear in general correlated gluon pairs, triplets etc. with an average distance between gluons of the order of $(\sigma)^{-1/2}$. These correlated exchanges are not taken into account during exponentiation and should be accounted for separately, yielding a Yukawa-type correction to the perturbative part of $E(R)$ starting with the order $O(g^4)$. We shall discuss these corrections in a separate publication, they do not change the results of this paper concerning the large distance logarithmic behaviour.

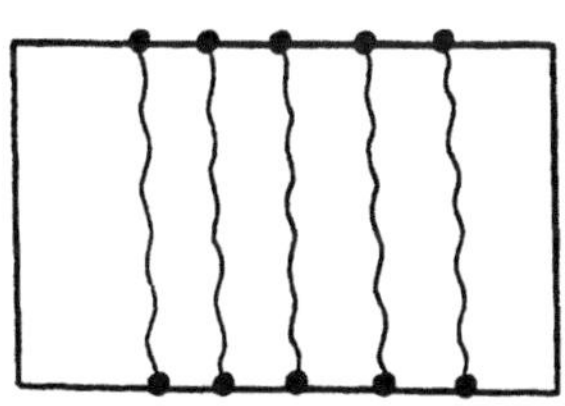

Fig. 15. Multiple iterative gluon exchanges.

Now we come to the fourth order diagrams, needed to renormalize α_s, shown in Figs. 9,10,12,13

First we discuss the structure of divergencies and logarithms in case of no background B_μ. We can, e.g., write the resulting integral for the diagram of Fig. 13

$$\left(\frac{f(R)}{R}\right)_{\text{Fig. 13}} = \frac{-3(1-\xi)}{16\pi^2} C_2 \int \frac{d^3x}{|\mathbf{x}||\bar{x} - R|^3} = \frac{-3(1-\xi)}{4\pi}\frac{C_2}{R} \ln \frac{Re}{\delta} . \tag{6.32}$$

In dimensional regularization one would have

$$\ln \frac{R}{\delta} \to \frac{C}{\varepsilon} + \ln \mu R + \text{const}. \tag{6.33}$$

For Figs. 9 and 10 one obtains (again $B_\mu = 0$)

$$\left(\frac{f(R)}{R}\right)_{\text{Figs. 9,10}} = (13-3\xi)C_2 \sim \frac{d^4r}{|\bar{R}-\bar{r}|}\Pi(r^2) \to \frac{13-3\xi}{12\pi}C_2 \ln \frac{R}{\delta}, \tag{6.34}$$

where $\Pi(r^2)$ is the trasverse self-energy part of gluon and ghost in coordinate space:

$$\Pi(r^2) = \frac{1}{(\bar{r}^2 + r_4^2)^2}. \tag{6.35}$$

Integrating over d^4r we obtain the same logarithm as in (6.32).

Some more care is needed for the graph of Fig. 12; we mention only the final result

$$\left(\frac{f(R)}{R}\right)_{\text{Fig. 12}} = \frac{(3-\xi)C_2}{2\pi R} \ln \frac{R}{\delta}. \tag{6.36}$$

Summing (6.32), (6.34), and (6.36) together we obtain

$$f(R) = \frac{b_0}{2\pi} \ln \frac{R}{\delta} \tag{6.37}$$

in agreement with (6.3) and (6.4).

Consider now the case of nonzero B_μ, and let us average each of the diagrams $W^{(4)}(B)$ in Figs. 9, 10, 12, 13 over $\{B_\mu\}$. We again use the limit $N_c \to \infty$ and replace each gluon line by a double fundamental line. Choosing the gauge fixing constant $\xi = 3$, we eliminate the logarithmic part (and divergence) of Fig. 12 and disregard it in what follows, since now we are only interested in the question: what happens with the logarithms in (6.32), (6.34), and (6.36) when the NP confining field is included.

Following the same line of reasoning as we used for the diagram of Fig. 14 and imposing the confining field B_μ, we obtain for the diagrams of Figs. 9, 10 and that of Fig. 13 three surfaces, depicted in Figs. 16 and 17, respectively, filled with a confining film due to the area law. As in (6.28), we must consider again the horizontal size T of the diagrams in Figs. 13,14 large, and as a result the surfaces S_1 and S_2 do not produce any force on the gluon line when $T \to \infty$. This is not true, however, for S_3 , which provides a strong confining force between the fundamental lines which border S_3.

As a result we must consider the behaviour of the self-energy part $\Pi(x,y)$ in Fig. 16, defined by the FS representation:

$$\Pi(x,y) \sim \int_0^\infty ds \int_0^\infty d\bar{s} Dz D\bar{z} \exp(-K - \bar{K} - \sigma S_3). \tag{6.38}$$

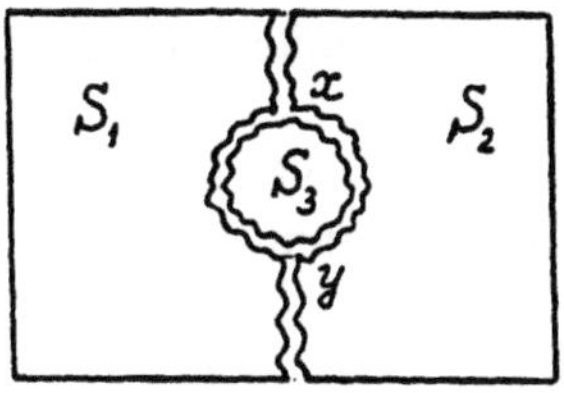

Fig. 16. At large N_c the diagrams of Figs. 9 and 10 contain three Wilson loops with areas S_1, S_2 and S_3.

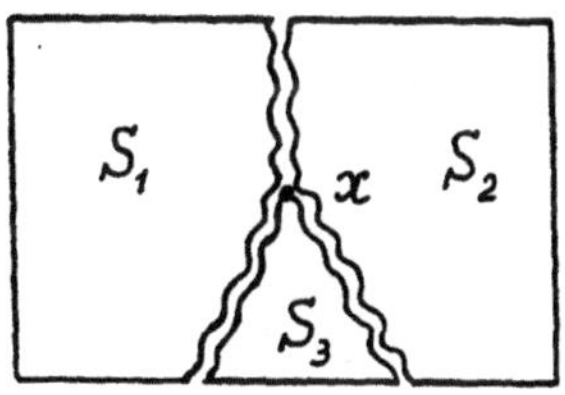

Fig. 17. At large N_c the diagram of Fig. 13 contains three Wilson loops with areas S_1, S_2, S_3.

It is easy to show that $\Pi(x,y)$ behaves as $\frac{1}{(x-y)^4}$ at small $|x-y|$, $|x-y| \ll T_g$, and has an exponential fall-off at large $|x-y|$

$$\Pi(x,y) \sim \exp(-m_1|x-y|), \quad |x-y| \to \infty . \tag{6.39}$$

Introducing this cut-off behaviour in (6.35) one obtains the following integral which replaces the original logarithmic expression in (6.34)

$$\frac{1}{R}\ln\frac{R}{\delta} \to I(R) \equiv \int_\delta^\infty \frac{dr}{r} e^{-m_1 r}\left(\frac{1}{R}\theta(R-r) + \frac{1}{r}\theta(r-R)\right). \tag{6.40}$$

There are two regimes of behaviour of $I(R)$: the region of asympthotic freedom,

$$m_1 R \ll 1, \quad I(R) \approx \frac{1}{R}\ln\frac{Re}{\delta}, \tag{6.41}$$

and the infrared region,

$$m_1 R \gg 1, \quad I(R) \approx \frac{1}{R}\int_\delta^R \frac{dr}{r} e^{-m_1 r} \approx \frac{1}{R}\ln\frac{1}{m_1\delta}. \tag{6.42}$$

The same situation occurs for the diagram of Fig. 17. Here the cut-off function $\exp(-m|\mathbf{R}-\mathbf{x}|)$ appears under the integral in (6.32), producing the same integral $I(R)$ as in (6.40).

As a final result of this section we formulate the new expression for $f(R)$ taking the NP confining into account (modulo constant (nonlogarithmic) terms)

$$\frac{1}{R} f_{\mathrm{NP}}(R) = \frac{b_0}{2\pi} I(R) \tag{6.43}$$

with $I(R)$ defined in (6.40).

Let us look more closely into the object corresponding to the mass m_1. To visualize it, we make a horizontal cross section of the Wilson loop with interval lines in Fig. 16 and 17 in such a way that we cross the surface S_3 – the crossing is shown as a broken line in Figs. 16 and 17. It is clear from Fig. 16 that the state corresponding to this cross section consists of infinitely heavy quarks on both ends of the fundamental string with the distance between quarks equal to T and two gluons "sitting" on the string with a piece of string from the surface S_3 between them. We are interested in the mass of this system when $T \to \infty$, i.e., we can associate the mass m_1 in (6.39) with this two–gluon hybrid state mass with the ends of the string fixed and tending to infinity,

$$m_1 = \lim_{T\to\infty} (M_{2g}(T) - \sigma T) \, . \tag{6.44}$$

The subtracted term σT corresponds to the definition in (6.13).

The information about the magnitude of m_1 can be obtained from three sources:

i) from the lattice calculations (see, e.g., the one–gluon excitations of the string with fixed ends measured in [52]); ii) from the string spectrum for strings with fixed ends, which was studied, e.g., in [53]; iii) from the Feynman–Schwinger representation of the diagram of Fig. 16, see (6.38).

Let us start with point iii). Applying to (6.38) the technique developed in [54] and neglecting the gluon spin interaction, one can find the relativistic 3d Hamiltonian for the hybrid system of Fig. 16 of the form

$$H = \sqrt{\mathbf{p}_1^2} + \sqrt{\mathbf{p}_2^2} + \sigma\{|\mathbf{r}_{21}| + r_{20} + r_{10} - |t_2 - t_1|\}, \tag{6.45}$$

$$\mathbf{r}_{21}^2 = (t_2 - t_1)^2 + (y_2 - y_1)^2 + z_2 - z_1)^2,$$

$$r_{20} = (T - t_2)^2 + y_2^2 + z_2^2,$$

$$r_{10}^2 = t_1^2 + y_1^2 + z_1^2 \, .$$

Here we consider x as the evolution parameter and H is the corresponding Hamiltonian. A glance at the Hamiltonian (6.45) leads to the immediate conclusion that the lowest mass configuration corresponds to large average values of $|T - t_2| \sim \frac{T}{3}, |t_1| \sim \frac{T}{3}, |H_2 - t_1| \sim \frac{T}{3}$. When $T \to \infty$ the average values of y_i^2, z_i^2 are of the order of σ^{-1}, hence expanding r_{21}, r_{20} and r_{10} in powers of $\frac{y_i^2+z_i^2}{T}$ one obtains an estimate

$$M_{2g}(T) = \sigma T + 0(\frac{1}{T}) . \tag{6.46}$$

A more explicit evaluation can be obtained from point ii) – the string spectrum.

Following [52-53] one can write the following spectrum of the string of length T with fixed ends

$$V_N = [\sigma^2 T^2 - \frac{\pi\sigma}{6} + 2\pi N\sigma]^{1/2}, \quad N = 0, 1, 2, 3, \dots . \tag{6.47}$$

For large T one can expand (6.47) to obtain

$$V_N \simeq \sigma T + \frac{\pi(N - \frac{1}{12})}{T} . \tag{6.48}$$

At this point we can make contact with the lattice calculations in [52], which generally agree with the spectrum (6.47) and its low-N approximation (6.48). In our specific case, corresponding to the diagrams in Figs. 16 and 17, we need the lowest two–gluon excitation of the string, i.e. the case $N = 2$ in (6.47) and (6.48).

Correspondingly we obtain

$$m_1 = m_1(T) \simeq \frac{2\pi}{T} . \tag{6.49}$$

Hence for the finite time extension T of the Wilson loop, the mass $m_1(T)$ is finite, e.g., for $T = 6a$ and $a \approx 0.2$fm one obtains $m_1(6a) \approx 1$GeV.

iv) Renormalization of α_s in the definition ii)

In the second nonperturbative definition of α_s through the two-point function $H(x, y)$ (6.11), depicted in Fig. 7, we can again exploit the $N_c \to \infty$ limit. In this case each adjoint line, including $\hat{\Phi}(x, y, B)$, is represented as double fundamental line [63], and each closed fundamental contour $W_c(B)$, after averaging over the fields $\{B_\mu\}$, acquires the area-law factor $\exp(-\sigma S_{\min}(C))$ – in agreement with our assumption (6.18). Thus each of the diagrams of Fig. 7 can be depicted as in Fig. 18 , the difference between (a) and (b) is that for the gluon Green's function there are in addition gluon spin operators inserted on the gluon line, leading to spin-dependent effects, which we disregard in the first approximation.

The diagrams of Fig. 7 can be written in a form similar to (6.38),

$$H_{\mu\nu}(x, y) = \frac{g^2}{(4\pi)^2} b_0 (\partial_\mu \partial_\nu - \partial^2 \delta_{\mu\nu}) \bar{\Pi}(x, y) , \tag{6.50}$$

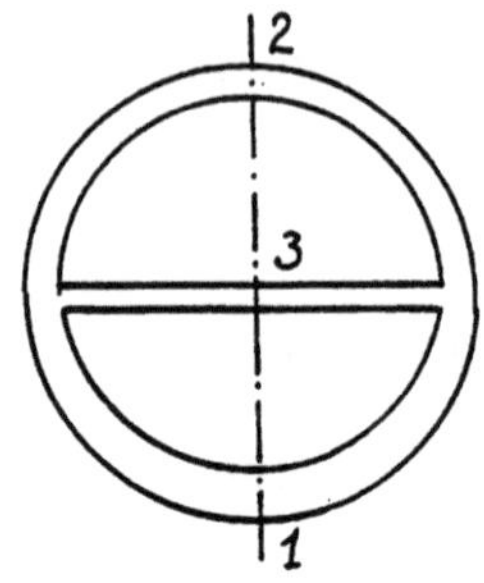

Fig. 18. The diagrams of Fig. 7 contain three Wilson loops at large N_c.

where $\bar{\Pi}$ is defined as

$$\bar{\Pi}(x,y) = \int_0^\infty ds \int_0^\infty d\bar{s}\, e^{-K-\bar{K}} (Dz)_{xy} (D\bar{z})_{xy}\, e^{-\bar{V}}, \tag{6.51}$$

with

$$\bar{V} = S^{(1)}_{\min} + S^{(2)}_{\min} + S^{(3)}_{\min}, \tag{6.52}$$

and $S^{(i)}_{\min}$ correspond to the minimal areas inside the three contours depicted in Fig. 18.

We have omitted in (6.52) for simplicity the gluon spin terms in the Green's function G in (6.24).

At small distances $|x-y|$ one can obtain that

$$\bar{\Pi}(x,y) \sim \frac{1}{(x-y)^{2d-4}}. \tag{6.53}$$

This leads to the familiar answer for the Fourier-transform of $H_{\mu\nu}$ [42]

$$H_{\mu\nu}(p) = \frac{g^2 b_0}{(4\pi)^2} (p_\mu p_\nu - p^2 \delta_{\mu\nu}) (N_\varepsilon - \ln \frac{p^2}{\mu^2} + \text{const}) \tag{6.54}$$

with

$$b_0 = \frac{11}{3} N_\varepsilon, \quad N_\varepsilon = \frac{1}{\varepsilon} - C + \ln 4\pi,$$

C is the Euler constant, $C = 0.577$.

At large $\mid x-y \mid$ one must take into account the confining dynamics present in $\bar{V}$ and find the corresponding mass. To this end one must look at the expressions (6.49), (6.51) more carefully.

To understand the structure of $H(x,y)$, one can dissect the graph in Fig. 18 along the dash-dotted line, and one obtains a system of two gluons 1 and 2 connected by a pair of strings, as shown in Fig. 19. One of the strings

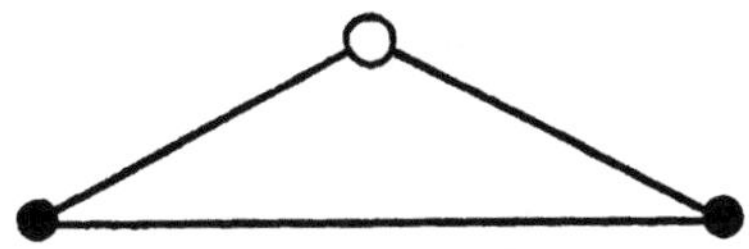

Fig. 19. Cutting the diagram of Fig. 18 along the dash–dotted line line reveals two gluons 1 and 2, connected by two fundamental strings.

goes through the fixed point 3 which appears due to the dissection of the straight-line parallel transporter $\Phi(x, y; B)$ – see Fig. 18.

One can show that the mass of the system of two gluons with two strings in the configuration of Fig. 19 is larger than the corresponding mass without the additional condition of the string passing through the point 3:

$$m(1,2;3) > m(1,2) \equiv m_2\,. \tag{6.55}$$

Thus we come to the conclusion that H describes the Green's function of a two-gluon glueball with a quantum number $J^{PC} = 1^{-+}$ (and with an additional "heavy gluon" propagating along the straight line of $\hat{\Phi}(x, y, B)$, if we keep the point 3 in Fig. 19, and without it if we calculate a lighter system according to the inequality (6.55)).

An estimate of $m(1,2)$ with $J^{PC} = 1^{-+}$ has been done within the proper time Hamiltonian in [64]

$$m(1,2) \cong 2\text{GeV} \pm 0.15\text{GeV}\,. \tag{6.56}$$

A similar estimate has been obtained in the lattice calculations, where usually two neighbouring members of the same multiplet $L = 1, S = 1$ are studied [65].

Using the arguments given in [30] and in the next section, one can approximate $\bar{\Pi}(x, y)$ as

$$\bar{\Pi}(x,y) \simeq C_0 \frac{\exp(-m_2 \mid x-y \mid)}{(x-y)^{2D-4}}\,, \tag{6.57}$$

which after renormalization yields the Fourier-transform

$$\bar{\Pi}(p) \approx C_0 \ln \frac{m_2^2 + p^2}{\mu^2} + \text{const}\,. \tag{6.58}$$

A comparison of (6.54) and (6.58) shows that up to power corrections the two expressions have the same μ dependence, and hence the same Z factors. Actually all the changes from (6.55) to (6.58) can be described as the introduction of a new (fifth) external momentum component $p_5^2 \equiv m_2^2$, so that the total momentum squared is equal to $p^2+m_2^2$ in case of a nonperturbative background. This fact does not change the Gell-Mann-Low equations which to one-loop order are [5]

$$\frac{dg}{d\ln\mu} = -b_0\frac{g^3}{16\pi^3}\,. \tag{6.59}$$

Solving (6.59) with boundary conditions as in (6.50) and (6.58) (neglecting power terms as before for large $(m^2+p^2)x_0^2 \gg 1$) one obtains

$$\alpha_s(p) = \frac{\alpha_s(\mu)}{1+\frac{b_0}{4\pi}\alpha_s(\mu)\ln\frac{m_2^2+p^2}{\mu^2}}\,. \tag{6.60}$$

Introducing the Λ parameter, (6.60) is rewritten as

$$\alpha_s(p) = \frac{4\pi}{b_0\ln\frac{m_2^2+p^2}{\Lambda_{\mathrm{NP}}^2}}\,. \tag{6.61}$$

We have used the notation Λ_{NP} to stress that this new Λ parameter may differ from the free vacuum Λ_{QCD} and needs a new determination from the experimental data. We stress here again that the appearence of m_2^2 together with p^2 is possible because of renormalization invariance of gB_μ [42], which in turn yields invariance of σ and $m_2 \cong \sqrt{\sigma}$.

At large p one obtains the familiar AF logarithm in (6.61). This logarithm is still large for $p=0$ if one chooses a low normalization parameter Λ_{NP}, $\Lambda_{\mathrm{NP}} \sim \Lambda_{\mathrm{QCD}}$. Thus the AF phenomenon can be maintained down to very low p in the Euclidean region. This is similar to the situation which we have observed in the Wilson-loop renormalization scheme, (6.42).

There, however, the mass m_1 corresponds to the one-Wilson-loop dynamics, while in our case actually a double film defines the dynamics for $m(1,2)$, with $\sigma_a \cong 2\sigma$, so that the mass $m_2 \cong \sqrt{2}m_1$. This fact demonstrates that the IR cutoff is scheme dependent. (Note that our definition of the renormalization scheme differs from the usual one, and the scheme dependence actually means the dependence on the physical process in question).

7 Determination of α_s via the Hadronic Ratio R

As we stated in Sect. 6, to define the renormalized quantity one needs, according to the common lore, first to choose the scheme and second to set the scale. The schemes mostly used in PQCD are $\overline{\mathrm{MS}}$ or MS, connected to the dimensional regularization, have no immediate physical meaning and do not correspond to a definite physical process. In principle all physical observables should be independent of the choice of the renormalization scheme and scale. However one is always dealing with a truncated perturbation series, so the scheme and scale dependence is intrinsic in PQCD and limits its accuracy and usefulness.

This problem becomes even more important in QCD when one enters the large distance region, since on one hand α_s is growing there, and on the other hand, the NP contribution intervenes and interferes with the perturbative series. In this region it is imperative from the very beginning to define the renormalized α_s in a gauge–invariant and renormalization–group invariant way, and this can be done through a given physical process. An example of this definition was presented in the previous section, when α_s, according to (6.1), is expressed through the (in principle physically measurable) energy of static quarks. This replacement of the scheme by the process is not new [55], and actually the study of the process–dependence instead of the scheme-dependence was done previously [56].

In the case of large α_s and NP environment, one has to deal with the process instead of the scheme, and, as we shall see, the process dependence in the determination of α_s will include, in addition to the usual perturbative one like that discussed in [56], also the NP process dependence entering into the coefficients and scales of the perturbative series.

As an example of this "process–renormalization" let us consider the perturbative series for the process $e^+e^- \to$ hadrons. To this end we consider the correlation function of electromagnetic currents

$$-i\int d^4x e^{iqx} < 0|T(j_\mu(x)j_\nu(0))|0 >= (q_\mu q_\nu - g_{\mu\nu}q^2)\Pi(q^2)\,. \tag{7.1}$$

The imaginary part of Π is related to the total hadronic cross section in e^+e^- annihilation by

$$R(q^2) = \frac{\sigma(e^+e^- \to \text{hadrons})}{\sigma(e^+e^- \to \mu^+\mu^-)} = 12\pi\Im\Pi(\frac{q^2}{\mu^2}, \alpha_{\rm QCD}(\mu))\,. \tag{7.2}$$

In (7.1) we have used the current j_μ,

$$j_\mu \equiv \sum_q Q_q : \bar{q}\gamma^\mu q : . \tag{7.3}$$

To lowest order of α_s one obtains from (7.1) the famous partonic result:

$$R^{(0)}(s) = 3\sum_{f=1}^{n_f} Q_f^2, \quad s = -Q^2 = +q^2\,. \tag{7.4}$$

The usual treatment of $\Pi(q^2)$ is based on perturbative calculations in the Euclidean region, i.e. for spacelike q^2, where $-q^2 = Q^2 > 0$.

It is convenient to introduce a real quantity $D(Q^2)$ free from the trivial divergence (which is regularized by the usual QED renormalization of $\Pi(q^2)$)

$$D(Q) = -Q^2\frac{\partial}{\partial q^2}\Pi(Q^2)\,. \tag{7.5}$$

Then the physical quantity $R(s)$ is obtained by a continuation to the Minkowskian region $Q^2 < 0$ and taking the imaginary part.

We shall consider here three approaches to the perturbative calculation of $\Pi(Q^2)$ which differ in their treatment of the NP contribution:

1) The purely perturbative approach, which can be also called the parton model approach. In this case one disregards completely all NP contributions and claims that they (if any) can be reestablished by the proper summation of the perturbative series.

By now $\Pi(Q^2)$ is known to $0(\alpha_s^3)$, and, e.g., one can write for $R(s)$ in the $\overline{\text{MS}}$ scheme [57]

$$R^{(3)}(s) = N_c \sum_{f=1}^{nf} Q_f^2 \{1 + \frac{\alpha_s(s)}{\pi} + r_2 (\frac{\alpha_s(s)}{\pi})^2 + r_3 (\frac{\alpha_s(s)}{\pi})^3\}, \tag{7.6}$$

where

$$r_2 \simeq 2.0 - 0.12 n_f ,$$

$$r_3 = 18.243 - 4.216 n_f + 0.086 n_f^2 - 1.240 \frac{(\sum Q_f)^2}{3 \sum Q_f^2} + r_3^{\text{an}} , \tag{7.7}$$

and the last term comes from the analytic continuation of $\alpha_s(s)$

$$r_3^{\text{an}} = -(\frac{121}{48} - \frac{11}{36} n_f + \frac{1}{108} n_f^2) \pi^2 . \tag{7.8}$$

It is important that in the limit of large N_c one has [57,5]

$$r_2 \simeq 0.28 N_c^2, \quad r_3 \simeq 1.05 N_c^3 . \tag{7.9}$$

Consequently, all terms in the expansion (7.6) yield nonvanishing and comparable contributions in the limit $N_c \to \infty$, which is clear from the fact that all planar diagrams without fermion loops give contributions to $R(s)$ of the same order $0(N_c)$.

2) In the second type of approach one admits nonperturbative contributions in the form of coefficients in the OPE [58,14], namely for light quarks one has an expansion in a series of operators of growing dimension d

$$\Pi(Q^2) = \sum_n C_n^{AB}(Q) < 0_n > = C_I I + C_G < 0_G > + C_M < 0_M > + \tag{7.10}$$

$$+ C_\sigma < 0_\sigma > + C_f < 0_f > + \ldots .$$

Here I is the unit operator, $< 0_M >$ includes the quark condensate, $< 0_G >$ the gluonic condensate, etc. For a detailed discussion the reader is referred to [14].

In particular, taking two light quarks of equal masses, $m_{\mathrm{u}} = m_{\mathrm{d}}$, one obtains the expansion [14]

$$\Pi(Q^2) = -\frac{1}{4\pi^2}(1+\frac{\alpha_s}{\pi}) \ln \frac{Q^2}{\mu^2} + \frac{6m^2}{Q^2} + \frac{2}{Q^4} m < \bar{q}q > + \frac{\alpha_s}{12\pi Q^4} < G^a_{\mu\nu} G^a_{\mu\nu} > - \tag{7.11}$$
$$-\frac{2\pi\alpha_s}{Q^6} < (\bar{q}\gamma_\mu\gamma_s\lambda^a q)(\bar{q}\gamma_\mu\gamma_s\lambda^a q) > -\frac{4\pi\alpha_s}{9Q^6} < (\bar{q}\gamma_\mu\lambda^a q)(\bar{q}\gamma_\mu\lambda^a q) > + \dots .$$

Let us stress that the perturbative series (7.6) appears in (7.10) -(7.11) in the coefficient function C_I of the unit operator, and in (7.11) one can see the first term of this expansion: $(1 + \frac{\alpha_s}{\pi})$. A similar perturbative series appears also in other coefficient functions (for the proper definition of the operators 0_n, normalization point dependence etc. see [14], p. 411).

Thus one can roughly visualize the OPE as the sum of a purely perturbative series $C_I I$ (as in the approach 1), and of power terms

$$\Pi_{\mathrm{OPE}} = \frac{6m^2}{Q^2} + C_G < 0_G > + C_M < 0_M > + \dots , \tag{7.12}$$

or in terms of the function $D(Q)$,

$$D(Q) = D^{\mathrm{pert}}(\alpha_s(Q)) + D_{\mathrm{OPE}}(Q) , \tag{7.13}$$

where $D_{\mathrm{OPE}}(Q)$ contains all terms (7.12).

In Sect. 9 we shall come back to the discussion of OPE and give a more refined discussion of separation of perturbative and NP contributions in OPE and QCD sum rules.

3) We now come to the third and the most complete approach, where the NP background is treated fully, and the perturbative series in α_s is calculated taking account of this background. To this end we calculate $\Pi(q^2)$ using the partition function (2.5)

$$Q^2\Pi(Q^2) = \int \frac{e^{iQx} d^4x}{N} \int DB\eta(B) Da e^{-S_{\mathrm{E}}(B+a)} tr(\gamma_\mu G_q(x,0)\gamma_\mu G_{\bar{q}}(0,x)) \times \tag{7.14}$$
$$\times \det(m + \hat{D}) ,$$

where G_q is the quark Green's function and $S_{\mathrm{E}}(B + a)$ contains only the gluon and ghost parts of the action,

$$G_q(x,y) =< x|(m + \hat{\partial} - ig(\hat{B} + \hat{a}))^{-1}|y > . \tag{7.15}$$

The expression (7.14) can be expanded in a perturbative series in powers of ga_μ. In lowest order one obtains a purely NP contribution. Note that the term $\det(m + \hat{D}(B))$ contributes a series of additional quark loops, which are connected to the valent loop by background field correlators.

To simplify our discussion we consider the limit of large N_c and neglect in what follows additional quark loops due to the term $\det(m+\hat{D})$. We shall also discard the term L_1 in $S_E(B+a)$ (see (4.8)), again for the reason of simplicity, the corresponding corrections can be easily restored at the end.

In next order $0(\alpha_s)$ one obtains the diagrams of Fig. 20, where the gluon exchange is shown by the wavy line. To order α_s^2 one has the diagrams shown in Fig. 21-22.

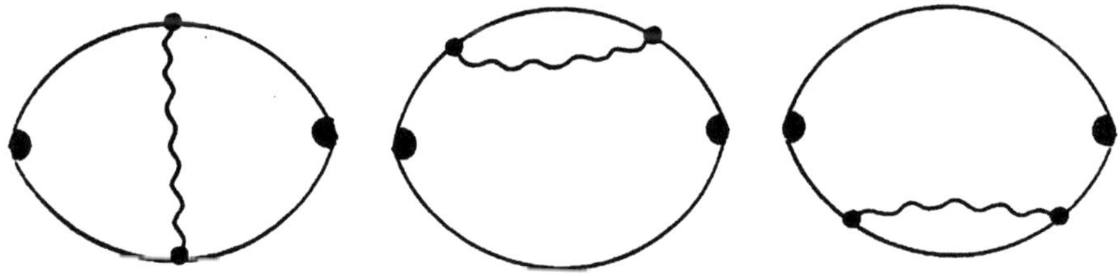

Fig. 20. Three diagrams of the the one–gluon exchange in the e^+e^- amplitude.

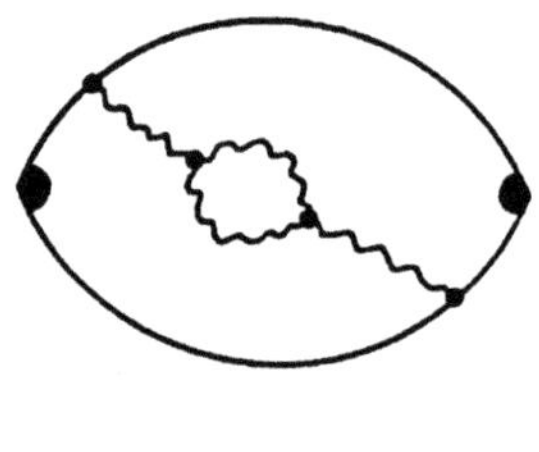

Fig. 21. The gluon loop diagram participating in the definition of the renormalized charge in the e^+e^- process.

Note that all broken lines correspond to the gluon Green's function in the background field B_μ, which one can take in the background Feynman gauge, (4.13).

Thus one can write the total contribution $\Pi(Q^2)$ as a series in the number of gluon exchanges, i.e., in powers of α_s:

$$\Pi(Q^2) = \Pi_{NP}^{(0)}(Q^2) + \alpha_s \Pi_{NP}^{(1)}(Q^2) + \alpha_s^2 \Pi_{NP}^{(2)}(Q^2) + \dots . \quad (7.16)$$

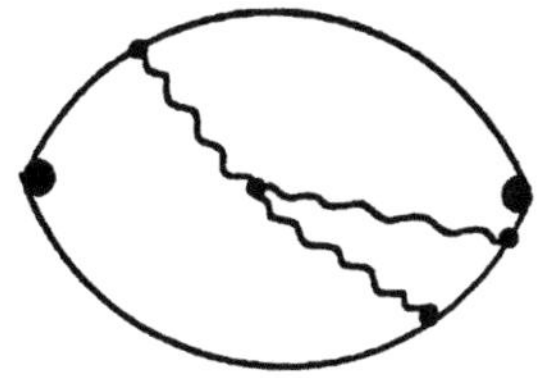

Fig. 22. The same as in Fig. 21 with the three-gluon vertex.

The series (7.16) does not look similar to the purely perturbative expansion (7.6) or the OPE–type expansion (7.11-7.13), since in all terms of (7.16) the NP interaction is present, which is implied by the subscript NP.

To find a connection between (7.6), (7.11-7.13) and (7.16) we have to look more closely at each of the terms $\Pi^{(n)}_{\rm NP}(Q^2)$, starting with $n=0$.

One can exploit the fact that in the limit of large N_c the spectrum of $\Pi^{(0)}_{\rm NP}(Q^2)$ consists of simple poles, and write for a given flavour f [66]

$$\Pi^{f(0)}_{\rm NP}(Q^2) = \frac{1}{12\pi^2}\sum_{n=0}^{\infty}\frac{C^f_n}{M^2_n(f)+Q^2}\,. \tag{7.17}$$

The hadronic ratio R_f for a given flavour is a sum of δ- functions

$$R_f(s) = \sum_{n=0}^{\infty} C^f_n \delta(s-M^2_n)\,. \tag{7.18}$$

To calculate c^f_n and $M^2_n(f)$ one needs dynamical computations. This can be done starting with the general expression for $\Pi^{(0)}(Q^2)$, obtained from (7.14)

$$Q^2\Pi^{(0)}_{\rm NP}(Q^2) = \frac{1}{N}\int d^4x e^{iQx} < {\rm tr}\gamma_\mu G^f_q(x,0)\gamma_\mu G^f_{\bar q}(0,x) >_B\,, \tag{7.19}$$

where now G^f_q is obtained from (7.15) by putting $m\to m_f, a_\mu\to 0$. The notation $<...>_B$ implies

$$<T>_B = \frac{1}{N}\int DB\eta(B)e^{-S_{\rm E}(B)}T(B)\,. \tag{7.20}$$

Inserting the quark Green's function $G^f_q(x,y)\equiv S(x,y)$ from (5.4) one obtains for the $q\bar q$ Green's function (the integrand of (7.19)) the following expression (cf (5.8))

$$G_{q\bar q}(x,0) = < {\rm tr}\gamma_\mu \int ds d\bar s Dz D\bar z(m_f-\hat D)\gamma_\mu(m_f-\hat D)e^{K-\bar K}W_F>\,, \tag{7.21}$$

where W_F is the Wilson loop with the spin insertions, see (5.5), which should be modified to the case of the closed contour, $x = y$.

In what follows we shall be interested in the dependence of $M_n^2(f)$ and C_n^f for large n, since this region is most important in the limit of large Q^2.

The nonperturbative bound states with masses $M_n(f)$ and decay constants C_n^f for large n have a nice property, i.e., that they are quasiclassical, have a large radius $r(n)$, and their spin dependence is irrelevant – spin interactions are short ranged and can be neglected at large distances (all these approximations can be tested a posteriori).

Correspondingly, one can neglect the spin factors in W_F, thus reducing it to the usual Wilson loops in (7.21), $W_F \to W$. Moreover, at large distances one can use the area law (6.18) for $< W >$ and introduce the Hamiltonian instead of the Green's function in (7.21) namely

$$G_{q\bar{q}}(x,0) = < x|e^{-HT}|0 >, \quad T = |x| \,. \tag{7.22}$$

Thus the problem boils down to writing down the Hamiltonian of spinless quarks with strings, and finding its eigenvalues and eigenfunctions. The Hamiltonian of the $q\bar{q}$ system in the c.m. system was found in [54] and on the light cone in the last two papers of [54].

In general it has a complicated structure, accounting for the relativistic motion of both quarks and the string, however in our specific problem of e^+e^- annihilation only small values of the orbital angular momentum L are involved, $L = 0, 2$. In this limit, when the radial quantum number n is much larger than the orbital one, $n \gg L$, the Hamiltonian simplifies considerably [54] and assumes the familiar potential form [24]

$$H^{(0)} = 2\sqrt{\mathbf{p}^2 + m_f^2} + \sigma r \,. \tag{7.23}$$

The eigenvalues $M_n(f)$ and decay contants C_n^f for the Hamiltonian $H^{(0)}$ have been found quasiclassically in [67]. In a leading large n approximation, when one can neglect the "current" quark mass m_f, one obtains [67]

$$M_n^2(f) = M_n^2(\infty) \equiv 4\pi\sigma n + \Delta \equiv m_0^2 n + \Delta \,, \tag{7.24}$$

where Δ is some constant, weakly depending on n [24, 67];

$$C_n^f(L=0) = \frac{2}{3} Q_f^2 N_c m_0^2 \,, \tag{7.25}$$

$$C_n^f(L=2) = \frac{1}{3} Q_f^2 N_c m_0^2 \,. \tag{7.26}$$

One can also introduce the radial quantum number n_r, such that $n = n_r + \frac{L}{2}$ and at large n_r

$$M_n^2 = 2\pi\sigma(2n_r + L) + \Delta \,. \tag{7.27}$$

Therefore one can see that at large n_r the states $n_r, L=0$ and $n_r-1, L+2$ are degenerate and correspond to the same value of n, and the decay contants add to

$$C_n^f(\infty) \equiv C_n^f(L=0, n_r=n) + C_n^f(L=2, n_r=n_1) = Q_f^2 N_c m_0^2 . \tag{7.28}$$

Introducing (7.28), (7.24) into (7.17) yields a diverging sum which can be represented as [66]

$$\sum_{n=n_0}^{\infty} \frac{1}{M_n^2+Q^2} = -\frac{1}{m_0^2}\psi(\frac{Q^2+\Delta+n_0 m_0^2}{m_0^2}) + \text{divergent const}, \tag{7.29}$$

where $\psi(z) = \frac{\Gamma'(z)}{\Gamma(z)}$, which has an asymptotic expansion

$$\psi(z)_{z\to\infty} = \ln z - \frac{1}{2z} - \sum_{k=1}^{\infty} \frac{B_{2k}}{2kz^{2k}} . \tag{7.30}$$

Here B_n are the Bernoulli numbers, $B_2 = \frac{1}{6}$. To obtain our final representation for $\Pi_{\mathrm{NP}}^{f(0)}(Q^2)$, we split the sum in (7.17) into the first n_0 states with $n = 0, \ldots n_0 - 1$, which do not follow the simple rule (7.24), and the rest with $n \geq n_0$, satisfying this rule. Using (7.29) we obtain

$$\Pi_{\mathrm{NP}}^{f(0)}(Q^2) = \frac{1}{12\pi^2}\sum_{n=0}^{n_0-1} \frac{C_n^f}{M_n^2(f)+Q^2} - \frac{Q_f^2 N_c}{12\pi^2}\psi(\frac{Q^2+\Delta+n_0 m^2}{m^2}) + \tag{7.31}$$
$$+\text{divergent const.}$$

The leading term in the asymptotics (7.30) can be now used in (7.31) and compared with the leading term in (7.11). One can see that both terms are equal to order $0(\alpha_s^0)$, namely

$$\Pi_{\mathrm{NP}}^{f(0)}(Q^2) = -\frac{Q_f^2 N_c}{12\pi^2}\ln\frac{Q^2+\Delta+n_0 m^2}{\mu^2} + 0(\frac{m^2}{Q^2}), \tag{7.32}$$

where we have regularized and renormalized (7.31), which amounts to the replacement $m^2 \to \mu^2$ in the argument of the logarithm in (7.32).

Taking the imaginary part of (7.32) for $Q^2 \to -s$ one obtains

$$12\pi \Im \Pi_{\mathrm{NP}}^{f(0)}(-s) = R^f(s) = N_c Q_f^2 , \tag{7.33}$$

in agreement with the partonic result (7.4). This property of the spectrum (7.23)-(7.28) was first noticed in [67]. It corresponds to the quark-hadron duality, which can be formulated as follows. Let us integrate $R^f(s)$ over some interval Δs, which is large enough so that it contains many resonances. We have from (7.18), (7.28), (7.24)

$$\int_{\delta s} R^f(s)ds = \sum_{n<\frac{\Delta s}{m^2}} C_n^f = N_c Q_f^2 m^2 \Delta n = N_c Q_f^2 \Delta s . \tag{7.34}$$

This can be also represented as a result of the closure theorem for the (almost) complete set of functions $\varphi_n \equiv < n|$, which are eigenvalues of the Hamiltonian H in (7.22), namely

$$\sum_n < p_1|n >< n|p_2 >= (2\pi)^4\delta(p_1 - p_2) \approx \sum_{k_1k_2} < p_1|k_1k_2 >< k_1k_2|p_2 >, \tag{7.35}$$

where $|k_1k_2 >$ is the quark-antiquark set of free states.

Since $C_n^f \sim | < 0|j_\mu|n > |^2$, one can derive that

$$\sum_n C_n^f \sim \sum_{k_1k_2} | < 0|j_\mu|k_1k_2 > |^2 \sim \Im \Pi^{(0)}(s), \tag{7.36}$$

where $\Pi^{(0)}$ is the free quark loop diagram depicted in Fig. 23. The equality of (7.36) and (7.34) give an exact meaning to the notion of the quark-hadron duality.

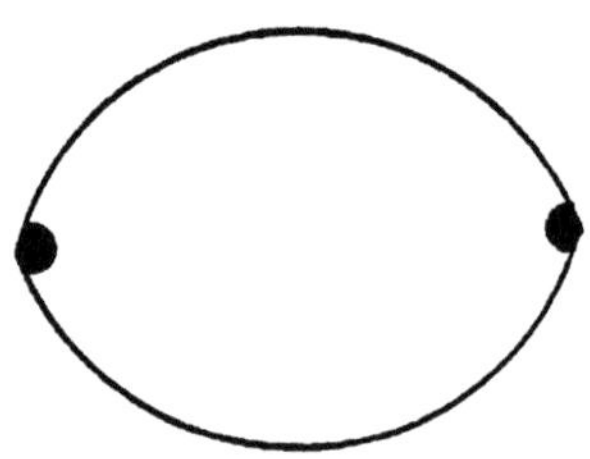

Fig. 23. The free quark loop diagram reproducing the partonic limit at large s.

In a similar manner one can analyze all other terms in (7.16) and conclude that each of them, at large Q^2, reduces to the free q$\bar{\text{q}}$ diagram with gluon exchanges contributing to order $0(\alpha_s^n)$ plus some power terms as in (7.32), i.e.,

$$\alpha_s^n \Pi_{\rm NP}^{(n)}(Q^2)_{Q^2\to\infty} \to \alpha_s^n \Pi_{free}^{(n)}(Q^2) + 0(\frac{m^2}{Q^2}). \tag{7.37}$$

Therefore one can define the renormalized α_s at large Q^2, where one can neglect power terms, in the following manner:

$$D(Q^2) = -Q^2\frac{\partial}{\partial Q^2}\Pi(Q^2) = D_{\rm NP}^{(0)}(Q^2) + \frac{\sum_f Q_f^2}{4\pi^2}\frac{\alpha_s^R(Q^2)}{\pi} + 0(\frac{m^2}{Q^2}). \tag{7.38}$$

The reader can notice that to the same order in $\frac{m^2}{Q^2}$ one can also replace $D_{\rm NP}^{(0)}(Q^2)$ by its free asymptotic value, which according to (7.32) is equal to

$$D_{\rm NP}^{(0)} = \frac{\sum_f Q_f^2}{4\pi^2}. \tag{7.39}$$

But we shall keep (7.38) for the moment, and generalize it later on to be able to describe the low Q region.

At large Q^2 one has for $\alpha_s^R(Q^2)$ to one loop

$$\alpha_s^R(Q^2) = \alpha_s^{(0)}(\mu) - \frac{b_0}{4\pi}(\alpha_s^{(0)}(\mu))^2 \ln\frac{Q^2}{\mu^2} + \dots . \tag{7.40}$$

Let us now generalize this expression in order to extend it to the region of smaller Q^2. To get the answer in the simplest way without computations, one can compare the logarithmic terms in (7.40) and (7.32). Both have the same physical meaning – in (7.40) the logarithm appears from the gluon loops, see Fig. 21 and Fig. 22, while in (7.32) the logarithm is due to the quark loop. In both cases this is the result of the integration over all two-particle states

$$\Pi(Q^2) - \Pi(\mu^2) \sim \int_0^\infty \frac{d\lambda^2}{Q^2+\lambda^2} - \int_0^\infty \frac{d\lambda^2}{\mu^2+\lambda^2} . \tag{7.41}$$

For the nonperturbative background case the integral in (7.41) is replaced by the sum over all resonances M_n^2

$$\Pi_{\rm NP}(Q^2) - \Pi(\mu^2) \sim \sum_{n=0}^{\infty}\left(\frac{1}{Q^2+M_n^2} - \frac{1}{\mu^2+M_n^2}\right) . \tag{7.42}$$

The sum on the r.h.s. of (7.42) is actually the Euler ψ–function (7.29). Whenever the following inequality is fulfilled

$$Q^2 + M_0^2 \equiv Q^2 + \Delta \gg m^2, \quad m \equiv m_0 , \tag{7.43}$$

one can use the logarithm in (7.30).

Hence the generalization of the logarithm in (7.40) in this case is

$$\ln\frac{Q^2}{\mu^2} \to \psi\left(\frac{Q^2+\Delta}{m^2}\right) - \psi\left(\frac{\mu^2}{m^2}\right) \approx \ln\frac{Q^2+\Delta}{\mu^2} , \tag{7.44}$$

and the generalization of the definition of α_s to one loop is the same as in Sect. 6, (6.60)-(6.61)

$$\alpha_s^R(Q^2+\Delta) = \frac{4\pi}{b_0 \ln\frac{Q^2+\Delta}{\Lambda^2}} . \tag{7.45}$$

In the next section we shall discuss two–loop corrections to (7.45) and phenomenological implications of the appearence of Δ in the logarithm, but now let us turn to the physical meaning of Δ. From (7.42) it is clear that $\Delta \equiv M_0^2$ is the physical threshold for the creation of two gluons inside the string, which has q and q̄ of flavour f at its ends. Making a cross section in Fig. 21 and Fig. 22 shown by the dash–dotted line one discovers the two–gluon hybrid state, shown in Fig. 24

The sum in (7.42) corresponding to the two–gluon renormalization of α_s, is actually over two–gluon hybrid states, and Δ is the ground state mass squared

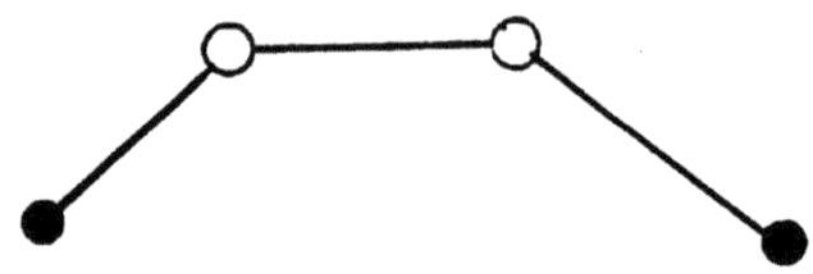

Fig. 24. The intermediate state, obtained by cutting the renormalization diagrams of Figs. 21 and 22 across two gluon lines.

of those states. The Hamiltonian analysis can be done in the same way as in (6.45), but now with quarks of finite mass m_f, so that the quark kinetic terms are present. For the one-gluon hybrids this analysis and calculations were done in [66], and yield masses in the interval m (1 gluon hybrid) $\approx 1.3 - 1.5\text{GeV}$.

An estimate of the lowest two gluon hybrid mass can be obtained in a similar way and gives

$$\Delta = (1.8 - 2\text{GeV})^2 . \tag{7.46}$$

Finally one obtains the following representation for $D(Q^2)$ in (7.38)

$$D(Q^2) = D^{(0)}_{\text{NP}}(Q^2) + \frac{1}{4\pi^2}\sum_f Q_f^2 \frac{\alpha_s(Q^2+\Delta)}{\pi} + 0(\frac{m^2}{Q^2+\Delta}) . \tag{7.47}$$

Note that on the r.h.s. of (7.47) one can explicitly take into account the lowest two-gluon hybrid states by adding

$$-Q^2\frac{\partial}{\partial Q^2}\sum_{n=0}^{n_0}\frac{C_n(2g)}{Q^2+M_n^2(2g)} . \tag{7.48}$$

In this case Δ should be changed to a larger value:

$$\Delta \to M^2_{n_0} . \tag{7.49}$$

Let us now apply the Borel transform as in [14,68] to (7.47) and compare the result with the experiment of [68] and with the sum rule analysis of [14, 69]. We define the quantities:

$$I_0^{\text{exp}}(M) = \int_{4m_\pi^2}^{\infty} e^{-s/M^2} R^{I=1}_{\text{exp}}(s) ds , \tag{7.50}$$

$$I_0^{\text{NP}}(M) = \int_{4m_\pi^2}^{\infty} e^{-s/M^2} R^{I=1}_{\text{NP}}(s) ds , \tag{7.51}$$

where $R^{I=1}_{\rm exp}(s)$ is the experimental hadronic ratio with isospin $I = 1$ reported in [68], and $R^{I=1}_{\rm NP}(s)$ refers to (7.47). More explicitly one can write, using (7.17),

$$I_0^{\rm NP}(M) = \frac{1}{12\pi}\sum_{n=0}^{\infty}\sum_f C_n^f e^{-M_n^2/M^2} + \frac{1}{4\pi}\sum Q_f^2 \frac{\alpha_s(M+\Delta)}{\pi}. \tag{7.52}$$

In Fig. 25 we plot three curves for $I_0^{\rm exp}(M), I_0^{rmNP}(M)$ and $I_0^{\rm SR}(M)$. The latter curve corresponds to the sum rule formalism [14] and is equal to

$$I_0^{\rm SR}(M) = \frac{3}{2}M^2[1 + \frac{\alpha_s(M)}{\pi} + \frac{\pi^2}{3}\frac{\frac{\alpha_s}{\pi} < G^a G^a >}{M^4} - \frac{448\pi^3\alpha_s}{81}\frac{|<\bar{q}q>|^2}{M^6}]. \tag{7.53}$$

One can notice that $I_0^{\rm NP}(M)$ behaves regularly in the whole region and is close to $I_0^{\rm exp}$, while $I_0^{\rm SR}(M)$ explodes at small M.

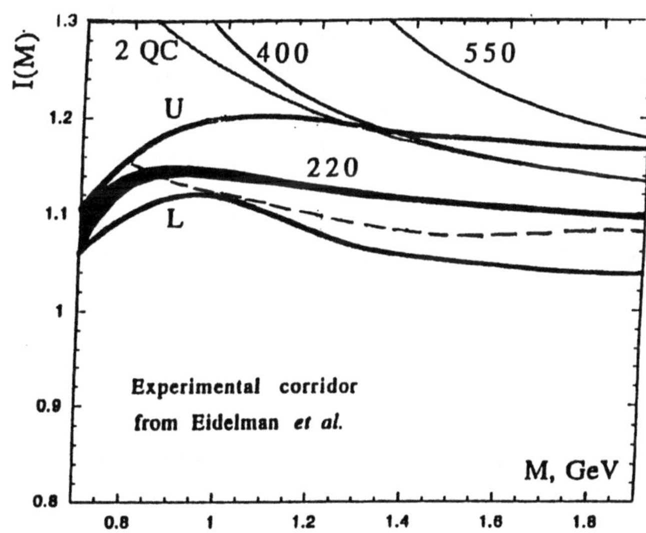

Fig. 25. The QCD sum rule in the ρ meson channel for different values of the scale parameter $\Lambda^{(3)}_{\overline{\rm MS}}$. The curves marked by "U" and "L" are the upper and lower side of the experimetal corridor (see text and Ref. [69]). The curve marked by QCD corresponds to $\Lambda^{(3)}_{\overline{\rm MS}} = 400$ MeV and the doubled value of the quark condensate. (From Ref. [70], except for the broken curve).

This fact gives additional confirmation to the stability of our formalism, (7.47) in the Euclidean region.

8 Freezing of α_s at Large Distances

As was discussed in Sect. 6, the background field in the combination gB_μ is not renormalized, since the Z factors of B_μ enter as $Z_g Z_B^{1/2}$ and the latter combination is equal to unity [42]. As a consequence, gB_μ enters in the physical amplitudes on the same ground as the external momenta p_μ, and consequently one can use for them the same Ovsyannikov–Callan–Simanzik equations (OCSE). Thus also the β–function is expressed through the same coefficients (1.6), and in the equation defining the scale dependence of α_s

$$\mu\frac{\partial\alpha_s}{\partial\mu} = -\frac{b_0}{2\pi}\alpha_s^2 - \frac{b_1}{8\pi^2}\alpha_s^2 - \dots, \tag{8.1}$$

$$b_0 = 11 - \frac{2}{3}n_f,\ \ b_1 = 102 - \frac{38}{3}n_f\,,$$

one should consider μ as any of the full-scale parameters p_μ, gB_μ. Going over to gauge–invariant and Lorentz–invariant quantities, one considers as a minimal set Q^2, $\Delta = \text{const}\,\sigma$, where Δ is the minimal mass in the "renormalization channel"- in absence of quarks this is the 2-gluon-hybrid mass, discussed in Sects. 6 and 7. Therefore in the two–loop solution to (8.1)

$$\alpha_s(\tilde\mu) = \frac{4\pi}{b_0 \ln\tilde\mu^2/\Lambda^2}[1 - \frac{b_1}{b_0^2}\frac{\ln\ln\tilde\mu^2/\Lambda^2}{\ln\tilde\mu^2/\Lambda^2}], \tag{8.2}$$

one should consider $\tilde\mu^2$ as any combination of Q^2 and Δ (multiplied by a dimensionless function of Q^2/Δ). The solution (8.2) applies when the ratio

$$\frac{\tilde\mu^2}{\Lambda^2} = \frac{c_1 Q^2 + C_2\Delta}{\Lambda^2} f(\frac{Q^2}{\Delta})$$

is large.

In the limiting case when $Q^2 \gg \Delta$ one should recover the usual one–scale–parameter formula (8.2) with $\tilde\mu^2 = Q^2$, which gives $f(\frac{Q^2}{\Delta}) \to \text{const}$. Moreover the analysis of the last two sections for $N_c \to \infty$ yields in the Euclidean region

$$\frac{\tilde\mu^2}{\Lambda^2} \simeq \frac{Q^2+\Delta}{\Lambda^2}\,. \tag{8.3}$$

In x–space one obtains for the static potential between heavy quarks an expression similar to (8.2):

$$\alpha_g^{(2)}(R) = \frac{4\pi}{b_0\ln(\frac{m_g^2+R^{-2}}{\Lambda_R^2})}\{1 - \frac{b_1}{b_0^2}\frac{\ln\ln(\frac{m_g^2+R^{-2}}{\Lambda_R^2})}{\ln(\frac{m_g^2+R^{-2}}{\Lambda_R^2})}\} \tag{8.4}$$

The most important property of (8.2-8.4) is that α_s remains finite at large distances R or small Euclidean momenta Q^2; we say that α_s is <u>frozen</u> at large distances. As we have discussed above, this property does not contradict the renormalization group, and, more specifically, OCSE.

More than that, we shall now demonstrate that the phenomenon of the freezing of the coupling constant is most common in QED and the Standard Model [3,71]. Consider, e.g., the renormalization of the photon or gluon self-energy part $\Pi^{\mu\nu}$ ([4] p.58) due to the transition into a quark-antiquark pair with masses m,

$$\Pi^{\mu\nu}(Q) \sim (\delta_{\mu\nu}Q^2 - Q_\mu Q_\nu)\{\text{div. part} - 4\int_0^1 x(1-x)\ln\frac{m^2+x(1-x)Q^2}{\nu_0^2}\}, \tag{8.5}$$

where ν_0 is the mass of dimensional regularization. It is easy to see that for $m^2 = 0$ one has a divergent contribution both at $Q^2 \to \infty$ and $Q^2 \to 0$, while for $m^2 \neq 0$ the quantity $\Pi^{\mu\nu}(Q^2)$ is finite for $Q^2 \to 0$. If quarks had been antiscreening rather than screening, the presence of m^2 would bring us the freezing of α_s, the corresponding logarithm in (8.2) would look like

$$\ln\frac{\tilde{\mu}^2}{\Lambda^2} \to \ln\frac{m^2+\frac{1}{4}Q^2}{\nu_0^2},$$

implying that the closest threshold in the "renormalization channel" is $4m^2$ – the channel of the $q\bar{q}$ pair creation. An even more spectacular example is that of the electroweak theory, where the self–energy part of Z (W) due to WW (WZ) is antiscreening; its logarithmic part has the form

$$\Pi_{\text{WW}}(Q^2) \sim \int dx \ln\frac{x(1-x)Q^2+m_{\text{W}}^2}{m_{\text{W}}^2}. \tag{8.6}$$

For $Q^2 \lesssim 4m_{\text{W}}^2$ the Q^2 dependence of (8.6) can be neglected, and the corresponding coupling constants are frozen. For $Q^2 \gg 4m_{\text{W}}^2$ one recovers the usual asymptotic freedom of the SU(2) part of the electroweak theory.

Thus the freezing of the coupling constant is a common phenomenon when the nonperturbative mechanism of mass generation is at work at large distances, as it happens in QCD due to confinement and in the electroweak theory due to the Higgs mechanism.

Let us now discuss whether the α_s freezing is supported (i) by lattice measurements, and (ii) by experiment.

Concerning the first item, one can take a recent analysis [72] of the lattice calculation [38] of the static potential $V(R)$, which we already discussed in subsection 3.2. Let us note, following [72], that the results [38] have been fitted differently in two regions of R: for $R \geq 4a$ the fit (3.11) with constant charge (E = const) Coulomb force and linear potential was used, while for $R \leq 4a$ another fit, (3.12) without nonperturbative interaction at all and free two–loop α_s (3.13) was done. When we have discussed this work in subsection 3.2, we only stressed the possibility of relating the lattice perturbative scale (Λ_{QCD}) and the NP one (σ).

Now we consider the same data [38] to obtain much more delicate information: whether the freezing of α_s is preferred (or rejected) by the data.

Looking closely at the data of [38] (in particular table 3), one notices that at the matching point $R_0 = 4a = 0.95\text{GeV}^{-1}$ the two–loop value of $\alpha_s^{(2)}(R_0) = 0.305$, while the constant $E = \frac{4}{3}\alpha_s(as)$ from the large R region yields the asymptotic value $\alpha_s(\text{as}) = 0.2085$, which is significantly smaller. Hence the matching is not so good, and one needs better input to appropriate a) the NP force in the whole region $a \le R \le 23a$, and b) to use the freezing α_s in that region. That program was fulfilled in [72]. The NP force was used in two variants : A) as that of a linear potential with the same σ as in the fit (3.11) but now in the whole interval $a \le R \le 23a$; B) in the second case the finite value of T_g was taken into account, which brings about the gradual increase of the NP potential from zero to an asymptotically linear form [18]. For the concrete analysis the correlator $D(x)$ measured in [25] was used and found to have the exponential form

$$D(x) = D(0)\exp(-x/T_g), \quad T_g = 0.2\text{fm}\,. \tag{8.7}$$

The static potential $\varepsilon(R)$ is found from this correlator, using [23],

$$\frac{d\varepsilon(R)}{dR} = \frac{2\sigma}{\pi}\int_0^{R/T_g} tK_1(t)dt\,, \tag{8.8}$$

where $K_1(t)$ is the McDonald function.

Note that $\varepsilon(R) \sim R^2$ at small R and asymptotically (when $R \gg T_g$) it grows linearly like σR. Here we have kept only the bilocal correlator $D(x)$, since the quartic correlator contributes $\Delta\varepsilon(R) \sim R^4$ at small R to $\varepsilon(R)$, and can be neglected.

In addition to the NP potential $\varepsilon(R)$ the freezing $\alpha_s(R)$ was used, so that

$$V_i(R) = -\frac{4}{3}\frac{\alpha_g^{(2)}(R)}{R} + \varepsilon_i(R), \quad i = A, B\,, \tag{8.9}$$

where (8.4) is used for $\alpha_g^{(2)}(R)$ and $\varepsilon_A(R) = \sigma R$ while $\varepsilon_B(R)$ corresponds to (8.8).

In Fig. 26 three curves of $\frac{dV}{dR}$ are shown, the solid line for the case B with $\Lambda_R = 0.32\text{GeV}$, $m_g = 0.96\text{GeV}$, the broken line for the case A with $\Lambda_R = 0.30\text{GeV}$, $m_g = 1.5\text{GeV}$ and the dotted line for the standard two–loop potential (the first term in (8.9) on the r.h.s.) with $m_g = 0$ and $\Lambda_R = 0.289\text{GeV}$. The "experimental" points are from [38] and corrected by the authors to remove lattice artefacts at small R. One can see that both fits A and B are very good in the whole region $a \le R \le 23a$, while the free two–loop curve diverges at $R \approx 10a$, where the Landau ghost pole appears.

From this fact one can conclude that the freezing α_s is indeed necessary to explain the data. A similar analysis of other lattice data is now in progress.

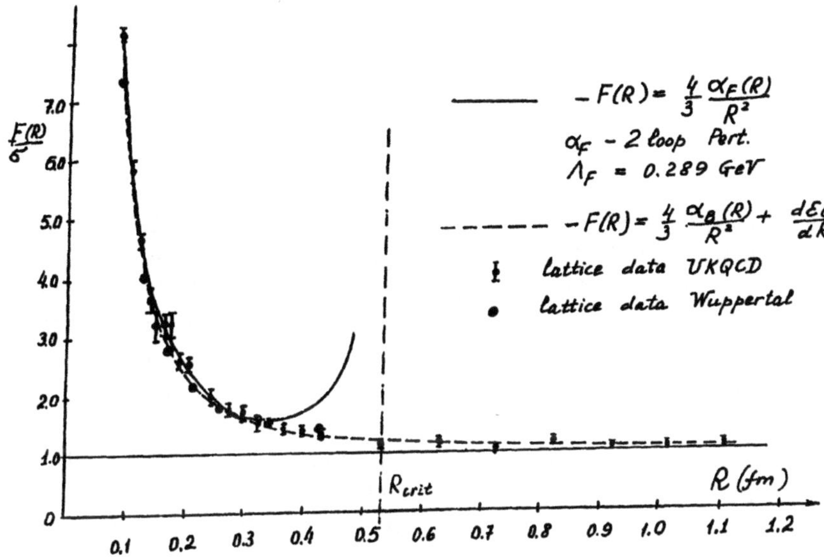

Fig. 26. The fit to the lattice measurements of the static force with the two–loop perturbative formula (solid line) and with the background–corrected formula (broken line).

Let us come now to other evidences in favour of a freezing α_s. In the previous section we have plotted the zeroth moment of $R(s)$ with standard nonfreezing α_s and two condensates from OPE (analysis of [69],[70]), and our result for $R(s)$ with the freezing $\alpha_s(Q)$ and $\Pi^{(0)}_{\mathrm{NP}}$. From Fig. 25, where the experimental points are shown, one can deduce that the freezing of α_s is certainly preferable, but the value of Λ_{QCD} cannot be defined here with the same accuracy as for higher Q^2 data – e.g. from the Z peak.

In [70] it was argued that data from the Z peak (high Q^2) which give

$$\alpha_s(M_Z) = 0.125 \pm 0.005 \tag{8.10}$$

correspond to $\Lambda^{(5)}_{\overline{\mathrm{MS}}} = 250 - 300$ MeV (or $\Lambda^{(4)}_{\overline{\mathrm{MS}}} = 450$ MeV) while the low-Q data prefer smaller Λ, say $\Lambda^{(4)} \approx 200$ MeV, which would give

$$\alpha_s(M_Z) \approx 0.110\,. \tag{8.11}$$

From that the author in [70] deduces that the low Q data are more reliable, and the discrepancy at high Q around the M_Z peak can be explained by new physics.

Our calculations in Fig. 25 of the sum rule, on which the conclusions of [70] are mostly founded, show that the freezing of α_s, with Λ from the high Q

region, well describes the experimental data, thus removing the discrepancy between the high Q and low Q data. To understand qualitatively why the freezing of Λ fits well at small and high Q values, let us take for simplicity the one-loop expression

$$\alpha_s(Q) = \frac{4\pi}{b_0 \ln \frac{Q^2+m^2}{\Lambda^2}} . \tag{8.12}$$

It is clear, that at $Q \sim 100\text{GeV}$ one can neglect $m \sim 1\text{GeV}$ and make a clean estimate of $\Lambda^{(5)}$, $\Lambda^{(5)} \sim 300$ MeV. If one would continue $\alpha_s(q)$ to the region $Q \sim 1\text{GeV}$ with the free $\alpha_s(Q)$, i.e., with $m = 0$ in (8.12), the resulting value $\alpha_s(1\text{GeV}) = \frac{4\pi}{b_0 \ln(\frac{1}{0.4})^2} \approx 0.7$ is too large to explain the low–energy data. However, with $m \approx 1.8\text{GeV}$ (according to our estimates of the two–gluon hybrid in the previous section) one obtains a reasonable value in the correct ballpark $\alpha_s(1\text{GeV}) \sim 0.4$ with the same large Λ_{QCD}.

Thus the discrepancy between high and low Q data can be easily reconciled by the freezing of $\alpha_s(Q)$. In concluding this section we illustrate this point in more detail. At the same time it should be stressed that the low–Q data are less sensitive to the value of Λ when the freezing of α_s is used, and the in principle less accurate low Q data therefore enter with less weight in our definition of α_s. (In this respect the conclusion of high accuracy of the sum rules for bottomonium in [73] seems to be exaggerated).

There are a lot of arguments in favour of the freezing of α_s from different processes; for a review see [74], and the recent papers [75], [76]. We hope to come to a systematic analysis of all these data in the nearest future.

9 Convergence of Perturbative Series. Renormalons

Let us take the example of the process $e^+e^- \to$ hadrons and consider as in Sect. 7 the regularized function $D(Q)$ defined as in (7.5) in the Euclidean region. Assume that the perturbation theory alone can describe the process, i.e. put

$$D(Q) \equiv D^{\text{pert}}(\alpha_s(Q)) , \tag{9.1}$$

and study the properties of a perturbative series for $D(Q)$ following [77]. $D(Q)$ can be expanded in the perturbative series, of which the first four terms are explicitly known [57]

$$D(Q) = \sum_{n=0}^{\infty} D_n \alpha^n(Q) . \tag{9.2}$$

It is understood by now that the series (9.2) can be, at best, an asymptotic one, so that the sum of the series may not exist. It has been suggested by 'tHooft [29] to use the Borel transform $\tilde{D}(b)$ and its properties in the complex

b-plane to characterize the convergence of the series (9.2). To this end one defines

$$D(\alpha) - D(0) = \int_0^\infty db e^{-b/\alpha(Q)} \tilde{D}(b) \,, \tag{9.3}$$

where $\tilde{D}(b)$ is expanded as

$$\tilde{D}(b) = \sum_{n=0}^{\infty} D_{n+1} \frac{b^n}{n!} \,. \tag{9.4}$$

Problems with (9.3) come i) from the convergence of the integral at large b, and ii) from the possible singularities of $\tilde{D}(b)$ on the real axis.

Usually point i) occurs when one tries to account for the nonperturbative singularities in $D(Q)$ via (9.3), e.g., the resonance masses in $D(Q)$ at $Q^2 = -M^2_{\text{res}}$ are due to confinement. Trying to obtain them from (9.3) one should require that $D(Q)$ has a singularity when $\alpha = \alpha_{\text{th}}(Q^2 = -M^2_{\text{res}})$, and consequently the integral (9.3) does not converge in b when $\alpha = \alpha_{\text{th}}$ [29]. From our point of view $D^{\text{pert}}(Q)$ is only a part of the general expression (7.38), and there $D^{(0)}_{\text{NP}}(Q^2)$ is responsible for singularities at thresholds and resonance masses, so that $D^{\text{pert}}(Q)$ does not have these singularities.

Point ii) is more relevant. Indeed it was shown [29] that $\tilde{D}(b)$ has branch points in the b-plane, as shown in Fig. 27, ultraviolet (UV) renormalons

$$b = -\frac{m}{\beta_0} = -\frac{m4\pi}{b_0} \quad , \quad m = 1, 2, \ldots \,, \tag{9.5}$$

infrared (IR) renormalons at

$$b = \frac{(m+1)4\pi}{b_0} \quad , \quad m = 1, 2, \ldots \,, \tag{9.6}$$

and branch points due to instanton–antiinstanton ($I\bar{I}$) pairs

$$b = 4m\pi \quad , \quad m = 1, 2, \ldots \,. \tag{9.7}$$

Actually the singularities on the positive axis (IR and $I\bar{I}$) make the integral (9.3) meaningless, and require redefinition of the theory, as we shall do.

The authors of [77, 78] suggest instead to interpret the nearest IR singularity at $b = \frac{8\pi}{b_0}$ as an ambiguity in the definition of the gluon condensate, $\frac{<G_a G_a>}{Q^4}$, since this term has a singularity of the same order (Q^{-4}) as resulting from $b = \frac{8\pi}{b_0}$.

We proceed below to show that in the nonperturbative vacuum the problems with IR singularities are absent – the perturbative series in the NP background has no IR renormalons and there are no obstacles in defining the

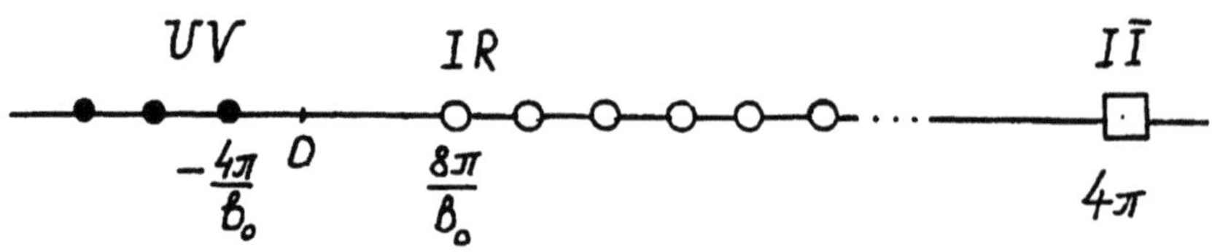

Fig. 27. The Borel parameter plane singularities.

integral (9.3) and in doing the Borel summation. (The problem with instantons is also solved by the explicit inclusion of $I\bar{I}$ pairs into the NP vacuum and by the definition of the modified perturbative series in such a vacuum.)

In this section we study the consequences of the freezing of α_s for the properties of the PTH series. We shall concentrate, for definiteness, on the Euclidean correlator of e.m. currents $\Pi(Q^2)$ as in [7,1] and shall write it as in [79] in a form exhibiting the contribution of the renormalon [29]

$$\Pi(Q^2) = \frac{1}{2\pi^2 b_0}\left(\frac{1}{\alpha_s(\mu^2)} - \frac{1}{\alpha_s(Q^2)}\right) + \frac{1}{2\pi^2 b_0}\ln\left(\frac{\alpha_s(Q^2)}{\alpha_s(\mu^2)}\right) + \sum \tilde{p}_n \alpha_s^n(Q^2) + \Delta\Pi\,, \tag{9.8}$$

where we have separated out of the PTH series the contribution of the set of diagrams shown in Fig. 28, denoted by $\Delta\Pi$. We shall first discuss the infrared (IR) renormalon, [29], i.e. the contribution from the graphs of Fig. 17 from the domain of integration over k with $k \ll Q$. If the coupling constant α_s is normalized at Q^2 (we shall take it at $Q^2 + m^2$) and (8.12) is used for α_s, we have

$$\Delta\Pi = \frac{\alpha_s(Q^2)}{8\pi^3} \sum_n \left(\frac{b_0 \alpha_s(Q^2)}{4\pi}\right)^n \int_0^{Q^2} \frac{k^2 dk^2}{Q^4} \ln^n\left(\frac{Q^2+m^2}{k^2+m^2}\right). \tag{9.9}$$

The analysis of the integral (9.9), made in the appendix to this section, yields

$$I = \int_0^{Q^2} \frac{k^2 dk^2}{Q^4} \ln^n\left(\frac{Q^2+m^2}{k^2+m^2}\right) = \begin{cases} \frac{n!}{2^{n+1}}\,, & n < 2n_0 \\ \eta \frac{n_0^{n+1}}{n+1}\,, & n > 2n_0,\ \ \eta < 1 \end{cases} \tag{9.10}$$

with the notation $n_0 = \ln\frac{Q^2+m^2}{m^2}$.

As a result we obtain the following contribution of the graphs in Fig. 28

$$\Delta\Pi = \frac{4\pi}{8 b_0 \pi^3} \sum_{n \gg 1}^{\infty} \left(\frac{\alpha_s(Q^2) b_0}{8\pi}\right)^n q_n\,, \tag{9.11}$$

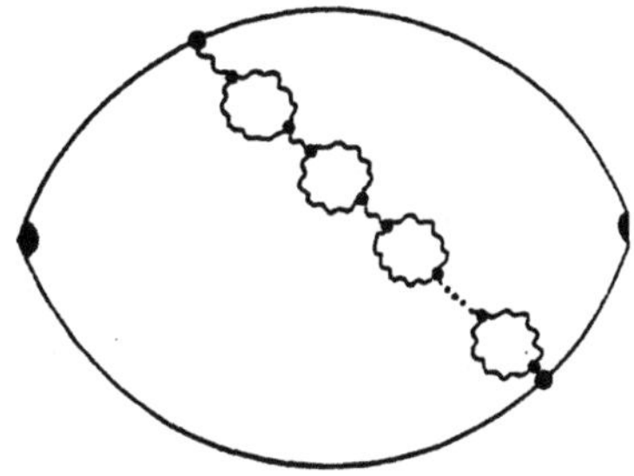

Fig. 28. The set of graphs contributing to the renormalons in the absence of a background.

where

$$q_n = \begin{cases} (n-1)! & , n < 2n_0 \\ \eta \frac{(2n_0)^n}{n} & , n > 2n_0 \end{cases} . \tag{9.12}$$

We remind that the estimate (9.12) is an upper limit for q_n.

First we discuss the connection with the previous results [29,79] on the IR renormalons. In this case $m^2 \to 0$ (no background field) and $n_0 \to \infty$; q_n is defined as in (9.12) for any n and the Borel transform of $\Delta \Pi$ is as in [29,79]

$$\Delta \Pi = (16\pi^3)^{-1} \int_0^\infty \frac{\exp(-\frac{t}{\alpha_s(Q^2)})dt}{1 - \frac{tb_0}{8\pi}} . \tag{9.13}$$

The Borel integral (9.13) exhibits an IR renormalon pole at $t = \frac{8\pi}{b_0}$ which makes the Borel summation of the PTH series prohibitive.

In our case, when n_0 is finite, the situation is drastically changed. Namely, having in mind that the estimate (9.12) is an upper limit of q_n both for $n > 2n_0$ and for $n < 2n_0$, we can take the sum in (9.11) which does not diverge and even does not need Borelization. Leaving details for the Appendix we quote only the final result

$$\Delta \Pi = \frac{\alpha_s}{8\pi^3} \sum_{n=1}^{\infty} \int_0^{n_0} \frac{\alpha_s(Q^2)tb_0}{4\pi})^n e^{-2t} dt \approx \frac{b_0 \alpha_s^2(Q^2)}{32\pi^4} . \tag{9.14}$$

There are no poles in the plane of the Borel variable, as can be easily seen from the absense of the factorial growth of the coefficient q_n at arbitrarily large n. We thus conclude that the convergence of the PTH series is radically improved when the strong nonperturbative background is taken into account, and in particular the IR renormalons disappear.

We now turn to the ultraviolet (UV) renormalons [29]. They come from the same series of graphs, Fig. 28, but from the integration region with $k^2 \gg$

Q^2. It is easy to understand that the UV renormalons do not change due to the modified logarithm (8.12), since in this kinematical region $k^2 \gg m^2$ and the change is immaterial. Hence we have, as in [79], the estimate

$$\Delta\Pi(UV) = -\frac{1}{6\pi^3}\sum_n \alpha_s^n(-\frac{b_0}{4\pi})^{n-1}(n-1)!\,, \qquad (9.15)$$

which brings about a pole at a negative value of the Borel parameter, $t = -\frac{4\pi}{b_0}$, and hence the Borel sum of the PTH series is well defined.

Appendix

The integral in (9.9) can be rewritten as

$$I \equiv \int_0^{Q^2} \frac{k^2dk^2}{Q^4}\ln^n\frac{Q^2+m^2}{k^2+m^2} = \int_0^{n_0} t^n e^{-2t}dt - \frac{m^2}{Q^2}\int_0^{n_0} t^n e^{-t}dt\,, \qquad (9.16)$$

where $n_0 = \ln\frac{Q^2+m^2}{m^2}$.

The sum in (9.9) can now be calculated easily (the interchange of the order of summation and integration is possible because the integral is between finite limits, and the sum is converging uniformly for $t \le n_0$). One has

$$\sum_{n=1}^{\infty}(\alpha_s(Q^2)t\frac{b_0}{4\pi})^n = \frac{\alpha_s(Q^2)t\cdot b_0/4\pi}{1-\alpha_s(Q^2)\frac{b_0}{4\pi}t}\,, \qquad (9.17)$$

and the contribution of I in the sum over n is

$$\frac{b_0}{4\pi}\alpha_s(Q^2)\int_0^{n_0}\frac{e^{-2t}tdt}{1-\alpha_s(Q^2)\frac{b_0}{4\pi}t} = \int_0^{n_0}\frac{tdte^{-2t}}{t_p - t} \sim \frac{1}{t_p}\,, \qquad (9.18)$$

where

$$t_p = \frac{4\pi}{b_0\alpha_s(Q^2)} = \ln\frac{Q^2+m^2}{\Lambda^2} > n_0\ , \qquad (9.19)$$

since $m > \Lambda$.

Insertion of (9.18) into (9.9) finally yields

$$\Delta\Pi = \frac{\alpha_s(Q^2)}{8\pi^3}\cdot\frac{1}{t_p} \cong \frac{b_0\alpha_s^2(Q^2)}{32\pi^4} \qquad (9.20)$$

This result is used in (9.14) of the main text.

Thus one can see that the inclusion of the nonperturbative background radically changes the situation with respect to the convergence of the perturbative series in the whole series

$$\Pi(Q) = \Pi_{\rm NP}^{(Q)} + \Pi_{\rm pert}^{(Q)}\,.$$

The first term yields singularities at $Q^2 = -M^2_{\rm res}$ and at the thresholds, which are nonperturbative, while the second term can be treated via the Borel summation.

10 Conclusions

The methods and results reported in these lectures may lead to a better understanding of old and new problems which exist in applications of QCD.

Below we shall demonstrate how the results of the new NP background formalism can explain and resolve the conflicts in the determination of α_s from different processes. As we discussed in Sect. 8, there seems to be a systematic discrepancy between the determination of α_s from low-Q physics and high-Q physics [70]. Whereas the low-Q e^+e^- experiment [69] and bottomonium data [71] prefer $\alpha_s(M_Z) \approx 0.110$ (the two-loop free perturbative evolution is used for α_s here and below), the high-Q data on Γ_h yield $\alpha_s(M_Z) = 0.125 \pm 0.05$. Fig. 29 displays the known determinations of α_s with the corresponding errors. Being seemingly small, this discrepancy becomes crucial for the region of $Q \approx 1$ GeV, where most hadron physics is involved: if one relies on Γ_h and takes $\Lambda^{(5)} = 0.295$ GeV, then one obtains $\alpha_s(1\text{GeV}, n_f = 3) = 0.967$, a value which is too large and contradicts both sum rules and potential models. On the other hand, if one starts with $\Lambda^{(3)} = 0.3$ GeV (equivalent to $\alpha_s(1\text{GeV}, n_f = 3) = 0.421$), which is reasonable from the considerations of [70], then one ends up with $\alpha_s(M_Z, n_f = 5) = 0.110$, in contradiction with Γ_h.

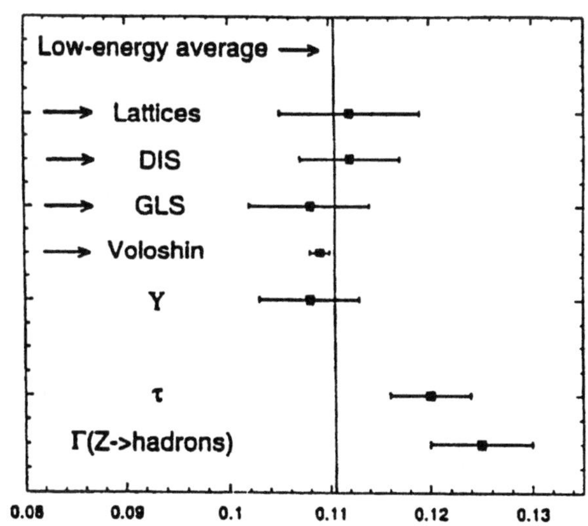

Fig. 29. Experimental data on $\alpha_s(M_Z)$ (udapted from Ref. [13]). The vertical line is a naive average of four low–energy points marked by the arrows (from Ref. [70]).

To check the formalism presented in the lectures, let us take the expansion for the freezing of α_s, (8.2), and start from the M_Z point, choosing $m_g = 1.5$GeV, $\Lambda^{(5)} = 0.295$GeV (this corresponds to $\alpha \equiv \alpha_g(M_Z) = 0.123$). Consider now the evolution to smaller values of Q, as shown in Fig. 30. The freezing of α_s evolves at 1GeV to a very reasonable value,

$$\alpha_g(1\text{GeV}, n_f = 3) = 0.42\,, \tag{10.1}$$

while the free α_s is equal to 0.967 at this point.

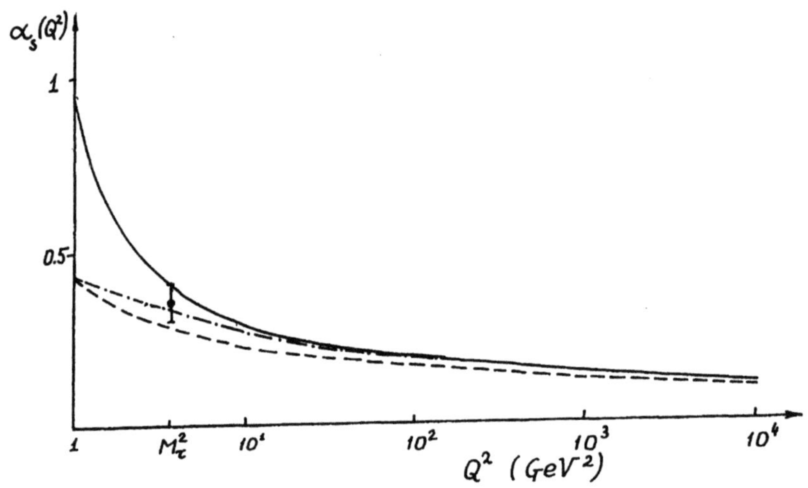

Fig. 30. The "universal" curve of the running α_s from the energy of the Z peak to 1 GeV. The standard perturbative two-loop evolution from the value $\alpha_s = .123$ (solid line). The background-corrected evolution from the same Z-peak value (dash-dotted line). The standard evolution of α_s from the lower value of .42 at 1 GeV (dashed line).

Let us now do another check and evolve α_s from the value (10.1), which is phenomenologically reasonable, to M_Z using the free, nonfreezing evolution. Then one obtains $\alpha_s(M_Z, n_f = 5) = 0.110$, in agreement with the results of [70].

Therefore the freezing evolution of $\alpha_s(Q)$ with $\Lambda^{(5)} = 0.295$GeV satisfies both high-Q data and low-Q data, as one can see from Fig. 30.

Comparison of Fig. 29 and Fig. 30 demonstrates clearly how the conflict between the data is resolved due to the freezing of α_s. In a recent analysis of $\sqrt{s} = 130 - 136$GeV data at LEP [80], a smaller value of $\alpha_s(M_Z)$ was quoted, $\alpha_s(M_Z) = 0.113 \pm 0.006 \pm 0.007$. A freezing evolution from this value yields also a reasonable $\alpha_s(1\text{GeV}) \approx 0.4$.

Acknowledgements

The author is indebted to A.M. Badalian for numerous discussions and computations used for Figs. 26 and 30, and in the text.

The partial support in the framework of the INTAS grant 93-79 and from the Netherlands grant NWO 07-30-148 is gratefully acknowledged.

References

1. D.J. Gross and F. Wilczek, Phys. Rev. **D8** (1973) 3633, **D9** (1974) 980
2. H.D. Politzer, Phys. Rep. **14C** (1974) 129
3. G. 'tHooft, Nucl. Phys. **B33** (1971) 173, **B35** (1971) 167;
   G. 'tHooft and M. Veltman, Nucl. Phys. **B44** (1972) 189;
   G. 'tHooft, talk at the Marseille Int. Conf. (1972)
4. F.J. Yndurain, *Theory of Quark and Gluon Interactions* (Springer, New York, 1993)
5. See talks in *QCD: 20 years later*, eds. P.M. Zerwas and H.A. Kastrup (World Scientific, Singapore, 1993)
6. Yu. Dokshitzer et al., *Basics of Perturbative QCD* (Editions Frontieres, 1991)
7. L.V. Gribov, E.V. Levin, and M.G. Ryskin, Phys. Rep. **100** (1983) 1
8. G. Sterman et al., Rev. Mod. Phys. **67** (1995) 157
9. G. Sterman, these proceedings
10. P. Lepage, these proceedings
11. K.G. Wilson, Phys. Rev. **D10** (1974) 2445
12. I.J. Ford et al., Phys. Lett. **B208** (1988) 286;
    V.M. Heller et al., Phys. Lett. **B335** (1994) 71
13. M. Creutz, *Quarks, gluons and lattices* (Cambridge Univ. Press, Cambridge, 1983)
14. M.A. Shifman, A.I. Vainshtein, and V.I. Zakharov, Nucl. Phys. **B147** (1979) 385
15. R. Crewther, Phys. Rev. Lett. **28** (1972) 1421;
    M. Chanowitz and J. Ellis, Phys. Lett. **B40** (1972) 397;
    J. Collins, L. Duncan, and S. Joglecar, Phys. Rev. **D16** (1977) 438
16. V.A. Novikov, M.A. Shifman, A.I. Vainshtein, and V.I. Zakharov, Uspekhi Fiz. Nauk. **136** (1982) 554
17. O.V. Tarasov, A.A. Vladimirov, and A. Zharkov, Phys. Lett. **B93** (1980) 429
18. H.G. Dosch, Phys. Lett. **B190** (1987) 177;
    H.G. Dosch and Yu.A. Simonov, Phys. Lett. **B205** (1988) 339
19. Yu.A. Simonov, Nucl. Phys. **B307** (1988) 512
20. V. Marquard and H.G. Dosch, Phys. Rev. **D 35** (1987) 2238
21. Yu.A. Simonov, Yad. Fiz. **54** (1991) 192
22. L. Del Debbio, A. Di Giacomo, and Yu.A. Simonov, Phys. Lett. **B332** (1994) 111
23. Yu.A. Simonov, Nucl. Phys. **B324** (1989) 67
24. D.R. Stanley and D. Robson, Phys. Lett. **B45** (1980) 235;
    J. Carlson et al., Phys. Rev. **D27** (1983) 233;
    N. Isgur and S. Godfrey, Phys. Rev. **D32** (1985) 189;
    J.L. Basdevant and S. Boukraa, Z. Phys. **C28** (1985) 413;
    W. Lucha, F.F. Schoeberl, and D. Gromes, Phys. Rep. **200** (1991) 127

25. A. Di Giacomo and H. Panagopoulos, Phys. Lett. **B285** (1992) 133;
A. Di Giacomo, E. Meggiolaro, and H. Panagopoulos, hep-lat/9603017, hep-lat/9603018
26. H.G. Dosch, E. Ferreira, and A. Kramer, Phys. Rev. **D50** (1994) 1992
27. C.G. Callan, R. Dashen, and D.J. Gross, Phys. Rev. **D17** (1978) 2717;
E.V. Shuryak, Nucl. Phys. **B203** (1982) 93, 116, 140;
D.I. Dyakonov and V.Yu. Petrov, Nucl. Phys. **B245** (1984) 259
28. E.V. Shuryak, Rev. Mod. Phys. **65** (1993) 1;
M.-C. Chu et al., Phys. Rev. **D49** (1994) 6039
29. G. 'tHooft, in *The Whys of Subnuclear Physics*, ed. A. Zichichi, (NY Plenum, 1977);
G. Parisi, Phys. Lett. **B76** (1977) 65;
B. Lautrup, Phys. Lett. **B69** (1977) 109
30. Yu.A. Simonov, Yad. Fiz. **58** (1995) 113, hep-ph/9311247;
Yu.A. Simonov, JETP Lett. **57** (1993) 513
31. Yu.A. Simonov, Yad. Fiz. **58** (1995) 357, hep-ph/93 11216, JETP Lett. **55** (1992) 605
32. E.L. Gubankova and Yu.A. Simonov, Phys. Lett. **B360** (1995) 93
33. O. Nachtmann, these proceedings
34. R.P. Feynman, *Statistical Mechanics* (W.A. Benjamin, Reading, 1972)
35. G. 'tHooft, Phys. Rev. **D14** (1976) 3432
36. I.I. Balitsky and A.V. Yung, Phys. Lett. **B168** (1986) 113
37. Yu.A. Simonov, Yad. Fiz. **46** (1987) 317, Lectures at the E. Fermi School, Varenna, 1995 (in press)
38. S.P. Booth et al. (UKQCD collaboration), Phys. Lett. **B294** (1992) 385
39. A. Billoire, Phys. Lett. **B104** (1981) 472
40. B.S. De Witt, Phys. Rev. **162** (1967) 1195, 1239;
J. Honerkamp, Nucl. Phys. **B48** (1972) 269
41. G. 'tHooft, Nucl. Phys. **B62** (1973) 444, Lectures at Karpacz, in: Acta Univ. Wratislaviensis **368** (1976) 345
42. L.F. Abbot, Nucl. Phys. **B185** (1981) 189
43. Yu.A. Simonov , S. Titard, and F.J. Yndurain, Phys. Lett. **B354** (1995) 435
44. S. Titard and F.J. Yndurain, Phys. Rev. **D49** (1994) 6007, Phys. Lett. **B351** (1995) 541
45. Yu.A. Simonov and J.A. Tjon, Ann. Phys. **228** (1993) 1
46. M.B. Voloshin, Nucl. Phys. **B154** (1979) 365;
M.B. Voloshin, Sov. J. Nucl. Phys. **36** (1982) 143;
H. Leutwyler, Phys. Lett. **B98** (1981) 447
47. I.I. Balitsky, Nucl. Phys. **B254** (1985) 166
48. M. Campostrini, A. Di Giacomo, and S. Olejnik, Z. Phys. **C31** (1986) 577
49. A. Kramer, H.G. Dosch, and R.A. Bertlmann, Phys. Lett. **B223** (1989) 105
50. A.M. Badalian and V.P. Yurov, Yad. Fiz. **51** (1990) 1368, Phys. Rev. **D42** (1990) 3138
51. S.J. Brodsky, G.P. Lepage, and P.B. Mac Kenzie, Phys. Rev. **D28** (1983) 228
52. S. Perantonis and C. Michael, Nucl. Phys. **B347** (1990) 854
53. M. Luescher, Nucl. Phys. **B180** (1981) 317
54. A. Dubin, A.B. Kaidalov, and Yu.A. Simonov, hep-ph 93//344, Yad. Fiz. **56** (1993) 213, **58** (1995) 348, Phys. Lett. **B323** (1994) 41, **B343** (1995) 310

55. S. Brodsky and Hung Ju Lu, SLAC-PUB-6389 Nov. 1993
56. Hung Ju Lu, S.J. Brodsky, Phys. Rev. **D48** (1993) 3310
57. S.G. Gorishny, A.L. Kataev, and S.A. Larin, Phys. Lett. **B259** (1991) 144; M.A. Samuel and L.R. Surguladze, Phys. Rev. Lett. **66** (1991) 560
58. K. Wilson, Phys. Rev. **179** (1969) 1499
59. M. Schiestl and H.G. Dosh, Phys. Lett. **B209** (1988) 85; E. Eichten and Feinberg, Phys. Rev. **D23** (1981) 2724
60. V.S. Dotsenko and S.N. Vergeles, Nucl. Phys. **169** (1980) 527; R.A. Brandt et al., Phys. Rev. **D24** (1981) 879, **D 26** (1982) 3611
61. R. Barbieri et al., Phys. Lett. **B57** (1975) 455
62. S.N. Gupta, S.F. Redford, and W.W. Repko, Phys. Rev. **D24** (1981) 2309, **D26** (1982) 3305
63. L.D. Landau, A.A. Abrikosov, and I.M. Khalatnikov, Dokl. ANSSSR **95** (1954) 773
64. G.'tHooft, Nucl. Phys. **B72**, (1974) 461
65. Yu.A. Simonov, Preprint TPI-MINN-90/19-T (unpubl.), Phys. Lett. **B249** (1990) 514
66. G. Bali et al., Phys. Lett. **B309** (1993) 378
67. Yu.A. Simonov, in: *Hadron '93*, eds. T. Bressani, A. Feliciello, G. Preparata, and P.G. Ratcliffe (Centro *A. Volta*, Como, 1993) pp.2629
68. P. Cea, G. Nardulli, and G. Preparata, Z. Phys. **C16** (1982) 135, Phys. Lett. **115** (1982) 310
69. S.I. Eidelman, L.M. Kurdadze, and A.I. Vainshtein, Phys. Lett. **B82** (1979) 278
70. M. Shifman, hep-ph/9511469, Talk at the XVIII Kazimierz Meeting on Particle Physics
71. V.A. Novikov, L.B. Okun, and M.I. Vysotsky, Mod. Phys. Lett. **A9** (1994) 1489;
G. Passarino and M. Veltman, Nucl. Phys. **B160** (1979) 151;
A. Denner et al., Nucl. Phys. **B440** (1995) 95
72. A.M. Badalian, in preparation
73. M.B. Voloshin, Int. J. Mod. Phys. **A10** (1995) 2865
74. A.C. Mattingly and P.M. Stevenson, Phys. Rev. **49** (1994) 437
75. V.A. Abramowski, in preparation
76. B.R. Webber, Phys. Lett. **B339** (1994) 148;
Yu.L. Dokshitzer and B.R. Webber, hep-ph/954219
77. A.H. Mueller, in: *QCD–20 Years Later*, eds . H.A. Kastrup and P.M. Zerwas (World Scientific, Singapore, 1993)
78. F. David, Nucl. Phys. **B234** (1984) 237
79. V.I. Zakharov, Nucl. Phys. **B385** (1992) 452
80. The L3 Collaboration, CERN-PPE/95-1192 December 21, 1995

Factorization and Resummation

George Sterman

Institute for Theoretical Physics, State University of New York at Stony Brook, Stony Brook NY 11794-3840, USA

Abstract. The basic methods of perturbative quantum chromodynamics are introduced and developed through an analysis of short- and long-time dependence in time-ordered perturbation theory for quantum fields. Jet cross sections are shown to be dominated by short times, and hence to be infrared safe. Next, the factorization of short- and long-time dependence for inclusive hard cross sections involving hadrons in the initial state is discussed. The basic results of the parton model, the evolution of deeply inelastic scattering structure functions and the resummation of Sudakov corrections in e^+e^- annihilation are derived from factorization. The lectures conclude with a brief account of the relation between high orders in perturbation theory and power corrections in e^+e^- annihilation.

1 Introductory Comments

In these lectures I will discuss the application of some very general methods in quantum field theory to the gauge theory of the strong interactions, quantum chromodynamics (QCD). Quantum chromodynamics should be thought of as a theory that is "off to a good start", rather than as an area closed to research. It first gained wide acceptance because its property of asymptotic freedom elegantly accounted for scaling in deeply inelastic scattering (see below). I like to draw an analogy between this success and the most famous of all clean explanations from fundamental principles, Newton's derivation of elliptical orbits from a $1/r^2$ force law. Although an extraordinary result, it is exact for the two-body problem only, and by itself is inadequate to establish the validity of Newtonian gravity. Indeed, much of the mathematical physics that we use today was developed (over several centuries) to test gravitation in a larger framework. We are now, perhaps, in an analogous stage with respect to the quantum field theories of the standard model and its extensions.

Among the components of the standard model, quantum chromodynamics shows, at currently accessible energies, the widest range of field theory dynamics. Examples include asymptotic freedom, but also phase structure and confinement, spontaneous symmetry breaking, topological structure, and the coexistence of perturbative and nonperturbative regimes. Thus, although it may not be the cleanest road to "new physics", in the sense of new elementary particles, it may be the most convenient arena for testing the ideas of field theory itself. In these lectures, we shall concentrate on how to use the simplicity of the theory at short distances, even in the presence of its rich structure at long distances. As we shall see, asymptotic freedom makes a perturbative treatment of short distances possible. Most of our work will be to match this perturbative picture to long-distance nonperturbative physics.

My philosophy will be to push perturbation theory as far as it can go. I will begin, in Sections 2-4, by using very general properties of field theory to identify a set of observables, primarily in e^+e^- annihilation experiments, that are dominated directly by short-distance behavior. These observables require, to good approximation, no reconciliation of perturbative and nonperturbative information. Here I will introduce the concept of, and criteria for, "infrared safety".

In Sections 5 and 6, I will show how the technique of factorization may be used to separate systematically long- and short-distance contributions to hard inclusive cross sections that involve one or more hadrons in the initial state. Here approximate scaling, and the quantitative treatment of scale breaking and evolution, will appear as consequences of our search for infrared safety through factorization. In Section 7, I will discuss the treatment of multiple perturbative scales, using Sudakov exponentiation as an example. Finally, Section 8 will introduce power corrections to the infrared safe cross sections of Section 4, and an aspect of the perturbative-nonperturbative interface that has gotten a lot of attention lately, the "infrared renormalons", and their relation to power corrections.

Space and time conspire to limit these lectures to a broad discussion of these results, without explicit calculations. The basic results described here are, of course, already in the literature. Such novelty as may be found in these lectures is in the extensive use of time-ordered perturbation theory as a guide, particularly in the use of stationary phase to identify physical processes as the source of long-time dependence (Section 3).

A few fundamental theorems can be demonstrated explicitly here, especially those related to infrared safety. More complete arguments for factorization theorems must be found elsewhere. Extensive discussions of many of the points introduced here, along with many references to the original literature may be found in a number of useful reviews of QCD. The field-theoretic basis of perturbative QCD has been discussed in books by Ynduráin (1983) and Muta (1987), although some developments are given below which may be less widely familiar (Sterman (1993)). Details of some related elementary calculations are in my lectures at the 1991 TASI Summer School (Sterman (1992)). My 1995 lectures at the same school (Sterman (1996)) cover much of the same material as the present lectures, and the final three sections here are adapted from them.

Very useful recent reviews that treat the material in these lectures in a more directly phenomenological manner include the 1994 TASI lectures of Ellis (1995), and a "Handbook" by the CTEQ Collaboration (Sterman et al. (1995)). As a review of perturbative QCD, the collection of monographs edited by Mueller (1989) is extremely helpful. Finally, for a theoretical introduction from a complementary point of view, see the book of Dokshitser, Khoze, Mueller and Troyan (1991).

2 Time-Ordered Perturbation Theory for QCD

We shall begin with a highly condensed introduction to "old-fashioned", or time-ordered perturbation theory for quantum fields. In using this method, we give up some of the beautiful Lorentz invariance that Feynman made manifest with his diagrammatic techniques, but we will gain insight into the roles of short and long times in field theory.

Let's begin by recalling the classical Lagrangian for quantum chromodynamics,

$$\mathrm{L}_{\mathrm{QCD}} = \int d^3x \left\{ \sum_q \bar{q}(x)\, \gamma \cdot iD[A]\, q(x) \;-\; \frac{1}{2}\mathrm{Tr}\, [F_{\mu\nu}(x)F^{\mu\nu}(x)] \right\}. \quad (1)$$

The full, quantized version of this Lagrangian requires several important ingredients, like gauge fixing and ghosts. For simplicity of presentation, we shall ignore them for the most part. Suffice it to say that a more careful analysis including ghost fields does not change our basic results. To set the stage, let me itemize the basic notation, with which I suppose you are already familiar,

- $A^\mu = \sum_{a=1}^{N^2-1} A_a^\mu\, T_a^{(F)}$: gluon field (vector potential)
- q: quark field
- $\bar{q} = q^\dagger \gamma_0$
- $D_\mu(A) = \partial_\mu + igA^\mu$: covariant derivative
- $F_{\mu\nu} = \partial_\mu A_\nu - \partial_\nu A_\mu + ig[A_\mu, A_\nu]$: field strength
- $\sum_a T_a^{(F)2} = C_{\mathrm{F}}\, I,\ C_{\mathrm{F}} = (N^2-1)/2N$ in SU(N),

where the $T_a^{(F)}$ are matrix generators of SU(N) in its defining representation.

We shall outline the development of perturbative quantum field theory in terms of the Hamiltonian, which for QCD is of the general form

$$H = H^{(0)} + gV_1 + g^2V_2\,, \quad (2)$$

with g the coupling, and V_i functionals of the fields and their conjugate momenta. $H^{(0)}$ consists of all the quadratic terms in H, and if that were all there were, the system would be described in terms of free quark-gluon states. Each term in $V = gV_1 + g^2V_2$ defines an elementary process that mixes free states. Let's see how, in a quantum mechanical interlude.

2.1 From $H^{(0)}$ to Free Quark-Gluon States

The quantization of any system begins with the identification of its degrees of freedom, which for free field theory we may identify with the spatial Fourier transforms of the fields,

$$q(x),\ A_a^\mu(x)\ \to \tilde{q}(\mathbf{k},x^0),\ \tilde{A}_a^\mu(\mathbf{k},x^0)\,. \tag{3}$$

The Hamiltonian is then a functional of these fields (letting the fields stand implicitly for conjugate momenta as well)

$$H = H^{(0)}(\tilde{q},\tilde{A}_\mu) + V(\tilde{q},\tilde{A}_\mu)\,. \tag{4}$$

With the free Hamiltonian $H^{(0)}$, the classical fields obey the wave equation, and hence superposition, and the quantum system is correspondingly described by free-particle states. These states are found by acting on a ground (vacuum) state $|0>$ with the (negative frequency parts of) the momentum-space free fields,

$$|m>=|\{k_i\},\{q_j\}>\sim \prod_i \tilde{q}^\dagger(\mathbf{k}_i,0)^{(-)} \prod_j \tilde{A}_a^\mu(\mathbf{q}_j,0)^{(-)}|0>\ . \tag{5}$$

Their energies are simply the sums of on-shell energies $\omega(\mathbf{k})$ corresponding to the momenta $\mathbf{k}$,

$$H^{(0)}|m>=\left(\sum_i \omega_i(\mathbf{k}) + \sum_j \omega(\mathbf{q}_j)\right)|m>\equiv S_m|m>\ . \tag{6}$$

This of course, is just for the free theory. We now introduce the potential terms as perturbations, which appear as interactions between the particles of the free states. In these terms:

2.2 The Interaction Mixes the Free States

To compute this mixing, we start with the Schrödinger picture of time development, in which the states of the theory inherit all the time dependence and obey the Schrödinger equation, in quantum field theory just as in quantum mechanics,

$$i\frac{\partial}{\partial t}|\psi(t)>=H|\psi(t)>\,, \tag{7}$$

with H the full Hamiltonian. A convenient way to solve this equation perturbatively is to separate the time development of the free part of H from the states and give it to the operators. This defines the interaction picture of time development, whose states are derived from Schrödinger picture states by

$$|\psi_{\rm I}(t)>=e^{iH^{(0)}t}|\psi(t)>\,, \tag{8}$$

or, in (7),

$$i\frac{\partial}{\partial t}\left(e^{-iH^{(0)}t}|\psi_{\rm I}(t)>\right) = \left(H^{(0)}+V\right)e^{-iH^{(0)}t}|\psi_{\rm I}(t)> \,. \tag{9}$$

This gives a new equation for the time development of interaction-picture states

$$i\frac{\partial}{\partial t}|\psi_{\rm I}(t)>= V_{\rm I}(t)\,|\psi_{\rm I}(t)>\,, \tag{10}$$

in term of a new potential, whose matrix elements between free states are easily computed,

$$< m'|V_{\rm I}(t)|m>=< m'|e^{iH^{(0)}t}\;V\;e^{-iH^{(0)}t}|m>= e^{-i(S_m-S_{m'})t}V_{m'm}\,, \tag{11}$$

with S_m the energy of free state m as in Eq. (6).

We are now interested in solutions to Eq. (10) with boundary conditions $|\psi_{\rm I}(-\infty)>=|m_0>$. This will tell us about how states that are effectively free far in the past evolve into the future. A bit of thought shows that this is just what we want for scattering experiments, in which pure states describing incoming isolated particles in the distant past scatter to give superpositions of pure states describing outgoing isolated particles in the distant future. The overlap of the second sort of states ("out states") with the first sort ("in states") defines the S-matrix.

The solution to Eq. (10) with this boundary condition is easily expressed as a perturbative series,

$$\begin{aligned}|m_{0,{\rm I}}(t)>=\sum_{n=0}^{\infty}(-i)^n\int_{-\infty}^{t}d\tau_n V_{\rm I}(\tau_n)\int_{-\infty}^{\tau_n}d\tau_{n-1}V_{\rm I}(\tau_{n-1})\\ \times\cdots\times\int_{-\infty}^{\tau_2}d\tau_1 V_{\rm I}(\tau_1)\,|m_0>\,.\end{aligned} \tag{12}$$

Then matrix elements between free states and $|m_{0,{\rm I}}(t)>$ are found by using (11), and inserting complete sets of free states between the $V_{\rm I}$'s,

$$\begin{aligned}< m_n|m_{0,{\rm I}}(t)>=\sum_{n=0}^{\infty}\sum_{m_1\ldots m_n}(-i)^n\,V_{n,n-1}\,V_{n-1,n-2}\ldots\times V_{1,0}\\ \times\int_{-\infty}^{t}d\tau_n\,e^{-i(S_{n-1}-S_n)\tau_n}\int_{-\infty}^{\tau_n}d\tau_{n-1}e^{-i(S_{n-2}-S_{n-1})\tau_{n-1}}\\ \times\cdots\times\int_{-\infty}^{\tau_2}d\tau_1 e^{-i(S_0-S_1)\tau_1}\,.\end{aligned} \tag{13}$$

In this form, time dependence is quite simple, occurring only in exponentials, a feature that we shall exploit in a moment. Now, the matrix element $< m_n|m_{0,{\rm I}}(t)>$ describes the amplitude for a state that begins as a free state m_0 at $t=-\infty$ to evolve into free state m_n at a time t, in terms of sums over all possible intermediate virtual free states through which the system may pass. The matrix elements $V_{i+1,i}$ which govern the transformation of each state into the next are the "vertices" of the theory, which are in general polynomials of the momenta of the free particles of the incoming and outgoing states that mix. They are given in their usual form in Fig. 1, which I include primarily to emphasize that these vertices are the normal ones in QCD perturbation theory, including ghost interactions. The only subtlety here is that the momenta at each vertex is "on-shell", corresponding to one of the on-shell gluons or quarks of the free states.

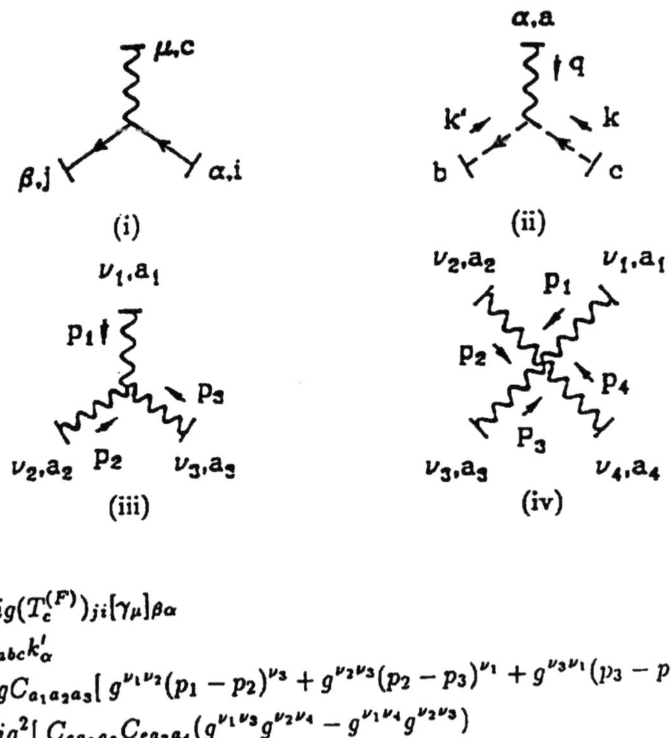

(i) $-ig(T_c^{(F)})_{ji}[\gamma_\mu]_{\beta\alpha}$

(ii) $gC_{abc}k'_\alpha$

(iii) $-gC_{a_1a_2a_3}[\,g^{\nu_1\nu_2}(p_1-p_2)^{\nu_3}+g^{\nu_2\nu_3}(p_2-p_3)^{\nu_1}+g^{\nu_3\nu_1}(p_3-p_1)^{\nu_2}\,]$

(iv) $$-ig^2[\,C_{ea_1a_2}C_{ea_3a_4}(g^{\nu_1\nu_3}g^{\nu_2\nu_4}-g^{\nu_1\nu_4}g^{\nu_2\nu_3}) + C_{ea_1a_3}C_{ea_4a_2}(g^{\nu_1\nu_4}g^{\nu_3\nu_2}-g^{\nu_1\nu_2}g^{\nu_3\nu_4}) + C_{ea_1a_4}C_{ea_2a_3}(g^{\nu_1\nu_2}g^{\nu_4\nu_3}-g^{\nu_1\nu_3}g^{\nu_4\nu_2})\,]$$

Fig. 1. The vertices of QCD.

2.3 Old-Fashioned Perturbation Theory, Energy Deficits

We now return to Eq. (13) and carry out the time integrals, which is easy because all time dependence is, as noted above, in linear exponentials. These time integrals give what is known as time-ordered, or old-fashioned perturbation theory (TOPT). Taking the final time t in (13) to infinity results in an overall energy conservation delta function. Suppressing this factor and the overall momentum conservation delta function, we find

$$\Gamma_{n0}(p) = -i \prod_{\text{loops } i} \int \frac{d^3\ell_i}{(2\pi)^3} \prod_{\text{lines } j} \frac{1}{2\omega_j(p,\ell_i)} \prod_{\text{states } a} \frac{1}{E_a - S_a + i\epsilon} N_{\rm V}(p,\ell_i)\,, \tag{14}$$

where p represents collectively the momenta of the incoming state m_0 and the final state m_n. The characteristic feature of this expression is the product of denominators $1/(E_a - S_a + i\epsilon)$, in which the $i\epsilon$ has been inserted to define the time integrals at minus infinity. Here E_a is the total energy which has entered the graph before time t_a, and S_a is, as in Eq. (6), the (on-shell) energy of the state that follows the action of the ath vertex. It is easy to verify that the normalized integrals over loops in Eq. (14) define precisely a sum over all the intermediate states m_a in (13),

$$\sum_{\text{states}} = \prod_{\text{loops } i} \int \frac{d^3\ell_i}{(2\pi)^3} \prod_{\text{lines } j} \frac{1}{2\omega_j(p,\ell_i)}\,. \tag{15}$$

Thus, TOPT in field theory has the usual form of a sum over states weighted by energy deficits, familiar from nonrelativistic perturbation theory. Finally, we have collected all momentum, group and coupling factors in $N_{\rm V}(p,\ell_i)$, which depends on the external and loop momenta.

The simplest nontrivial example of this formalism is the one-loop self-energy of a single scalar line, shown in Fig. 2. Two time orderings (the right-hand side of the figure) are possible, and together they give

$$\begin{aligned}\Gamma_1 + \Gamma_2 = -i \sum_{2-\text{particle states}} \Bigg(&\frac{1}{p_0 - \omega(k_1) - \omega(k_2) + i\epsilon} \\ &+ \frac{1}{-p_0 - \omega(k_1) - \omega(k_2) + i\epsilon}\Bigg)\,.\end{aligned} \tag{16}$$

As indicated in the figure, the sum over time-ordered diagrams gives the covariant diagram. In this case, it is easily verified that

$$\Gamma_1 + \Gamma_2 = \int \frac{d^4k}{(2\pi)^4} \frac{1}{k^2 - m^2 + i\epsilon} \frac{1}{(p-k)^2 - m^2 + i\epsilon}\,. \tag{17}$$

The price of Lorentz covariance is to attribute off-shell momenta to the lines, and to lose the direct correspondence with physical intermediate states. More generally, for any diagram, we have

$$\Gamma(p) = \sum_{\Gamma_i} \Gamma_i = \prod_{\text{loops } i} \int \frac{d^4\ell_i}{(2\pi)^4} \prod_{\text{lines} j} \frac{1}{k_j(p,\ell_i) - m_j^2} \tilde{N}_{\mathrm{V}}(p,\ell)\,, \qquad (18)$$

where now the momenta in the function $\tilde{N}$ are no longer fixed to be on-shell.

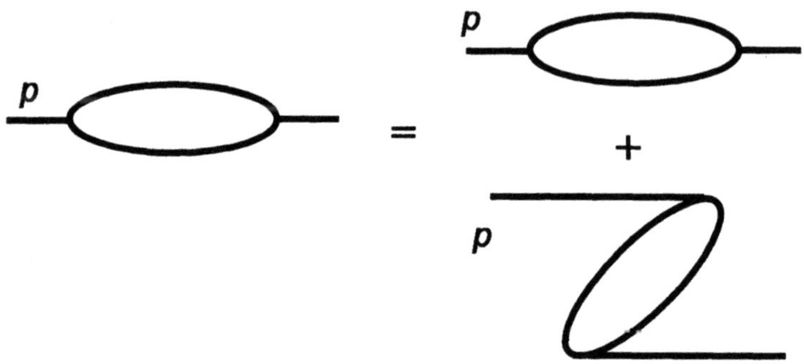

Fig. 2. The two TO diagrams corresponding to the scalar self-energy.

2.4 Asymptotic Freedom and the Running Coupling

From their correspondence with covariant perturbation theory, it is clear that time-ordered diagrams suffer from the usual divergence problems in field theory, both ultraviolet and infrared. In a theory like QCD, there is an extra problem, and an extra benefit, associated with the property called asymptotic freedom. This property implies that the ultraviolet behavior of QCD is simpler, and the infrared behavior more complicated, than we might otherwise expect.

Perhaps the simplest way to introduce asymptotic freedom is to contrast the measurement of the isolated charge in quantum electrodynamics with the corresponding process in quantum chromodynamics. In either case, the answer we get depends upon the resolution with which the measurement is made. This is because any charge is surrounded by virtual quanta, corresponding to the states which are mixed with the isolated charge by the interactions.

For instance, in QED a charge is "dressed" by the emission of virtual photons, which produce in turn virtual electron-positron pairs. Now a probe that seeks to measure any charge does so by testing the field in the neighborhood of that charge, for instance, by absorbing a photon. This photon may originate directly from the original charge, but also from any of the virtual pairs produced by it. In quantum field theory, it is always such a "net" charge that is measured. Its value, not surprisingly, depends on the momentum of the photon that is absorbed by the probe, a dependence that varies from theory to theory. It is readily computed, however, from one-loop diagrams. Qualitatively, in QED electron-positron pairs tend to screen the original point charge, and the latter looks larger at larger photon energy, corresponding to shorter wavelength and better resolution. In QCD, just the opposite is the case: virtual processes increase the net charge seen at larger distance, and the observed charge decreases as the energy of the probing gluon increases. In summary, short wavelength gluons couple more weakly, and long wavelength gluons couple more strongly.

These considerations are incorporated quantitatively in perturbation theory though a scale-dependent coupling, $g(\mu)$, which satisfies an evolution in terms of the QCD beta function

$$\beta(g(\mu)) = \frac{\partial g}{\partial \ln \mu} = \frac{g^3}{16\pi^2} b_2 + \ldots ,$$
$$b_2 = 11 - 2n_f/3 , \tag{19}$$

whose solution is

$$g^2(\mu) = \frac{16\pi^2}{b_2 \ln(\mu^2/\Lambda^2_{\rm QCD})} = \frac{4\pi\alpha_s(\mu_0)}{1 + \frac{\alpha_s b_2}{4\pi} \ln(\mu^2/\mu_0^2)} . \tag{20}$$

Here we use the conventional notation $\alpha_s = g^2/4\pi$. For scale $\mu = \Lambda_{\rm QCD} \equiv \Lambda$ the perturbative coupling diverges, and perturbation theory becomes ill-defined.

In perturbation theory, we are free to expand in a coupling defined at any scale that we like; that is, we may choose any value of μ. Because the theory is independent of this choice, however, μ must also appear in ratios with the other dimensional quantities, in particular in arguments of logarithms. Such quantities include particle masses, but also momentum or space-time invariants.

Suppose, for simplicity, that we have found a physical quantity that depends on single time, T only. Then all logarithms are of the form $\ln(\mu T)$. To keep this logarithm small, the appropriate scale of the coupling is $1/T$. For short times, much less that $1/\Lambda$, the corresponding perturbative coupling $g(1/T)$ is small in QCD. On the other hand, for processes that depend on large times, such that $1/T \sim \Lambda$, the relevant coupling is large, even infinite. Thus, *perturbative* QCD works for short times only, and certainly not for the infinite times that are necessary to define a scattering matrix in TOPT. We are left wondering how we can use it at all. Of course, we can, and to this end the concepts of infrared safety and factorization have been developed. The next section deals with the first of these.

3 Narrowing Down the Time: Infrared Safety

In QCD, in fact in any field theory, it is possible to find physical quantities that *don't* depend on large times, at least when some energy or momentum transfer becomes large enough. Such quantities are said to be *infrared safe*. In identifying infrared safe quantities, for which a perturbative expansion may be useful, we will find that unitarity plays a central role. We will also see how to construct a generalized form of unitarity, from which will follow the ideas of jets associated with the quark-gluon content of QCD. Later, pushing the perturbative picture even further, we will introduce the concept of parton-hadron duality, to describe some of the details of high energy final states.

3.1 Large Times and Physical Pictures

To identify infrared safe quantities, we begin by finding out how long-time dependence arises. Let's recall that the general TOPT structure arises from a "universal" time integral in Eq. (13), of the form,

$$\int_{-\infty}^{t} d\tau_n \, e^{-i(S_{n-1}-S_n)\tau_n} \int_{-\infty}^{\tau_n} d\tau_{n-1} e^{-i(S_{n-2}-S_{n-1})\tau_{n-1}} \times \cdots \times \int_{-\infty}^{\tau_2} d\tau_1 e^{-i(S_0-S_1)\tau_1} \,. \tag{21}$$

Although the time integrals are always unbounded, oscillating phases generally suppress large times. This is obvious when energy denominators, as in Eq. (14), are large, but the vanishing of energy denominators, or of differences $S_a - S_{a-1}$ in (21), is not enough to generate sensitivity to the long-time structure of the theory. Vanishing denominators correspond to vanishing phases, which may occur accidentally in the sum over intermediate states. True long time dependence only arises from points where the phase is *stationary*. Otherwise, the contributions of vanishing phase will simply cancel out.

Now the phase in Eq. (21) has a very simple interpretation, which makes the identification of the sources of large-time dependence much simpler than we might expect. This is summarized in the equation,

$$\begin{aligned} \text{PHASE} &= \sum_{\text{states } m=1}^{n} S_m(\tau_m - \tau_{m-1}) \\ &= \sum_{\text{states } m=1}^{n} \left(\sum_{\text{particle } j \text{ in } m} \omega(\mathbf{p}_j) \right) (\tau_m - \tau_{m-1}) \,. \end{aligned} \tag{22}$$

The conditions of stationary phase with respect to the loop momentum variables are simply

$$\frac{\partial}{\partial \ell_i^\mu} \sum_m S_m(\ell_i)(\tau_m - \tau_{m-1}) = 0 \,, \tag{23}$$

or,

$$\sum_{m} \sum_{\text{lines } j \text{ in } m} v_j^\mu (\tau_m - \tau_{m-1}) \epsilon_{ij}^{(m)} = 0\,, \tag{24}$$

where v_j is the relativistic velocity,

$$v_j^\mu = \frac{p_j^\mu}{\omega_j}\,. \tag{25}$$

$\epsilon_{ij}^{(m)} = +1$ for ℓ_i flowing in the same sense as the momentum p_j in state m; it equals -1 when the sense of flow is opposite with p_j in state m, and it is zero for p_j independent of ℓ_i and/or p_j not in state m.

The conditions (24) are equivalent to the famous Landau equations (Eden et al. (1966), Bjorken and Drell (1965), Sterman (1993)). The normal derivation of the Landau equations involves the analysis of momentum-space integrals in covariant Feynman diagrams, where the singularities appear as pinches between coalescing poles.

Eq. (24) may be interpreted as consistency conditions on the velocities, and hence on the four-momenta of the particles in on-shell intermediate states. If we associate each vertex with a point in space-time, these points are separated by vectors proportional to the velocities of particles times their total time of propagation. For each choice of times τ_i, the equations (24) require that if two vertices are connected by more than one set of lines, the vector separation between the vertices is the same, no matter which set of lines we use to compute it. This means that the time-ordered diagram represents a *consistent physical process*, in which quarks and gluons propagate freely between points in space-time. That is, the regions in momentum space that are troublesome to perturbation theory are those in which quarks and gluons propagate and interact as if they were real, classical particles. When they do not, oscillations due to phases keep correlations over short times. In terms of covariant perturbation theory, the physical processes correspond to "pinch surfaces", where four-dimensional momentum integrals are trapped (Sterman (1993)). The power of these simple considerations is well-illustrated in an important example, QCD corrections to the decay width of the Z electroweak vector boson.

3.2 Z Decay: Collinear and Soft Gluons

The Z vector boson is very heavy compared to the scale Λ at which the QCD coupling becomes strong, $m_Z \sim 90$ GeV. It couples to quarks via weak interactions, so that even at lowest order it can decay to a quark pair. Each of its decay amplitudes, of course, is corrected by QCD itself. We want to identify long-distance behavior (stationary phases) in these amplitudes due to such corrections. The vertex correction, Fig. 3a to the amplitude $A(\mathrm{Z} \to \mathrm{q\bar{q}})$ at order α_s illustrates what happens when the masses of the quarks and gluon are negligible compared to m_Z.

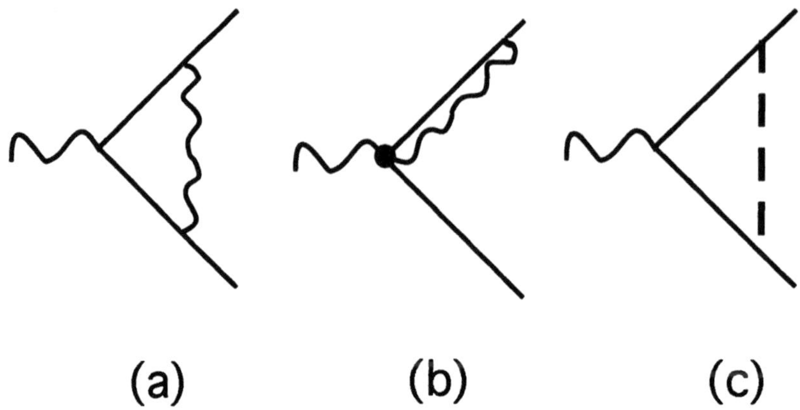

Fig. 3. (a) Vertex correction for Z decay, (b) collinear gluon configuration, (c) soft gluon.

The basic physical processes that involve free propagation of virtual particles in the vertex correction are illustrated in Figs. 3b and c. In the former, the gluon's momentum is proportional to the momentum of the outgoing quark, and two particles are created together at the decay vertex and travel freely at the speed of light for an arbitrary time, after which the gluon is absorbed. In such a configuration, called *collinear*, two of the three virtual lines are on-shell, with the third off-shell. It is common to refer to a set of collinear on-shell particles as a "jet", and we shall use this terminology to describe intermediate as well as final states.

All three lines are on-shell when the gluon's momentum actually vanishes, Fig. 3c. In this case, the gluon may be thought of as possessing an infinite wavelength, and is thus able to connect fast-moving fermions that are separated by spacelike distances.

Collinear and soft configurations are the only physical states that can propagate for finite times, and give rise to long-distance behavior in the vertex function. They correspond to only a small subset of degenerate intermediate states. Any momentum configuration in which the quark pair is moving in directions other than the directions in the final state does not correspond to a point of stationary phase (pinch surface), even though the pair may have the same total energy as the final state. This is because once the quark and antiquark start moving in different directions away from the original decay vertex, they can never meet again at a point in space-time to exchange momentum and produce the final state.

The generalization of this reasoning to all orders for processes such as Z decay and the total e^+e^- annihilation cross section is straightforward. Long-distance behavior comes entirely from processes in which two or more jets of particles are produced at short distances, and propagate into the final state, exchanging soft quanta as they go. This is illustrated in Fig. 4 for two particles. ***Exclusive amplitudes*** are thus a combination of short- and long-time contributions, and cannot be computed reliably in perturbation theory. In certain ***inclusive cross sections***, however, long-time contributions cancel. The simplest of these is the totally inclusive Z decay rate, or more generally the total cross section for e^+e^- annihilation. Our discussion so far will make the proof of their infrared safety to all orders in perturbation theory a simple matter.

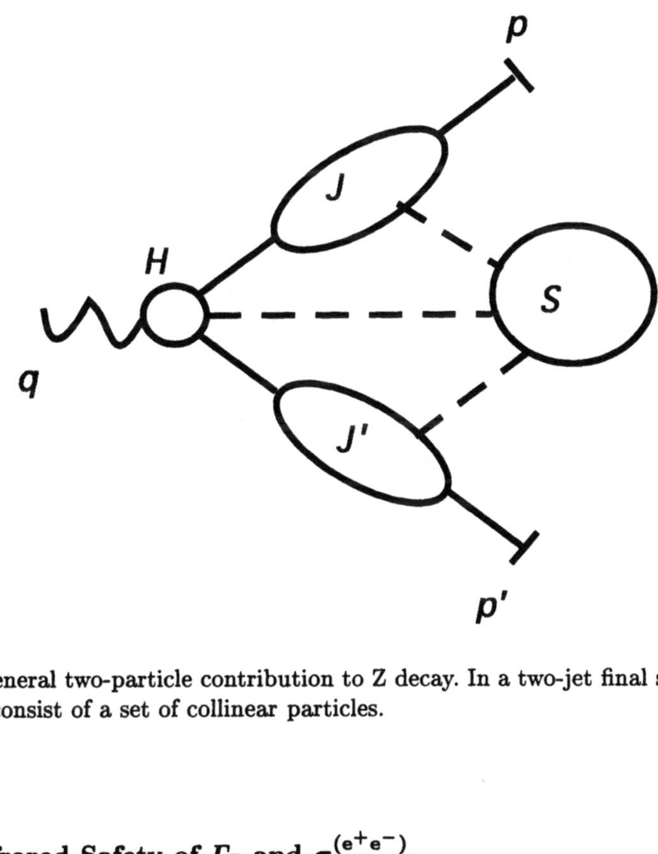

Fig. 4. General two-particle contribution to Z decay. In a two-jet final state, p and $p\prime$ would consist of a set of collinear particles.

3.3 Infrared Safety of Γ_{Z} and $\sigma_{\mathrm{tot}}^{(\mathrm{e}^+\mathrm{e}^-)}$

The infrared safety of the Z total decay rate, Γ_{Z} is a consequence of unitarity in the form of the optical theorem,

$$\Gamma_{\mathrm{Z}} \sim \mathrm{Im}\, \Pi_{\mathrm{Z}}(q^2 = m_{\mathrm{Z}}^2)\,, \tag{26}$$

which relates Γ_{Z} to the imaginary part of the hadronic component of the Z self-energy, Π_{Z}. Long-time behavior in Γ_{Z} can only come from Π_{Z}. To find long-time behavior in Π_Z, we again look for physical pictures. But there are none! We have already seen that the only relevant physical processes involve the decay of the Z into a set of jets. If these jets propagate for a finite amount of time, they can never meet again and annihilate to reform the Z. The details of their propagation into the future, which determines exclusive Z amplitudes, thus washes out in the total rate. This is again unitarity, as expressed in the KLN theorem (Muta (1987), Sterman (1993)): processes that occur long after the decay cannot interfere with the decay itself, and their effects must cancel in the total probability for that decay.

We have now proved that Γ_Z is infrared safe. As such it may be computed with all quark masses m_q set to zero, and is guaranteed to give an expansion in $\alpha_s(M_Z)$ which is free of infrared divergences and of any logarithms of M_Z/m_q,

$$\Gamma_Z = \Gamma_Z^{(EW)} \sum_{n=0}^{\infty} c_n \alpha_s^n(M_Z) . \tag{27}$$

The same reasoning applies quite generally to $\sigma_{tot}^{(e^+e^-)}$ at invariant mass squared Q^2, given as an expansion in $\alpha_s(Q)$. This computation has actually been carried out to third order in α_s for annihilation through both virtual photon and Z. The explicit form for the former is (Ellis (1995))

$$\begin{aligned}\sigma_{tot}^{(\gamma^*)} &= \left(\frac{4\pi\alpha^2}{3s}\right) \sum_q Q_q^2 \sum_{n=0}^{\infty} c_n \alpha_s^n(Q) \\ &= \left(\frac{4\pi\alpha^2}{3s}\right) \sum_q Q_q^2 \left(1 + \frac{\alpha_s(Q)}{\pi} + (1.986 - 0.115 n_q)\left(\frac{\alpha_s(Q)}{\pi}\right)^2 \right. \\ &\quad + \left(-6.637 - 1.200 n_q - 0.005 n_f^2 - 1.240 \frac{(\sum_q Q_q)^2}{3\sum_q Q^2}\right)\left(\frac{\alpha_s(Q)}{\pi}\right)^3 \\ &\quad \left. + \ldots \right) . \end{aligned} \tag{28}$$

Since $\alpha_s(Q)$ decreases as Q grows, we expect QCD effects almost to disappear at high energy, and that is just what happens, as illustrated in Fig. 5 (Particle Data Group (1994)). The rise at the high end of the energy scale is due to the Z. Because QCD effects are small, the shape of the Z peak is predicted by a combination of electroweak and strong perturbation theory with impressive precision.

It is possible to use this precision to measure $\alpha_s(Q)$ by comparing Eq. (28) to experiment (Takeuchi (1996)). It turns out to be of the order of 0.1 at $Q \sim M_Z$. This is encouraging, but we'd like to do more. In particular, a somewhat closer look will enable us to see that the quarks and gluons of QCD have a direct influence on what final states look like.

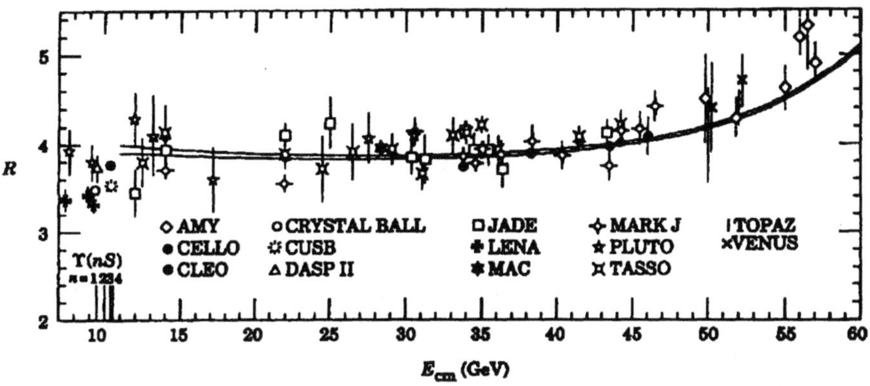

Fig. 5. Ratio R of the total e^+e^- annihilation cross section to hadrons to the total cross section for $e^+e^- \to \mu^+\mu^-$, as a function of center of mass energy.

4 Jets, Quarks and Gluons

4.1 Unitarity and Its Generalization

We have already used unitarity in the form of the optical theorem. Its general expression in field theory may be represented as

$$\sum_m A^\dagger_{fm} A_{mi} = 2\mathrm{Im}\,(-iA_{fi})\,, \tag{29}$$

where A_{mi} is the amplitude for some "initial" state i to mix with intermediate state m, and similarly for A_{fm} with "final" state f. The proof of (29) is particularly simple in TOPT. Schematically, the left-hand side of (29) may be written as

$$\sum_m A^\dagger_{fm} A_{mi} = \sum_{m=1}^{N} \prod_{j=m+1}^{N} \frac{1}{E_j - S_j - i\epsilon}(2\pi)\delta(E_m - S_m) \prod_{k=1}^{m-1} \frac{1}{E_k - S_k + i\epsilon}\,, \tag{30}$$

where we have suppressed numerator factors and momentum integrals (*i.e.* sums over intermediate states). On the other hand, the right-hand side, simplified in the same manner, is

$$2\mathrm{Im}\,(-iA_{fi}) = -i\left[-\prod_{j=1}^{N} \frac{1}{E_j - S_j + i\epsilon} + \prod_{j=1}^{N} \frac{1}{E_j - S_j - i\epsilon}\right]. \tag{31}$$

The equality of these two expressions follows by repeated use of the distribution identity

$$i\left(\frac{1}{x + i\epsilon} - \frac{1}{x - i\epsilon}\right) = 2\pi\delta(x)\,. \tag{32}$$

What I want to emphasize is that to prove unitarity *we didn't need to integrate over momenta*. In technical terms, unitarity follows from the hermiticity of the Hamiltonian, even in a system for which the Hilbert space is truncated. From this observation of "generalized" unitarity, it is a simple step to jet cross sections.

4.2 Infrared Safety of Jet Cross Sections

Suppose we consider a particular exclusive final state consisting of a number of jets of collinear, or nearly collinear, particles. We already know that long-distance behavior in the amplitudes for the production of this final state comes only from intermediate states with the same set of jets, each with the same momentum, although with different particle content. These states will, in general produce infrared divergences and large logarithms in perturbative amplitudes, but if we can cancel these divergences and logs, we will have eliminated long-time behavior. To achieve this, all we have to do is to sum over all final states with the same set of jets. Generalized unitarity, Eq. (29) then allows us to combine this restricted set of states, which produce long-time sensitivity in each others' amplitudes. Their sum equals the imaginary part of a two-point function, just as for the total cross section, but now with all intermediate states required to have the specified jet content. Nevertheless, there are no physical pictures for this two-point function, and we are again assured that long-time behavior cancels in the sum.

Such "jet cross sections" must be defined by restrictions on phase space, the simplest of which is to introduce a set of cones in the annihilation center of mass, and then to specify the energy flowing into each cone, up to some accuracy. For instance, for a two-jet cross section in Z decay, we may demand that the energy in two back-to-back cones of angular width δ be greater than or equal to $(1-\epsilon)m_{\mathrm{Z}}$,

$$\sigma_{2\mathrm{J}} = \sigma(E_{\mathrm{cones}} \geq (1-\epsilon)m_{\mathrm{Z}})\,. \tag{33}$$

By analogy to the treatment of the infrared problem in QED, we consider ϵm_{Z} as an energy "resolution". That is, in perturbative QCD, as in QED, we must account for gluons that are so soft that our detector might miss them. Similarly, δ may be regarded as an angular resolution, beyond which we cannot resolve parallel-moving quarks and gluons. In any case, the details of energy flow at accuracy ϵm_{Z} depend on time scales $1/(\epsilon m_{\mathrm{Z}})$, so we do not want to choose ϵ (or δ) to be too small.

The perturbative cross section for the production of such states is readily computed. The result is, as expected, infrared finite, and, in the limit of zero quark mass is proportional to $1+\cos^2\theta$, with θ the angle between the incoming electron and the outgoing quark. This is direct evidence, strikingly confirmed by experiment, that the quark has spin-1/2.

4.3 "Seeing" Quarks and Gluons with Jet Cross Sections

There are many variations on the jet theme. The general criteria for infrared safety in e^+e^- annihilation are that the parameters that define a cross section be insensitive to the emission of soft gluons and/or the rearrangement of momenta through the interaction of collinear particles.

There is no unique jet definition, rather each event has a sum of possible histories, combined according to quantum mechanical rules. Thus the relationship of jets to quarks and gluons is always approximate. It remains substantive, however, because corrections are computable. The general form of a jet cross section for SU(N) gauge theory in e^+e^- annihilation at center of mass energy Q is a power series in $\alpha_s(Q)$ with coefficients that are functions of group parameters and of a set of variables, denoted y_i, that define the jets,

$$\sigma_{\text{jet}} = \sigma_0 \sum_{n=0}^{\infty} c_n(y_i, N, C_F)\alpha_s^n(Q)\,. \tag{34}$$

The simplest y_i is a cone opening angle, δ. Others are so-called shape variables, whose value is a measure of the distribution of energy in the final state. For example, the "thrust", T is defined by

$$T = \frac{1}{\sqrt{s}}\text{max}_{\hat{n}} \sum_i |\hat{n}\cdot \mathbf{p}_i|\,, \tag{35}$$

where the sum is over all particles in the final state, and the maximum is taken over all unit vectors ("thrust axes") $\hat{n}$. For two perfectly back-to-back jets of massless particles, $T = 1$, so that the thrust of an event is a two-jet measure. We easily check that the thrust cross section is infrared safe according to the criteria introduced above.

Other infrared safe jet cross sections are defined by cluster algorithms. In a cluster algorithm, for each pair of particles, of energy E_i and E_j, we compute a measure y_{ij} that vanishes when the particles are collinear, or when one of their energies vanishes. The most useful of these measures defines the "k_T-algorithm", for which

$$y_{ij} = \min\left(E_i^2, E_j^2\right)\left(1-\cos\theta_{ij}\right)\,, \tag{36}$$

where $\cos\theta_{ij}$ is the angle between their momenta. The momenta of the pair of particles with minimum y is combined according to some linear rule, and the resulting cluster momentum is treated as a particle momentum for the next step. This procedure is continued until all y_{ij} are larger than some predetermined value y_{cut}. If there are n clusters left at this stage, the event is counted as an n-jet event.

Constructions such as these complement the "anecdotal" evidence for jets available from selected event portraits, like those in Fig. 6, from the OPAL Collaboration at LEP, taken at the Z mass (OPAL Collaboration

(1995b)). Thus, as the energy increases in e^+e^- annihilation, we expect the average thrust to approach its two-jet value, $T = 1$, since all higher orders in the thrust cross section vanish as $Q \to \infty$ in Eq. (34). This expectation is satisfied, as is shown by data from TASSO (Wu (1984)), in Fig. 7, taken at PETRA energies, far below the Z mass.

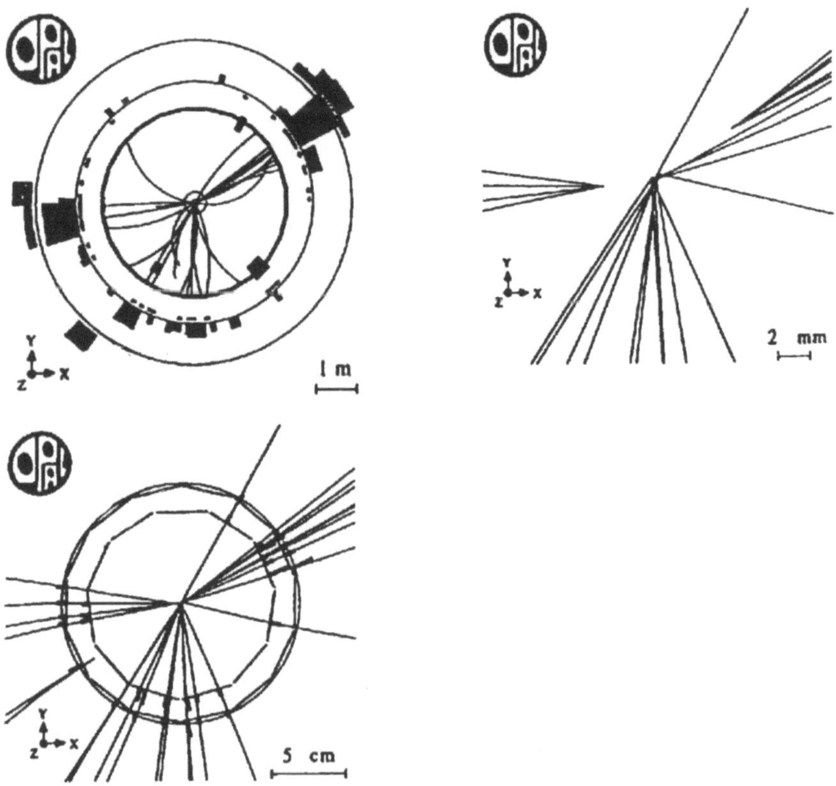

Fig. 6. Jet events from the OPAL Collaboration at LEP.

Aside from their intrinsic interest, jet cross sections may be used to verify the group structure of QCD, for instance by solving (34) for C_F, thus affording direct evidence that QCD with $N = 3$ is indeed the theory of the strong interactions at high energy (OPAL Collaboration (1995a)). Similarly, we may use jet cross sections to determine the strong coupling (Takeuchi (1996)). In practice, we start with a jet cross section $\hat{\sigma}$, computed to some order $n_{\max}$ in α_s. If Δ represents all uncomputed contributions to $\hat{\sigma}$, our determination of

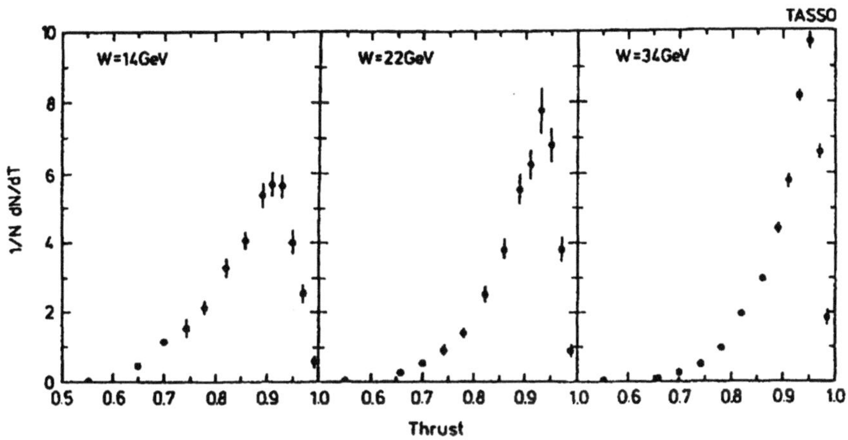

Fig. 7. Experimental thrust profiles measured by TASSO.

$\alpha_s(\mu)$ is schematically,

$$\hat{\sigma}(\alpha_s(Q)) = \sum_{n=0}^{n_{\max}} C_n(Q)\, \alpha_s^n(Q) + \Delta \rightarrow \alpha_s(Q) = f(\hat{\sigma}(Q), C_n(Q), \Delta)\,, \quad (37)$$

for some particular function f. Any experimental determination of α_s must include an estimate of the uncertainty in Δ. When values of α_s determined from various energy scales in this fashion are combined, there is little doubt that the QCD coupling runs with the scale (Particle Data Group (1994)). There is debate, however, as to whether the measured values are consistent with the running predicted by Eq. (20) (Shifman (1995)), given the number of quark flavors in the standard model. Nevertheless, it is worth noting that between the scales 1 and 90 GeV, α_s changes by a factor of roughly two, which takes the running coupling to the edge of the formal radius of convergence of the expansion according to Eq. (20) of α_s at the higher scale in terms of its value at the lower.

4.4 Quark-Gluon Dynamics: Perturbative Formation of the Final State

In principle, when the center of mass energy Q in e^+e^- annihilation is large, all times t,

$$1/Q \leq t \leq 1/\Lambda_{\rm QCD}\,, \quad (38)$$

are available for perturbation theory. If we take this viewpoint, then we can try to calculate a much more fine-grained picture of the final state than simply counting jets. The assumption that hadrons in the final state follow the distribution of partons whose virtuality is slightly larger than Λ is termed *parton-hadron duality.*

Following this line of reasoning allows us to test certain details of QCD dynamics, particularly those relating to "coherence" phenomena. One of the characteristic features of gluon emission as it cascades from highly off-shell to nearly on-shell gluons is *angular ordering*, by which gluons emitted within a jet tend to be increasingly softer and in more forward directions within the jet (Dokshitzer et al. (1988)). We will show how angular ordering arises naturally in the discussion of factorization in Section 6 below. Angular ordering in perturbation theory leads to a somewhat suppressed emission of soft gluons. Parton-hadron duality then suggests a distribution of *hadrons* in jets that also falls off for very small momenta. This prediction has been verified experimentally (OPAL Collaboration (1990)).

On the other hand, in a classical gauge theory, an accelerated charge always radiates. Then the correspondence principle requires some radiation of soft gluons, and thus some soft hadrons. This is evident in the suppression of thrust distributions near $T = 1$ (see Fig. 7), which persists to the highest energies. We will return to this suppression and its analysis in Section 7 below.

5 Forgetting the Past: Factorization

So far, we have only treated cross sections without strongly-interacting particles in the *initial* state. In e^+e^- annihilation, hadronic matter first appears at a definite time, and we have seen that unitarity suppresses contributions from large times after that, at least for infrared safe quantities. It is now natural to ask whether an analogous treatment is possible for large times in the *past*. We still look for quantities that are sensitive to short times, so we consider large momentum transfer processes, inclusive, to control dependence on large positive times. This class of processes is very large, including first of all deeply inelastic scattering,

$$\mathrm{e}(k) + \mathrm{N}(p) \to \mathrm{e}(k-q) + X_{\mathbf{hadronic}}\,, \tag{39}$$

in some sense the classic case. At the next step in complexity are hard hadron-hadron scattering cross sections that are induced by electroweak processes, including

$$\begin{aligned}
&\text{Drell} - \text{Yan} \quad h(p) + h'(p') \to \gamma^*(q),\ \mathrm{Z},\ \mathrm{W} + X_{\mathbf{hadronic}} \\
&\text{Direct Photon} \quad h(p) + h'(p') \to \gamma(q_{\mathrm{T}}) + X_{\mathbf{hadronic}} \\
&\text{Higgs Production} \quad h(p) + h'(p') \to \mathrm{H} + X_{\mathbf{hadronic}}\,.
\end{aligned} \tag{40}$$

Next are hadron-hadron cross sections induced by QCD itself,

$$\begin{aligned}
&\text{Jet Production} \quad h(p) + h'(p') \to J(p_J) + X_{\mathbf{hadronic}} \\
&\text{Heavy} - \text{Quark Production} \quad h(p) + h'(p') \to Q\bar{Q} + X_{\mathbf{hadronic}}\,.
\end{aligned} \tag{41}$$

Indeed, the methods that we will describe apply to the production of any heavy system in hadronic scattering, including those involving extensions of the standard model. Unlike e^+e^- annihilation, it will not be possible to compute these cross sections from first principles in perturbation theory. The consolation is that from the nonperturbative input we will learn about the partonic substurcture of hadrons.

5.1 Factorization for Deeply Inelastic Scattering

Turning first to deeply inelastic scattering, we start by identifying long-distance behavior in perturbative amplitudes from the consistent physical processes that are possible in the zero-mass limit. These are illustrated in Fig. 8. Here the initial-state hadron, of momentum p, scatters from a lepton by absorbing an electroweak boson of spacelike momentum q, with $q^2 = -Q^2$. Physical processes begin with the virtual states of the incoming hadron, consisting of arbitrary numbers of virtual quarks and gluons ("partons"), all collinear to p. One of these partons, of momentum ξp absorbs the boson, producing a hadronic system of momentum $\xi p + q$. This consists of a set of jets, all of which propagate into the final state. By the kinematics of free propagation, there can be no further scattering between the produced jets and the remaining, unscattered fragments of p, whose total momentum is $(1-\xi)p$.

For a given Q^2, there is a minimum value x of ξ, at which $(\xi p + q)^2 = 0$, given by

$$x = \frac{Q^2}{2p \cdot q}\,. \tag{42}$$

This value is often referred to as the Bjorken scaling variable, or as simply "Bjorken x".

Because deeply inelastic scattering can proceed through different polarizations, and through the exchange of different vector bosons, it is conventional to separate this dependence into overall factors, and to define dimensionless "structure functions", $F_i(x, Q^2)$,

$$\sigma_{\mathrm{DIS}} = (\text{Kinematic and EW Factors}) \times F_i(x, Q^2)\,. \tag{43}$$

For electroweak deeply inelastic scattering, three independent structure functions F_i, $i = 1, 2, 3$, are necessary.

Now unitarity arguments may be applied to the deeply inelastic scattering cross section to cancel dependence on long times in the final state. Formally, the inclusive cross section is proportional to the imaginary part of the forward Compton scattering amplitude of a hadron and the off-shell boson. The only remaining long-time dependence is associated with the incoming jet, and its propagation through the past, relative to the hard scattering. This dependence is illustrated in Fig. 9, in which the incoming jet J is connected

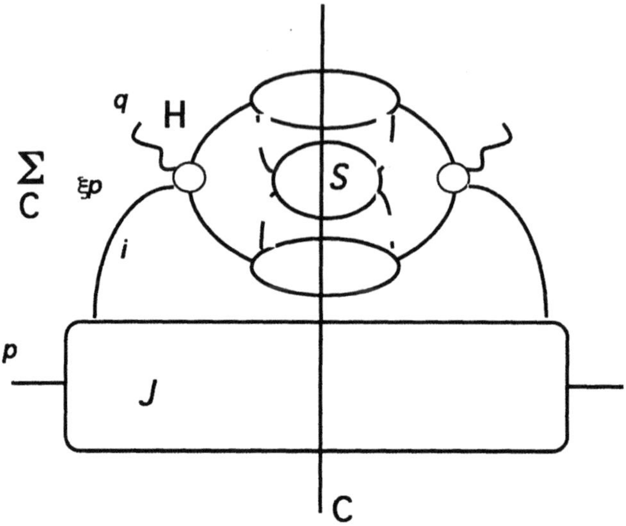

Fig. 8. Consistent physical pictures (pinch surfaces) for deeply inelastic scattering. Final states are labeled by C.

to a hard-scattering function H by a virtual line, whose momentum is nearly equal to ξp, with

$$x \leq \xi \leq 1 . \tag{44}$$

The line connecting J and H is of parton type $i =$ quark, antiquark or gluon. With these observations, we recognize that Fig. 9 is equivalent to the parton model picture of deeply inelastic scattering.

To complete the connection to the parton model, we note that the hard-scattering portion of the picture in Fig. 9, H, consists of lines that are generally far off the mass shell. In computing H, we may, therefore, neglect all components of the momentum of parton i other than ξp, in particular the components of p_i transverse to p and q. An important, and perhaps surprising consequence of this approximation is that the integrals over the transverse momenta of line i *no longer converge in the ultraviolet.* As a result, we must cut them off at some scale, which we shall label μ below. The choice of μ, and our freedom to choose it, will have important consequences. For unpolarized scattering, we can also separate the spin degrees of freedom in J from H, and average over the spins of line i. The result of these technical steps is the

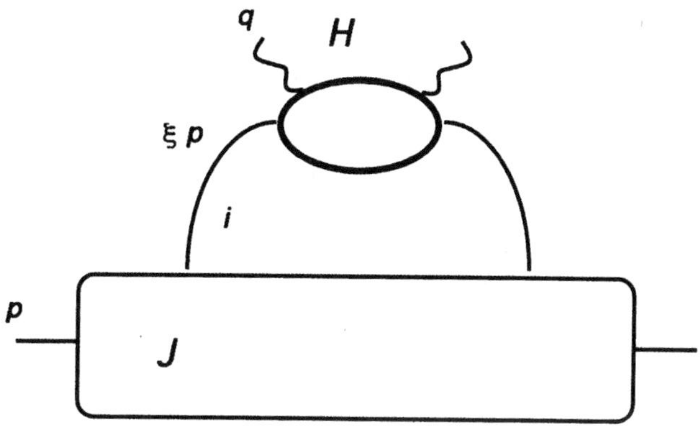

Fig. 9. Physical process for inclusive deeply inelastic scattering after sum over final states (forward Compton scattering).

following "factorized" expression, given for a typical structure function F_2,

$$F_2{}^{(h)}(x,Q^2) = \sum_{i=f,\bar{f},G} \int_x^1 d\xi \; C_2{}^{(i)}\left(\frac{x}{\xi},\frac{Q^2}{\mu^2},\alpha_s(\mu^2)\right)\,\phi_{i/h}(\xi,\mu^2)\,. \tag{45}$$

The ingredients in this expression are:

- $C_2{}^{(i)}\left(\frac{x}{\xi},\frac{Q^2}{\mu^2},\alpha_s(\mu^2)\right)$ is the "coefficient function", which is what becomes of H, after the approximations outlined above. It is process-dependent and computable in perturbation theory.
- $\phi_{i/h}(\xi,\mu^2)$ is the "distribution of parton i in hadron h". It is nonperturbative, and must be inferred from experiment by measuring F_2 experimentally. It is, however, universal among various hard-scattering processes.
- μ is the factorization scale. It serves as an IR Cutoff in C, and a UV cutoff in ϕ. It thus separates short- and long-distance dependence in the structure function.

The parton distributions may be thought of as field-theoretic matrix elements in QCD. In fact, they may be given a natural interpretation as the expectation values of gauge-invariant operators in the single-particle state of the incoming hadron. For the quark, for instance, we may write

$$\phi_{q/h}(x,\mu^2) = \int \frac{dy^-}{4\pi} e^{-ixp^+y^-} \times \langle h(p)|\bar{q}(y^-)\mathrm{P}\, \mathrm{e}^{-ig\int_0^{y^-} dz A^+(z)} \gamma^+ q(0)|h(p)\rangle\,. \quad (46)$$

The ultraviolet cutoff noted above, the need for which results from the factorization of long and short distances, corresponds to the renormalization of this composite operator. These matrix elements are reminiscent of those encountered in the effective field theory for heavy quarks, discussed elsewhere in this school. In particular, the propagator that corresponds to the ordered exponential in (46) is of the form $1/u \cdot k$, with u^μ a lightlike vector, closely related to the propagator $1/v \cdot k$ for a heavy quark of velocity v^μ in heavy quark effective theory.

In the parton model, parton distributions are taken to be independent of μ, and C is computed without QCD corrections, an approximation that makes sense in view of asymptotic freedom. Under these assumptions, the structure functions become independent of Q entirely, and depend on x only. This is known as *scaling*, and explains the use of the term "scaling variable" for x, Eq. (42). We have thus found scaling as an approximate result of asymptotic freedom, applied in connection with the factorization of short- and long-distance (time) dependence based on the identification of physical processes in perturbation theory. Historically, the experimental observation of approximate scaling in deeply inelastic scattering at the SLAC experiments in the late nineteen sixties led to a reevaluation of strong interaction theory, and eventually to the development of QCD. In the meantime, especially since the turn-on of HERA, many details of deeply inelastic scattering final states have become accessible to study. The jet associated with the scattered quark may often be resolved at HERA, along with coherence effects in its development. Of paramount interest, however, is the quantitative analysis of scaling violation.

5.2 Evolution

Perhaps the most striking consequence of Eq. (45) is that it allows us to calculate the μ-dependence of parton distributions. This is particularly useful because once these distributions are computed at one scale, μ_0 say, they can be used to predict cross sections at arbitrary scale μ so long as both $\alpha_s(\mu_0)$ and $\alpha_s(\mu)$ are small. This remarkable result is easy to prove.

Let us denote the convolution in (45) by $\int d\xi \leftrightarrow \otimes$, in terms of which it becomes

$$F = C \otimes \phi\,. \quad (47)$$

Now the physical quantity F must be independent of the arbitrary factorization scale μ,

$$\frac{dF}{d\ln\mu} = 0\,. \quad (48)$$

To this expression we apply separation of variable arguments, observing that the only μ-dependence that C and ϕ have in common is through $\alpha_s(\mu)$. Therefore, the only way that Eq. (48) can be satisfied is for the derivatives of C and ϕ to be convolutions of these functions with functions of $\alpha_s(\mu)$ only,

$$\begin{aligned} \frac{d\phi_{a/A}}{d\ln\mu^2} &= \sum_b P_{ab}(z,\alpha_s)\otimes\phi_{b/A} \\ \frac{dC_c}{d\ln\mu^2} &= -C_c\otimes P_{bc}(z,\alpha_s)\,, \end{aligned} \tag{49}$$

where, because the C_c are perturbative, the $P_{ab}(z,\alpha_s)$ may be computed from perturbation theory. The P_{ab}, which turn out to be distributions in the convolution variable z, are the famous evolution kernels of QCD, and Eq. (49) is known (nowadays) as the DGLAP equation. An example, for $a = b = q$, is the one-loop quark kernel,

$$P^{(1)}_{\rm qq}(\xi,\alpha_s) \equiv \frac{\alpha_s}{2\pi}C_{\rm F}\left\{(1+\xi^2)\left[\frac{1}{1-\xi}\right]_+ + \frac{3}{2}\,\delta(1-\xi)\right\}. \tag{50}$$

Here, the "plus distribution" is defined by its integrals with smooth functions $f(\xi)$ by

$$\int_x^1 d\xi\; f(\xi)\,\left[\frac{1}{1-\xi}\right]_+ = \int_0^1 d\xi\;\left(\frac{f(\xi)-f(1)}{1-\xi}\right) - \int_0^x d\xi\;\frac{f(\xi)}{1-\xi}\,. \tag{51}$$

Generically, the $f(1)$ term in the plus distribution comes from virtual corrections, and the rest from gluon emission.

Moments with respect to x further factorize the DGLAP equation, leading to simple scale breaking expressions in terms of anomalous dimensions,

$$\int_0^1 dx\; x^{n-1}\phi(x,Q^2) \sim e^{-\int^{Q^2}\frac{d\mu^2}{\mu^2}\gamma(n,\alpha_s(\mu^2))}\,, \tag{52}$$

which may themselves be computed from the evolution kernels,

$$\gamma_n(\alpha_s) = -\int_0^1 dz\; z^{n-1}P(z,\alpha_s)\,. \tag{53}$$

Here we have ignored partonic labels a, b; taking them into account leads to qualitatively similar matrix anomalous dimensions.

Asymptotic freedom implies that Q^2-dependence, *i.e.* scale breaking, is relatively weak (logarithmic). This approximate scaling in QCD accounts quantitatively for the many successes of the parton model.

Scale breaking in deeply inelastic scattering according to the DGLAP equation is now firmly established experimentally. Of particular current interest are the limit $x \to 1$, in which the struck parton takes nearly all of the momentum of the incoming hadron, and the limit $x \to 0$, in which it takes a vanishing fraction. The total invariant mass of the hadronic final state is

$$W^2 = Q^2 \frac{1-x}{x} \,. \tag{54}$$

The $x \to 1$ limit thus corresponds to vanishing final-state mass, in general a single well-collimated jet, while $x \to 0$ corresponds to the "Regge" limit: fixed momentum transfer t and large center of mass energy squared s. The connection between small-x deeply inelastic scattering structure functions and the total hadronic cross section at high energy, which is dominated by small momentum transfers, has been recognized for some time. This limit may afford a perturbative approach to some of the most important nonperturbative features of the strong interactions.

5.3 Factorization in Hadron-Hadron Scattering

The program of factorization can be extended beyond deeply inelastic scattering to inclusive hard-scattering cross sections with two hadrons in the initial state. We do not have the time to develop the details of this formalism, but its general form is a direct generalization of Eq. (47),

$$E_C \frac{d\sigma_{ABC}}{d^3 p_C} = \sum_{a,b,c} \phi_{a/A} \otimes \phi_{b/B} \otimes \frac{d\hat{\sigma}_{abc}}{d^3 k_c} \otimes D_{C/c} \,, \tag{55}$$

illustrated by Fig. 10. The cross section is expressed as a sum over parton types a and b from each hadron. The contribution of each term in the sum is factorized into four functions, two parton distributions ϕ, a perturbatively-computable "partonic cross section" $\hat{\sigma}$, and a "fragmentation" function $D_{C/c}$ which describes the probability for a high-p_T or very massive parton c to decay into the observed final state, which may itself include a high-p_T particle C, or a jet, an electroweak heavy boson, a pair of heavy quarks, or a Higgs or even the superpartner of a standard model particle. The fragmentation functions are in many ways analogous to parton distributions, and obey evolution equations similar to Eq. (49). They are also universal among hard-scattering cross sections for given parton c and daughter particle C. They are less well known, however, since their determination requires even more delicate measurements than for parton distributions (Chiappetta et al. (1994), Binnewies et al. (1995)).

Factorized cross sections like Eq. (55) have found many applications at high energy accelerators operating in both fixed-target and collider modes. Their success, when supplemented by the evolution of parton distributions, over extraordinary ranges of momentum transfer is a compelling justification for the perturbative treatment of high-energy QCD, and indeed, for the quantum field theory picture of elementary processes at the shortest accessible distance scales.

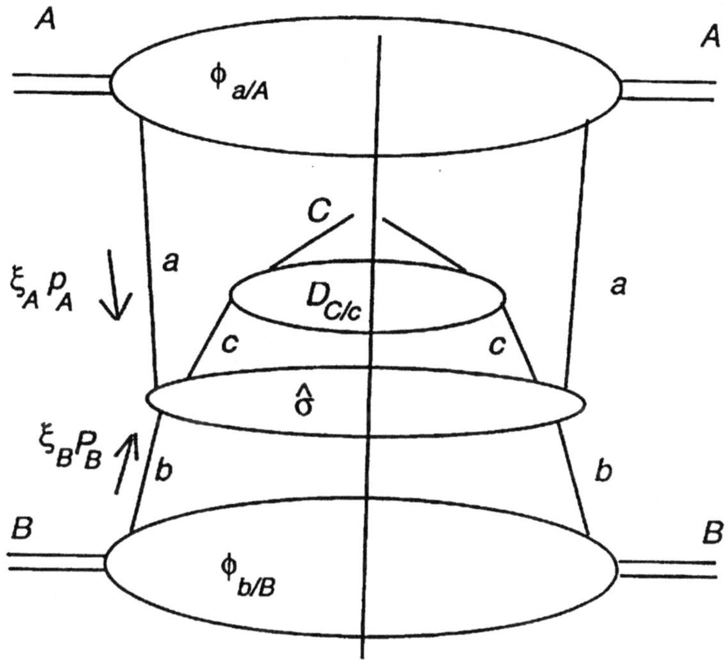

Fig. 10. Factorized hard hadron-hadron scattering.

The data which has become available in the 1990's from LEP, HERA and the Tevatron has not only confirmed the broad outlines of perturbative QCD, it has also presented the challenge to lift perturbative QCD to a new level of quantitative accuracy. Nowhere is the potential for, and the difficulties in, this project better illustrated than by high-p_T jet events at the Tevatron, for which the cross section changes by more than eight orders of magnitude between jet transverse momenta of twenty and five hundred GeV. Given specific parton distributions, the factorization formalism provides absolute predictions for these cross sections, with hard-scattering functions known to next-to-leading order in α_s. Most recent parton distributions give predictions that are accurate to some tens of percent in the full experimental range. At the high end, for p_T's over 200 GeV, considerable interest has been generated by a possible excess of events (CDF Collaboration (1996)). These predictions depend on many variables, both theoretical and experimental, and the highest-p_T jets depend upon parton distributions at relatively large values of x, for which they, particularly the gluon distribution, are not so well known (Huston et al. (1995)). The final interpretation of the highest-p_T jet cross section is therefore as yet not clear.

6 The Nature of Factorization Proofs

Since factorization is so central to the use of perturbative QCD at high energy, it is worthwhile to spend some time discussing the arguments that justify it (Collins, Soper and Sterman (1989)). Without going far into details, I will review first simple heuristic arguments, and then go on to describe a few technical considerations.

6.1 Heuristics of Factorization

Consider deeply inelastic scattering at momentum transfer Q. We assume that the processes that bind the nucleon occur on fixed time scales, $T \gg 1/Q$. This suggests that these interactions cannot interfere quantum mechanically with the large momentum transfer (electroweak) process that initiated the scattering. As seen by the initial-state electron in its rest frame, the lifetimes of all relevant virtual states are highly dilated, and the electron may be regarded as encountering a nucleon state with a definite number of partons, each with a well-defined fraction of the nucleon's momentum. These partons are randomly distributed over the nucleon, which is also seen as highly Lorentz-contracted. At large Q, the electron must (by accident) encounter one of the partons, whose momentum is ξp, $0 \leq \xi \leq 1$, at a transverse distance of no more than $1/Q$ (by the uncertainty principle). The probability for encountering more than one parton at this distance scale is further suppressed by an extra (geometrical) factor of $1/\pi R^2 Q^2$ with R the nucleon radius. Thus the cross section is an integral over ξ of the probability ($\phi(\xi)$) for finding a parton with fraction ξ of the hadron's momentum times a single-parton hard-scattering cross section, just as in Eq. (45). This line of reasoning also suggests that scatterings involving more than one parton are power-suppressed compared to (45).

A very similar heuristic description of hadron-hadron scattering may be given, in which the partons of each colliding hadron play the role of the electron in deeply inelastic scattering for the partons of the other hadron. There is an important difference, however, since in principle the partons in different hadrons can interact before the hard scattering takes place. Such interactions could produce a polarization of the partons in each hadron by the other's partons. They could in this way modify the distribution of partons as seen at the hard scattering. This would have important consequences for hadron-hadron scattering, since we would lose universality for the functions $\phi_{a/A}$, which could be different in Eq. (55) and in (45). This universality, however, seems to be respected by experiment.

A (similarly heuristic) resolution to this problem may be found in the Lorentz transformation properties of classical Coulomb fields. Qualitatively, the vector field of an isolated charge transforms as a Lorentz vector, and appears, in a frame in which the charge is rapidly moving, to be proportional to the four-velocity of the charge. The field strengths, which produce sensible forces, however, behave very differently. Field strengths, in fact, are Lorentz contracted more strongly than rulers! Their magnitude at any fixed time before the collision shrinks as $1/\gamma^2$, with $\gamma = 1/\sqrt{1-v^2/c^2}$ the familiar parameter of special relativity. Now $\gamma \sim E_{\mathrm{projectile}}/m_{\mathrm{projectile}}$ in a target rest frame, which suggests that nonfactoring corrections from the initial state decrease as $1/\gamma^2 \sim 1/s^2$, with s the center of mass energy squared. Arguments for this strong suppression of nonfactoring corrections have been extended to perturbative field theory (Qiu and Sterman (1991)).

6.2 Factoring Soft Gluons from Jets in Perturbation Theory

The proof of factorization theorems in hadron-hadron scattering (Collins, Soper and Sterman (1989)) is highly nontrivial in perturbation theory, and has not reached the sophistication of treatments of renormalization. Many technical issues, however, correspond closely to the heuristic arguments which are sketched above.

Consider, for instance, the physical pictures relevant to the Drell-Yan process, illustrated by Fig. 11. The jets of the two incoming hadrons are connected by soft gluons, corresponding to the initial-state scattering of partons from either hadron due to the Coulomb fields of partons in the other hadron.

In our heuristic discussion, we observed that although the vector potential of a moving charge may be large, its field strength is highly Lorentz-contracted. In perturbation theory, the corresponding phenomena is manifested by the unphysical degrees of freedom of gauge fields A^μ. The large unphysical polarization of a gluon turns out to be proportional to its momentum, and thus may grow under Lorentz boosts. We expect the effects of such unphysical polarizations to contribute in individual diagrams in perturbation theory, but to cancel in a gauge invariant sum over diagrams.

To see how this works out in perturbation theory requires application of Ward identities that decouple unphysical gluons from physical processes. Let us sketch how this can be done, taking as an example connections of soft lines to a final-state jet, as shown in Fig. 12.

At a typical pinch surface, soft lines, q to the left of the cut, and q' to the right, are attached to the jet, which we may take in the plus direction, with large momentum component p^+. We may define both soft momenta to flow into the jet, and then along jet lines away from the final state and toward the hard scattering functions H. The momentum of each jet line along which a soft momentum flows is then of the form $\ell - q$ for lines to the left of the cut (amplitude) and $\ell + q$ to the right (complex conjugate amplitude). Here ℓ is the momentum of a jet line in the absence of the extra soft momentum, and is hence naturally parameterized as

$$\ell = (xp^+, \ell^-, \ell_\perp)\,, \tag{56}$$

with $0 < x \leq 1$. For ℓ a jet momentum, we can assume that xp^+ is its largest component. We can now make two approximations that will enable us to

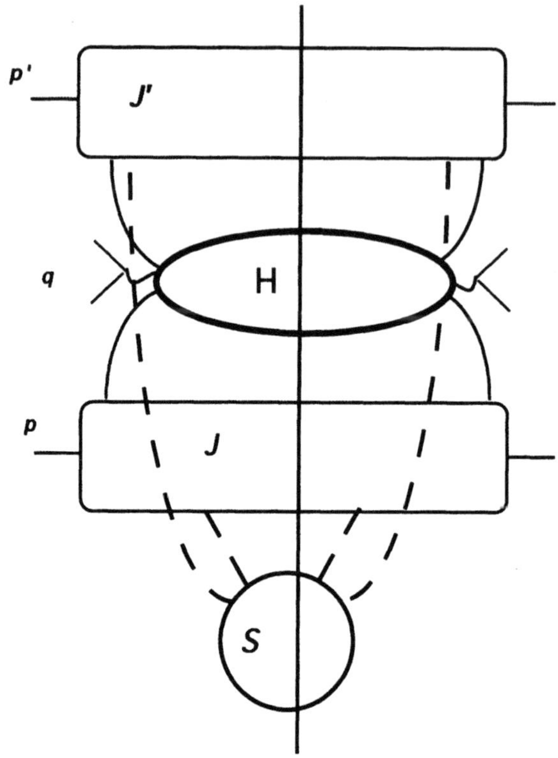

Fig. 11. General physical process in the Drell-Yan cross section. The subdiagram H includes possible final-state jets, and for lack of space the Drell-Yan pair is reversed in direction.

factor the soft gluons from the jet, replacing their couplings by an effective interaction. The first is that the coupling of the gluons' propagators $G_{\nu\lambda}(q)$ to the jet is always through the combination $p^\nu G_{\nu\lambda}(q)$, that is, that the soft gluons couple only to the large component of the current. The second is that propagator denominators $(\ell+q)^2$, depend only upon the minus component of the soft gluon's momenta, *i.e.* the components in the direction opposite to the jet direction,

$$(\ell \pm q)^2 \sim \pm 2xp^+q^- + \ell^2 . \tag{57}$$

We shall return to this condition in a moment.

Once these approximations are made, the jet depends only on one component of each soft gluons' momentum, and on the same component of its polarization. Thus the coupling of soft gluons to a jet moving in the plus direction is equivalent to the coupling of a set of gluons with *only* minus momenta and *only* minus polarizations. Such gluons are unphysical, and are

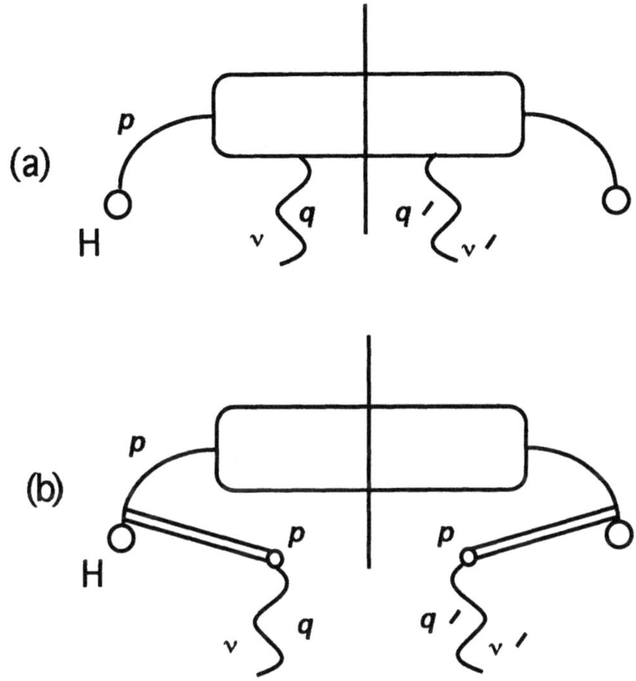

Fig. 12. (a) Diagram in which soft lines connect to a final-state jet. (b) Factorization of soft lines.

equivalent to a phase rotation on the external lines of the jet, which attach to the hard scatterings. In an abelian theory, for instance, an arbitrary jet with one vector (photon) attached to each side of the cut, is equivalent to a jet with no photons, multiplied by two linearized propagators ("eikonal lines"), one for each of the photons,

$$J_2^{\nu'\nu}(p,q,q') \sim J_2^{++}(p,\hat{q},\hat{q}') = g^2 \, \frac{p^{\nu'}}{p^+q'^- - i\epsilon} \, \frac{p^{\nu}}{-p^+q^- + i\epsilon} \, J_0(p) \,, \qquad (58)$$

where $J_0(p)$ is the jet with no soft photon connections and where

$$\hat{q}^\mu = (0, q^-, \mathbf{0}) \,. \qquad (59)$$

The factors on the right of (58) are of the form of an eikonal propagator and vertex, illustrated in Fig. 12b. In Eq. (58), g is the total charge of the jet.

In a nonabelian theory, multiple gluons still factor, and their color interactions are summarized by connections to eikonal Wilson lines, which have the same perturbation theory rules for lines and vertices as for deeply inelastic scattering parton distributions Eq. (46). Schematically, we write

$$J_n^{+\cdots+}(p,\hat{q}_i,\hat{q}'_j) = E^*(\{q'_j\})E(\{q_i\})J_0(p) \,, \qquad (60)$$

where the E's are generated from the operators

$$U(A) = \exp\left[-ig\int_0^\infty d\lambda\, p\cdot A(\lambda p)\right]\,. \tag{61}$$

Here A is in the color representation carried by the hard parton that initiates the jet. The soft divergences factorized in this manner cancel due to the unitarity of the Wilson lines,

$$U[A]^\dagger\, U[A] = I\,. \tag{62}$$

To realize this cancellation, it is necessary to sum over all final states that differ by soft gluon emission. This, of course, is exactly what we do in inclusive hard-scattering cross sections. Given this cancellation, the remaining jets are independent of each other, and may be factored from the hard scattering as in deeply inelastic scattering.

The argument that we have given above is, of course, quite rough. In particular, the approximation (57) requires justification. It is clearly not true for every soft momentum q^μ, since it fails whenever q^- vanishes compared to the other components of q^μ. This is fine if $q^+ \gg q_\perp$, because then the soft gluon may be thought of as part of the jet. Even for gluons whose momenta are not in the jet direction, however, the approximation will hold if the q^- integral is not trapped between singularities in the complex plane at or very near the origin. In this case, we may use Cauchy's theorem to deform the q^- contour into a region where (57) holds.

Arguments that the q^- contour is not pinched are relatively easy to give for many cross sections in e^+e^- annihilation, where we verify that all the poles in q^- from lines within the jet, $(\ell \pm q)^2 \mp i\epsilon$, are in the upper half-plane for soft gluon momenta routed as above. The situation is much more difficult when the jet originates from an incoming hadron, as in the Drell-Yan process. Here, singularities may pinch momentum contours on a diagram-by-diagram basis, as in Fig. 13, in regions where the eikonal approximation fails, and Ward identities do not help. In the figure, for instance, the lines $xp+q$ and $(1-x)p-q$ have q^- poles in opposite half-planes. The resolution of this problem is again found in TOPT, where upper half-plane poles are seen to arise from times after the hard scattering, and lower half-plane poles from times before it. In sufficiently inclusive cross sections, however, final-state interactions, and hence upper half-plane poles, cancel for the physical reasons that we have outlined above. The q^- contours of soft gluons that are not in the jet direction are thus no longer pinched after the sum over final states. The decoupling of soft gluons from incoming as well as outgoing jets may then be demonstrated (Collins, Soper and Sterman (1989)).

We now ask when a soft gluon is close enough to a jet's direction to be part of that jet. Let us consider the case when the two momenta in Eq. (57) are both on-shell, *i.e.*, $q^2 = \ell^2 = 0$. To factor q from the p-jet we require, in particular, that

$$xp^+q^- > |\ell_\perp\cdot q_\perp|\,. \tag{63}$$

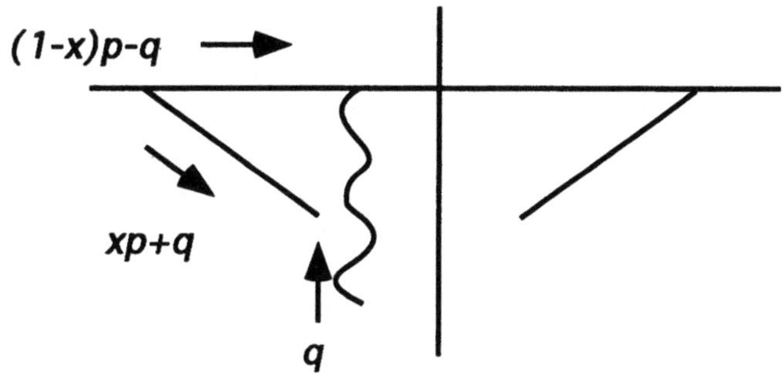

Fig. 13. Diagram for which q^- is pinched close to zero.

This condition has a natural geometrical interpretation, which itself has important physical consequences. For a well-collimated jet, $\theta_\ell = |\ell_\perp|/(xp^+)$ is the angle of ℓ^μ to the overall jet direction. Defining θ_q similarly, and parameterizing $q^+ = yp^+$, we have

$$|\ell_\perp| = \theta_\ell xp^+, \quad |q_\perp| = \theta_q yp^+, \quad q^- = \frac{q_\perp^2}{2yp^+} . \tag{64}$$

Using these relations it is easy to show that the soft gluon q will factorize from the jet, *i.e.* Eq. (57) will hold, *unless*

$$\theta_q < \theta_\ell , \tag{65}$$

that is, unless q is emitted closer to the jet axis than ℓ, the particle that emitted it. This "angular ordering", limits the region of phase space into which soft gluons may be emitted late in jet evolution, and hence suppresses the multiplicity of very soft gluons, since gluons that factor from the jet do not get enhancements from physical processes (pinch surfaces) where jet lines are near the mass shell. This is one of the many consequences of the coherence of gluon emission in QCD (Dokshitzer et al. (1988)).

7 Resummation Near the Edge of Phase Space

The factorization theorems of the previous section go far toward connecting perturbative QCD to experiment. Their application, however, is limited somewhat by our assumption that there is only a single large scale in the problem, for instance, Q^2 in deeply inelastic scattering. In this case, it is natural to choose the factorization scale to be of order of Q, and to use the evolution of the parton distributions discussed above. The coefficient function is then of the general form

$$C(x/\xi) = \sum_n d_n(x/\xi)\, \alpha_s^n , \tag{66}$$

where $d_n(z)$ is a distribution in z. Then, if $\alpha_s(Q^2)$ is small, we may hope that the effect of higher-order terms in C is small as well.

Unfortunately, this is not always the case even in deeply inelastic scattering; the $d_n(z)$ will generally include factors like $\alpha_s^n \ln^n(z)/z$, or even $\alpha_s^n \times [\ln^{2n-1}(1-z)/(1-z)]_+$. When x is very small, or very close to unity, these logarithms may produce large corrections at every order. Recalling that $(p+q)^2/Q^2 = (1-x)/x$, we see that in both cases the arguments of the logarithms are ratios of kinematic invariants. There are many other such examples, all with two hard scales, Q_1^2 and Q_2^2, which satisfy

$$Q_1^2 \gg Q_2^2 \gg \Lambda^2 \,. \tag{67}$$

The first inequality ensures the presence of a large ratio in the hard scattering function. The second ensures that parton densities evolve perturbatively, and that the running coupling in the hard scattering function, $\alpha_s(Q_2^2)$ is small, even if the combination $\alpha_s(Q_2^2)\ln(Q_1^2/Q_2^2)$ is not. In this case, we may try to "resum" the series in $\ln(Q_1^2/Q_2^2)$ to all orders in perturbation theory. As Q_2 approaches Λ, we smoothly connect to the idea of parton-hadron duality. Time allows the treatment of only one example, the "Sudakov" double logarithms that occur in many cross sections when they approach the kinematics of elastic scattering. The other canonical application of resummation is illustrated by the treatment of small-x logarithms (Lipatov (1989), Del Duca (1995)), which are related, as noted above, to the Regge limit of field theory.

7.1 Sudakov Double Logarithms

A classic example of Sudakov double logs is in the electromagnetic form factor for pair creation, which in $n = 4 - 2\epsilon$ dimensions takes the form,

$$\Gamma_\mu(p_1, p_2, \epsilon) = -ie\mu^\epsilon \; \bar{u}\,(p_1)\gamma_\mu v(p_2)\; \rho(Q^2, \epsilon)\,, \tag{68}$$

with $Q^2 = (p_1 + p_2)^2$. For massless quarks, explicit calculation shows that double logarithms, and poles in dimensional regularization, exponentiate as

$$\rho(Q^2, \epsilon) = \exp\left[-\frac{\alpha_s}{4\pi} C_F \left(\frac{2}{\epsilon^2} + \ln^2(Q^2/\mu^2) + \ldots\right)\right] . \tag{69}$$

Double logarithms in momentum transfer are a reflection of the overlap of collinear and soft singularities in the vertex function, and are generally referred to as Sudakov double logarithms (Collins (1989)). Although the electromagnetic vertex is not infrared safe, it is an important input to the calculation of any process whose kinematics approaches those of elastic scattering.

As an example of resummation applied to an infrared safe cross section, I will discuss the thrust near $T = 1$ (Catani et al. (1993)). Here Sudakov double logarithms appear not in Q^2 directly, but in $(1 - T)$. Consulting the definition of thrust, Eq. (35), we see that at $T = 1$ the final state consists of two back-to-back massless jets. A little kinematics also shows that when both jet masses p_i^2 are small compared to Q^2 they are related to the thrust by

$$1 - T = \frac{p_1^2 + p_2^2}{Q^2} \,. \tag{70}$$

At order α_s^n the leading logarithm in $1 - T$ is given by an exponential of Sudakov logarithms,

$$\frac{1}{\sigma_{\text{tot}}} \frac{d\sigma}{dT} = -2C_{\text{F}} \frac{\alpha_s}{\pi} \frac{\ln(1-T)}{1-T} \exp\left\{ -C_{\text{F}} \frac{\alpha_s}{\pi} \ln^2(1-T) \right\} \,. \tag{71}$$

As T approaches one, $\frac{d\sigma}{dT}$ vanishes. This is a quantum-mechanical reflection (already noted in Section 4) of the classical radiation field that must accompany any process in which a charged particle is accelerated (quark pair creation being an extreme example). Quantum field theory assembles this classical field out of many soft and collinear gluons (the correspondence principle). Cross sections in which gluon emission is forbidden in part of phase space are correspondingly suppressed.

7.2 Factorization for $T \to 1$

Our goal in the following is to rederive Eq. (71), and to extend it to include nonleading logarithms and the effects of the running coupling. Our arguments apply to a large class of cross sections. In fact, the resummation of Sudakov logarithms follows from the factorization properties of the cross section in the regions of momentum space that give rise to the logarithms. The factorization of soft glue from jets has already been illustrated at low order in Fig. 12. The infrared divergences associated with such configurations cancel, according to the arguments of Section 4, but for $T \sim 1$, the cancellation between diagrams with virtual and real gluons is constrained by the requirements that the real gluons be either very soft or emitted very close to the quark or antiquark directions. This leaves large finite remainders in the cancellation of divergences. These are the logarithms of $1 - T$.

The form of Fig. 12, combined with the factorization of soft gluons from jets, suggests that the thrust cross section factorizes into functions that describe the two jets, the hard, and the soft subdiagrams. We will simplify the expression even further, and assume that the effects of soft exchanges may be absorbed into redefinitions of the jets. In fact, this is true to next-to-leading logarithm in a axial gauge,

$$\xi \cdot A = 0 \,, \tag{72}$$

where we assume $\xi^2 \neq 0$. A more general treatment, which is accurate to all logarithmic order (Contopanagos et al. (1996)) gives essentially equivalent results. At fixed values of jet masses p_i^2, we shall therefore begin with the factorized expression

$$\frac{d\sigma}{dp_1^2 dp_2^2} = J_1(p_1, \mu, \xi) J_2(p_2, \mu, \xi) H(p_1, p_2, \mu, \xi) \,, \tag{73}$$

where the J's represent the jets, and H the hard-scattering factor. We shall suppress particle labels, but it is relatively easy to show from power counting that to leading *power* in $1-T$ the jets are connected to the hard scattering by a quark and antiquark only. Eq. (73) is enough to derive the exponentiated double logarithms of Eq. (71) above, with corrections due to the running of the coupling.

In axial gauge (72) the jets depend, not only on their invariant momenta, but also on their energies through the products $p_i \cdot \xi$. The jets and H are thus not individually Lorentz invariant. For a general gauge vector ξ^μ, the precise arguments are $p_i \cdot \xi / \sqrt{\xi^2}$ because the gluon propagator, and hence J, is invariant under rescalings of ξ^μ. (In the following, we set $\xi^2 = 1$.)

In view of Eqs. (70) and (73), the cross section at fixed $1 - T \sim 0$ is of a convolution form

$$\frac{1}{\sigma_0}\frac{d\sigma}{dT} \simeq \int_0^{Q^2} dp_1^2 dp_2^2\, \delta\left(1 - T - \frac{p_2^2}{Q^2} - \frac{p_2^2}{Q^2}\right) H\left(p_1 \cdot \xi/\mu, p_2 \cdot \xi/\mu, \alpha_s(\mu^2)\right)$$
$$\times J_1\left(p_1^2/\mu^2, p_1 \cdot \xi/\mu, \alpha_s(\mu^2)\right) J_2\left(p_2^2/\mu^2, p_2 \cdot \xi/\mu, \alpha_s(\mu^2)\right) . \tag{74}$$

Dividing by the Born cross section σ_0 gives $H = 1$ at lowest order. Because we are interested in the limit $p_i^2/Q^2 \to 0$, the p_i may be expanded about back-to-back lightlike momenta. We choose axes so that p_1^μ is in the plus direction and p_2^μ is in the minus direction. Then p_1^+ and p_2^- are both nearly equal to $Q/\sqrt{2}$, while

$$p_1^- \sim \frac{p_1^2}{\sqrt{2}Q}, \quad p_2^+ \sim \frac{p_2^2}{\sqrt{2}Q} . \tag{75}$$

For ξ in an arbitrary direction, we may take $p_1 \cdot \xi \sim p_1^+ \xi^-$ and $p_2 \cdot \xi \sim p_2^- \xi^+$ to leading power in Q. Then to leading power in $1 - T$, the p_i^2 integrals are independent of $p_i \cdot \xi$.

We want to identify singular $\ln^m(1-T)/(1-T)$ behavior for $T \to 1$, and for this purpose moments with respect to T are particularly useful,

$$\tilde{\sigma}(n) = \frac{1}{\sigma_0}\int_0^1 dT\, T^{\,n} \frac{d\sigma}{dT} , \tag{76}$$

since the moments of any function that is finite at $T = 1$ falls off as $1/n$ for $n \to \infty$. In particular, logarithms of $1 - T$ are transformed into logarithms of n by

$$\int_0^1 dT\, \frac{T^{\,n} - 1}{1-T} \ln^m(1 - T) = \frac{(-1)^m}{m+1} \ln^{m+1} n + \ldots . \tag{77}$$

Keeping only terms that are finite or grow as $n \to \infty$, and using the relation $T^{\,n} \sim e^{-n(1-T)}$, which holds in this approximation, we find that the convolution in (74) factors into a simple product under moments,

$$\tilde{\sigma}(n) = \tilde{J}_1\left(Q^2/n\mu^2, p_1\cdot\xi/\mu, \alpha_s(\mu^2)\right)\ \tilde{J}_2\left(Q^2/n\mu^2, p_2\cdot\xi/\mu, \alpha_s(\mu^2)\right) \\ \times H\left(p_1\cdot\xi/\mu, p_2\cdot\xi/\mu, \alpha_s(\mu^2)\right)\,, \tag{78}$$

up to corrections that vanish as $1/n$. Note that n now appears only in the combination $Q^2/n\mu^2$. By analogy to the derivation of evolution for deeply inelastic scattering structure functions, we shall use this factorized expression, coupled with renormalization group arguments, to derive a resummed cross section. The new feature in our Sudakov factorization is the dependence on the axial gauge vector ξ. Although each of the factors that makes up $\tilde{\sigma}$ is ξ-dependent, the physical quantity $\tilde{\sigma}$ must be gauge-invariant. This invariance will drive the resummation (Collins and Soper (1981)).

7.3 Resummation for $T \to 1$

We start with the renormalization group behavior of J, which is simple, since it has only two external (quark or antiquark) lines,

$$\left[\mu\frac{\partial}{\partial\mu} + \beta\frac{\partial}{\partial g}\right]\ln\tilde{J} = -2\gamma_q\,, \tag{79}$$

with q the quark anomalous dimension. Since the cross section is independent of the renormalization scale μ, H must behave in a corresponding fashion,

$$\left[\mu\frac{\partial}{\partial\mu} + \beta\frac{\partial}{\partial g}\right]\ln H = 4\gamma_q\,. \tag{80}$$

We now observe that very similar reasoning may be applied to the gauge-fixing vector ξ.

Let us change (boost) ξ^+ and ξ^- in a manner that leaves $\xi^2 = 1$. The independence of $\tilde{\sigma}(n)$ from ξ^μ may be expressed as $\partial\tilde{\sigma}/\partial\ln\xi^- = -\partial\tilde{\sigma}/\partial\ln\xi^+$. The chain rule then gives

$$\frac{\partial\ln\tilde{J}_2(Q^2/n\mu^2, p_2^-/\mu)}{\partial\ln p_2^-} + \frac{\partial\ln H(p_1^+/\mu, p_2^-/\mu)}{\partial\ln p_2^-} \\ = -\frac{\partial\ln\tilde{J}_1(Q^2/n\mu^2, p_1^+/\mu)}{\partial\ln p_1^+} - \frac{\partial\ln H(p_1^+/\mu, p_2^-/\mu)}{\partial\ln p_1^+}\,, \tag{81}$$

where we have made components explicit and have suppressed α_s. This relation is surprisingly powerful, because J_1, J_2 and H depend on different sets of arguments. J_2, for instance, depends on both p_2^- and $Q^2/n\mu^2$. Its derivative with respect to p_2^- may depend upon either of these arguments, but must be canceled in (81) by the derivatives of H and J_1, which contribute additively. Thus, its dependence on p_2^- and $Q^2/n\mu^2$ can only be additive after the derivative,

$$\frac{\partial}{\partial \ln p_2^-} \ln \tilde{J}_2 \left(\frac{Q^2}{n\mu^2}, \frac{p_2^-}{\mu}, \alpha_s(\mu^2) \right) = K\left(\frac{Q^2}{n\mu^2}, \alpha_s(\mu^2) \right) + G\left(\frac{p_2^-}{\mu}, \alpha_s(\mu^2) \right). \tag{82}$$

The function K cancels a corresponding term from J_1 while G cancels the contribution from H, whose derivatives must satisfy,

$$\frac{\partial \ln H}{\partial \ln p_2'^-} + \frac{\partial \ln H}{\partial \ln p_1^+} = -G\left(\frac{p_2^-}{\mu}\right) - G\left(\frac{p_1^+}{\mu}\right). \tag{83}$$

This separation of short-distance and long-distance dependence in jets is characteristic of Sudakov factorization. As the gauge changes, the jets exchange contributions with each other (via K) and with the hard part (via G). Here we find a strong analogy to the "matching conditions" of effective field theory.

We now have at our disposal two evolution equations, the first relying on invariance under renormalization group rescalings, the other on gauge invariance, but both based on factorization. Combining the two, we shall find enough information to determine all logarithmic n-dependence.

By Eq. (79), $(d \ln J_i/d\mu)$ is independent of momenta, so, for instance

$$\frac{d^2}{d\mu\, dp_1^+} \ln J_1 = 0. \tag{84}$$

Applying this result to (82), we conclude that the combination $K+G$ is itself a renormalization group invariant

$$\mu \frac{d}{d\mu}(K+G) = 0, \tag{85}$$

which implies that yet another anomalous dimension relates K and G (Collins and Soper (1981)),

$$\mu \frac{d}{d\mu} K = -\gamma_K(\alpha_s) = -\mu \frac{d}{d\mu} G. \tag{86}$$

Now we can relate the moment-dependence of K and G through

$$K\left(\frac{Q^2}{n\mu^2}, \alpha_s(\mu^2)\right) + G\left(\frac{Q}{\mu}, \alpha_s(\mu^2)\right) = K\left(1, \alpha_s(Q^2/n)\right) + G\left(1, \alpha_s(Q^2)\right) - \frac{1}{2}\int_{Q^2/n}^{Q^2} \frac{d\mu'^2}{\mu'^2} \gamma_K\left(\alpha_s(\mu'^2)\right), \tag{87}$$

in which all logarithms of n are generated either through the running coupling and/or the explicit μ' integral. There are only two steps left, to solve for the

n-dependence of the J's and to combine everything together in the cross section.

Combining Eqs. (82) and (87), we derive the full evolution of J_2 in terms of p_2^-, an exactly similar equation holding for J_1 in terms of p_1^+,

$$\frac{\partial}{\partial \ln p_2^-} \ln J_2 = K(1, \alpha_s(Q^2)) + G(1, \alpha_s(Q^2)) - \frac{1}{2} \int_{Q^2/n}^{Q^2} \frac{d\lambda^2}{\lambda^2} \Gamma_J(\alpha_s(\lambda^2)) . \tag{88}$$

Here Γ_J combines γ_K with a term that allow us to have the same running coupling in K and G,

$$\begin{aligned}\Gamma_J(\alpha_s) &= \gamma_K(\alpha_s) + \beta(g)\frac{\partial}{\partial g}K(1,\alpha_s) \\ &= \left(\frac{\alpha_s}{\pi}\right) 2C_F + \left(\frac{\alpha_s}{\pi}\right)^2 \left[\left(\frac{67}{18} - \frac{\pi^2}{6}\right) C_F C_A - \left(\frac{5}{9}\right) n_f C_F\right] .\end{aligned} \tag{89}$$

In the second line, we have given the two-loop expression for Γ_J, where as usual n_f is the number of quark flavors.

It is now a simple application of the chain rule to derive a differential equation for the n-dependence of both jets,

$$\begin{aligned}\left[\frac{\partial}{\partial \ln n} + \frac{1}{2}\beta\frac{\partial}{\partial g}\right] \ln J\left(\frac{Q^2}{n\mu^2}, \frac{Q}{\mu}, \alpha_s(\mu^2)\right) &= \frac{1}{2}\, \Gamma'_J\left(\alpha_s(\mu^2)\right) \\ &\quad - \frac{1}{2}\int_{Q^2/n}^{Q^2} \frac{d\lambda^2}{\lambda^2} \Gamma_J\left(\alpha_s(\lambda^2)\right) ,\end{aligned} \tag{90}$$

where we have set $p_i \cdot \xi = Q$ and where

$$\Gamma'_J\left(\alpha_s(\mu^2)\right) \equiv G_J\left(1, \alpha_s(Q^2)\right) + K_J\left(1, \alpha_s(Q^2)\right) - 2\gamma_q\left(\alpha_s(\mu^2)\right) . \tag{91}$$

The solution of Eq. (91) relates J at large n to J at $n = 1$ [1],

$$\begin{aligned}\ln J\left(\frac{Q^2}{n\mu^2}, \frac{p'^-}{\mu}, \alpha_s(\mu^2)\right) &= \ln J\left(\frac{Q^2}{\mu^2}, \frac{p'^-}{\mu}, \alpha_s(\mu^2/n)\right) \\ &\quad - \frac{1}{2}\int_{Q^2/n}^{Q^2} \frac{d\lambda^2}{\lambda^2}\left[\ln\frac{\mu}{\lambda}\Gamma_J\left(\alpha_s(\lambda^2)\right) - \Gamma_J'\left(\alpha_s(\lambda^2)\right)\right] .\end{aligned} \tag{92}$$

If we now set $\mu = Q$, all logarithms of n are generated by the integrals of the two anomalous dimensions Γ and Γ' and the expansion of $\alpha_s(Q^2/n)$, and, as promised, exponentiate in the moments of J.

The resummed expression (92) for the jets organizes all logarithms of n in the moments of the cross section, since the hard function H has no

[1] One way to verify this result is to observe that $\beta(g)\partial\Gamma(\alpha_s(\lambda^2))/\partial g = \lambda\partial\Gamma(\alpha_s(\lambda^2))/\partial\lambda$.

$\ln n$-dependence. The inverse transform of $\tilde{J}(n)$ then gives the singular $1-T$-dependence. To be explicit, the inverse Mellin transform of $\tilde{J}$ in (92) is given by (Contopanagos and Sterman (1993), Catani et al. (1993))

$$J\left((1-x)\frac{Q^2}{\mu^2}, \frac{Q}{\mu}, \alpha_s(\mu^2)\right)\Big|_{\mu=Q} = \left[\frac{e^{E(1-x,\alpha_s(Q^2))}\left[\frac{1}{\pi}\sin(\pi_1 P_1)\Gamma(1+P_1)+\ldots\right]}{1-x}\right]_+ , \tag{93}$$

where $1-x \equiv p^2/Q^2$, and where the exponent E is given by the right-hand side of (92) with $\mu = Q$,

$$E = -\frac{1}{2}\int_{(1-x)Q^2}^{Q^2} \frac{d\lambda^2}{\lambda^2}\left[\ \ln\left(\frac{\mu}{\lambda^2}\right)\Gamma_J(\alpha_s(\lambda^2)) - \Gamma'_J(\alpha_s(\lambda^2))\ \right] + \ln \tilde{J}\left(1,1,\alpha_s((1-x)Q^2)\right) , \tag{94}$$

while P_1 is related to the exponent by

$$P_1 \equiv -\frac{dE(1-x,\alpha_s(Q^2))}{d\ln(1-x)} . \tag{95}$$

Terms omitted in Eq. (93) are suppressed by powers of p^2/Q^2. It is a straightforward matter to verify that the leading logarithms in $1-T$ in Eq. (71) are indeed generated by this form. Here, however, we see the essential role of the running coupling for nonleading logarithms. In particular, because of asymptotic freedom, the exponent receives relatively larger contributions from long distances, and relatively smaller contributions from short distances than in the case of a fixed coupling.

Another important aspect of Eq. (94) is a singularity in the λ^2 integral due to the divergence of the perturbative running coupling (20) at $\lambda^2 = \Lambda^2$. When the invariant mass of the jet is sufficiently small, perturbation theory cannot give a consistent account, and must be supplemented by nonperturbative information. This is a very general feature of resummed expressions, which brings us to the final topic of these lectures.

8 High Orders in Perturbation Theory

Throughout these lectures, we have used the singularities of perturbation theory as a diagnostic for long-distance behavior, and as a guide for organizing the relation of short to long distances. In this section, we shall briefly discuss yet another aspect of perturbation theory that gives hints of nonperturbative structure, its behavior at high orders (Mueller (1993)).

Recall the relation between the total e^+e^- annihilation cross section and the imaginary part of the two-current correlation function,

$$\sigma^{(\rm tot)}_{e^+e^-} \sim {\rm Im} \int d^4x\, e^{iqx} < 0 \,|\, T\ (J^\mu(x) J_\mu(0))\,|\,0 > , \tag{96}$$

with J_μ an electromagnetic current. Here we can apply the operator product expansion, but, because there is no "external" momentum in the matrix element, only the expectation values of scalar operators can contribute, and only a few operators appear with singularities at $x^2 = 0$,

$$J^\mu(x)J_\mu(0) \sim \frac{1}{x^6} C_0(x^2\mu^2) I + \frac{m}{x^2} C_q(x^2\mu^2)\bar{q}q(0) \\ + \frac{1}{x^2} C_F(x^2\mu^2) F_{\mu\nu}F^{\mu\nu}(0) + \dots , \tag{97}$$

with I the identity operator. Perturbation theory with all masses set to zero contributes at any finite order to $C_0(x^2\mu^2)$ only. Yet, as we shall now see, there is a problem with C_0 from high orders, which suggests the presence of the higher terms in the operator product expansion, even in the absence of explicit quark masses.

Our reasoning begins with the sources of long-distance behavior in Eq. (96). As we have pointed out above in Section 3, there are no physically realizable processes in this matrix element with on-shell free particles that carry nonzero momenta. There is, however, another sort of physical picture, which we have ignored up until now because its contribution is highly nonleading (Sterman (1993)). This is one in which a set of internal lines all have *vanishing* momentum. This sort of picture, illustrated in Fig. 14, involves a subdiagram S, consisting entirely of a "cloud" of zero-momentum lines, attached to a single hard subdiagram H. Near the physical process, all soft momenta may be neglected in H, which may therefore be treated effectively as a gauge-invariant local vertex. Consider any single gluon internal to subdiagram S, whose momentum we label k. S may formally be written as

$$S = \int d^4k\, \frac{g^{\alpha\beta}}{k^2}\, T_{\alpha\beta}\,(k, \mu, \alpha_s(\mu))\ , \tag{98}$$

where we have isolated the k propagator (in Feynman gauge) and where $T_{\alpha\beta}$ is the remainder of the subdiagram, as in Fig. 15.

Dimensional counting and gauge invariance then require $T_{\alpha\beta}$ to have the form

$$T_{\alpha\beta}(k, \mu, \alpha_s(\mu)) = \left(k^2\ G_2(k^2)\right)\ (k_\alpha k_\beta - k^2 g_{\alpha\beta})\, t\left(k^2/\mu^2, \alpha_s(\mu^2)\right)\ , \tag{99}$$

where $G_2(k)$ is the trace of the full gluon propagator. Neglecting for simplicity any renormalization of the vertex H, the function t is renormalization-group invariant,

$$\mu \frac{d}{d\mu}\, t(k^2/\mu^2, \alpha_s(\mu^2)) = 0\,, \tag{100}$$

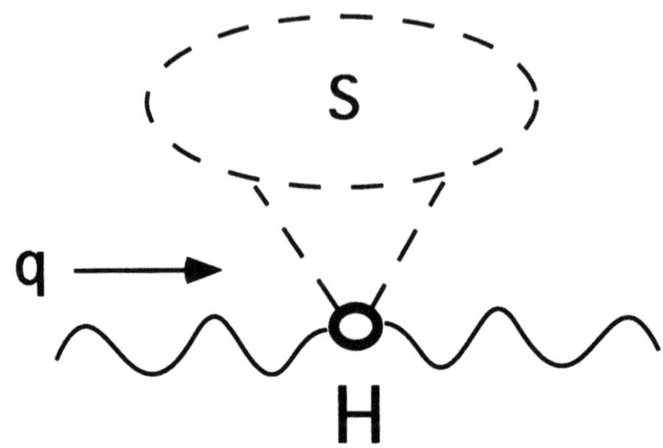

Fig. 14. Process with only zero-momentum lines (subdiagram S) coupled to a point-like hard part H.

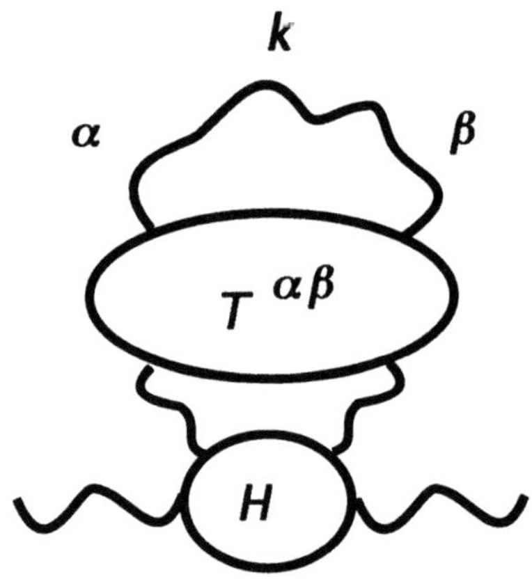

Fig. 15. Decomposition of Fig. 14.

from which we conclude that we may choose $\mu^2 = k^2$ as we integrate over k, at the cost of running the coupling to the scale k^2,

$$
\begin{aligned}
t\left(k^2/\mu^2, \alpha_s(\mu^2)\right) &= t\left(1, \alpha_s(k^2)\right) \\
&= \sum_i a_i {\alpha_s}^i(k^2)\,,
\end{aligned}
\tag{101}
$$

with the a_i numbers. We can already see something funny by looking at the resulting a_1 term for S. Using the one-loop running coupling, we find

$$\begin{aligned}
S^{(1)} &= -3\int_0^Q d^4k\, k^2 \frac{\alpha_s(k^2)}{k^2} \\
&= -12\pi \int_0^{Q^2} dk^2 k^2 \frac{\alpha_s(Q^2)}{1+\frac{\alpha_s(Q^2)}{4\pi} b_2 \ln \frac{k^2}{Q^2}} \\
&= -12\pi Q^4 \alpha_s \sum_{n=0}^{\infty} \left(\alpha_s \frac{b_2}{4\pi}\right)^n \int_0^1 dx\, x \ln^n \frac{1}{x} \\
&= -6\pi Q^4 \alpha_s \sum_{n=0}^{\infty} \left(\alpha_s \frac{b_2}{8\pi}\right)^n \Gamma(n+1)\,,
\end{aligned} \tag{102}$$

where in the last two lines we have reexpanded the running coupling in terms of $\alpha_s(Q^2) \equiv \alpha_s$. As a result, the nth order in α_s has a coefficient that grows like $\Gamma(n+1) = n!$. This is despite the infrared safety of the matrix element. Evidently, this uncontrolled growth in perturbative coefficients is a direct reflection of the singularity in the running coupling. All is not lost, however, although this behavior will require us to reevaluate how we regard the perturbative expansion in QCD.

A very useful conceptual tool for treating high orders in a perturbative series is the Borel transform. Consider a general power series in an expansion variable, in this case α_s,

$$\Pi(\alpha_s) = \sum_{n=0}^{\infty} c_n {\alpha_s}^n\,, \tag{103}$$

with the c_n constants. If $\Pi(\alpha_s)$ is analytic at $\alpha_s = 0$, the c_n are coefficients in a Taylor series expansion. This need not always be the case, however. The c_n may grow with n, and Π may possess no radius of convergence at all about $\alpha_s = 0$. Nevertheless, there can be a good deal of information in the expansion (103). To see how, we define the Borel transform of $\Pi(\alpha_s)$ by

$$\tilde{\pi}(b) = \sum_{n=0}^{\infty} \frac{c_n}{n!} b^n\,. \tag{104}$$

It is an expansion in a conjugate variable b, whose coefficients are simply $c_n/n!$. $\tilde{\pi}(b)$ is thus much more convergent than $\Pi(\alpha_s)$, and has a finite radius of convergence about $b = 0$ even when the c_n grow as fast as $n!$. Formally, the inverse transform from $\tilde{\pi}$ back to Π is

$$\Pi(\alpha_s) = \alpha_s^{-1} \int_0^{\infty} db\, e^{-b/\alpha_s} \tilde{\pi}(b)\,, \tag{105}$$

since the integral over b precisely generates $n!\alpha_s^{n+1}$ from the b^n term. Factorial growth in the expansion coefficients of Π now shows up as a singularity in $\tilde{\pi}$.

If this singularity is on the real axis, the inverse transform is an ambiguous integral. Even if the singularity is off the real axis, its presence indicates contributions to Π that cannot be described fully by the series (103), *i.e.*, nonperturbative contributions.

Returning to our example $S^{(1)}$ above, we observe that a simple change of variables,

$$b' = \frac{\alpha_s(Q^2)}{4\pi} \ln(Q^2/k^2), \tag{106}$$

in the second line of Eq. (102) leads to an expression for $S^{(1)}$ that is precisely of the inverse Borel form,

$$S^{(1)}(Q^2) = -48\pi^2\alpha_s^{-1}Q^4 \int_0^\infty db' \frac{e^{-8\pi b'/\alpha_s(Q^2)}}{(1-b_2b')}. \tag{107}$$

Here $1/(1-b_2b')$ plays the role of the Borel transform of $S^{(1)}$, and its singularity is a direct reflection of the singularity in the perturbative running coupling at $k^2 = \Lambda^2$. Any such singularity in the plane of the Borel variable due to the infrared behavior of the running coupling is called an infrared renormalon.

Although (107) is ill-defined, we have gained something by reexpressing the integral in this fashion, if we regard the singularity as an ambiguity in the inverse Borel transform, which is well-defined in the full theory, although not in perturbation theory alone. We imagine that the Borel transform is well approximated by perturbation theory up to $b' = 1/b_2$. Although the perturbative integral is not well-defined beyond this point, the integrand is already suppressed at $b' = 1/b_2$ by a factor $\exp[-8\pi/b_2\alpha_s(Q^2)] = (\Lambda^2/Q^2)^2$. Thinking back to the operator product expansion, Eq. (97), we recognize that this is the power corresponding to the gluon condensate, the vacuum expectation of the operator F^2. Perturbation theory itself thus signals its own incompleteness by generating an infrared renormalon ambiguity at precisely the leading nonperturbative power of the operator product expansion (at zero quark mass, in this approximation).

It is a widely accepted viewpoint that the correct way to treat the perturbative expansion is to *define* perturbation theory by regulating the inverse Borel transform in such a way that it introduces a new nonperturbative parameter that may be associated with the vacuum expectation value $\langle 0|F^2|0\rangle$. The theory in principle then gives a consistent picture of the function Π up to corrections of order Q^{-6} relative to the leading power, and up to the next uncalculated order in perturbation theory (Mueller (1993)).

In closing, we may note that infrared renormalons appear not only in the total cross section for e^+e^- annihilation, but in many other cross sections as well. They are particularly interesting in resummed quantities, for example the jet of Eq. (93), where integrals over running couplings, analogous to those just encountered, appear in the organization of large corrections. This scenario has been explored in a number of recent papers, among them: Akhoury and Zakharov (1995), Beneke and Braun (1995), Dokshitser and Webber (1995), Korchemsky and Sterman (1995), Manohar and Wise (1994), Nason and Seymour (1995) and Webber (1994). It is natural to ask whether here, as above, perturbation theory is signaling a new set of nonperturbative parameters, which probably cannot be reduced to the operator product expansion. The full answer to this intriguing question is not, to my knowledge, available at present.

References

Akhoury, R. and Zakharov,V.I. (1995): Phys. Lett. B357, 646.

Beneke, M. and Braun, V.M. (1995): Nucl. Phys. B454, 253.

Binnewies, J., Kniehl, B.A. and Kramer, G. (1995): Z. Phys. C65, 471.

Bjorken, J.D., and Drell, S.D. (1965): *Relativistic Quantum Fields* (McGraw-Hill, New York).

Catani, S., Trentadue, L., Turnock, G. and Webber, B.R. (1993): Nucl. Phys. B407, 3.

CDF Collaboration (F. Abe et al.) (1996): FERMILAB-PUB-96-020-E, Jan. 1996.; e-Print Archive: hep-ex/9601008 .

Chiappetta, P., Greco, M., Guillet, J.P., Rolli, S. and Werlin, M. (1994): Nucl. Phys. B412, 3.

Collins, J.C. (1989): in *Perturbative Quantum Chromodynamics*, ed. A.H. Mueller (World Scientific, Singapore), p. 573.

Collins, J.C. and Soper, D.E. (1981): Nucl. Phys. B193, 381.

Collins, J.C., Soper, D.E. and Sterman, G. (1989): in *Perturbative Quantum Chromodynamics*, ed. A.H. Mueller (World Scientific, Singapore), p. 1.

Contopanagos, H. Laenen, E. and Sterman, G. (1996): Stony Brook preprint ITP-SB-96-17; e-Print Archive hep-ph/9604313.

Contopanagos, H. and Sterman, G. (1993): Nucl. Phys. B400, 211.

Del Duca, V. (1995): DESY preprint DESY 95-023, Feb. 1995, Scientifica Acta, INFN Pavia, X, 91; e-Print Archive hep-ph/9503226.

Dokshitser, Yu.L. and Webber, B.R. (1995) Phys. Lett. B352, 451.

Dokshitzer, Yu.L., Khoze, V.A., Troian, S.I. and Mueller, A.H. (1988): Rev. Mod. Phys. 60, 373.

Dokshitzer, Yu.L., Khoze, V.A., Mueller, A.H. and Troian, S.I. (1991): *Basics of Perturbative QCD* (Editions Frontières, Gif-sur-Yvette).

Eden, R.J., Landshoff, P.V., Olive, D.I. and Polkinghorne, J.C. (1966): *The Analytic S-Matrix* (Cambridge University Press, Cambridge).

Ellis, R.K. (1995): in *CP Violation and the Limits of the Standard Model*, proceedings of the 1994 Theoretical Advanced Study Institute, ed. J.F. Donoghue (World Scientific, Singapore, 1995).

Huston, J., Kovacs, E., Kuhlmann, S., Lai, H.L., Owens, J.F., Soper, D. and Tung, W.K. (1995) Michigan State preprint MSU-HEP-50812, Nov 1995; e-Print Archive: hep-ph/9511386 .

Korchemsky, G.P. and Sterman, G. (1995): Nucl. Phys. B437, 415.

Lipatov, L.N. (1989): in *Perturbative Quantum Chromodynamcis*, ed. A.H. Mueller (World Scientific, Singapore), p. 411.

Manohar, A.V. and Wise, M.B. (1995): Phys. Lett. B344, 407.
Mueller, A.H. ed. (1989): *Perturbative Quantum Chromodynamics*, (World Scientific, Singapore).
Mueller, A.H. (1993): in *QCD 20 Years Later*, Aachen, 1992, ed. P.M. Zerwas and H.A. Kastrup (World Scientific, Singapore) v.1, p.162.
Muta, T. (1987): *Foundations of Quantum Chromodynamics* (World Scientific, Singapore).
Nason, P. and Seymour, M.H. (1995): Nucl. Phys. B454, 291.
OPAL Collaboration (M.Z. Akrawy et al.) (1990): Phys. Lett. B247, 617.
OPAL Collaboration (R. Akers et al.) (1995a): Z. Phys. C65, 367.
OPAL Collaboration (R. Akers et al.) (1995b): Z. Phys. C68, 179.
Particle Data Group (1994): *Review of Particle Properties*, Phys. Rev. D50, 1171.
Qiu, J. and Sterman, G. (1991): Nucl. Phys. B353, 105, 137.
Shifman, M. (1995): Mod. Phys. Lett. A10, 605; e-Print Archive: hep-ph/9501222.
Sterman, G. (1992): in *Perspectives in the Standard Model*, proceedings of the 1991 Theoretical Advanced Study Institute, ed. R.K. Ellis, C.T. Hill and J.D. Lykken (World Scientific, Singapore), p. 475.
Sterman, G. (1993), *An Introduction to Quantum Field Theory* (Cambridge University Press, Cambridge).
Sterman, G. (1996): Stony Brook preprint ITP-SB-96-4, to appear in *QCD and Beyond*, proceedings of the 1995 Theoretical Advanced Study Institute, ed. D.E. Soper (World Scientific, Singapore, in press).
Sterman, G. et al. (CTEQ Collaboration) (1995): Rev. Mod. Phys. 67, 157.
Takeuchi, T. (1996): CERN preprint CERN-TH/96-79, March 1996, e-Print Archive: hep-ph/9603415.
Webber, B.R. (1994): Phys. Lett. B339, 148.
Wu, S.L. (1984): Phys. Rep. 107, 59.
Ynduráin, F.J. (1983): *Quantum Chromodynamics* (Springer-Verlag, New York).

The Use of Computer Algebra in QCD

Jos Vermaseren

NIKHEF, P.O.Box 41882,
NL-1009 DB Amsterdam, Netherland

Abstract. These lectures will focus on the use of computer algebra in many problems that arise in perturbative QCD calculations. First we look at some mathematical/combinatorial problems such as non-commutative algebras. Next we consider color traces. After an intermezzo with determinants we will attack several aspects of the computation of Feynman diagrams, both at the tree level and at two loops. This will show useful techniques for when things become complicated. All examples are worked out in the language of FORM [1], which is quite suitable for this kind of problems.

1 Introduction

Computer Algebra is a field of science that is nowadays becoming part of mathematics. Its original development in the early 60's was more inspired by its users, many or even most of whom were physicists. Hence a number of the well known packages were either created by physicists (Reduce, Schoonschip) or for physicists (Macsyma) be it not exclusively. The next generation of programs involved a larger number of mathematicians, because it was recognized that these programs can do to the average use of mathematics what hand-held calculators did to arithmetic. Even so Mathematica comes from the physics corner (and so does FORM), while Maple comes from computer science. Nevertheless the mathematicians have become very much interested in the algorithms that are needed for the automatic treatment of certain problems. One should realize that computer algorithms are essentially different from 'hand algorithms'. Some examples are:

- Automatic symbolic integration (Risch algorithm),
- Solving sets of non-linear equations (Gröbner bases),
- Finite sums,

and of course much more. The possibility to solve problems by a massive number of simple manipulations rather than by having several excellent ideas at the conceptual level, has opened new areas of science. A good example could be the evaluation of a formula which might take the expert maybe an hour, using many symmetry principles, tricks, short cuts etc., while a 'dumb' computer program might evaluate the thing into a million terms that afterwards add and cancel to give the same answer after 5 minutes (and 5 minutes typing and 10 minutes debugging). To illustrate this on a simplified level:

```
    S    a,b;
    L    F = (a+b)*(a+b)*(a+b)*(a+b)*(a+b)*(a+b)*(a+b)*(a+b)*
             (a+b)*(a+b)*(a+b)*(a+b)*(a+b)*(a+b)*(a+b);
    Print;
    .end

Time =        4.64 sec    Generated terms =      32768
                 F        Terms in output =         16
                          Bytes used      =        260

   F =
      15*a*b^14 + 105*a^2*b^13 + 455*a^3*b^12 + 1365*a^4*b^11 +
      3003*a^5*b^10 + 5005*a^6*b^9 + 6435*a^7*b^8 + 6435*a^8*b^7 +
      5005*a^9*b^6 + 3003*a^10 *b^5 + 1365*a^11*b^4 + 455*a^12*b^3
       + 105*a^13*b^2 + 15*a^14*b + a^15 + b^15;
```

In this case the program was made intentionally stupid to force it to expand into 2^{15} terms. Even so, one must be either a calculational prodigy, or know the binomials by heart to beat the time needed to type and run the program.

In addition computer algebra has made large scale calculations more reliable. A hundred page hand calculation in which one finds an error on page 1 can be redone completely (some tricks might still be useful though), while in the equivalent of a symbolic computer manipulation a single editor command and a short rerunning might save the day. Moreover one has a much nicer record of what has been done (depends on programming style of course). Also, for simple types of operations, one may assume that the computer and the algebra system do not make mistakes. For more complicated operations one still has to check. A field in which CA systems (Computer Algebra systems) are notoriously weak is Rieman sheets. Hence many physics integrals are either not done at all, result in a crash or in a wrong answer (which is of course worst). Therefore a good approach is to not only try to solve your problems, but also build in safeguards that can give some idea about the reliability of the answer. In particle physics such a safeguard is often the inclusion of a gauge parameter. It is not 100% safe because the hardest integrals are usually in the terms without the gauge parameter. This is due to the simplifications that come from the Ward identities. But it will draw attention to missing graphs, sign and combinatoric errors, and the more trivial mistakes. Other good checks are numerical evaluation (if possible of course) before and after the interesting symbolic steps. This is for instance very easy when one has to do difficult summations in which the boundaries are symbolic. There are however no general rules, and the safeties are part of the ingenuity that goes into making the programs.

Actually there is a common confusion in the nomenclature. We distinguish the term Symbolic Manipulation and the term Computer Algebra. The first is just the capability to manipulate classes of formulae. The second indicates additional mathematical capabilities that have been built in. CA systems can be very handy. They provide us with much mathematical knowledge and very powerful tools. Yet they have a drawback. Their design has been optimized for many of these operations and hence for whole classes of big problems they may be unnecessarily slow. In addition they need enormous amounts of CPU memory. This puts a practical limitation on the size of the problems that can be handled. Hence sometimes one has to resort to a program that is a pure symbolic manipulation program that has been optimized for speed and size of its expressions at the cost of built in generality. FORM is such a program. The user has to provide all the knowledge, but in exchange one can attack much larger problems. This is not as horrible as it may seem. For large problems usually general solutions are not sufficient anyway. Therefore also in a CA system one would have to do extensive programming to get any type of success.

In the current lectures we will run our examples with FORM. Maybe the output will not be quite as pretty, but it will focus our attention on the algorithms. In addition it will allow us to do some things that might otherwise be quite difficult to do. Also the instruction set of FORM is not as limited as it may seem at first sight. Because this course is not intended for learning the language in the first place, we will skip over many details, assuming that some high level examples that are properly explained will clarify what is going on.

The first example concerns non-commutative algebras. This is a problem that that comes back in much theoretical research, be it in different guises.

2 Campbell Baker Hausdorf Expansions

Working in quantum mechanics and/or field theory it is rather hard to avoid the use of the Campbell Baker Hausdorf (CBH) relation. Usually one needs only very few terms in it. Suppose however that one needs more terms. Suppose one would like to go to order 10. This is what we will do in this section. In the end we will have a little glimpse of what future versions of FORM may do here.

The problem of the expansion is rather old. Actually Dynkin has given a solution in terms of a formula, but this formula is rather useless. It still contains many sums in it and the coefficients of many objects turn out to be zero after summing many terms explicitly. Our target is to get a practical expansion that can be used. For this we need two steps. First we want to rewrite $\log e^A e^B$ as an expansion in terms of A and B. Next we want to rewrite this expansion in terms of commutators. One way of doing this is given in the manual of version 2 of FORM. Because there is always room for improvement we give a faster method here. And in the end we will see how the future versions of FORM will do such rewriting in a very concise way.

Let us first look at the expansion. If we define $\alpha = e^A - 1$ and $\beta = e^B - 1$, then we want the expansion of $\log 1+z$ with $z = \alpha+\beta+\alpha\beta$. The expansion in terms of A and B can then be obtained by substituting the power expansions for α and β. In order to cut the series off in time we use an auxiliary variable x of which the power is limited to the desired depth of the expansion:

```
Functions A,B,al,be,ga;
Symbols   z(:5),x(:5),i;
Local     ABexp5 = sum_(i,1,5,-(-1)^i*z^i/i);
identify  z = x*al+x*be+x^2*al*be;
.sort

Time =        0.04 sec      Generated terms =        118
            ABexp5          Terms in output =         62
                            Bytes used      =       1362
```

The program begins with some declarations. By default functions are non-commuting and they do not have to have arguments. Regular algebraic objects are called symbols. They do commute with everything. In the case of x and z we have put a power restriction on them. Powers higher than 5 are discarded. Then we define the local expression (global expressions can be stored for later use) 'ABexp5' which is a sum over i. The function sum_ is a built in object. It is a general property of FORM that all built in objects have a name that ends in _ while the user is not allowed to use the _ in the definition of names. This avoids conflicts. Hence the imaginary symbol i is in the language of FORM i_. We will see more of these objects while continuing. In this case the statement says

$$AB\text{exp}5 \leftarrow \sum_{i=1}^{5} -(-1)^i \frac{z^i}{i} \tag{1}$$

which is of course the expansion of $\ln 1 + z$. The last statement tells FORM to substitute z by $x\alpha + x\beta + x^2\alpha\beta$. Finally the funny instruction that starts with a period tells FORM to get busy and sort the result into some normal form. The lines that start with "Time" are comments by the system. They are the running statistics and they give us an idea of what is going on. For small programs this may be bothersome, but for big programs these are vital. They can be very early indicators that something is terribly wrong. After the .sort the program contains 5 more lines. Hence after executing the previous step, FORM continues:

```
id   al = sum_(i,1,5,x^i/x*(A)^i/fac_(i));
id   be = sum_(i,1,5,x^i/x*(B)^i/fac_(i));
Bracket x;
Print;
.end

Time =       16.33 sec      Generated terms =      20363
            ABexp5          Terms in output =         44
                            Bytes used      =        854

   ABexp5 =
       + x * ( A + B )

       + x^2 * ( 1/2*A*B - 1/2*B*A )

       + x^3 * ( 1/12*A*A*B - 1/6*A*B*A + 1/12*A*B*B + 1/12*B*A*A - 1/6*B*A*B
          + 1/12*B*B*A )

       + x^4 * ( 1/24*A*A*B*B - 1/12*A*B*A*B + 1/12*B*A*B*A - 1/24*B*B*A*A )
```

```
+ x^5 * (  - 1/720*A*A*A*A*B + 1/180*A*A*A*B*A + 1/180*A*A*A*B*B - 1/
  120*A*A*B*A*A - 1/120*A*A*B*A*B - 1/120*A*A*B*B*A + 1/180*A*A*B*B*B
   + 1/180*A*B*A*A*A - 1/120*A*B*A*A*B + 1/30*A*B*A*B*A - 1/120*A*B*A*B
  *B - 1/120*A*B*B*A*A - 1/120*A*B*B*A*B + 1/180*A*B*B*B*A - 1/720*A*B*
  B*B*B - 1/720*B*A*A*A*A + 1/180*B*A*A*A*B - 1/120*B*A*A*B*A - 1/120*B
  *A*A*B*B - 1/120*B*A*B*A*A + 1/30*B*A*B*A*B - 1/120*B*A*B*B*A + 1/180
  *B*A*B*B*B + 1/180*B*B*A*A*A - 1/120*B*B*A*A*B - 1/120*B*B*A*B*A - 1/
  120*B*B*A*B*B + 1/180*B*B*B*A*A + 1/180*B*B*B*A*B - 1/720*B*B*B*B*A )
  ;
```

The first two statements are also identifications. In this case they are indicated by id only. FORM will accept shorter versions for many of its commands. We will slowly start using them during the lectures. The bracket statement is a statement that tells FORM that in the printout we want x to be pulled outside brackets. Then the print statement causes a printout of the result after the next normal ordering. The `.end` forces such a normal ordering by sorting the results and adding identical terms. In addition it terminates the program, but we do get our printout first. Of course the execution time is not very impressive and it becomes even worse. For the 6-th order the program needs 267 sec. for generating 72 terms. The reason is simple: If a term has six powers of α and we substitute α in one go, we generate 6^6 terms, all of which are thrown away with the exception of one.

In our next attempt we go a little bit slower. The substitution of one power series into another one is an art by itself. We will introduce one deeper power at each step. This gives the program a chance to eliminate terms in as early a stage as possible.

```
#do MAX = 5,10
F   A,B,al,be,ga;
S   z(:'MAX'),x(:'MAX'),i;
L   ABexp'MAX' = sum_(i,1,'MAX',-(-1)^i*z^i/i);
id,z = x*al+x*be+x^2*al*be;
nwrite statistics;
.sort
id al=A*al;
id be=B*be;
#do i = 2,'MAX'
nwrite statistics;
id al = (1+x*A*al/'i');
id be = (1+x*B*be/'i');
.sort
#enddo
write statistics;
.store

Time =        0.46 sec    Generated terms =          48
            ABexp5        Terms in output =          48
                          Bytes used      =        1118
    #enddo

Time =        1.75 sec    Generated terms =          76
            ABexp6        Terms in output =          76
                          Bytes used      =        1742

Time =        5.96 sec    Generated terms =         202
            ABexp7        Terms in output =         202
                          Bytes used      =        4658
```

```
Time =       19.15 sec    Generated terms =        326
              ABexp8      Terms in output =        326
                          Bytes used      =       7402

Time =       62.31 sec    Generated terms =        716
              ABexp9      Terms in output =        716
                          Bytes used      =      16502

Time =      199.70 sec    Generated terms =       1104
             ABexp10      Terms in output =       1104
                          Bytes used      =      25662
   .end
```

The 'nwrite statistics' statement switches the writing of the statistics off. At a later stage we turn it on again because we are curious about the final statistics. We also see a new type of instruction here. Instructions that start with a # character are for the preprocessor. This is some kind of editor that treats the incoming text before it is sent to the compiler. The preprocessor has also variables. These variables contain character strings and they are referred to by putting their name between quotes. Because of this they can be concatenated to other variables or used for composing names like is done here with the names of the expressions. The expansion here is done by providing each time one more term in the expansion of the exponent. Because this is done one statement at a time and FORM brings individual terms to normal form after each statement, terms that are zero are discarded before the expansion goes too deep. Hence the program is much faster already. But we can do even better. So let us try a different approach:

```
   #define MAX "10"
   F   A,B,al,be,ga;
   S   z(:'MAX'),x(:'MAX'),i,n;
   L   ABexp'MAX' = sum_(i,1,'MAX',-(-1)^i*z^i/i);
   id,z = x*al+x*be+x^2*al*be;
   .sort

Time =       17.54 sec    Generated terms =       9799
             ABexp10      Terms in output =       2046
                          Bytes used      =      45010

   repeat;
     id,once,x^n?*al=x^n*A*sump_(i,1,{'MAX'+1}-n,x*A/i);
     id,once,x^n?*be=x^n*B*sump_(i,1,{'MAX'+1}-n,x*B/i);
   endrepeat;
   .end

Time =      148.71 sec    Generated terms =     291148
             ABexp10      Terms in output =       1100
                          Bytes used      =      25550
```

We notice here the **#define** instruction to give a value to a preprocessor variable. We also see the sump_ function which is a special type of sum. Each term is the previous term times the last argument. The first term is always 1, no matter what the summation boundaries are. In this case it gives a quick way to define the expansion of the exponential. There is also the remark

',once,' in the id-statement. This means that the pattern should be applied only once, even if it occurs more times in each term. We also see "n?" which is a so called wildcard. FORM looks what value the power of x has (this is allowed to be zero) and then in the rhs it gives n this value. In this case it means that we expand the exponent to just enough terms and we never generate terms with a power of x that is too high. The repeat–endrepeat combination tells FORM to keep doing what is between them until there are no more substitutions.

We see an enormous improvement over method one, but it is not really better than method 2. What we also see in the statistics is that the first step also starts taking up resources. What we want now is a hybrid of the second and the third methods. In this case we will practise with the first step only.

```
    #define MAX "10"
    F   A,B,al,be,[al+be];
    S   z(:'MAX'),x(:'MAX'),i,n;
    L   ABexp'MAX' = sum_(i,1,'MAX',-(-1)^i*z^i/i);
    repeat;
        id,once,z = x*[al+be]+x^2*al*be;
        if ( count(x,1,z,1) > 'MAX' ) discard;
    endrepeat;
    .sort

Time =        0.19 sec    Generated terms =        231
         ABexp10          Terms in output =        231
                          Bytes used      =       5432

    #do i = 'MAX',1,-1
        if ( count([al+be],1) >= 'i' ) id,once,[al+be]=al+be;
        .sort: [al+be] reduction 'i';

Time =        0.26 sec    Generated terms =        232
         ABexp10          Terms in output =        232
    [al+be] reduction 1   Bytes used      =       5520

    #enddo

Time =        0.32 sec    Generated terms =        235
         ABexp10          Terms in output =        234
    [al+be] reduction 9   Bytes used      =       5652


Time =        0.39 sec    Generated terms =        249
         ABexp10          Terms in output =        246
    [al+be] reduction 8   Bytes used      =       6336

Time =        2.20 sec    Generated terms =       1802
         ABexp10          Terms in output =       1364
    [al+be] reduction 2   Bytes used      =      35420
                      .
                      .
                      .
                      .
Time =        3.26 sec    Generated terms =       2723
         ABexp10          Terms in output =       2046
    [al+be] reduction 1   Bytes used      =      45010

    .end
```

```
Time =        3.76 sec    Generated terms =       2046
            ABexp10       Terms in output =       2046
                          Bytes used      =      45010
```

The new feature here is the composite name. A name is allowed to start with the character `[` and end with the character `]`. In that case all characters between this pair are accepted, provided the braces remain matching braces. This notation can make programs more readable. From the name we can see what we want with this variable, even though it is a single symbol for FORM. We also notice an if statement. Of course one may wonder what criteria can be used in an if-statement. In this case we use 'count' which does power counting. It has an even number of arguments, composed of pairs of objects and their weights. Based on this weight, each term gets a combined weight and depending on its size the if may be taken or not. The discard statement throws the term away if the if is taken. In the second if the substitution is made when the if is taken.

We notice a very good improvement. It is actually even better than one might believe at this point. The old expansion would fill up the disk rather quickly for larger values of MAX. This expansion method keeps the usage of disk space almost at a minimum. The reason why we work the loop backwards is that we give the highest order terms first a chance to cancel, before we substitute the next power. This we can try also with the substitution of α and β. This gives the final program:

```
    #define MAX "10"
    F   A,B,al,be,ga,[al+be];
    S   z(:'MAX'),x(:'MAX'),i,j,n;
    Set ab:A,B;
    Set albe:al,be;
    G   ABexp'MAX' = sum_(i,1,'MAX',-(-1)^i*z^i/i);
    repeat;
        id,once,z = x*[al+be]+x^2*al*be;
        if ( count(x,1,z,1) > 'MAX' ) discard;
    endrepeat;
    .sort

Time =        0.18 sec    Generated terms =        231
            ABexp10       Terms in output =        231
                          Bytes used      =       5432

    #do i = 'MAX',1,-1
        if ( count([al+be],1) >= 'i' ) id,once,[al+be]=al+be;
        .sort: [al+be] reduction 'i';

                    .

                    .

Time =        3.21 sec    Generated terms =       2723
            ABexp10       Terms in output =       2046
    [al+be] reduction 1   Bytes used      =      45010

    #do i = 'MAX',1,-1
        if ( count(x,1) >= 'i' ) id,once,x^n?*ga?albe[i] =
                 x^n*ab[i]*sump_(j,1,{'MAX'+1}-n,ab[i]*x/j);
        .sort: i = 'i';

Time =        4.72 sec    Generated terms =       4094
```

```
        ABexp10          Terms in output =       2046
   i = 10                Bytes used      =      49094

   #enddo

Time =      6.99 sec     Generated terms =       5630
        ABexp10          Terms in output =       2036
   i = 9                 Bytes used      =      50902
                    .
                    .
                    .
Time =     31.02 sec     Generated terms =       5300
        ABexp10          Terms in output =       1100
   i = 1                 Bytes used      =      25550

   B   x;
   .store

Time =     31.33 sec     Generated terms =       1100
        ABexp10          Terms in output =       1100
                         Bytes used      =      23470

   Save ABexp'MAX'.sav;
   .end
```

Here we notice the definition of two sets and their use. A set is a collection of objects of the same type that can be used to restrict wildcarding to matches to objects in a set. In this case `ga?albe[i]` means a wildcard function in set `albe`. In addition it gives i the value of the number of the set element that gave the match. Hence a match with al will give i the value 1 and be will give it the value 2. In the rhs this is used to obtain A or B in their place. The final statement saves the resulting expression into a file ABexp10.sav.

The program has become significantly faster. This method also prevents a massive buildup of intermediate expressions and is rather friendly with respect to the disk. It has been used to evaluate the expansion to order 19. To that order there are 867882 terms in the answer.

We have now mastered the first stage of the project. The answer can be found in a file. Hence we do not have to rerun the first part all the time. Rewriting the expression in terms of commutators requires some extra care. It is not quite as trivial as one would like it to be. The first thing is to establish a notation for the commutators. We define:

$$\begin{aligned} C &= [A,B] \\ C(x) &= [x,C] \\ &= [x,[A,B]] \\ C(x,?) &= [x,C(.)] \end{aligned}$$

This notation can make things very compact. Because we must have always at least one A and one B in the deepest commutator, it is only necessary to mention all the other commutators. Hence a proper 8-th order term in the perfect output will be a C with 6 arguments. The notation with the question mark is one of the wildcard notations of FORM. A single or a multiple question mark for an argument of a function in the left hand side means that it will match a whole set of arguments. Hence $C(x,?)$ means a function C of which the first argument is called x. All other arguments together are represented by this special variable. In the right hand side the wildcard variable ? is represented by a single period, the variable indicated by ?? is represented by two periods, etc. We can use up to ten of these variables simultaneously.

Our first attempt is to commute all B's to the right, and all C's to the left. This is done with

```
   #define MAX "8"
   F   A,B,C;
   S   x;
   .global

   Load ABexp10.sav;
 ABexp10 loaded
   L  F = ABexp10;
   if ( count(x,1) > 8 );
     discard;
   endif;
   .sort
Warning: Conflicting power restrictions for x

Time =        0.29 sec     Generated terms =       322
                 F          Terms in output =       322
                            Bytes used      =      7306
   repeat;
     id B*A = A*B-C;
   endrepeat;
   repeat;
     id A*C(?) = C(A,.)+C(.)*A;
     id B*C(?) = C(B,.)+C(.)*B;
   endrepeat;
   Print +f;
   B  x;
   .end

Time =       37.74 sec     Generated terms =     52971
                 F          Terms in output =       511
                            Bytes used      =      9270

   F =
       + x * ( A + B )

       + x^2 * ( 1/2*C )

       + x^3 * ( 1/12*C(A) - 1/12*C(B) )

       + x^4 * (  - 1/24*C*A*B + 1/24*C*B*A + 1/24*C*C - 1/24*C(A,B) )

       + x^5 * (  - 1/180*C*A*A*B + 1/90*C*A*B*A - 1/180*C*A*B*B - 1/180*C*B*A
         *A + 1/90*C*B*A*B - 1/180*C*B*B*A + 1/360*C*C(A) - 1/72*C*C(B) - 1/90
         *C(A)*A*B + 1/90*C(A)*B*A + 1/72*C(A)*C - 1/720*C(A,A,A) - 1/180*C(A,
         A,B) + 1/180*C(A,B,B) + 1/90*C(B)*A*B - 1/90*C(B)*B*A - 1/360*C(B)*C
          + 1/720*C(B,B,B) )
   etc
```

Unfortunately this program still has many terms with loose A's and B's in the output. In addition, not all B's are to the right of A's. This is annoying, because we wanted only terms with a single C. But getting rid of the A's and B's is our first worry. This is done in the next program. Here we do it meticulously:

```
#procedure commute(size)
#do j = 1,'size'/2
Multiply,right,D;
#do i = 1,'size'
repeat;
  id,B*D = D*B;
endrepeat;
repeat;
  id A*D = D*A;
endrepeat;
repeat;
  id B*D*A = A*B*D - C;
endrepeat;
.sort
#enddo
id D=1;
repeat;
  id A*C(?) = C(A,.)+C(.)*A;
  id B*C(?) = C(B,.)+C(.)*B;
endrepeat;
.sort
#enddo
#endprocedure
```

This is an example of a procedure. Such a procedure is like a gigantic macro. It is processed by the preprocessor. Its variables are text strings and they become preprocessor variables. Such a procedure can be put in a file which must in this case have the name "commute.prc". If FORM does not find the procedure in the current file (before the first call) it will look in the current directory for such a file. If the file is not in the current directory it can also be looked for in the 'procedure directory' which is something that can be defined as an environment variable in the unix shell.
We want to replace the rightmost pair BA, provided that that term did not pick up a C in this round yet. To this end we multiply by the right by the object D, and move it between the rightmost BA pair. Then we commute until there is no more D or no more BDA combination. After that we do not need D any longer in this round. Now we can move the new commutators to the left. By doing this a number of times we get terms with the product of more than one commutator. This is unfortunate but it cannot be helped.

Next we 'pack' commutators in terms with more than one commutator.

```
#procedure simplify
repeat id,disorder,C(?)*C(??) = C(..)*C(.)+C(C(.),..);
.sort
repeat id C(?,C(x?,??),???) = C(.,x,C(..),...)-C(.,C(..),x,...);
.sort
repeat id C(?,C,??) = C(.,A,B,..)-C(.,B,A,..);
.sort
if ( count(C,1) > 1 );
 repeat id C(x?,?) = x*C(.)-C(.)*x;
endif;
.sort
```

```
if ( count(C,1) > 1 ) id C = A*B-B*A;
.sort
#endprocedure
```

In this step we pack the commutators into each other. Of course, putting just an id statement (without the option 'disorder') would get FORM into an infinite loop. The disorder only makes the substitution if the two functions are out of order. This is to say: if they would be commuting functions, FORM would have exchanged them. Now we have complicated commutators, but they can be simplified. This is done in the next statements. Try to understand what they do. If, after this there are still terms with more than one commutator, these are of a type that turns out to be zero (empirical fact). Hence we expand them and at the A,B level all their terms cancel. This standardizes the commutators. The full program gives now:

```
    #define MAX "10"
    F   A,B,C,D;
    F   A1,A2,A3,A4,A5,A6;
    nwrite statistics;
    .global
    Load ABexp10.sav;
 ABexp10 loaded
    L  F = ABexp10;
    .sort
    if ( count(x,1) > 'MAX' );
      discard;
    endif;
    #call commute{'MAX'}
    #call simplify
    Write statistics;
    .sort

Time =      138.90 sec     Generated terms =        400
                 F         Terms in output =        400
                           Bytes used      =      12364

    id C(?,B,A) = C(.,A,B);
    Print +f;
    if ( count(x,1) > 6 );
      discard;
    endif;
    B x;
    .end

Time =      139.06 sec     Generated terms =         19
                 F         Terms in output =         18
                           Bytes used      =        324

   F =
       + x * ( A + B )

       + x^2 * ( 1/2*C )

       + x^3 * ( 1/12*C(A) - 1/12*C(B) )

       + x^4 * (  - 1/24*C(A,B) )

       + x^5 * (  - 1/720*C(A,A,A) - 1/120*C(A,A,B) - 1/360*C(A,B,B) + 1/360*
         C(B,A,A) + 1/120*C(B,A,B) + 1/720*C(B,B,B) )
```

```
+ x^6 * ( 1/720*C(A,B,B,B) + 1/1440*C(B,A,A,A) + 1/240*C(B,A,A,B) - 1/
  720*C(B,A,B,B) - 1/720*C(B,B,A,A) + 1/1440*C(B,B,A,B) );
```

In the run presented here we print only the terms up to the sixth power. The statistics show one of the reasons. The other reason is more interesting from the scientific viewpoint. There are still relations between the commutators. We show one of these relations:

```
C(B,A) = C(A,B)
```

which means simultaneously that

```
C(?,B,A) = C(.,A,B)
```

More relations can be found by considering that

```
0 = [C(B),C(B)] = C(C(B),B) = C(B,C,B)-C(C,B,B)
                = - C(A,B,B,B) + 2*C(B,A,B,B) - C(B,B,A,B)
```

Applying such relations can simplify the expressions considerably, especially the even orders. It is an interesting problem to find the shortest expression in a given order. The sixth order can be reduced to 4 terms, the seventh order to 18 terms and the eight order to at most 13 terms. The ninth order needs at most 54 terms, and the tenth order is known to be writable with at most 41 terms.

In future versions of FORM the program will be shorter and faster. There will be facilities to define sets of equations. In that case we construct all commutators in a given order that have to be zero, work them out, and consider these expressions as a set of equations that we solve. This gives automatically the relations between the commutators. If we give the expression in terms of A's and B's as an extra equation, FORM will apply all these simplifications automatically, and in the end we get a rather short expression. This way short expressions have already been obtained up to order 12. An example of the 8-th order result:

```
Time =       19.92 sec    Generated terms =        151
               REL        Terms in output =        151
                          Bytes used      =       3158

   REL =
      D(8) + 1/60480*C(B,A,A,A,A,A) + 1/10080*C(B,A,A,B,A,A) - 1/20160*C(B,A,A
      ,B,B,A) + 1/20160*C(B,A,B,A,A,A) + 1/2016*C(B,A,B,A,B,A) + 1/10080*C(B,A
      ,B,A,B,B) - 1/15120*C(B,A,B,B,A,A) + 1/6720*C(B,A,B,B,B,A) + 1/20160*C(B
      ,A,B,B,B,B) - 1/20160*C(B,B,A,A,A,A) - 1/2520*C(B,B,A,B,A,A) - 1/3360*C(
      B,B,A,B,B,A) + 5/24192*C(B,B,B,A,A,A) + 1/5040*C(B,B,B,A,B,A) - 1/12096*
      C(B,B,B,A,B,B) - 1/20160*C(B,B,B,B,A,A) + 1/20160*C(B,B,B,B,B,A)
      ,
      C(A,A,A,A,A,B) - 5*C(A,A,B,A,A,A) + 6*C(A,B,A,A,A,A) - 2*C(B,A,A,A,A,A)
      ,
   etc.
```

We see here the expression in which one should read that `D(8)+rest=0` and `D(8)` represents the 8-th order result. The computer time includes computing all the lower identities also. The second expression is one of the many identities that FORM derived while doing this. We suppressed the others. Unfortunately the expressions do not quite have the minimal size. Anyway there does not seem to be a good theory about how to obtain the shortest expression.

3 Color Traces in SU(N)

When one is computing diagrams in non-trivial gauge groups, one is faced with the evaluation of color traces. Such traces can be rather nasty at times. A computer however can evaluate these traces for most semi simple Lie algebras. We follow the paper by Cvitanovic [2]. In the current case we look only at SU(N). If other Lie algebras are needed, it should be considered as a nice exercise to make an equivalent program.

The first problem one is confronted with, is to choose a name for the routine. We call it SU. It will have a single parameter N which will indicate the specific SU group. Of course we have to take a few more decisions. There has to be input and output. We have to have a notation for the group objects. In this matter we follow the article by Cvitanovic. We define:

$$T(a,b,i) = \frac{1}{2}\lambda_i^{ab}$$

$$[T_i, T_j] = \frac{i}{2} f_{ijk} T_k$$

Because the character f is so popular in many programs, we will represent this f by the name `fff`. In addition, when we have the T-matrices, we represent them almost like the gamma matrices: We can omit the indices a and b if we are indicating a trace, and we can string the i,j indices together into one T:

```
T(a,b,i,j) = T(a,c,i)*T(c,b,j)
T(a,a,i,j) = T(i,j)
```

For the input we accept only traces of T's without the a,b indices and the function `fff`. If we want 'open' a,b indices we will use a function named TT for which this 'trace' convention does not apply.

In addition to the above objects which should be declared as (commuting) functions in the calling program, we need a number of indices. These indices must have been declared in N dimensions, and in addition, at the time of the call, the default dimension should be N. We will see why this is important. The indices are `i1,i2,i3,i4`. Additionally we need the indices `i,j,k`. Their dimension is not important, provided it is not zero. It should be noted that if the dimension of the `i,j` type indices is important, it should be a dimension that is equivalent to $N^2 - 1$.

The central relation, which is particular to SU(N), is

$$T_i^{ab}T_i^{cd} = \frac{1}{2}\delta^{ad}\delta^{bc} - \frac{1}{2N}\delta^{ab}\delta^{cd}$$

This means that all objects should be transformed into the function TT, and then we should apply this simplifying relation as many times as possible. Hence we have to start with the transformation. First we express the traces in terms of TT:

```
repeat;
    id,once,T(??) = TT(i1,i1,..)*nf;
    sum i1;
endrepeat;
```

We have to make sure that the indices are contracted properly. Hence we should make sure that no more than two indices i_1 are introduced simultaneously. We sum over them immediately. This changes the index i_1 into an internal index which has the default dimension. Hence it is important that the default dimension is N. For these indices FORM checks that it does not use one that occurs already inside the term. Also FORM can rename these indices, if that brings the term to a more sensible representation. This happens for instance when the lower numbered indices disappear. This trick of the summing to avoid more than two occurrences of indices in the same term can be quite useful at times.

Next we take the TT functions apart into individual matrices. This is done with the statements

```
repeat;
    id,once,TT(i1?,i2?,i?,j?,?) = TT(i1,i3,i)*TT(i3,i2,j,.);
    sum i3;
endrepeat;
```

Again we see that the new index that is introduced, is immediately summed over to make the name i_3 available again for the next substitution. Note that it is therefore important that we use the 'once' option of the id statement, insuring that only one pair of the index i_3 is introduced simultaneously.

After the above statements we have only individual matrices TT and the functions `fff`. Therefore we have to attack the `fff`'s now. We can use their defining relation in reverse when one of the indices of the `fff` is contracted with an index of one of the TT. We apply this rule as many times as possible, but it is not enough by itself. If, for instance, we started with only functions `fff`, we cannot apply this relation. Fortunately there is also another relation for the `fff`:

$$f_{ijk} = \frac{2}{i}T_i^{ab}T_j^{bc}T_k^{ca} - \frac{2}{i}T_k^{ab}T_j^{bc}T_i^{ca} \tag{2}$$

In total the code we construct is

```
repeat;
  repeat;
    id,once,fff(i?,j?,k?)*TT(i1?,i2?,i?) = TT(i1,i3,j)*TT(i3,i2,k)/i_
                                          -TT(i1,i3,k)*TT(i3,i2,j)/i_;
    sum i3;
    id,once,fff(k?,i?,j?)*TT(i1?,i2?,i?) = TT(i1,i3,j)*TT(i3,i2,k)/i_
                                          -TT(i1,i3,k)*TT(i3,i2,j)/i_;
    sum i3;
    id,once,fff(j?,k?,i?)*TT(i1?,i2?,i?) = TT(i1,i3,j)*TT(i3,i2,k)/i_
                                          -TT(i1,i3,k)*TT(i3,i2,j)/i_;
    sum i3;
  endrepeat;
  id,once,fff(i?,j?,k?) = TT(i1,i2,i)*TT(i2,i3,j)*TT(i3,i1,k)*2/i_
                         -TT(i1,i2,k)*TT(i2,i3,j)*TT(i3,i1,i)*2/i_;
  sum i1,i2,i3;
endrepeat;
```

One may ask why the first statements are necessary. The last one should be sufficient. This is a matter of economy. The last statement introduces three dummy indices for each `fff` and three TT's. The first statements give only one dummy index and one TT for each `fff`. This means that fewer TT's have to be eliminated later. We see also a good example of nested repeat loops. In this case we try to apply the simpler relation whenever possible. If this does not work any more, we apply the more complicated relation once, after which we try the simpler relation again as many times as possible. Etc. This gives a good efficiency.

Finally we can start applying the simplifying relation for pairs of TT with the same i-index:

```
repeat;
    id,once,TT(i1?,i2?,i?)*TT(i3?,i4?,i?) =
            (d_(i1,i4)*d_(i2,i3)-d_(i1,i2)*d_(i3,i4)/'N')/2;
endrepeat;
```

Note that this is the only place where the parameter N enters our equations explicitly. But of course, when there are functions `d_` with two identical indices, these indices are usually dummy indices that were introduced by a sum statement. Hence such a `d_` will be replaced by the default dimension. And things will go wrong if this is not N.

After this there rests only cleaning up. We remove objects that are obviously zero or something else that is simple:

```
id TT(i1?,i1?,i?) = 0;
id TT(i1?,i2?,i?)*TT(i2?,i1?,j?) = d_(i,j)/2;
```

Considering that no two TT's were left with identical i-type indices, we will never have to replace a `d_(i,i)` by the dimension of i. Hence the statement at the beginning that the dimension of these indices is not relevant, unless later stages in the users part of the program make it important.

Finally we string the TT's back together and introduce traces again:

```
repeat;
    id TT(i1?,i2?,?)*TT(i2?,i3?,??) = TT(i1,i3,.,..);
endrepeat;
id TT(i1?,i1?,?) = T(.)/nf;
```

This finishes the procedure. Of course it is wise to put some commentary at the beginning, especially to indicate how to use the routine and what external declarations are expected. Let us run the program now:

```
Symbols N,nf,[N^2-1];
CFunctions T,TT,fff;
Dimension [N^2-1];
Indices i,j,k;
I a1,a2,a3,a4,a5,a6,a7,a8,a,b;
Dimension N;
Indices i1,i2,i3,i4;

L [2loop]   = fff(a,a2,a3)*fff(a2,a1,a5)*fff(a5,a4,a3)*fff(a1,b,a4);
L [3loopLA] = fff(a,a3,a4)*fff(a3,a2,a8)*fff(a2,a1,a7)*fff(a1,b,a6)
        *fff(a6,a5,a7)*fff(a5,a4,a8);
L [3loopNO] = fff(a,a3,a4)*fff(a3,a2,a8)*fff(a2,a1,a7)*fff(a1,b,a6)
        *fff(a6,a5,a8)*fff(a5,a4,a7);
#call SU{N}
print;
.end

Time =       0.21 sec    Generated terms =        42
          [2loop]        Terms in output =         1
                         Bytes used      =        26

Time =       1.48 sec    Generated terms =       306
        [3loopLA]        Terms in output =         1
                         Bytes used      =        26

Time =       2.79 sec    Generated terms =       300
        [3loopNO]        Terms in output =         0
                         Bytes used      =         2

   [2loop] =
      1/2*d_(a,b)*N^2;

   [3loopLA] =
      - 1/4*d_(a,b)*N^3;

   [3loopNO] = 0;
```

There are several new features here: CFunctions are commuting functions. These functions are assumed to be commuting with everything else. The other new type of variables are the indices. Indices have a dimension connected to them. This is relevant in objects like $\delta^{\mu\mu}$. This dimension is by default 4 unless it is changed with the dimension statement. It must be either a positive integer, a single symbol or zero in which case the index is not automatically summed over. The dimension statement influences the dimension of all indices that are declared after it and the dimension of dummy indices that are generated internally.

We see the evaluation of three color traces. The first is the color factor of the most complicated diagram in the two loop gluon propagator. The next one is a three loop diagram of the gluon propagator with the ladder topology. The last one is a so called non-planar diagram in the three loop gluon propagator. This diagram is rather nasty. Mercifully it is zero, making the evaluation of the Lorenz part of the diagram unnecessary.

If the only relation that is used to eliminate the `fff`'s is the one that gives three TTT's we get:

```
Time =         0.87 sec    Generated terms =        210
          [2loop]          Terms in output =          1
                           Bytes used      =         26

Time =        26.49 sec    Generated terms =       4170
        [3loopLA]          Terms in output =          1
                           Bytes used      =         26

Time =        52.21 sec    Generated terms =       5152
        [3loopNO]          Terms in output =          0
                           Bytes used      =          2

   [2loop] =
      1/2*d_(a,b)*N^2;

   [3loopLA] =
       - 1/4*d_(a,b)*N^3;

   [3loopNO] = 0;
```

The answers are still the same, but the routine SU2 that we used here which does not have the simpler relations, is considerably slower.

One can obtain a speedup by having the routine do some intermediate sorts. The problem with this is that one does not know how many of these sorts are beneficial. For fun we added three sorts in the simplifying relations:

```
#do i = 1,3
  id,once,TT(i1?,i2?,i?)*TT(i3?,i4?,i?) =
               (d_(i1,i4)*d_(i2,i3)-d_(i1,i2)*d_(i3,i4)/'N')/2;
  id TT(i1?,i1?,i?) = 0;
  .sort
#enddo
```

This gives a faster program:

```
         .
         .
   #call SU3{N}
   print;
   .end

Time =         1.93 sec    Generated terms =          4
          [2loop]          Terms in output =          1
                           Bytes used      =         26

Time =         1.94 sec    Generated terms =          4
        [3loopLA]          Terms in output =          1
                           Bytes used      =         26

Time =         1.95 sec    Generated terms =          4
        [3loopNO]          Terms in output =          0
                           Bytes used      =          2

   [2loop] =
      1/2*d_(a,b)*N^2;

   [3loopLA] =
```

```
      - 1/4*d_(a,b)*N^3;

   [3loopNO] = 0;
```

Of course it has lost in elegance. We also suppressed the intermediate statistics.

Let us now construct a procedure SU4(N,s) which has 's' of these sorts as a parameter. We will use it to compute the traces of the maximally non-planar graph in a purely gluonic diagram in the n-loop gluon propagator. We construct a program that can do these diagrams up to a given number of loops. The result can be seen:

```
    #define MAX "6"
    Symbols N,nf,[N^2-1];
    CFunctions T,TT,fff;
    Dimension [N^2-1];
    Indices i,j,k;
    I
    #do i = 1,3*'MAX'
        a'i'
    #enddo
        ;
    I a,b;
    Dimension N;
    Indices i1,i2,i3,i4;
    Nwrite Statistics;
    .global
    #do LOOPS = 2,'MAX'
    L ['LOOPS'loopNO] =
         fff(a{'LOOPS'+1},a,a'LOOPS')*fff(a1,b,a{2*'LOOPS'})
    #do i = 1,'LOOPS'-1
        *fff(a{'i'+1},a'i',a{2*'LOOPS'+'i'})
        *fff(a{'LOOPS'+'i'+1},a{'LOOPS'+'i'},a{2*'LOOPS'+'i'})
    #enddo
        ;
    #call SU4{N|'LOOPS'}
    Write statistics;
    print;
    .store

   [2loopNO] =
      1/2*d_(a,b)*N^2;

    #enddo

   [3loopNO] = 0;

   [4loopNO] =
      3/2*d_(a,b)*N^2;

   [5loopNO] =
      d_(a,b)*N^3;

Time =      209.80 sec    Generated terms =          10
        [6loopNO]         Terms in output =           2
                          Bytes used      =          50

   [6loopNO] =
      d_(a,b)*N^4 + 3/2*d_(a,b)*N^2;

     end
```

Of course, this way we would like to make the program even faster. In addition one could notice (if the statistics had been turned on) that the program starts to use much disk space. The build up of terms starts with the elimination of the fff's, and the shrinking occurs with the simplifying relation. Hence we should simplify as quickly as possible. We can do this with

```
id,once,fff(i?,j?,k?) = TT(i1,i2,i)*TT(i2,i3,j)*TT(i3,i1,k)*2/i_
                        -TT(i1,i2,k)*TT(i2,i3,j)*TT(i3,i1,i)*2/i_;
sum i1,i2,i3;
#do i = 1,3*'s'-2
  if ( match(fff(i?,j?,k?)*TT(i1?,i2?,i?)) );
    id,once,fff(i?,j?,k?)*TT(i1?,i2?,i?) = TT(i1,i3,j)*TT(i3,i2,k)/i_
                                          -TT(i1,i3,k)*TT(i3,i2,j)/i_;
    sum i3;
  else;
   if ( match(fff(k?,i?,j?)*TT(i1?,i2?,i?)) );
     id,once,fff(k?,i?,j?)*TT(i1?,i2?,i?) = TT(i1,i3,j)*TT(i3,i2,k)/i_
                                           -TT(i1,i3,k)*TT(i3,i2,j)/i_;
     sum i3;
   else;
    if ( match(fff(j?,k?,i?)*TT(i1?,i2?,i?)) );
      id,once,fff(j?,k?,i?)*TT(i1?,i2?,i?) = TT(i1,i3,j)*TT(i3,i2,k)/i_
                                            -TT(i1,i3,k)*TT(i3,i2,j)/i_;
      sum i3;
    endif;
   endif;
  endif;
  id,once,TT(i1?,i2?,i?)*TT(i3?,i4?,i?) =
             (d_(i1,i4)*d_(i2,i3)-d_(i1,i2)*d_(i3,i4)/'N')/2;
  id TT(i1?,i1?,i?) = 0;
  .sort
#enddo
```

We replace one fff by TT's and then after each introduction of a new TT we try to eliminate a TT. The result is a much faster program:

```
          .
          .
   #call SU5{N|'LOOPS'}
   Write statistics;
   print;
   .store

  [2loopNO] =
     1/2*d_(a,b)*N^2;

   #enddo

  [3loopNO] = 0;

  [4loopNO] =
     3/2*d_(a,b)*N^2;

  [5loopNO] =
     d_(a,b)*N^3;

Time =       31.97 sec    Generated terms =          10
        [6loopNO]         Terms in output =           2
                          Bytes used      =          50

  [6loopNO] =
```

```
      d_(a,b)*N^4 + 3/2*d_(a,b)*N^2;

   [7loopNO] =
      d_(a,b)*N^5 + d_(a,b)*N^3;

   [8loopNO] =
      d_(a,b)*N^6 + d_(a,b)*N^4 + 3/2*d_(a,b)*N^2;

   [9loopNO] =
      d_(a,b)*N^7 + d_(a,b)*N^5 + d_(a,b)*N^3;

Time =     3183.44 sec    Generated terms =         18
       [10loopNO]         Terms in output =          4
                          Bytes used      =         98

   [10loopNO] =
      d_(a,b)*N^8 + d_(a,b)*N^6 + d_(a,b)*N^4 + 3/2*d_(a,b)*N^2;

    .end
```

The drawback is a much more specialized program. Hence we keep the original program as a general color procedure, and we keep the special versions for special occasions.

4 Determinants of Large Matrices

One of the things one encounters at times is the symbolic evaluation of determinants. If $n \times n$ is the size of the matrix, this is typically a problem that involves $n!$ steps, provided that all elements are non-zero and they all consist of a single symbol. In reality the situation is usually different. There may be many zeroes. The entries may be expressions. Not all symbols have to be different, causing many cancellations in intermediate steps, provided things are done in a given order, etc. Hence it pays to have a good look at the problem. One very simple way to generate a determinant in FORM is by the contraction of two Levi-Civita tensors. These are built-in objects in FORM. Hence

```
    CFunction M;
    Indices i1,i2,i3,j1,j2,j3;

    Local F = e_(i1,i2,i3)*e_(j1,j2,j3)/6
              *M(i1,j1)*M(i2,j2)*M(i3,j3);
    sum i1,1,2,3;
    sum i2,1,2,3;
    sum i3,1,2,3;
    Contract;
    Print +s;
    .end

Time =         0.01 sec    Generated terms =         36
                  F        Terms in output =          6
                           Bytes used      =        190

   F =
       + M(1,1)*M(2,2)*M(3,3)
       - M(1,1)*M(2,3)*M(3,2)
```

```
    - M(1,2)*M(2,1)*M(3,3)
    + M(1,2)*M(2,3)*M(3,1)
    + M(1,3)*M(2,1)*M(3,2)
    - M(1,3)*M(2,2)*M(3,1)
  ;
```

We note some new feature. The Levi-Civita tensor is indicated by e_. We can sum over symbolic indices explicitly. In that case one has to give the values that the index should take. These must be positive integers. The contract statement contracts pairs of Levi-Civita tensors. This program is not very smart, but it works. The next approach is already slightly better.

```
   Vectors p1,p2,p3,q1,q2,q3;

   Local F = e_(p1,p2,p3)*e_(q1,q2,q3);
   Contract;
   Print +s;
   .end

Time =          0.00 sec    Generated terms =          6
                   F        Terms in output =          6
                            Bytes used      =        144

   F =
       + p1.q1*p2.q2*p3.q3
       - p1.q1*p2.q3*p3.q2
       - p1.q2*p2.q1*p3.q3
       + p1.q2*p2.q3*p3.q1
       + p1.q3*p2.q1*p3.q2
       - p1.q3*p2.q2*p3.q1
      ;
```

Here we use vectors. If one puts a vector in a position that would normally be taken by an index, one should assume that there was an index, but it has been contracted. Hence $\epsilon^{p_1 p_2 p_3} = \epsilon^{\mu_1 \mu_2 \mu_3} p_1^{\mu_1} p_2^{\mu_2} p_3^{\mu_3}$. A contraction between two vectors is written as a dotproduct. FORM does not remember the name of the contracted index. In our example the dotproducts are like the elements of a matrix. One could substitute other values with id statements. Of course one could do this for a bigger matrix:

```
   Vectors p1,p2,p3,p4,p5,p6,p7,p8;
   Vectors q1,q2,q3,q4,q5,q6,q7,q8;

   Local F = e_(p1,p2,p3,p4,p5,p6,p7,p8)
            *e_(q1,q2,q3,q4,q5,q6,q7,q8);
   Contract;
   .end

Time =          5.69 sec    Generated terms =      40320
                   F        Terms in output =      40320
                            Bytes used      =     980166
```

Yet this is not a perfect algorithm for sparse matrices. Especially when it concerns sparse 15×15 matrices. The algorithm we are going to construct is going to solve just that. It is based on constructing first all the 2×2 minors of the last two rows. Then we construct the 3×3 minors of the last three rows, without keeping a memory of the previous 2 × 2 minors. We will not need them anymore. Many of these minors will be zero or have just one or two terms. Hence the expression does not grow too much usually (unless the determinant will be gigantic in the end anyway). But let us practise first with a smaller determinant to see what the algorithm does.

```
S  a,b,c,d,e;
CF  T;
Table T(1:4,1:4);
Fill T(1,1) =  a,0,0,b;
Fill T(2,1) =  0,c,d,0;
Fill T(3,1) =  0,e,c,0;
Fill T(4,1) =  b,0,0,a;

Local F = + e_(1)*T(4,1) + e_(2)*T(4,2)
          + e_(3)*T(4,3) + e_(4)*T(4,4);
#do k = 3,1,-1
id  e_(?) = e_(1,.)*T('k',1)+e_(2,.)*T('k',2)
           +e_(3,.)*T('k',3)+e_(4,.)*T('k',4);
B   e_;
Print;
.sort: determinant step 'k';
#enddo
id  e_(1,2,3,4) = 1;
Print;
.end
```

The program contains now a table to enter the matrix. That is much easier than substituting each element separately. A table is a special function. Hence tables can also be non-commuting if necessary. Tables have more properties than what is shown here, but it is important to know that it has to be filled with elements. When we run the program the output is

```
   F =
      + e_(1,2) * (  - b*e )

      + e_(1,3) * (  - b*c )

      + e_(2,4) * ( a*e )

      + e_(3,4) * ( a*c );

   F =
      + e_(1,2,3) * ( b*c^2 - b*d*e )

      + e_(2,3,4) * ( a*c^2 - a*d*e );

   F =
      + e_(1,2,3,4) * ( a^2*c^2 - a^2*d*e - b^2*c^2 + b^2*d*e );


    id  e_(1,2,3,4) = 1;
    Print;
    .end

   F =
      a^2*c^2 - a^2*d*e - b^2*c^2 + b^2*d*e;
```

The 1,2 minor in the last two rows is indicated by the factor `e_(1,2)` etc. Because FORM knows that the Levi-Civita tensor is fully antisymmetric it kills terms with two identical indices immediately. It also sorts the arguments, taking the sign into account. Let us go now to a bigger matrix which was provided by David Wood of the university of Delaware:

```
S   aa,ad,ah,ai,aj,ak,al,am,an,ao;
S   bb,be,bh,bi,bj,bk,bl,bm,bn,bo;
S   cc,cf,ch,ci,cj,ck,cl,cm,cn,co;
S   dd,dg,dh,di,dj,dk,dl,dm,dn,do;
S   ee,eh,ei,ej,ek,el,em,en,eo;
S   fa,ff,fh,fi,fj,fk,fl,fm,fn,fo;
S   gb,gg,gh,gi,gj,gk,gl,gm,gn,go;
S   hc,hh,hi,hj,hk,hl,hm,hn,ho;
S   ii,il,jj,jm,kk,kn,ll,lo,mh,mm,ni,nn,oj,oo;
CF  T;
Table T(1:15,1:15);
Fill T(1,1) = aa, 0, 0,ad, 0, 0, 0,ah,ai,aj,ak,al,am,an,ao;
Fill T(2,1) =  0,bb, 0, 0,be, 0, 0,bh,bi,bj,bk,bl,bm,bn,bo;
Fill T(3,1) =  0, 0,cc, 0, 0,cf, 0,ch,ci,cj,ck,cl,cm,cn,co;
Fill T(4,1) =  0, 0, 0,dd, 0, 0,dg,dh,di,dj,dk,dl,dm,dn,do;
Fill T(5,1) =  0, 0, 0, 0,ee, 0, 0,eh,ei,ej,ek,el,em,en,eo;
Fill T(6,1) = fa, 0, 0, 0, 0,ff, 0,fh,fi,fj,fk,fl,fm,fn,fo;
Fill T(7,1) =  0,gb, 0, 0, 0, 0,gg,gh,gi,gj,gk,gl,gm,gn,go;
Fill T(8,1) =  0, 0,hc, 0, 0, 0, 0,hh,hi,hj,hk,hl,hm,hn,ho;
Fill T(9,1) =  0, 0, 0, 0, 0, 0, 0, 0,ii, 0, 0,il, 0, 0, 0;
Fill T(10,1)=  0, 0, 0, 0, 0, 0, 0, 0, 0,jj, 0, 0,jm, 0, 0;
Fill T(11,1)=  0, 0, 0, 0, 0, 0, 0, 0, 0, 0,kk, 0, 0,kn, 0;
Fill T(12,1)=  0, 0, 0, 0, 0, 0, 0, 0, 0, 0, 0,ll, 0, 0,lo;
Fill T(13,1)=  0, 0, 0, 0, 0, 0, 0,mh, 0, 0, 0, 0,mm, 0, 0;
Fill T(14,1)=  0, 0, 0, 0, 0, 0, 0, 0,ni, 0, 0, 0, 0,nn, 0;
Fill T(15,1)=  0, 0, 0, 0, 0, 0, 0, 0, 0,oj, 0, 0, 0, 0,oo;

Local F =
#do i = 1,15
    +e_('i')*T(15,'i')
#enddo
    ;
#do k = 14,1,-1
id  e_(?) =
#do i = 1,15
    +e_('i',.)*T('k','i')
#enddo
    ;
.sort: determinant step 'k';

Time =        0.02 sec    Generated terms =          4
                 F        Terms in output =          4
    determinant step 14   Bytes used      =        106
                     .
                     .
Time =        1.89 sec    Generated terms =        136
                 F        Terms in output =        136
    determinant step 2    Bytes used      =       7914

Time =        2.20 sec    Generated terms =         64
                 F        Terms in output =         64
    determinant step 1    Bytes used      =       3698

    id  e_(1,2,3,4,5,6,7,8,9,10,11,12,13,14,15) = 1;
    .end

Time =        2.22 sec    Generated terms =         64
                 F        Terms in output =         64
                          Bytes used      =       3664
```

This seems to work rather well. But it can even be improved. When we extend the minors by adding the extra row we would like to exploit the zeroes in the Levi-Civita tensor and the zeroes in the table as quickly as possible. That is, before we multiply by the minors. This can be arranged, even though everything is residing in one expression. We use the bracket statement to get the Levi-Civita tensor outside the brackets. The expression is stored this way during the sorting. Then after the sorting we put a 'Keep Brackets' statement. This tells FORM not to look inside the brackets until the next end-of-module (like `.sort`) is reached. Just before the sorting the contents of the brackets are multiplied by whatever is the result of what happened outside. Hence the code looks like

```
#do i = 1,15
    +e_('i',.)*T('k','i')
#enddo
    ;
B   e_;
.sort: determinant step 'k';
Keep Brackets;
#enddo
```

This way we obtain for the final statistics

```
Time =        1.04 sec    Generated terms =        64
                 F        Terms in output =        64
                          Bytes used      =      3664
```

This is a clear improvement. Especially for bigger determinants the improvement can be of vital importance.

Of course the next step is to automatize this by putting the code in a procedure that can handle the size of the matrix and the name of the table as parameters. This has of course been done, but showing the result is outside the aim of the course. The algorithm was the most important thing. Finally an example of how important it is to deal with the matrix in the 'right way'. If we would have started with the minors in the last columns, rather than the last rows, this would have happened:

```
Local F =
#do i = 1,15
    +e_('i')*T('i',15)
#enddo
    ;
#do k = 14,1,-1
id  e_(?) =
#do i = 1,15
    +e_('i',.)*T('i','k')
#enddo
    ;
B   e_;
.sort: determinant step 'k';
```

```
Time =        0.05 sec    Generated terms =        92
                 F        Terms in output =        92
    determinant step 14   Bytes used      =      2450
```

```
    Keep Brackets;
    #enddo

Time =        0.33 sec    Generated terms =        776
                 F        Terms in output =        776
    determinant step 13   Bytes used      =      20514


Time =        2.53 sec    Generated terms =       5892
                 F        Terms in output =       5892
    determinant step 12   Bytes used      =     152746

Time =       14.78 sec    Generated terms =      34678
                 F        Terms in output =      34678
    determinant step 11   Bytes used      =     892264

Time =       85.62 sec    Generated terms =     207037
                 F        Terms in output =     207037
    determinant step 10   Bytes used      =    5448552

Time =      466.40 sec    Generated terms =    1100961
                 F        Terms in output =    1100961
    determinant step 9    Bytes used      =   30213478



Time =      648.81 sec    Generated terms =     559236
                 F 174242 Terms left      =     559236
    determinant step 8    Bytes used      =   15663026

Time =      650.67 sec    Generated terms =     565664
                 F 178562 Terms left      =     565664
    determinant step 8    Bytes used      =   15828242
```

At this point the program was killed because it was clear that the disk was not large enough (only 80 Mbytes and FORM needs twice the size of the expression). It was still generating terms and expectation was that at the end of step 8 the expression would be about 90 Mbytes. It is important to use the zeroes in the matrix as early as possible.

5 Feynman Diagrams

One of the subjects that people are interested in, is a program that can do automatic calculations. This is to say, we specify a Lagrangian and a reaction, and out comes either a full formula for the cross section or a program that can calculate such a cross section numerically. Of course one could try to write such a program completely from scratch, but the smarter approach is probably to make good use of existing possibilities. And part of that is the intensive use of computer algebra. After all, formulae have to be worked out. Of course, for some parts one may have to resort to numerical programs. Also sometimes it is better to have a single program control the action, and then some people prefer that to be a program like Mathematica or Maple. Some systems that can do these things are FeynCalc [3] (for Mathematica) and HIP [4] for Maple. Because the complexity of the inherent problems and the relatively little amount of manpower that went into them, these systems are far from perfect. They can handle relatively small problems nicely, but the big problems for which automatization would pay off most handsomely are beyond their reach. This is due both to the systems and to the underlying algebraic program. Hence it is believed that a very modular system in which the modules obey some form of standardization for their communication will create the most powerful and complete system. This will enable specialists to create something they are good at in the language that is suitable for it. Hence we have by now two good diagram generators, QGRAF [5] (written in FORTRAN) by Nogueiro and the program by Kaneko [6]. In addition FeynCalc has also a decent generator. There also exists another system: CompHEP [7], made at the university of Moscow. It has been programmed originally in Pascal and uses for its algebra mainly REDUCE. It has of course also a diagram generator.

Some of the problems connected to automatic diagram computations will never find an algebraic solution. The effect of experimental cuts is usually something that needs to be solved in terms of Monte Carlo integration. Hence it requires a numerical approach. Also the optimal distribution of the points in this Monte Carlo integration seems to be a numerical problem, but in this case the construction of a program that is tailored to a given reaction may involve symbolic techniques. This is not clear currently. Of course the evaluation of the formulae connected to the Feynman diagrams, and the loop integrals will always involve symbolic work. Some programs exist that evaluate some loop integrals numerically, but even then much preparatory symbolic work is needed to reduce the diagram to a standard set of integrals.

Let us have a short look at diagram generation from the viewpoint of computer algebra. When we have a course on field theory we learn usually that the proper way to derive the relevant Feynman graphs is by means of functional derivatives of the path integral. And indeed from a theoretical viewpoint it is quite elegant. Unfortunately it is also quite inefficient as we will see. Let us look at ϕ^m theory. This only illustrates some of the more obvious problems. After having a good look at the method, things come down to finding a way to couple the two ends of a given number of propagators to the lines of a number of vertices. If this can be done in more than one way that is equivalent, we want to keep only one of them but with an appropriate combinatorics factor. Let us have a look at the program to find all nontrivial 4-loop propagator diagrams in ϕ^4 theory.

```
I   z,z1=0,z2=0,z3=0,z4=0,z5=0,z6=0;
Set zz:z3,z4,z5,z6;
I   n,n1=0,n2=0,n3=0,n4=0,n5=0,n6=0;
Set nn:n3,n4,n5,n6;
S   i;
Tensors D,T,v;
L   F = T(z1)*T(z2)*v(z3)*v(z4)*v(z5)*v(z6)/24;
id,once,T(?) = D(.);
repeat id T(?)*D(??)  = D(..,.);
```

```
repeat id v(z?)*D(??) = D(..,z,z,z,z)/24;
id D(?)=dd_(.);
.sort
```

The specifications in the declaration of the indices indicate their dimensions. A dimension of 0 indicates that FORM should not apply automatic summation conventions. Tensors are special functions. Their arguments can only be indices or vectors. By default they are commuting (NTensors or NT indicates non-commuting tensors).

We define the two endpoints with z_1 and z_2 and then we attach 4 vertices for the 4-loop graphs. They are labeled z_3 to z_6. Then we put all legs into a single function D. This means that the endpoints get only one entry in D and the vertices get 4 entries each. Finally we put the whole thing inside the function dd_. This is a special function in FORM that generates products of Kronecker delta's:

$$dd\_(\mu_1, \mu_2) = \delta^{\mu_1\mu_2}$$
$$dd\_(\mu_1, \mu_2, \mu_3, \mu_4) = \delta^{\mu_1\mu_2}\delta^{\mu_3\mu_4} + \delta^{\mu_1\mu_3}\delta^{\mu_2\mu_4} + \delta^{\mu_1\mu_4}\delta^{\mu_2\mu_3}$$

etc. It is a fully symmetric tensor. Its strong point is that it detects identical arguments. It does not generate identical terms, but it does the combinatorics right and generates the proper coefficient. In this program it is quite necessary because it has 18 arguments and it would generate otherwise 17!! = 34459425 terms. In our case we obtain

```
Time =       0.43 sec    Generated terms =       1326
                F        Terms in output =       1326
                         Bytes used      =      30386
    TryReplace,z3,z4,z4,z3;
    .sort
Time =       0.95 sec    Generated terms =       1326
                F        Terms in output =        706
                         Bytes used      =      16574
    TryReplace,z3,z5,z5,z3;
    .sort
Time =       1.26 sec    Generated terms =        706
                F        Terms in output =        462
                         Bytes used      =      10866
    TryReplace,z3,z6,z6,z3;
    .sort
Time =       1.46 sec    Generated terms =        462
                F        Terms in output =        340
                         Bytes used      =       8050
    TryReplace,z4,z5,z5,z4;
    .sort
Time =       1.59 sec    Generated terms =        340
                F        Terms in output =        195
                         Bytes used      =       4662
    TryReplace,z4,z6,z6,z4;
    .sort
Time =       1.67 sec    Generated terms =        195
                F        Terms in output =        134
                         Bytes used      =       3290
    TryReplace,z5,z6,z6,z5;
    .sort
```

```
Time =        1.71 sec    Generated terms =        134
                 F        Terms in output =         86
                          Bytes used      =       2162
```

Next we see a very new statement. FORM has a special function `replace_`. It should have an even number of arguments. When this function is attached to a term, FORM will first normal order the term and then do as if the odd arguments are wildcards that still have to be replaced by the values in the even arguments. This can be very powerful. It is also the fastest way to exchange objects.

```
Multiply replace_(z3,z4,z4,z3);
```

will replace z3 by z4 and at the same time z4 by z3 for all occurrences in each term. The new statement TryReplace is related to this. FORM will see the arguments as arguments for a `replace_` function, make the replacements and then see whether the resulting term would come before the current term after sorting. If it would, it will keep the new term, otherwise it keeps the old term. Hence this can be used to bring all terms to some standard form by trying permutations of the vertices. In our case we do not have to try all permutations. All pairwise exchanges seem to be sufficient. And we see that the number of terms gets reduced quickly. The `.sort` instructions were inserted to have the terms cancel as quickly as possible.

After this we will eliminate the vacuum bubble graphs:

```
id  d_(z1,z2) = 0;
id  d_(z?,n?) = D(z,n);
id  D(z1,z?zz[i])=d_(z1,z)*replace_(z,nn[i]);
repeat;
 id,once,D(n?nn,z?zz[i])=d_(n,z)*replace_(z,nn[i]);
 id,once,D(z?zz[i],n?nn)=d_(z,n)*replace_(z,nn[i]);
endrepeat;
id  D(z2,n?nn) = d_(z2,n);
id  D(z?zz,n?zz) = 0;
id  d_(n1?,n2?) = D(n1,n2);
.sort
```

The replace_ function is used here to replace an element of the set zz by a corresponding element in the set nn. We start at z1 and its partner is replaced by one of the n indices. This is done for all its occurrences! Next we enter a loop in which partners of n-indices that are still z-indices are replaced by their corresponding n-indices. If, in the end, there are still z-indices left they must be in a part of the graph that is not connected to z1.

```
Time =        2.08 sec    Generated terms =         39
                 F        Terms in output =         39
                          Bytes used      =       1482
```

```
id D(n?,n?) = 0;
id D(z1,n?)*D(z2,n?) = 0;
id D(n1?,n2?)*D(n1?,n2?)*D(n1?,n2?)*
                            D(n1?,n3?)*D(n2?,n3?) = 0;
id D(n1?,n2?)*D(n1?,n2?)*D(n1?,n2?)*
```

```
                         D(n1?,n3?)*D(n3?,n2?) = 0;
id D(n1?,n2?)*D(n1?,n2?)*D(n1?,n2?)*
                         D(n3?,n1?)*D(n3?,n2?) = 0;
id  D(z1,n?)*D(n?,n1?)*D(n?,n1?)*D(n?,n1?) = 0;
Print +f +s;
.end
```

Finally we eliminate diagrams with tadpoles or diagrams that are one particle reducible. For a higher number of loops one might need more statements here.

```
Time =        2.22 sec    Generated terms =          4
                 F        Terms in output =          4
                          Bytes used      =        238

   F =
      + 1/12*D(z1,n3)*D(z2,n4)*D(n3,n4)*D(n3,n4)*D(n3,
     n5)*D(n4,n6)*D(n5,n6)*D(n5,n6)*D(n5,n6)
      + 1/8*D(z1,n3)*D(z2,n4)*D(n3,n4)*D(n3,n5)*D(n3,
     n5)*D(n4,n6)*D(n4,n6)*D(n5,n6)*D(n5,n6)
      + 1/4*D(z1,n3)*D(z2,n4)*D(n3,n4)*D(n3,n5)*D(n3,
     n6)*D(n4,n5)*D(n4,n6)*D(n5,n6)*D(n5,n6)
      + 1/4*D(z1,n3)*D(z2,n4)*D(n3,n5)*D(n3,n5)*D(n3,
     n6)*D(n4,n5)*D(n4,n6)*D(n4,n6)*D(n5,n6)
      ;
```

The answer consists of only 4 diagrams and amazingly enough this simple program gives even the correct combinatoric factors.

If we modify the above program for 5 loop diagrams we find for the final statistics

```
Time =       46.32 sec    Generated terms =         16
                 F        Terms in output =         16
                          Bytes used      =        886
```

It should be clear from the above example that the straight forward brute attack is not always useful. The diagrams we derived might have been written down by an expert in almost the same time that the computer needed for them. Things even become worse when three-point vertices are involved. This method needs more than 110 sec. for the three loop propagator graphs. An expert will easily beat that! Hence the good diagram generators try to construct the topologies in a direct way, without constructing the disconnected diagrams and without generating all the symmetrically equivalent diagrams. This is less trivial of course but necessary. Anyway, when the full standard model is considered, one should not only construct the topologies. That is the easy part. One has to assign a particle to each line and that may be done in many different ways. For $2 \to 4$ reactions at the tree level with only three-point vertices (like in QED) one has only two topologies. Yet the reaction $e^-e^+ \to e^-e^+e^-e^+$ has already 36 diagrams in QED, all based on one topology. The best thing here is to just use one of the existing generators. They can be obtained from the respective authors.

Of course diagram generation is not the only thing needed. After obtaining the diagrams we have to calculate matrix elements. A good example would be the reaction $e^-e^+ \to q\overline{q}q\overline{q}$. If we ignore the problems with fermi statistics because we might have identical quarks, there are only four diagrams. Labeling the quarks 1 and 3 and the antiquarks 2 and 4 we have (after some preparatory work in the electron line, using current conservation)

```
V q,pe,p1,p2,p3,p4,[q-p1],[q-p2],[q-p3],[q-p4];
I mu,nu,ka,la;
S s;

L F = (8*pe(mu)*pe(nu)+2*s*d_(mu,nu))*
      g_(1,p1)*g_(2,p3)*(
        +g_(1,ka)*(g_(1,q)-g_(1,p2))/[q-p2].[q-p2]*g_(1,mu)/2/p3.p4*g_(2,ka)
        +g_(1,mu)*(g_(1,p1)-g_(1,q))/[q-p1].[q-p1]*g_(1,ka)/2/p3.p4*g_(2,ka)
        +g_(2,ka)*(g_(2,q)-g_(2,p4))/[q-p4].[q-p4]*g_(2,mu)/2/p1.p2*g_(1,ka)
        +g_(2,mu)*(g_(2,p3)-g_(2,q))/[q-p3].[q-p3]*g_(2,ka)/2/p1.p2*g_(1,ka)
      )*g_(1,p2)*g_(2,p4)*(
        +g_(1,nu)*(g_(1,q)-g_(1,p2))/[q-p2].[q-p2]*g_(1,la)/2/p3.p4*g_(2,la)
        +g_(1,la)*(g_(1,p1)-g_(1,q))/[q-p1].[q-p1]*g_(1,nu)/2/p3.p4*g_(2,la)
        +g_(2,nu)*(g_(2,q)-g_(2,p4))/[q-p4].[q-p4]*g_(2,la)/2/p1.p2*g_(1,la)
        +g_(2,la)*(g_(2,p3)-g_(2,q))/[q-p3].[q-p3]*g_(2,nu)/2/p1.p2*g_(1,la)
      );
Trace4,2;
Trace4,1;
id  pe.pe = 0;
id  p1.p1 = 0;
id  p2.p2 = 0;
id  p3.p3 = 0;
id  p4.p4 = 0;
id  q.q = s;
.end

Time =        0.97 sec    Generated terms =       1693
                 F        Terms in output =        368
                          Bytes used      =      15454
```

The function g_ is the gamma matrix. Its first argument indicates to what spinline it belongs. This must be an index. If it has more than two arguments it is a string of gamma matrices. The argument 5_ is the matrix γ_5 while FORM also uses the schoonschip convention $\gamma_6 = 1+\gamma_5$ and $\gamma_7 = 1-\gamma 5$. The use of these objects can improve the speed of calculations enormously. The command Trace4 takes traces in 4 dimensions. There is a similar command for traces in n dimensions: Tracen. When we assume the quarks to be massless, the expression is easily manageable for a FORTRAN program, provided we use a different notation for the vectors `[q-p1]` etc. We have used this here to make the program more readable.

Of course there are ways to make the expression shorter, using all kinds of tricks for massless spinors. But that would be far more work than we have done here. And it would not work once we would add masses. Here we just want to see how easy it is to derive such traces. The main work is to type the input, and the main worry is that one has made a typing error. And clearly, if such an error would be there, it would take very little work to repair it, and it would take the computer less than 1 sec to redo the calculation.

And now a surprising result. Let us assume a hypothetical reaction, involving a charged and a neutral heavy lepton:

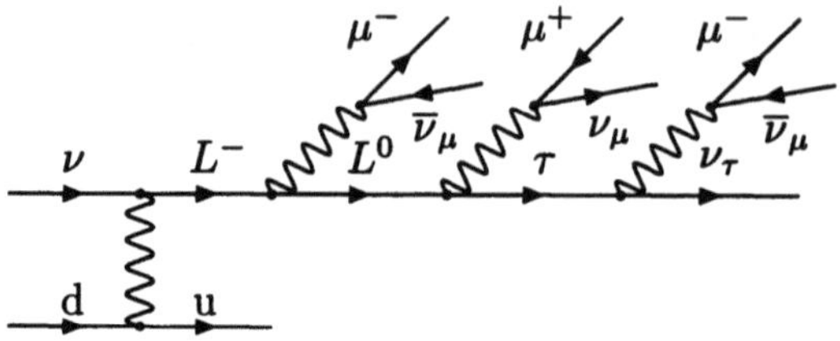

Assume we make the particle in a neutrino interaction via the diagram, and we observe events by its multi-lepton signal. Hence we want to know the distribution of the charged leptons. Fortunately we assume that all couplings are V-A and that simplifies the matrixelement enormously. The program would read:

```
S    mt,mLm,mLO,mmu;
V    pLm,pLO,pt,pm1,pm2,pm3,pd,pu,pn1,pn2,pn3,pn4,pn5;
I    i1,i2,i3,i4,j1,j2,j3,j4;

L    M2 = g_(1,pn1)*(g_(2,pm1)+mmu)*g_(3,pn4)*(g_(4,pm3)+mmu)*g_(5,pu)
        *g_(1,i1,7_)*(g_(1,pt)+mt)*g_(1,i2,7_)*(g_(1,pLO)+mLO)
        *g_(1,i3,7_)*(g_(1,pLm)+mLm)*g_(1,i4,7_)
        *g_(2,i1,7_)*g_(3,i2,7_)*g_(4,i3,7_)*g_(5,i4,7_)
        *g_(1,pn2)*g_(2,pn3)*(g_(3,pm2)-mmu)*g_(4,pn5)*g_(5,pd)
        *g_(1,j4,7_)*(g_(1,pLm)+mLm)*g_(1,j3,7_)*(g_(1,pLO)+mLO)
        *g_(1,j2,7_)*(g_(1,pt)+mt)*g_(1,j1,7_)
        *g_(2,j1,7_)*g_(3,j2,7_)*g_(4,j3,7_)*g_(5,j4,7_)
        ;
Trace4,5;
Trace4,4;
Trace4,3;
Trace4,2;
Trace4,1;
Multiply 1/2^20;
Print +f +s;
.end
```

The number of gamma matrices is very large in this expression. And in addition we have mass terms for the muons and the heavy leptons. Therefore the result is very surprising:

```
Time =       0.56 sec    Generated terms =       672
               M2        Terms in output =         8
                         Bytes used      =       364

   M2 =
       - 512*pLm.pLm*pLO.pLO*pt.pt*pm1.pn1*pm2.pm3*pd.pn2*pu.pn5*pn3.pn4
       + 1024*pLm.pLm*pLO.pLO*pt.pn3*pt.pn4*pm1.pn1*pm2.pm3*pd.pn2*pu.pn5
       + 1024*pLm.pLm*pLO.pm2*pLO.pm3*pt.pt*pm1.pn1*pd.pn2*pu.pn5*pn3.pn4
       - 2048*pLm.pLm*pLO.pm2*pLO.pm3*pt.pn3*pt.pn4*pm1.pn1*pd.pn2*pu.pn5
       + 1024*pLm.pu*pLm.pn5*pLO.pLO*pt.pt*pm1.pn1*pm2.pm3*pd.pn2*pn3.pn4
       - 2048*pLm.pu*pLm.pn5*pLO.pLO*pt.pn3*pt.pn4*pm1.pn1*pm2.pm3*pd.pn2
       - 2048*pLm.pu*pLm.pn5*pLO.pm2*pLO.pm3*pt.pt*pm1.pn1*pd.pn2*pn3.pn4
       + 4096*pLm.pu*pLm.pn5*pLO.pm2*pLO.pm3*pt.pn3*pt.pn4*pm1.pn1*pd.pn2
      ;
```

Yet it needs rather sophisticated trace algorithms to give the answer in a concise form without Levi-Civita tensors. Version 1 of FORM could not handle it, but version 2 is much better at it. The 8 terms in the output even factorize into 2x2x2 terms! And we did not ignore any of the masses! Actually even the inclusion of neutrino masses would have given the same result. Of course, if you are rather good at playing with gamma matrices, you can discover the trick by which you could do this even by hand in very little time. Nevertheless this little program shows one of the strong points of computer algebra: The experience of experts becomes available to the masses.

6 A Two Loop Gluon Diagram

In this section we are going to do a real calculation. We are going to work out the worst diagram one can encounter when computing the two loop gluon propagator. And we are not going to make it easy: we are going to include a gauge parameter as well. Once one can compute this diagram, the program can be used to calculate all other diagrams as well. We will leave this to the very diligent student(s). The diagram we are going to compute is given by

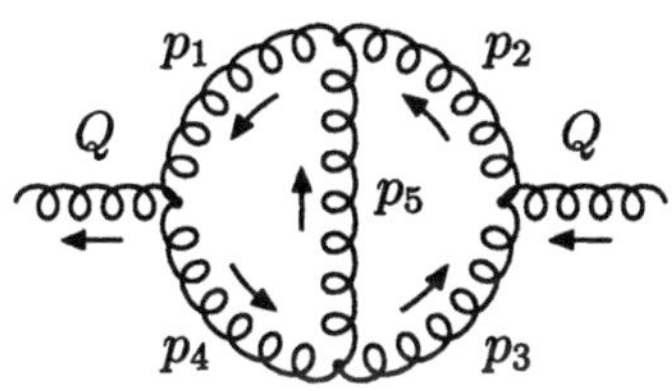

In principle it looks rather simple: no fermions. But beware: vector particles are much nastier than fermions when it comes down to the complexity of a calculation. Hence let us start with inspecting what we are getting us into when we just substitute vertices and propagators:

```
S   D,ep,xi;
Dimension D;
I   MU,NU,m1,m2,m3,i1,i2,i3,i4,i5,j1,j2,j3,j4,j5;
V   Q,p,p1,p2,p3,p4,p5;
CF  v3g,dg;

L   F = v3g(-p1,i1,p4,j4,Q,NU)
        *v3g(p1,j1,-p2,i2,-p5,i5)
        *v3g(p2,j2,-Q,MU,-p3,i3)
        *v3g(p3,j3,-p4,i4,p5,j5)
        *dg(i1,j1,p1)
        *dg(i2,j2,p2)
        *dg(i3,j3,p3)
        *dg(i4,j4,p4)
        *dg(i5,j5,p5);
id  dg(m1?,m2?,p?) = d_(m1,m2)/p.p-xi*p(m1)*p(m2)/p.p^2;
id  v3g(p1?,m1?,p2?,m2?,p3?,m3?) =
            +d_(m1,m2)*(p1(m3)-p2(m3))
            +d_(m2,m3)*(p2(m1)-p3(m1))
            +d_(m3,m1)*(p3(m2)-p1(m2));
Multiply d_(MU,NU)/Q.Q;
Print +f +s;
.end
```

```
Time =       16.71 sec    Generated terms =      41472
                 F        Terms in output =       2552
                          Bytes used      =     117890

   F =
       - Q.Q^-1*Q.p1*Q.p2*p1.p1^-2*p1.p3*p2.p2^-2*p2.p5*p3.p3^-2*p3.p4*
      p4.p4^-2*p4.p5*p5.p5^-1*xi^4
       + Q.Q^-1*Q.p1*Q.p2*p1.p1^-2*p1.p3*p2.p2^-2*p2.p5*p3.p3^-2*p3.p5*
      p4.p4^-1*p5.p5^-1*xi^3
  etc
```

We use the dimension statement to fix the dimension of all indices at $D = 4-2\epsilon$. Because each vertex gives 6 terms and each propagator 2 the expansion gives $6^4 2^5 = 41472$ terms. Of course many terms cancel or are identical, but the expression is still far from trivial. We give the first two terms, just to give an impression of what we have. Next we want to express all the mixed dotproducts in terms of squares of the momenta. This is possible and it is done by adding code to the original program. The execution time is not so long that we would like to write the intermediate result to a file.

```
    id  p4.p?!{p4} = p1.p-Q.p;
    id  p3.p?!{p3} = p2.p-Q.p;
    .sort

Time =       22.21 sec    Generated terms =      15789
                 F        Terms in output =       2575
                          Bytes used      =      86006

    id  p5.p?!{p5} = p1.p-p2.p;
    .sort

Time =       24.00 sec    Generated terms =       6230
                 F        Terms in output =       1905
                          Bytes used      =      52914

    id  Q.p1 = p1.p1/2+Q.Q/2-p4.p4/2;
    id  Q.p2 = p2.p2/2+Q.Q/2-p3.p3/2;
    .sort

Time =       26.82 sec    Generated terms =      12073
                 F        Terms in output =       2038
                          Bytes used      =      52358

    id  p1.p2 = p1.p1/2+p2.p2/2-p5.p5/2;
    .sort

Time =       28.15 sec    Generated terms =       6061
                 F        Terms in output =       1302
                          Bytes used      =      32038

    id  D = acc(4-2*ep);
    Print +f +s;
    .end(PolyFun=acc);

Time =       28.58 sec    Generated terms =       2050
                 F        Terms in output =        928
                          Bytes used      =      45924

   F =
```

```
      + Q.Q^-1*p1.p1^-2*p2.p2^-2*p3.p3^-1*p4.p4^-1*p5.p5^3*xi^2*acc(1/8)
      + Q.Q^-1*p1.p1^-2*p2.p2^-2*p3.p3^-1*p4.p4*p5.p5*xi^2*acc( - 1/4)
      + Q.Q^-1*p1.p1^-2*p2.p2^-2*p3.p3^-1*p4.p4^2*xi^3*acc(1/16)
      + Q.Q^-1*p1.p1^-2*p2.p2^-2*p3.p3^-1*p5.p5^2*xi^2*acc( - 1/8)
      + Q.Q^-1*p1.p1^-2*p2.p2^-2*p3.p3^-1*p5.p5^2*xi^3*acc( - 1/16)
      + Q.Q^-1*p1.p1^-2*p2.p2^-2*p3.p3*p4.p4^-1*p5.p5*xi^2*acc( - 1/4)
      + Q.Q^-1*p1.p1^-2*p2.p2^-2*p3.p3*xi^2*acc(9/8 - ep)
etc
```

The new feature here is the particular wildcarding. p?!{p5} gives us two new concepts. The first is that sets can be introduced locally. {p5} is a set with one element p5. One can use it in the same way as declared sets. The exclamationmark in p?!{p5} is a logical 'not'. Hence p will match all vectors with the exception of p5.

We notice now a slow convergence to an independent set of variables. The output becomes smaller this way. We also introduce another useful feature. It was invented for exactly this type of calculations in which expansions in terms of a parameter ϵ take place. One would like the coefficient of each term to be a polynomial in ϵ. In that case FORM will have far fewer terms to manipulate. FORM can handle that. We have to define a function (we call it acc for accumulator) and as an option in the end-of-module instructions we define it as the 'PolyFun'. This way FORM will see the argument of this function as the coefficient of the terms, provided the function has only one argument. In addition multiple occurrences of acc in one term are reduced to a single occurrence by multiplying the arguments. One has to be careful that the argument does not become too long. This is usually done by putting a power restriction on the parameter ep. We notice one such occurrence of acc in the few terms of the output that are shown. Two terms were added to give a more complicated argument.

```
    if ( count(p5.p5,1) >= 0 );
        if ( count(p1.p1,1) >= 0 ) discard;
        if ( count(p2.p2,1) >= 0 ) discard;
        if ( count(p3.p3,1) >= 0 ) discard;
        if ( count(p4.p4,1) >= 0 ) discard;
    endif;
    if ( ( count(p1.p1,1) >= 0 ) && ( count(p2.p2,1) >= 0 ) ) discard;
    if ( ( count(p3.p3,1) >= 0 ) && ( count(p4.p4,1) >= 0 ) ) discard;
    if ( ( count(p1.p1,1) >= 0 ) && ( count(p4.p4,1) >= 0 ) ) discard;
    if ( ( count(p2.p2,1) >= 0 ) && ( count(p3.p3,1) >= 0 ) ) discard;
    .sort(PolyFun=acc);

Time =       29.20 sec    Generated terms =        398
                 F        Terms in output =        398
                          Bytes used      =      19652


    while ( match(1/p1.p1/p2.p2/p3.p3/p4.p4/p5.p5) );
        id  1/p2.p2^n2?/p3.p3^n3?/p5.p5^n5? = 1/p2.p2^n2/p3.p3^n3/p5.p5^n5*(
                +n2*p5.p5/p2.p2-n2*p1.p1/p2.p2
                +n3*p5.p5/p3.p3-n3*p4.p4/p3.p3
                    )*den(4-2*n5-n2-n3,-2*ep);
    endwhile;

    if ( match(1/p1.p1/p2.p2/p3.p3/p5.p5) && ( count(p4.p4,1) >= 0 ) )
```

```
        Multiply,replace_(p4,-p1,p1,-p4,p2,-p3,p3,-p2,p5,-p5);
if ( match(1/p1.p1/p2.p2/p4.p4/p5.p5) && ( count(p3.p3,1) >= 0 ) )
        Multiply,replace_(p1,p3,p3,p1,p2,p4,p4,p2,p5,-p5,Q,-Q);
if ( match(1/p1.p1/p3.p3/p4.p4/p5.p5) && ( count(p2.p2,1) >= 0 ) )
        Multiply,replace_(p1,-p2,p2,-p1,p3,-p4,p4,-p3,Q,-Q);
if ( match(1/p1.p1/p3.p3/p5.p5) && ( count(p2.p2,1) >= 0 ) && (
                                       count(p4.p4,1) >= 0 ) )
        Multiply,replace_(p1,-p2,p2,-p1,p3,-p4,p4,-p3,Q,-Q);
Print +f +s;
.end(PolyFun=acc);

Time =      29.55 sec    Generated terms =       973
                F        Terms in output =       276
                         Bytes used      =     13324

   F =
      + Q.Q^-1*p1.p1^-2*p2.p2^-2*p3.p3^-1*p4.p4^-1*p5.p5^3*xi^2*acc(1/8)
      + Q.Q^-1*p1.p1^-2*p2.p2^-1*p3.p3^-2*p4.p4^-1*p5.p5^3*xi^2*acc(1/8)
      + Q.Q^-1*p1.p1^-2*p2.p2^-1*p3.p3^-1*p4.p4^-1*p5.p5^2*xi*acc(9/4 - 2*ep)
      + Q.Q^-1*p1.p1^-2*p2.p2^-1*p3.p3^-1*p4.p4^-1*p5.p5^2*xi^2*acc( - 3/8)
      + Q.Q^-1*p1.p1^-1*p2.p2^-2*p3.p3^-1*p4.p4^-2*p5.p5^3*xi^2*acc(1/8)
      + Q.Q^-1*p1.p1^-1*p2.p2^-2*p3.p3^-1*p4.p4^-1*p5.p5^2*xi*acc(9/4 - 2*ep)
      + den(-2, - 2*ep)*den(-1, - 2*ep)*den(0, - 2*ep)*Q.Q^3*p1.p1^-1*
     p2.p2^-3*p3.p3^-2*p4.p4^-1*xi^2*acc(1/2)
   etc
```

The next step kills a number of terms that will give zero after integration. This reduces the number of terms considerably. We will see soon why they are zero. In the second part of this example is the most important statement of the program. It reduces the most complicated terms with 5 different dotproducts in the denominator gradually into terms that have at most 4 different dotproducts in the denominator. Where does this statement come from?

Let us look at a so-called triangle graph. We have two of these in our diagram.

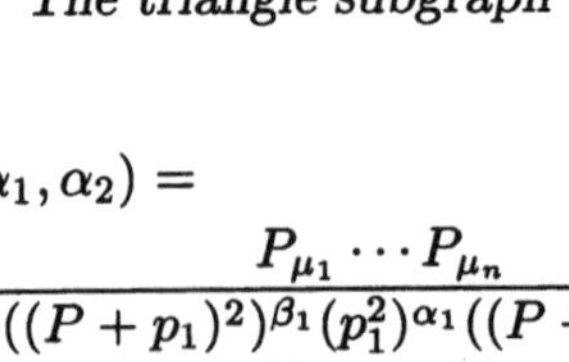

The triangle subgraph

We define the integral

$$I(n,\alpha_0,\beta_1,\beta_2,\alpha_1,\alpha_2) = \int d^DP \frac{P_{\mu_1}\cdots P_{\mu_n}}{(P^2)^{\alpha_0}((P+p_1)^2)^{\beta_1}(p_1^2)^{\alpha_1}((P+p_2)^2)^{\beta_2}(p_2^2)^{\alpha_2}}, \tag{3}$$

This integral can be studied by first considering the identity:

$$\int d^DP \left\{ \frac{\partial}{\partial P_\mu} \frac{P_\mu \quad \mathcal{P}_n(P)}{(P^2)^{\alpha_0}((P+p_1)^2)^{\beta_1}(p_1^2)^{\alpha_1}((P+p_2)^2)^{\beta_2}(p_2^2)^{\alpha_2}} \right\} = 0. \tag{4}$$

Working out the derivative one obtains the recursion relation [8]:

$$
\begin{aligned}
&I(n, \alpha_0, \beta_1, \beta_2, \alpha_1, \alpha_2) = (\\
&\quad +\beta_1(I(n, \alpha_0 - 1, \beta_1 + 1, \beta_2, \alpha_1, \alpha_2) - I(n, \alpha_0, \beta_1 + 1, \beta_2, \alpha_1 - 1, \alpha_2)) \\
&\quad +\beta_2(I(n, \alpha_0 - 1, \beta_1, \beta_2 + 1, \alpha_1, \alpha_2) - I(n, \alpha_0, \beta_1, \beta_2 + 1, \alpha_1, \alpha_2 - 1)) \\
&\quad)/(D + n - 2\alpha_0 - \beta_1 - \beta_2).
\end{aligned}
\tag{5}
$$

In many cases one cannot apply this identity in a direct way. The momenta in the numerator can make that the coefficient in the left hand side becomes proportional to ϵ more than once when one is reducing complicated three loop graphs. This is not very nice if one would like to keep the expansion in ϵ limited in depth to save computer time. In that case one has to resort to dirty tricks, like summing up the complete effect of the application of the formula. This has been done [9] and gives a rather ugly but very useful formula with three terms, each with three sums. In our case we have no loop momenta in the numerator and hence we are going to keep things easy. Notice that we keep the coefficient that involves ϵ inside a special function, named den for denominator. Its use will become clear rather soon.

Having killed one propagator, we use symmetry to diminish the number of terms somewhat. It also reduces the code we need to do the integration, because we move the missing propagator to the one-line. Of course p_1^2 can still have a positive power. And so can p_5^2 when it does not occur in the denominator anymore. This can be seen in the few output terms that are shown.

Next we take a good look at the den function. By splitting the fractions we can make sure that there is no more than a single occurrence of den in each term. This is done by

```
    repeat id den(n1?!{n2?},-2*ep)*den(n2?!{n1?},-2*ep) =
            den(n1,-2*ep)/(n2-n1)-den(n2,-2*ep)/(n2-n1);
    Print +f +s;
    .end(PolyFun=acc);

Time =       29.51 sec    Generated terms =        423
                 F        Terms in output =        255
                          Bytes used      =      11992

   F =
       + den(0, - 2*ep)*Q.Q*p1.p1^-1*p2.p2^-2*p3.p3^-1*p4.p4^-1*xi*acc( - 9/2
       + 4*ep)
       + den(0, - 2*ep)*Q.Q*p1.p1^-1*p2.p2^-2*p3.p3^-1*p4.p4^-1*xi^2*acc(1/4)
       + den(0, - 2*ep)*Q.Q*p1.p1^-1*p2.p2^-2*p3.p3^-1*p4.p4^-1*acc( - 3/2 + 2
      *ep)
       + den(0, - 2*ep)*Q.Q^3*p2.p2^-3*p3.p3^-1*p4.p4^-2*p5.p5^-1*xi^2*acc( -
      1/2)
      ;
```

The wildcarding reaches here a new level of complexity. We demand that the two wildcards are not equal to each other. But it works. Actually all terms with `den(-2,-2*ep)` and `den(-1,-2*ep)` cancel each other and these objects are removed from the expression.

At this point we are ready for integration. Each term corresponds now either to the product or to the convolution of two one loop integrals. Therefore let us study one loop integrals first.

The basic one loop diagram is given by:

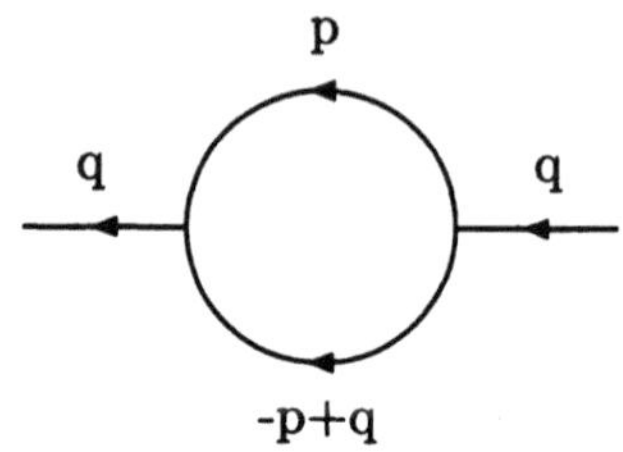

The basic one loop diagram

The integral is given by [10]

$$\int \frac{d^D p}{(2\pi)^D} \frac{\mathcal{P}_n(p)}{p^{2\alpha}(p-q)^{2\beta}} =$$

$$\frac{1}{(4\pi)^2}(q^2)^{D/2-\alpha-\beta} \sum_{\sigma\geq 0}^{[n/2]} G(\alpha,\beta,n,\sigma) q^{2\sigma} \left\{ \frac{1}{\sigma!}\left(\frac{\Box}{4}\right)^{\sigma} \mathcal{P}_n(p) \right\}_{p=q}, \tag{6}$$

in which

$$\mathcal{P}_n(p) = p_{\mu_1} p_{\mu_2} \cdots p_{\mu_n}. \tag{7}$$

D is the dimension of space-time and is also given by $D = 4 - 2\epsilon$, $\Box = \partial^2/\partial p_\mu \partial p_\mu$ and G can be expressed in terms of Γ-functions:

$$G(\alpha,\beta,n,\sigma) =$$

$$(4\pi)^\epsilon \frac{\Gamma(\alpha+\beta-\sigma-D/2)\Gamma(D/2-\alpha+n-\sigma)\Gamma(D/2-\beta+\sigma)}{\Gamma(\alpha)\Gamma(\beta)\Gamma(D-\alpha-\beta+n)}. \tag{8}$$

We see here quickly that one loop diagrams in which one line has a power that is zero or a positive integer must be zero. Basically such a diagram corresponds to a massless tadpole graph and those are zero in dimensional regularization. Because the above formulae are also valid if the power of the lines is not an integer we can use it also for convolutions of one loop graphs. We have such a convolution when the one-line is missing. First we do the one loop propagator subgraph with the 4-line and the 5-line. Their result gives (among others) $(p_3^2)^{-\epsilon}$. But that is no problem in the remaining diagram with only the 2-line and the 3-line remaining.

We do have to be careful with positive powers of the dotproducts. For a positive power of p_5^2 we write $p_5^2 = p_1^2 + p_2^2 - 2p_1{\cdot}p_2$. The term with the mixed dotproduct may get treated with the d'Alembertian. We have programmed it here for just a single mixed object to the power n. Of course it would be more elegant to make something general, and this has been done, but it is a rather intransparent piece of code and we do not need it here. It is very efficient though. Hence the next step becomes:

```
    if ( count(p5.p5,1) >= 0 );
        id p5.p5 = p1.p1+p2.p2-2*p1.p2;
        if ( count(p1.p1,1) >= 0 ) discard;
        if ( count(p2.p2,1) >= 0 ) discard;
        id p1.p2^n?/p2.p2^n2?/p3.p3^n3? = Q.Q^(2-n2-n3)*sum_(j,0,integer_(n/2),
            G(n2,n3,n,j)*Q.p1^(n-2*j)*Q.Q^j*p1.p1^j*fac_(n)/fac_(n-2*j)/4^j);
        id  Q.p1 = p1.p1/2+Q.Q/2-p4.p4/2;
        if ( count(p1.p1,1) >= 0 ) discard;
        if ( count(p4.p4,1) >= 0 ) discard;
        id 1/p1.p1^n1?/p4.p4^n4? = Q.Q^2/Q.Q^n1/Q.Q^n4*G(n1,n4,0,0);
    elseif ( count(p1.p1,1) >= 0 );
        id p1.p1 = Q.Q+p4.p4+2*Q.p4;
        if ( count(p4.p4,1) >= 0 ) discard;
        id Q.p4 = -Q.p4;
        id Q.p4^n?/p4.p4^n4?/p5.p5^n5? = p3.p3^(2-n4-n5)*sum_(j,0,integer_(n/2),
            G(n4,n5,n,j)*Q.p3^(n-2*j)*Q.Q^j*p3.p3^j*fac_(n)/fac_(n-2*j)/4^j);
        id  Q.p3 = -p2.p2/2+Q.Q/2+p3.p3/2;
        if ( count(p2.p2,1) >= 0 ) discard;
        id 1/p3.p3^n3?/p2.p2^n2? = Q.Q^2/Q.Q^n3/Q.Q^n2*G(n3+ep,n2,0,0);
    endif;
    Print +f +s;
    .end(PolyFun=acc);

Time =      32.11 sec    Generated terms =      5852
                F        Terms in output =       369
                         Bytes used      =     16840

   F =
       + den(0, - 2*ep)*G(1,1,0,0)*G(1 + ep,2,0,0)*xi*acc(9 - 8*ep)
       + den(0, - 2*ep)*G(1,1,0,0)*G(1 + ep,2,0,0)*xi^2*acc( - 1/2)
       + G(1,1,0,0)*G( - 1 + ep,2,0,0)*xi*acc(41/4 - 8*ep)
       + G(1,1,0,0)*G(1 + ep,1,0,0)*xi*acc(34 - 28*ep)
   etc
```

The integration has only two distinct cases. This is due to the symmetry relations. In the case that p_1^2 is missing from the denominator we flip the direction of the 4 and 3 momenta. This avoids problems with minus signs later. In the intermediate step we ignored the factor $(p_3^2)^{-\epsilon}$ and put it in the G function by hand. In the output we ignore the overall factor $(Q^2)^{-2\epsilon}$. At this point we only have to substitute the G functions and start playing with the Gamma functions:

```
    id  G(n1?int_,n2?,n3?,n4?) = G(1,1)
              *Gamma(n1+n2-n4-2,ep)*InvGamma(0,ep)
              *Gamma(2-n1+n3-n4,-ep)*InvGamma(1,-ep)
              *Gamma(2-n2+n4,-ep)*InvGamma(1,-ep)
              /fac_(n1-1)/fac_(n2-1)
              *InvGamma(4-n1-n2+n3,-2*ep)*Gamma(2,-2*ep);
    id  G(n1?,n2?,n3?,n4?) = G(1+ep,1)
              *Gamma(n1-ep+n2-n4-2,2*ep)*InvGamma(0,2*ep)
              *Gamma(2-n1+ep+n3-n4,-2*ep)*InvGamma(1,-2*ep)
              *Gamma(2-n2+n4,-ep)*InvGamma(1,-ep)
              *Gamma(1,ep)*InvGamma(n1-ep,ep)/fac_(n2-1)
              *InvGamma(4-n1+ep-n2+n3,-3*ep)*Gamma(2,-3*ep);
    id  Gamma(n1?,n2?)*InvGamma(n1?,n2?) = 1;
    .sort(PolyFun=acc);

Time =      39.20 sec    Generated terms =     18986
                F        Terms in output =       269
                         Bytes used      =     38580
```

```
repeat;
    id,once,Gamma(n1?,n?)*InvGamma(n2?,n?) = X(n1,n2,n1-n2,n);
    repeat id X(n1?,n2?,n3?pos_,n?) = X(n1-1,n2,n3-1,n)*acc(n1-1+n);
    repeat id X(n1?,n2?,n3?neg_,n?) = X(n1,n2-1,n3+1,n)*den(n2-1,n);
    id  X(n1?,n1?,0,n?) = 1;
endrepeat;
B   Gamma,InvGamma;
Print +f +s;
.end(PolyFun=acc);

Time =       45.34 sec    Generated terms =      16601
                 F        Terms in output =        122
                          Bytes used      =      13942

   F =
       + den(-1, - ep)*den(0, - ep)*den(0, - 2*ep)*den(0, - 2*ep)*den(1,ep)*G(
         1,1)*G(1 + ep,1)*xi^2*acc( - 3*ep^2 - 9*ep^3 + 21*ep^4 + 81*ep^5 + 54
         *ep^6)
          + den(-1, - ep)*den(0, - ep)*den(0, - 2*ep)*G(1,1)*G(1 + ep,1)*xi*
         acc(27*ep^2 - 51*ep^3 - 138*ep^4 + 144*ep^5)
          + den(-1, - ep)*den(0, - ep)*den(0, - 2*ep)*G(1,1)*G(1 + ep,1)*xi^2*
         acc( - 9/2*ep^2 + 9/2*ep^3 + 27*ep^4)
          + den(-1, - ep)*den(0, - ep)*den(0, - 2*ep)*G(1,1)*G(1,1)*xi*acc( -
         9*ep^2 + 17*ep^3 + 10*ep^4 - 16*ep^5)
   etc
```

When we introduce the gamma functions we normalize everything to two very simple versions of the G function. These can be evaluated separately when the need arises. We give the gamma functions two names: Gamma in the numerator and InvGamma= $1/\Gamma$. This is much easier for FORM, because it does tolerate composite denominators, but it is not very good with them. When a Gamma an a InvGamma have identical arguments they cancel each other. We give the functions two arguments. The first is an integer, and the second contains the ϵ part. This way we can eliminate the gamma's completely because when a Gamma and a InvGamma have an identical second argument the pair can be written as a polynomial in ϵ, either in the numerator or in the denominator. The ones in the denominator we leave inside the function den. We will split the fractions next.

```
id  den(ep) = acc(1/ep);
id  den(2*ep) = acc(1/2/ep);
id  den(3*ep) = acc(1/3/ep);
id  den(-ep) = acc(-1/ep);
id  den(-2*ep) = acc(-1/2/ep);
id  den(-3*ep) = acc(-1/3/ep);
id  den(n1?,2*ep) = den(n1/2,ep)/2;
id  den(n1?,3*ep) = den(n1/3,ep)/3;
id  den(n1?,-2*ep) = -den(-n1/2,ep)/2;
id  den(n1?,-3*ep) = -den(-n1/3,ep)/3;
id  den(n1?,-ep) = -den(-n1,ep);
id  den(0,ep) = acc(1/ep);
.sort(PolyFun=acc):normalize den;

Time =       45.18 sec    Generated terms =        894
                 F        Terms in output =         46
   normalize den          Bytes used      =       4920
```

```
    repeat id den(n1?!{n2?},ep)*den(n2?!{n1?},ep) =
              den(n1,ep)/(n2-n1)-den(n2,ep)/(n2-n1);
    B   Gamma,InvGamma;
    Print +f +s;
    .end(PolyFun=acc);

Time =      45.32 sec     Generated terms =       314
               F          Terms in output =        39
                          Bytes used      =      4190

   F =
      + den( - 3/2,ep)*G(1,1)*G(1 + ep,1)*xi*acc( - 351/5 + 3009/10*ep - 464*
        ep^2 + 619/2*ep^3 - 389/5*ep^4 + 8/5*ep^5)
         + den( - 3/2,ep)*G(1,1)*G(1 + ep,1)*xi^2*acc( - 21/5 + 29/2*ep - 219/
        10*ep^2 + 56/5*ep^3 + 18/5*ep^4 - 16/5*ep^5)
         + G(1,1)*G(1,1)*acc(32 - 3/2*ep^-1 - 22*ep)
   etc
```

We start with normalizing the contents of the den function in such a way that the second argument is always ep. This is not too much work because there is only a limited number of possibilities for the second argument.

The final splitting of the fractions has to be done after expanding the contents of the acc function.

```
    id  acc(n?) = n;
    repeat id ep*den(n1?,ep) = 1-n1*den(n1,ep);
    repeat den(n1?,ep)/ep = 1/ep/n1-den(n1,ep)/n1;
    B   xi;
    Print +f +s;
    .end;

Time =      46.08 sec     Generated terms =       570
               F          Terms in output =        37
                          Bytes used      =      1046

   F =
      + xi * (
         - 11/2*den( - 2/3,ep)*G(1,1)*G(1 + ep,1)
         - 27/2*G(1,1)*G(1 + ep,1)
         + 9*G(1,1)*G(1 + ep,1)*ep^-1
         - 28*G(1,1)*G(1 + ep,1)*ep
         + 24*G(1,1)*G(1 + ep,1)*ep^2
         - 99/8*G(1,1)*G(1,1)
         - 9/2*G(1,1)*G(1,1)*ep^-1
         - 5*G(1,1)*G(1,1)*ep
         + 12*G(1,1)*G(1,1)*ep^2
         )
   etc
```

This gives us an exact but still messy answer. Considering that we let $\epsilon \to 0$ sooner or later we may as well make some approximations now. By expanding the gamma function we can write

$$G(1+\epsilon,1) = \frac{1}{2}(1+\epsilon+3\epsilon^2+9\epsilon^3-6\epsilon^3\zeta_3)G(1,1) \tag{9}$$

In addition we take into account that $G(1,1)$ contains a pole in ϵ.

```
    id  G(1+ep,1) = G(1,1)*(1/2+ep/2+3*ep^2/2+9*ep^3/2-6*ep^3*z3/2);
    id  G(1,1) = G(1,1)/ep;
    repeat id ep*den(n1?,ep) = 1-n1*den(n1,ep);
    repeat den(n1?,ep)/ep = 1/ep/n1-den(n1,ep)/n1;
    id  ep = 0;
    id  den(n1?,ep) = 1/n1;
    id  1/ep = acc(1/ep);
    B   xi,G;
    Print +f +s;
    .end(PolyFun=acc);

Time =       45.82 sec    Generated terms =        231
                 F        Terms in output =          8
                          Bytes used      =        396

   F =
       + G(1,1)*G(1,1)*xi * (
          + z3*acc(-27)
          + acc(1859/32 - 21/2*ep^-2 - 31/16*ep^-1)
          )
   etc
```

The leftover $G(1,1)$ is finite. We can expand it in terms of ϵ to obtain the final answer in the MSbar scheme:

```
    id  G(1+ep,1) = G(1,1)*(1/2+ep/2+3*ep^2/2+9*ep^3/2-6*ep^3*z3/2);
    id  G(1,1) = G(1,1)/ep;
    *
    *   Conversion to MSbar:
    *
    id  G(1,1)*G(1,1) = 1+4*ep+12*ep^2+32*ep^3-14/3*ep^3*z3;
    repeat id ep*den(n1?,ep) = 1-n1*den(n1,ep);
    repeat den(n1?,ep)/ep = 1/ep/n1-den(n1,ep)/n1;
    id  ep = 0;
    id  den(n1?,ep) = 1/n1;
    id  1/ep = acc(1/ep);
    B   xi,G;
    Print +f +s;
    .end(PolyFun=acc);

Time =       46.19 sec    Generated terms =        568
                 F        Terms in output =          8
                          Bytes used      =        370

   F =
       + xi * (
          + z3*acc(-27)
          + acc( - 2421/32 - 21/2*ep^-2 - 703/16*ep^-1)
          )
       + xi^2 * (
          + z3*acc(3/2)
          + acc(1053/32 - 27/4*ep^-2 + 255/16*ep^-1)
          )
       + xi^3 * (
          + acc( - 61/4 - 21/4*ep^-1)
          )
       + xi^4 * (
          + acc(3/8)
          )
       + z3*acc(-9)
          + acc(1543/8 + 117/4*ep^-2 + 333/4*ep^-1)
         ;
```

Once we have this program it is easy to run also the other projection that one can make and which serves as a good check by the time all diagrams have been computed: the contraction of the external gluon indices with $Q^\mu Q^\nu$. We only have to change one statement at the beginning:

```
Multiply d_(MU,NU)/Q.Q;
```

becomes

```
Multiply Q(MU)*Q(NU)/Q.Q^2;
```

and we are immediately confronted with the enormous advantages of a good computer program. This calculation we get for free:

```
Time =       37.46 sec    Generated terms =        464
                 F        Terms in output =          5
                          Bytes used      =        252
   F =
       + xi * (
          + acc(543/16 + 9/8*ep^-2 + 61/8*ep^-1)
          )
       + xi^2 * (
          + acc( - 35/6 - ep^-1)
          )
       + xi^3 * (
          + acc(1/8)
          )
       + z3*acc(3/2)
          + acc( - 4801/96 - 13/8*ep^-2 - 547/48*ep^-1)
       ;
```

It is now rather easy to do all other diagrams. We only have to replace the input for the diagram, and maybe add some statements to deal with the fermion traces, the ghosts and/or 4-gluon vertices. But the integration program is general enough that it should be capable of handling all this.

At this point I would like to emphasize that this is the type of programs that is being used for doing actual three loop calculations. For those calculations one needs of course far more complicated reduction algorithms to reduce all terms to a very limited set of integrals. Two of these had to be computed by hand to sufficient orders in ϵ. There is however an extra class of computations that can be done. When one computes moments of structure functions one has in principle to evaluate four-point functions but they are of the type $P+Q \to P+Q$. Then one has to take a sufficient number of derivatives with respect to P and put P equal to zero in the remainder. But putting the momentum of a line equal to zero is like amputating that line. Hence we have effectively a propagator-like diagram left with the difference that many internal propagators can have quite high powers. This makes the reduction program quite slow and hence it had to be optimized to a very high degree. In the end the code takes thousands of lines and many tables which provide the program with shortcuts. Not all recursions have to be taken to the bitter end. In addition the number of diagrams becomes quite large (more than

10000). It took of the order of a year of CPU time on a big Silicon Graphics computer to do the 4 lowest moments (N=2, N=4, N=6 and N=8). One can see how important the optimizations are..

References

1. J.A.M. Vermaseren, Symbolic Manipulation with FORM, published by CAN (Computer Algebra Nederland), Kruislaan 413, 1098 SJ Amsterdam, 1991, ISBN 90-74116-01-9. An earlier version of FORM can be obtained for free by means of anonymous ftp from nikhefh.nikhef.nl. Version 1 is however insufficient to run the program described in this publication.
2. P. Cvitanovic, Phys. Rev. **D14** (1976) 1536
3. R. Mertig, M. Böhm and A. Denner, Comp. Phys. Comm. **64** (1991) 345
4. E. Yehudai, Fermilab-PUB-92-22-T
5. P. Nogueira, J. Comp. Phys. **105** (1993) 279
6. T. Kaneko, Comp. Phys. Comm. **92** (1995) 127
7. E. E. Boos et al., Preprint Moscow State University, MGU-89-63/140
8. K.G. Chetyrkin and F.V. Tkachov, Nucl. Phys. **B192** (1981) 159
9. F.V. Tkachov, Teor. Mat. Fiz. **56** (1983) 350
10. F.V.Tkachov, Phys. Lett. **100B** (1981) 65

The Analytical Value of the 6th-Order Electron (g-2) in QED

Ettore Remiddi

Dipartimento di Fisica, Università di Bologna, and INFN, Sezione di Bologna, I-40126 Bologna, Via Irnerio 46, Italy

As it happens that I have the opportunity of speaking just after a lecture of Jos Vermaseren dealing with `FORM`, his computer algebra program, I will report the results of an "exercise" just carried out by Stefano Laporta and myself [1] by using that program. The "exercise" is the complete analytic evaluation of the three loop radiative corrections to the anomalous magnetic moment of the electron in QED. Actually the exercise is quite a long one, as it was started by Goyi Mignaco and myself back in the late 60's [2], when we were both young fellows at CERN. In those years Vermaseren was presumably already born – but for sure not yet `FORM`; and in fact we embarked on the enterprise by using `SCHOONSCHIP`, the computer algebra program by Tini Veltman which can be considered, from any point of view, the direct ancestor of `FORM`.

When arriving at CERN in the fall of 1966 I was somehow expecting that computers could help not only in numerical calculations but also in some of the algebraic manipulations usually carried out with paper and pencil, and in fact I had already written a very primitive program for the numerical evaluation of $SU(3)$ Clebsch-Gordan coefficients. During a conversation with some friends, I proposed that something similar could also be tried for the traces of Dirac-matrices; immediately, I was told that Veltman had already written such a program, so I started looking for both (the author and the program). At that time Veltman was a little older than me, already much more authoritative (which is still true nowadays) – and commuting between CERN and Utrecht. At his next appearance at CERN I plucked up courage to ask him about the program. Contrary to my fears, his reaction was extremely kind; he gave me immediately his beloved program, which in those times had the form of four boxes filled with punched cards, with the customary threats (which I took however extremely seriously) of the worst punishments for any injury that the punched cards could ever suffer in my hands. He provided me with rudimentary verbal instructions on how to use it; again contrary to my expectations, the instructions turned out to be really *foolproof*, so that I immediately became an expert user – even if, due to the bias of my Italian origin, I never succeeded in pronouncing correctly the word `SCHOONSCHIP`, the (dutch) name of the program. Shortly afterwards, I was even able to propose a few additional features, such as the command `Ratio`, quickly implemented by Veltman in one of the periodical updatings of the program. In the notation of the latest versions, the command reads something like

```
id  Ratio,[x + a],[x + b],[b - a] ;
```

it forces the partial fractioning in x of the products of two arbitrary powers of $(x+a)$ and $(x+b)$, such as for instance

$$\frac{1}{x+a}\,\frac{1}{x+b} = \frac{1}{b-a}\left(\frac{1}{x+a} - \frac{1}{x+b}\right) ,$$

an algebraic manipulation of great use when "cleaning" an expression containing x in view of a subsequent integration in that variable.

The first versions of SCHOONSCHIP ran on the CDC mainframes, the CERN computers of those days; to improve the performance, Veltman had the "obvious" idea of writing SCHOONSCHIP in plain assembler language. At the beginning of the 80's, the physics community started moving towards the proliferation of the midframes (later on followed by the workstations and personal computers which we have now); unfortunately for me, and despite all my vain efforts to convince him of the contrary, Veltman never accepted the idea of rewriting SCHOONSCHIP in a "conventional" language (say C). I tried for a while to follow him and his Motorola versions of SCHOONSCHIP, but eventually I had to give up and to switch to something else. The obvious choice was FORM; Vermaseren was in fact another user of SCHOONSCHIP, he shared my point of view on the opportunity of having a C-version of SCHOONSCHIP, but rather than limiting himself to propose and complain, as I was doing, much more effectually he actually undertook the task of "rewriting SCHOONSCHIP in C". Not surprisingly, even if he had in mind at the beginning the structure of SCHOONSCHIP, in the course of the work his project evolved into FORM, acquiring a growing independence from the reference model; but yet the similarity of the underlying structure remained, making particularly easy for me and my collaborators the transition from SCHOONSCHIP to FORM.

To complete this digression on computer algebra, I have to add that I have been in various occasions at the Carnegie Mellon University of Pittsburgh; there, in the course of my collaboration with Michael Levine and Ralph Roskies (of the University of Pittsburgh), I started using ASHMEDI, the algebraic program written by Levine. That program, as I was told by both authors, was originated at CERN in the early 60's at the same time as SCHOONSCHIP, after a number of discussions on computer algebra at the CERN Cafeteria. ASHMEDI is very "essential"; it lacks for instance negative powers, so that if you have a variable X and you need its inverse you must create a new name for it, say IX and then supply repeatedly the instruction SUBST X*IX=1. Despite this apparent shortcoming, ASHMEDI is very efficient and robust, it allows user-defined macros and recursive insertion of files, which speeds up remarkably the writing of the input code and the execution; furthermore, ASHMEDI is written in FORTRAN, with very few machine dependent instructions, so that in the course of the years it was easily moved to a few machines with different CPU's.

In this short review of the computer algebra programs which have been used so far, I have already referred to my first work on the (g-2) of the electron in the late 60's with J.A. Mignaco; the work continued in the 70's with R. Barbieri at Pisa, with M. Caffo and S. Turrini at Bologna and then, omitting short term students or collaborators, with M.J. Levine and R. Roskies at Pittsburgh [3]. In 1983-84 I spent a sabbatic year at CMU (the Carnegie Mellon University of Pittsburgh), trying to figure out how to complete the light-light scattering graphs, whose scalar part was already known from previous work. Back in Bologna, the light-light graphs became the *Ph.D.* thesis of Stefano Laporta. He quickly went through the various stratifications of the code, grasping the key points – and at last completing that calculation (1991) [4]. The next step was the extension to the μ-meson (g-2) light-light graphs with an electron loop (1992) [5]; the extension provided the opportunity of trying a slightly different approach (another way of combining dispersion relations and hyperspherical variables, already used in the electron light-light graphs), which turned out to be of direct applicability to the second-last class of graphs still to be evaluated, the so-called corner-ladder graphs – which in fact were quickly completed by Laporta [6].

Still to evaluate were the triple cross graphs, the only non planar graphs among all the three-loop QED vertex graphs, from any point of view the most difficult to deal with. I had already attacked them during my previous sabbatic leave at CMU, but as one too many square root was appearing at the very beginning of the calculation, heavily jeopardizing the possibility of continuing, on that occasion I decided to invest more safely my time in the light-light graphs, for which I had reasonable hope of being able to progress towards the conclusion. Once the light-light graphs were completed and the corner-ladder graphs in the progress of being evaluated too, the problem of the triple-cross graphs could not be ignored any more. In 1993-94 I spent another sabbatic year, this time at CERN, fully devoted to processing the extra square root that appeared ten years before at CMU. With a judicious choice of the integration variables and of the integration pattern (an appropriate form of integration by parts) the extra root eventually disappeared! The path to follow was found; slightly afterwards we obtained one of the simplest (but still significant!) terms appearing in the calculation (1995) [7] – and by January 1996 the whole analytical calculation was completed [1]: the value of the triple-cross graphs to the electron (g-2) was found to be

$$\begin{aligned} a_e(3-\text{cross}) &= \frac{1}{2}\pi^2\zeta(3) - \frac{55}{12}\zeta(5) - \frac{16}{135}\pi^4 + \frac{32}{3}\left(a_4 + \frac{1}{24}\ln^4 2\right) \\ &+ \frac{14}{9}\pi^2\ln^2 2 - \frac{1}{3}\zeta(3) + \frac{23}{3}\pi^2\ln 2 - \frac{47}{9}\pi^2 - \frac{113}{48} + \ln\lambda \\ &= 0.785\ 401\ 156\ \ldots\ + \ln\lambda \end{aligned}$$

As usual in this kind of calculations, λ (appearing as $\ln\lambda$), is the infrared regularizator (the triple-cross graphs, when taken separately in the conventional on mass-shell renormalization scheme used here, are known to possess an infrared divergence), $\zeta(p)$ is the Riemann ζ-function, $\zeta(p) = \sum_{n=1}^{\infty} 1/n^p$, with $\zeta(2) = \pi^2/6$, while a_4 stands for the $p = 4$ case of the mathematical constants $a_p = \sum_{n=1}^{\infty} 1/(2^n n^p)$, with $a_4 = 0.517\ 479\ 061\ 673\ 899 \ldots$. Note in this respect that $a_1 = \ln 2$, while for $p = 2$ and $p = 3$ one can show the identities $a_2 = \frac{1}{2}\zeta(2) - \frac{1}{2}\ln^2 2$ and $a_3 = \frac{7}{8}\zeta(3) - \frac{1}{2}\zeta(2)\ln 2 + \frac{1}{6}\ln^3 2$, while no similar identity is known for $p = 4$; that is the reason why a_2, a_3 do not appear explicitly in the results, while a_4 does (in fact it always appears in the combination $a_4 + \frac{1}{24}\ln^4 2$).

The above result fits perfectly with the value 0.785 419 (40), recently obtained by T. Kinoshita [8] by means of numerical methods (or rather semi-numerical methods, as the most singular part is isolated and evaluated analytically, and only the remaining well behaved part is evaluated by high precision numerical integration).

By summing the previously obtained analytical values of the contributions of all the other graphs, the expression of the complete (and of course infrared finite) three-loop QED corrections to the electron (g-2) is [1]

$$\begin{aligned} a_e(3-\text{loop}) = \left(\frac{\alpha}{\pi}\right)^3 \Bigg\{ & \frac{83}{72}\pi^2\zeta(3) - \frac{215}{24}\zeta(5) \\ & + \frac{100}{3}\left[\left(a_4 + \frac{1}{24}\ln^4 2\right) - \frac{1}{24}\pi^2\ln^2 2\right] \\ & - \frac{239}{2160}\pi^4 + \frac{139}{18}\zeta(3) - \frac{298}{9}\pi^2\ln 2 + \frac{17101}{810}\pi^2 + \frac{28259}{5184} \Bigg\} \\ = \left(\frac{\alpha}{\pi}\right)^3 & (1.181241456\ldots) \, . \end{aligned}$$

Having obtained the result, one may ask what use can be made of it. Let us recall that the current experimental value of the electron (g-2) is [9]

$$a_e(\text{exp}) = 1\ 159\ 652\ 188.4\ (\pm 4.3) \times 10^{-12} \ ,$$

with a relative error smaller than 4 ppb (parts per billion; it is convenient to observe that an absolute error 10^{-12} in the electron (g-2) corresponds almost exactly to 1 ppb in the relative error), while another experiment is underway and sooner or later the 1 ppb barrier will be overcome. The role of the electron (g-2) in the development of QED does not need to be recalled here; let us just say that the theoretical prediction for it must be at least as precise as the experiment in order to have a significant test (in fact the most stringent one of QED, radiative corrections, renormalization procedures and the like).

Quite in general, one can write

$$a_e(\text{th}) = a_e(\text{QED, electrons}) + \delta a_e(\text{th}) \ ,$$

where a_e(QED, electrons) is due to the pure QED contributions involving only electrons, while δa_e(th) stands for all the rest (such as QED contributions from the μ-meson, or the contributions from the strong or the weak interactions). It turns out that δa_e(th) is small (less than 5×10^{-12}) and, what is more important here, it has an error which lies well below 1×10^{-12}, *i.e.* below 1ppb. Therefore only a_e(QED, electrons) needs to be considered for the discussion of the 1ppb precision; according to perturbative QED, it is given by

$$a_e(\mathrm{QED, electrons}) = \frac{1}{2}\left(\frac{\alpha}{\pi}\right) + c_2\left(\frac{\alpha}{\pi}\right)^2 + c_3\left(\frac{\alpha}{\pi}\right)^3 + c_4\left(\frac{\alpha}{\pi}\right)^4 + \ \dots ,$$

where 1/2 is the Schwinger result,

$$c_2 = \frac{197}{144} + \frac{1}{12}\pi^2 - \frac{1}{2}\pi^2 \ln 2 + \frac{3}{4}\zeta(3) = -0.328\ 478\ 965\ 579\ \dots$$

is the two loop value, c_3 was given in the previous formula, $c_4 = -1.557$ (70) according to a numerical calculation by T. Kinoshita [10] – and the coefficient of $\left(\frac{\alpha}{\pi}\right)^5$ is fortunately not needed as $\left(\frac{\alpha}{\pi}\right)^5$ is just equal to 0.1×10^{-12}. As $\left(\frac{\alpha}{\pi}\right) \simeq 2 \times 10^{-3}$, in order to have a 1ppb prediction one needs relative precisions on the value of c_3 and c_4 respectively better than 80 parts per million and 3%. For c_3 the condition is surely fulfilled (as c_3 is known analytically, the corresponding numerical error is zero), while the numerical error in c_4 should be still reduced by at least a factor 2. The difficulty in such a task cannot be underestimated; yet, as the current error ($70/1557 \simeq 4.5\%$) is still relatively large, one may hope in first approximation that just increasing the statistics of the Monte Carlo numerical integration by a factor 5 or 10 will do the job without giving rise to the much less tractable problems of loss of numerical precision around integrand spikes and the like.

A main problem is still left: the problem of the precise value of the fine structure constant α. It is obvious that the degree of accuracy of QED can be tested only when a sufficiently precise value of α is given. One may of course trust QED and use the experimental value of the electron (g-2) together with the theoretical prediction just for obtaining a precise value of α, but that would not be a test; for a real test the value of α must be provided in an independent way. We are so lead to a short discussion of the current status of the non-QED measurements of α, which will imply a short digression into the "exact effects" of solid state physics, the Josephson effect and the Quantized Hall Effect (QHE), and, as will be immediately apparent, the related methodical problems.

According to the QHE, in special circumstances suitable semiconductor devices show a Hall resistance whose value, when expressed in Ohm, is exactly equal to a known rational fraction of $R_H \equiv h/e^2 = \mu_0 c/2\alpha$ (where μ_0, the permeability of the vacuum, in SI units, is by definition exactly equal to $4\pi \times 10^{-7} N/A^2$). A precise measure of R_H would then give a precise measurement of α; but R_H, being a resistance, must be measured in Ohm –

so that the precision in the measurement of $R_{\rm H}$ is anyhow limited by the precision in the knowledge of the Ohm, which is much worse than 1ppb! Similarly, the Josephson effect states that in a Josephson junction a frequency ν is associated to an electrostatic potential V by the relation $2eV = h\nu$, *i.e.* $2e/h = \nu/V$; again, that provides a way of measuring e/h with the same precision as reached in the knowledge of the Volt (which is still far from 1ppb). A clever way of skipping the ignorance in the definition of the measuring units consists in choosing a suitable product of precisely measured quantities which is independent of the units themselves, whose value is α (or a power of it). A good choice is, for instance, given by

$$\alpha^{-1} = \left[\frac{(\mu'_{\rm p}/\mu_{\rm B})\ R_{\rm H}\ (2e/h)}{2\mu_0\ R_\infty\ \gamma'_{\rm p}}\right]^{1/3} ,$$

where $R_\infty = m_{\rm e}c\alpha^2/2h$ is the Rydberg constant (measured with very high precision), $\mu_{\rm B} = e\hbar/2m_{\rm e}$ (the Bohr magneton), $\mu_{\rm p}$ the proton magnetic moment and $\gamma_{\rm p} = (2/\hbar)\mu_{\rm p}$ the proton gyromagnetic ratio (the primes on $\mu_{\rm p}, \gamma_{\rm p}$ refer to low field measurements in water). It is easily seen that when the explicit values of those physical constants are used in the above equation, α^{-1} is immediately obtained; at the same time, recalling that $\gamma'_{\rm p}$ is measured as the ratio of an NMR frequency to a magnetic field, one checks also that the above product is indeed independent of the choice of the measuring units. In the above relation, R_∞ is the best measured constant (the error is of the order of 0.04 ppb or better!), the least precise is $\gamma'_{\rm p}$, whose error determines in fact the final precision in α (but fortunately the exponent 1/3 reduces the relative error by a factor 3). By suitably combining the best values available for the various constants, one obtains the so-called "quantum Hall" value of α [11]

$$1/\alpha(\text{q.Hall}) = 137.035\ 997\ 9(32)\quad (24\text{ppb})\ .$$

With the above value of α, the already referred to values of the perturbative QED coefficients, and the known value of $\delta a_{\rm e}(\text{th})$ one obtains the prediction

$$a_{\rm e}(\text{th}) = 1\ 159\ 652\ 201.2\ (2.1)\ (27.1)\ \times 10^{-12}\ ,$$

where (2.1) comes from the numerical uncertainty in the four loop coefficient c_4 and (27.1) from the experimental error in α. The difference $a_{\rm e}(\text{th}) - a_{\rm e}(\text{exp}) = 12.8\ \times 10^{-12}$ is well within the error of the prediction. One can conversely obtain α by using $a_{\rm e}(\text{th})$ and $a_{\rm e}(\text{exp})$; the result is

$$1/\alpha(a_{\rm e}) = 137.035\ 999\ 41(56);$$

in this case the difference is $1/\alpha(a_{\rm e}) - 1/\alpha(\text{q.Hall}) = 1.51\ \times 10^{-6}$, again well within the error in $1/\alpha(\text{q.Hall})$, which is $3.2\ \times 10^{-6}$.

Obviously, other independent and precise values of α are welcome. A new and very promising experiment has been proposed by S. Chu [12]: an "atomic fountain" emits slow atoms upright, which rise for a while and then fall in the gravitational field of the earth (an atom emitted with a velocity of 2 m/s will reach the height of 20 cm and then fall back in a few tenth of a second); during this long time (on an atomic scale) an atom, of mass M and originally in the ground state, is brought to absorb a photon, corresponding to some very narrow internal transition to an excited state of mass M_*, from a suitable laser beam of frequency ν_1 coming from the left – and as a consequence of the absorption it recoils to the left; the atom is then stimulated by another laser beam coming from the right to make the opposite transition to the fundamental state – and as a consequence it recoils for the second time to the left, so that the emitted photon must have the frequency ν_2 corresponding to the suitable Doppler shift of the absorption frequency ν_1. More exactly, an elementary exercise of relativistic kinematics gives $\nu_2 = \nu_1(1 - 2h\nu_1/M_*c^2)$ *i.e.*

$$\frac{h}{M_*} = \frac{1}{2}c^2\frac{\nu_1 - \nu_2}{\nu_1} ,$$

expressing h/M_* in terms of known constants and spectroscopical frequencies, which can be measured with great precision. Once a good value of h/M_* is available, one can obtain α by the relation

$$\alpha^2 = R_\infty \frac{2}{c^2}\left(\frac{\hbar}{M_*}\right)\left(\frac{M_*}{m_\mathrm{p}}\right)\left(\frac{m_\mathrm{p}}{m_\mathrm{e}}\right) ,$$

where the ratios of the masses are either already known or easy to measure.

That new approach to the measurement of α, although not yet competitive with the previous ones (it has reached so far only the precision of 10 ppm), is however of the greatest interest, as it relies upon almost no assumptions – or, in any case, on assumptions different from those underlying the exact solid state effects, so that it can provide a really independent cross check of QED, Josephson, and Quantum Hall effects.

As a last subject, let me shortly discuss the problem of the analytical integration. First of all, it requires a definition: it is reasonable to say that a definite integral is analytically evaluated when it is shown to be equal to a sum of as few as possible terms, each having the form of a rational fraction times some known mathematical constant; here known means really known, such as π (in fact some of the first even powers of it, as it turns out to be the case in (g-2) calculations), $\zeta(3)$, $\zeta(5)$, $\ln 2$ or a_4, as discussed above, so that the actual numerical value of the integral, once analytically evaluated, can then be obtained without effort up to any required precision. Each definite and mathematically well defined integral is by itself a mathematical constant, but that does not help in obtaining its numerical value, and in general it is not *a priori* known whether it can be expressed in terms of already known constants times simple rational coefficients; so one has to embark on a very long work without knowing for a long while whether the sought result does exist at all, at variance with the case, say, of numerical integration, where one is almost sure that some approximate numerical value will be anyhow obtained (what is not trivial even in this case is whether the result has the required precision).

Having that in mind, let me sketch the main lines of the algorithms used in the analytical integration. Given any Feynman graph amplitude for a QED graph referring to a vertex with on mass-shell electrons and a photon with momentum Δ, off the mass shell, one can extract the contribution to the $(g-2)$ of the electron by means of standard algebra (some traces of γ-matrices) in the $\Delta \to 0$ limit. When that is done, in the case of a 3-loop graph one obtains typically a few hundred terms of the form n_i/d_i, where the numerators n_i are combinations of all the possible scalar products which can be formed with the external electron momentum and the 3 internal loop momenta, and any of the denominators d_i is a product of the 8 denominators corresponding to the 8 propagators present in the 3-loop vertex graph. More exactly, any denominator d_i is of the form $D_1^{k_1} D_2^{k_2} ... D_8^{k_8}$, where each of the 8 D_j is raised to a power k_j whose possible values are $0, 1, 2$; 1 is, so to speak, the "default" value, the value 2 can arise as a consequence of the $\Delta \to 0$ expansion, while 0 corresponds actually to the absence of that denominator, due to its compensation by some suitable factor generated by the trace in the numerator. Terms with less denominators have in general a simpler structure, however, very often they cannot be evaluated directly as they are not *u.v.* convergent when isolated from all the other terms. Likewise, a few terms – those requiring renormalization of inserted self-mass or vertex amplitudes – are genuinely *u.v.* divergent; but in an analytical calculation those divergences are more a nuisance than a real problem. More troublesome are in general the *i.r.* divergences, which can also appear in single Feynman graphs, even if they cancel in the final result, especially when they are regularized by the traditional method of giving a fictitious small mass λ to the photon.

It turns out that the analytical integration of all the various terms generated in the trace is very much the same, so that considering a few typical terms can be sufficient for understanding whether the approach to the analytical integration will work. The "most typical" term is the term with the first power of all the 8 denominators and 1 or a simple scalar product in the numerator. Any 3-loop graph is given by the Feynman rules as an integral over the 12 components of the 3 loop momenta; in order to start the analytical integration, it is convenient to look for more convenient integration variables. In practice, it is convenient to use a suitable combination of hyperspherical variables for some loop (that amounts to replacing the 4 components of the loop momentum by their hyperspherical counterparts, say $\int d^4q \to i \int Q^3 dQ\; d\Omega(\hat{q})$, where $d\Omega(\hat{q})$ runs over the 3 hyperspherical angles) and dispersion relations for the remaining loops, and then carrying out explicitly the trivial angular integrations. At that point the integral has become a 4-dimensional definite integral, the integrand being just a logarithm (of some more or less complicated argument) times a factor, which is a fraction involving one or more square roots in the integration variables. A further explicit integration is sometimes possible – but in general of no particular use.

It is convenient here to open a digression on the dilogarithm of Euler,

$$S_2(x) = -\int_0^1 \frac{dt}{t} \ln(1-xt)$$

and its generalization, the polylogarithms of Nielsen [13]

$$S_{n,p}(x) = \frac{(-1)^{n+p-1}}{(n-1)!p!} \int_0^1 \frac{dt}{t} \ln^{n-1} t \ln^p(1-xt).$$

In this notation, $S_2(x) = S_{1,1}(x)$, and more generally $S_n(x) = S_{n-1,1}(x)$; $(n+p)$ is called the degree of the polylogarithm. Two remarkable properties of the polylogarithms are worth to be recalled: *i)* they take special values at $x = 1, \frac{1}{2}, -1$, such as for instance $S_n(1) = \zeta(n)$ (that is in fact the reason why the $\zeta(n)$'s and the other constants appear in the final expression of the (g-2) coefficients), and *ii)* their derivatives are always expressed in terms of polylogarithms of smaller degree (and therefore simpler)

$$\frac{d}{dx} S_{n,p} = \frac{1}{x} S_{n-1,p}(x)$$

(incidentally, we can consider the logarithm as a particular polylogarithm of degree 1).

It is time to come back to the definite integration. Let us single out the last integration variable, and take for granted that by some suitable change of variable it can be taken to be a variable x ranging from 0 to 1; assume further that we can guess that the whole integral is, for instance, of the form

$$\int_0^1 \frac{dx}{x} F(x) ,$$

where $F(x)$ is a polylogarithm or a combination of polylogarithms of degree 4. Remember that at this stage we know only that $F(x)$ is a triple definite integral, whose integrand contains a logarithm; if the guess on the nature of $F(x)$ is correct, its x-derivative will consist of a combination of simpler polylogarithms of degree 3, to be indicated by $G(x)$ in the following. An expression for the derivative can be obtained by direct action under the integral signs, without trying any explicit integration; after some manipulation it will take the form of a 3-dimensional definite integral, whose integrand is just a fraction, not anymore a fraction times a logarithm, as in the case of $F(x)$. Only at this point one can carry out explicitly one of the integrations, so that the degree-3 $G(x)$ will take the form of a double definite integral (not anymore a triple integral!) whose integrand contains a logarithm. If the procedure can be repeated two more times (and that was actually the case in the just completed (g-2) calculations), no more integrations are left and a simple logarithm, depending only on x, is obtained. At this point one reconstructs the desired $F(x)$ by means of 3 successive quadratures – relatively straightforward, as a rule, within the family of the polylogarithms. One obtains in this way a definite integral on x from 0 to 1, whose integrand is a combination of polylogarithms of degree 4 (or lower); that last definite integral gives the nice mathematical constants appearing in the final result. Two comments are in order: first, the success of the procedure is not guaranteed *a priori*, at every step complicated algebraic structures are generated (including a number of square roots and double radicals), only their almost "miraculous" disappearance (such as double radicals actually found to be the square root of a perfect square and therefore reduced to "standard" single radicals) allows the calculation to proceed; second, at every step an enormous number of intermediate terms is generated (even if the results are usually much simpler than any of the intermediate expressions leading to them), which can be dealt with only by relying on a robust, "heavy duty" symbolic manipulation program (such as the already mentioned `SCHHONSCHIP`, `ASHMEDI`, `FORM`).

The method used for the typical simplest terms can be painstakingly extended to the calculation of all the other terms as well; the success of the approach is almost granted, but the procedure can be really long and time consuming (that approach was indeed followed for the light-light vertex graphs, so we know it by direct experience). In the case of the triple cross graphs, after the direct calculation of a few typical terms, it was found more convenient to implement another method, based on the continuous dimensional integration by parts algorithm [14]. The idea is that the integral of a total derivative vanishes (provided that the end point contributions are absent); more exactly one considers the integral, on all the loop momenta, of the divergence $\frac{\partial}{\partial q_\mu}$, q_μ being one of the loop momenta, of any of the vectors obtained by suitably combining the external and loop momenta, their scalar products and the denominators of the various Feynman propagators. By carrying out explicitly the derivatives implied in the evaluation of the divergence, one generates a combination of terms which are of the same kind as those appearing in the initial trace which gives the contribution to the electron (g-2) by the considered graph. As the integral of such a total divergence is zero, one has obtained in this way an identity which can be used to express one of the generated terms – say the "most complicated" – as a combination of the other "simpler" terms. By systematically investigating all the possible identities which can be generated by considering all the divergences of virtually all the vectors which can be built with the scalar products and the denominators of the various Feynman propagators, one reduces all the integrals occurring in the original trace to a combination of as few as possible "basic" integrals.

The actual situation is slightly more complicated; to guarantee the convergence of all the integrals it is convenient to switch from 4 to n continuous dimensions, and then take the $n \to 4$ limit in a later stage. Also the initial trace must be evaluated in n (rather than 4) dimensions, but that is not a problem. In the case of the triple-cross graphs, everything was expressed in terms of just 18 independent "basic" integrals; the integration by parts identities in n dimensions in the $n \to 4$ limit generate a number of apparent singularities in $(n-4)$ (which of course cancel in the final result), so that also the 18 "basic" integrals must be evaluated for n different from 4 and then expanded in $(n-4)$ (the 18 basic integrals, which can in any case be chosen in various ways, can be either evaluated directly, or obtained by inverting the system expressing a sufficient number of previously evaluated typical integrals in terms of the basic integrals). It was very encouraging to see that a subset of the 18 integrals, necessary for the triple-cross graphs, was also sufficient for the complete evaluation of other classes of graphs, already obtained by different methods.

It is tempting to consider the possibility of extending the approach to the 4 loop (g-2); one might think of expressing everything as the combination of "few" (a few hundred ?) basic and relatively simple integrals, to be later evaluated either analytically or – more realistically – by numerical methods. But that is still in a less than preliminary stage.

References

1. S. Laporta and E. Remiddi, *Phys.Lett.* (to be published)
2. J. A. Mignaco and E. Remiddi, *Nuovo Cimento* **60**A, 519 (1968)
3. An almost complete list of references is given in M. J. Levine, E. Remiddi and R. Roskies, in *Quantum Electrodynamics*, edited by T. Kinoshita, Advanced series on Directions in High Energy Physics, Vol. 7, (World Scientific, Singapore, 1990), 162, p.214-216
4. S. Laporta and E. Remiddi, *Phys.Lett.* **B265**, 181 (1991)
5. S. Laporta and E. Remiddi, *Phys.Lett.* **B301**, 440 (1993)
6. S. Laporta, *Phys.Lett.* **B343**, 421 (1995)
7. S. Laporta and E. Remiddi, *Phys.Lett.* **B356**, 390 (1995)
8. T. Kinoshita, *Phys.Rev.Lett.* **75**, 4728 (1995)
9. R. S. Van Dick, Jr., P. B. Schwinberg, and H. G. Dehmelt, *Phys.Rev.Lett.* **59**, 26 (1987)
10. T. Kinoshita, *IEEE Trans. Instrum. Meas.* **44**, 498 (1995)
11. M. E. Cage *et al.*, *IEEE Trans. Instrum. Meas.* **38**, 284 (1989)
12. M. Weitz, B. C. Young, and S. Chu, *Phys.Rev.Lett.* **70**, 2706 (1993)
13. N. Nielsen, *Der Eulersche Dilogarithmus und seine Verallgemeinerungen*, Nova Acta Naturae Curiosorum (Halle) **90**, 125 (1909)
14. K. G. Chetyrkin and F. V. Tkachov, *Nucl. Phys.* **B192**, 159 (1981);
F. V. Tkachov, *Phys.Lett.* **B100**, 65 (1981)

Effective Field Theories

Aneesh V. Manohar

Physics Department, University of California, San Diego,
La Jolla CA 92093, 9500 Gilman Drive, USA

Abstract. These lectures introduce some of the basic ideas of effective field theories. The topics discussed include: relevant and irrelevant operators and scaling, renormalization in effective field theories, decoupling of heavy particles, power counting, and naive dimensional analysis. Effective Lagrangians are used to study the $\Delta S = 2$ weak interactions and chiral perturbation theory.

1 Introduction

An important idea that is implicit in all descriptions of physical phenomena is that of an effective theory. The basic premise of effective theories is that dynamics at low energies (or large distances) does not depend on the details of the dynamics at high energies (or short distances). As a result, low energy physics can be described using an effective Lagrangian that contains only a few degrees of freedom, ignoring additional degrees of freedom present at higher energies. One of the main purposes of these lectures is to make these qualitative statements quantitative.

First a simple example: The energy levels of the Hydrogen atom are calculated in textbooks using the Schrödinger equation for an electron bound to a proton by a Coulomb potential. To a good approximation, the only properties of the proton that are relevant for the computation are its mass and charge. An understanding of the quark substructure of the proton (let alone quantum gravity) is not necessary to compute the energy levels of the Hydrogen states. This is true provided an answer which has some theoretical uncertainty is sufficient. A more accurate calculation of the energy levels, for example including the hyperfine splitting, requires that we also know that the proton has spin-1/2, and a magnetic moment of 2.793 nuclear magnetons. An even more accurate calculation of the energy levels requires some knowledge of the proton charge radius, etc. More details of the proton structure are needed as we require a more accurate answer for the energy levels.

When we discuss effective theories, we will frequently talk about momentum scales characteristic of a given problem. The typical length scale characteristic of the Hydrogen atom is the Bohr radius $a_0 = 1/(m_e\alpha)$, and the typical momentum scale is of order $\hbar/a_0 \sim 1/a_0 = m_e\alpha$, using units in which $\hbar = 1$. The typical energy scale characteristic of Hydrogen is the Rydberg $\sim m_e\alpha^2$, and the typical time scale is $1/(m_e\alpha^2)$. The Hydrogen atom is more complicated than many relativistic bound states because it has two characteristic scales, $m_e\alpha$ and $m_e\alpha^2$. We can now give a quantitative estimate of the error caused by neglected interactions: the energy levels of Hydrogen can be computed by ignoring all dynamics on momentum scales Λ much larger than $m_e\alpha$, with an error of order $m_e\alpha/\Lambda$. As the desired accuracy increases, the scale Λ of the interactions that can be ignored, also increases.

The relevant interactions in an effective theory also depend on the question being studied. In the Hydrogen atom, the energy levels can be computed to an accuracy $(m_e\alpha/M_W)^2$ while ignoring the weak interactions, but if we are interested in atomic parity violation, the weak interactions are the leading contribution since strong and electromagnetic interactions conserve parity. Atomic parity violation will still be a very small effect, because the weak scale is much larger than the atomic scale.

An effective field theory describes low energy physics in terms of a few parameters. These low energy parameters can be computed in terms of (hopefully fewer) parameters in a more fundamental high energy theory. This computation can be done explicitly when the high energy theory is weakly coupled. In QED, for example, one can predict low energy parameters such as the magnetic moment of the electron which can be used in the Schrödinger equation. If the high energy theory is strong coupled, as in QCD, one usually treats the low energy parameters (such as the magnetic moment of the proton) as free parameters that are fit to experiment. We will deal with both cases in these lectures when we study the Fermi theory of weak interactions, and chiral perturbation theory.

We have said that high energy dynamics can be ignored in the study of processes at low energies. The precise form of this statement is subtle. It is not true that parameters in the high energy theory do not affect the low energy dynamics in any way. The precise statement is that the only effect of the high energy theory is to modify coupling constants in the low energy theory, or to put symmetry constraints on the low energy theory. The energy levels of Hydrogen should not depend on the masses of heavy particles such as the top quark. This is not true: changing the top quark mass while keeping the electromagnetic coupling constant at high energies fixed, changes the electromagnetic coupling constant at low energies,

$$m_t \frac{d}{dm_t}\left(\frac{1}{\alpha}\right) = -\frac{1}{3\pi}. \tag{1}$$

The proton mass also depends on the top quark mass,

$$m_p \propto m_t^{2/27}. \tag{2}$$

Despite this dependence, the value of m_t is irrelevant for studying the Hydrogen atom. The reason is that α and m_p are parameters of the Schrödinger equation for the Hydrogen atom. Fitting to the observed energy levels determines the value of α at low energies to be 1/137.036, and the proton mass to be 938.27 MeV. *The value of m_t is irrelevant for atomic physics if the Schrödinger equation is treated as a low energy theory whose parameters α, m_e, m_p are determined from low energy experiments.* The value of m_t is relevant if one studies how atomic physics changes as a function of m_t while keeping the high energy parameters constant.

High energy dynamics places non-trivial symmetry constraints on a low energy effective theory. An interesting example of such a constraint is the spin-statistics theorem. Non-relativistic quantum mechanics is a perfectly satisfactory theory, regardless of whether electrons are quantized using Bose, Fermi or Boltzmann statistics. However, a consistent relativistic formulation of the theory requires that electrons obey Fermi statistics, which is a constraint on non-relativistic quantum mechanics that follows from causality in quantum electrodynamics. The spin-statistics theorem is a statement about symmetry, and holds regardless of whether there is a simple connection between the high energy and low energy theories. In low energy QCD, the spin-statistics theorem implies that baryons are fermions and mesons are bosons.

The effective field theory technique is powerful precisely because one can compute low energy dynamics without any knowledge of the details of high energy interactions. This also has an unfortunate consequence – information about high energy interactions cannot be obtained using low energy measurements. Luckily, the last statement is not quite true. There are some vestiges of the high energy interactions in the symmetry constraints on the low energy theory, and in small corrections to low energy dynamics. Thus high precision low energy experiments can be used to probe high energy dynamics, and provide an alternative to high energy experiments.

2 The Renormalization Group and Scaling

Effective actions were used by Wilson, Fisher, and Kadanoff to study critical phenomena in condensed matter systems, and many of the ideas of effective theories were developed in this context. Consider the classic example of an Ising spin system on a square lattice with lattice spacing a, in an external magnetic field. The partition function is

$$Z = \sum_{s_i=\pm} \exp\left[K \sum_{\langle ij\rangle} s_i s_j + B \sum_i s_i\right], \tag{3}$$

where $\langle ij\rangle$ is a sum over nearest neighbors. At a second order phase transition, the correlation length ξ of the system becomes infinite. Intuitively, one expects that the properties of the Ising system near its critical point should not depend on the details of the system and on the scale of the lattice spacing a. It took a decade of inspired work to convert this intuitive statement into equations.

To study the Ising model at its critical point, it is not necessary to retain all the information in the partition function (3). The idea of Kadanoff was to reduce the degrees of freedom by introducing a block spin. Divide the lattice of spins into blocks of four spins each (see Fig. 1). The block spin s' is defined for each block to be the average of the spins at the four corners of the block

$$s'_B = \frac{s_{B1} + s_{B2} + s_{B3} + s_{B4}}{4}, \tag{4}$$

where s'_B is the block spin for the block B, and s_{Bi} are the original spins at the four corners of block B . One can write the partition function Z as

$$\begin{aligned} Z &= \sum_{s_i=\pm} \exp\left[K\sum_{\langle ij\rangle} s_i s_j + B\sum_i s_i\right] \\ &= \int \sum_{s_i=\pm} \prod_B ds'_B \delta\left(s'_B - \frac{s_{B1} + s_{B2} + s_{B3} + s_{B4}}{4}\right) \\ &\qquad \times \exp\left[K\sum_{\langle ij\rangle} s_i s_j + B\sum_i s_i\right], \end{aligned} \tag{5}$$

where the product is over all blocks B. Performing the sum over s_i leads to

$$Z = \int \prod_B ds'_B \; e^{S[s'_B]}, \tag{6}$$

where

$$e^{S[s'_B]} = \sum_{s_i=\pm} \prod_B \delta\left(s'_B - \frac{s_{B1} + s_{B2} + s_{B3} + s_{B4}}{4}\right) \exp\left[K\sum_{\langle ij\rangle} s_i s_j + B\sum_i s_i\right]. \tag{7}$$

This is an exact renormalization group transformation (called a Kadanoff block spin transformation), that expresses the partition function in terms of a new action $S[s'_B]$ with a quarter the number of degrees of freedom and twice the lattice spacing as the original action.

The new variable s'_B is the average of four spins with values ± 1 (4), and can have the values ± 1, $\pm 1/2$, 0. The new action $S[s'_B]$ is much more complicated than the original action $K\sum_{\langle ij\rangle} s_i s_j + B\sum_i s_i$, but can in principle be computed using (7). Now repeat the block spin transformation an infinite number of times. At each step, the number of degrees of freedom is reduced by four. The block spin s eventually becomes a continuous variable, which is usually denoted by ϕ. The only problem is that the action becomes more and more complicated, and more and more non-local at each step. This was the difficulty that prevented the Kadanoff block spin method from being used for a long time. What is needed is a way to truncate the effective action in a systematic and controlled manner. It is also important (particularly in field theory) to have an effective action that is local.

A technical difficulty with the block spin transformation is that it is discrete; it is much easier to deal with continuous transformations. Wilson suggested studying the Ising model in momentum space. The variables in momentum space are Fourier transformed variables $s(k)$, where the momentum k is restricted to the Brillouin zone, $|k| \le k_{\max} = \pi/a$. The Kadanoff block spin transformation

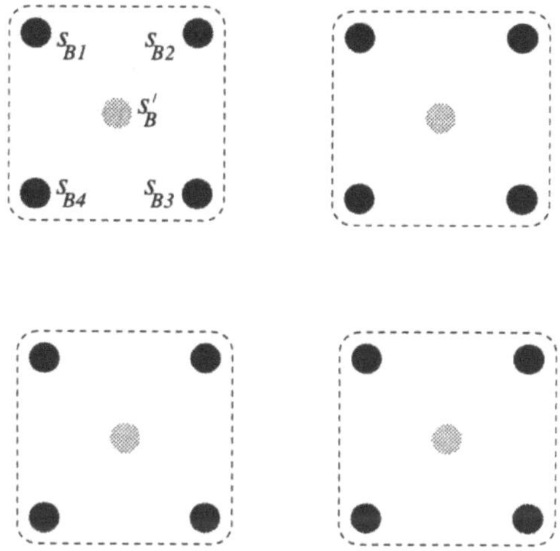

Fig. 1. The Kadanoff block spin transformation. The four spins at the corner of each block are replaced by an average spin at the center.

takes $a \to 2a \to 4a$, etc., which corresponds to letting $k_{\text{max}} \to k_{\text{max}}/2 \to k_{\text{max}}/4$, etc. k is a continuous variable, so instead consider decreasing k_{max} continuously. The partition function transformation formula becomes (using ϕ, Λ instead of s, k_{max})

$$Z = \int_{k \le \Lambda} \mathcal{D}\phi_k \; e^{-S_\Lambda[\phi_k]} = \int_{k \le \Lambda'} \mathcal{D}\phi'_k \; e^{-S'_{\Lambda'}[\phi'_k]}, \tag{8}$$

which is the momentum space analog of (7). The original action $S_\Lambda[\phi_k]$ contains all momentum modes up to some maximum value Λ, whereas the new action $S'_{\Lambda'}[\phi'_k]$ contains momentum modes up to Λ', where $\Lambda' < \Lambda$. The idea is to take $\Lambda' = \Lambda - \delta\Lambda$ infinitesimally different from Λ, so that S' is infinitesimally different from S. In the limit $\delta\Lambda \to 0$, the effective action satisfies a differential equation,

$$\frac{\partial S_\Lambda}{\partial \Lambda} = F[S_\Lambda], \tag{9}$$

where F is a functional of the action that can be determined from (8). Think of S_Λ as a set of actions, so that (9) gives the change in action as a function of cutoff. This is usually referred to as the renormalization group flow of the action. The action can be written as

$$\sum_i c_i \, \mathcal{O}_i, \tag{10}$$

in terms of coefficients c_i and some operator basis $\mathcal{O}_i$. The differential equation (9) is then a differential equation for the couplings,

$$\frac{\partial c_i}{\partial \Lambda} = F[\{c_i\}], \tag{11}$$

so that the renormalization group equation gives a flow in coupling constant space.

Finally, an extremely important point: the renormalization group equations are obtained by integrating out variables with momenta between $\Lambda - \delta\Lambda$ and Λ. There is both an infrared ($\Lambda - \delta\Lambda$) and ultraviolet (Λ) cutoff on the integration, so the renormalization group equations are local and non-singular.

Free Field Theory

To explicitly study the renormalization group equations, it is helpful to consider first a free scalar field in D dimensions, with action

$$S = \int d^D x \; \frac{1}{2}\partial_\mu \phi \partial^\mu \phi - \frac{1}{2} m^2 \phi^2 . \tag{12}$$

The action S is dimensionless, so the dimension of $\phi(x)$ is determined from the kinetic term to be $[\phi] = (D-2)/2$, and the dimension of m^2 is $[m^2] = 2$.

We would like to study correlation functions

$$G_n(x_1, \ldots, x_n) = \langle \phi(x_1) \ldots \phi(x_n) \rangle_S , \tag{13}$$

computed using the action S at long distances (i.e. low momentum). It is convenient to make the change of variables,

$$x = s x', \qquad \phi(x) = s^{(2-D)/2} \phi'(x'), \tag{14}$$

so that

$$S' = \int d^D x' \; \frac{1}{2}\partial'_\mu \phi'(x') \partial'^\mu \phi'(x') - \frac{1}{2} m^2 s^2 \phi'(x')^2 . \tag{15}$$

Correlation functions of $\phi(x)$ with action S are related to correlation functions of $\phi'(x')$ with action S' by

$$\langle \phi(s x_1) \ldots \phi(s x_n) \rangle_S = s^{n(2-D)/2} \langle \phi'(x_1) \ldots \phi'(x_n) \rangle_{S'} . \tag{16}$$

The long distance (low momentum) limit of correlation functions with action S is obtained by letting $s \to \infty$. These can be obtained by studying correlation functions at a fixed distance (fixed momentum) of the action S' as $s \to \infty$. The mass term in S' is $s^2 m^2$. Clearly, in the limit $s \to \infty$, the mass term becomes more and more important. The mass m^2 is called a relevant coupling, and dominates the long distance behavior of the correlation functions. Equivalently, ϕ^2 is called a relevant operator.

What about integrating out momentum shells to lower the cutoff? The original action had an implicit cutoff Λ. We should have integrated out momentum modes and lowered the cutoff to Λ/s, so that the rescaling transformation (14) restored the cutoff to its original value Λ. In free field theory, there is no coupling between the different modes. Thus integrating out a momentum shell produces an overall multiplicative factor in Z, i.e. an additive constant to the effective action. This shifts the cosmological constant, but does not affect the dynamics of ϕ.

Interactions

Next, add the interaction terms $\lambda\phi^4/4! + \lambda_6\phi^6/6!$ to the free Lagrangian. The dimensions of the coefficients are $[\lambda] = 0$, $[\lambda_6] = -2$. Rescaling the field as before (and ignoring, for the moment, integrating out momenta between Λ and Λ/s) gives the rescaled action

$$S' = \int d^D x' \ \frac{1}{2}\partial'_\mu\phi'\partial'^\mu\phi' - \frac{1}{2}m^2 s^2 \phi'^2 - \frac{\lambda}{4!}\phi'^4 - \frac{\lambda_6}{6! s^2}\phi'^6, \qquad (17)$$

with the implicit rescaled cutoff Λ. In the limit $s \to \infty$, the ϕ^6 term vanishes as $1/s^2$, so ϕ^6 is called an irrelevant operator, and λ_6 is called an irrelevant coupling. The ϕ^4 term remains unchanged under rescaling, so it is equally important at all length scales. For this reason, ϕ^4 is known as a marginal operator, and λ_4 is called a marginal coupling.

In effective field theories, we are usually interested in studying the dynamics at low energies, but not exactly at zero energy. For example, we will be studying hadron dynamics at a scale of order 1 GeV, which is much smaller than the weak interaction scale of $M_{\mathrm{W}} \sim 80$ GeV. In this case, the scale factor s between the weak and strong scales is $s = 80$, which is large but finite. Irrelevant operators (despite their name) then produce small corrections. In our scalar field theory example, the ϕ^6 operator produces corrections of order $1/s^2$, ϕ^8 produces corrections of order $1/s^4$, etc.

The alert reader will have noticed that the above results follow from dimensional analysis. The counting can trivially be generalized to an arbitrary Lagrangian:

1. Determine the canonical dimensions of the fields using the kinetic term.
2. Determine the (mass) dimensions of all the couplings.
3. Terms with a coupling constant with dimension d scale as s^d, so that the coupling is relevant, irrelevant, or marginal depending on whether $d > 0$, $d < 0$ or $d = 0$. Equivalently, the operator is relevant, irrelevant, or marginal depending on whether its dimension is less than, greater than, or equal to the space-time dimension D.
4. To include all corrections up to order $1/s^r$, one should include all operators with dimension $\leq D + r$, i.e. all terms with coefficients of dimension $\geq -r$.

Let us now turn on the interactions. The first problem is that there are divergences in the quantum theory. These are handled by the standard regularization and renormalization procedure. In scalar field theory, for example, one can introduce a cutoff Λ to regulate the functional integral. In the presence of a cutoff, the relation (16) between correlation functions becomes

$$G_n\left(\{sx\}; m^2, \lambda_4, \lambda_6; \Lambda\right) = s^{n(2-D)/2} G_n\left(\{x\}; s^2 m^2, \lambda_4, s^{-2}\lambda_6; s\Lambda\right), \qquad (18)$$

where we have explicitly included the cutoff dependence, and $\{x\}$ denotes $x_1, \ldots, x_n$. The left hand side is the desired correlation function. To get the infrared behavior of the left hand side, we need to replace the cutoff $s\Lambda$ by Λ on the right hand side. This is the hard part of the calculation which we have ignored so far, but one that you have all seen before – it is the standard renormalization group equation of quantum field theory:

$$\left[\Lambda\frac{\partial}{\partial\Lambda}+\beta_i\frac{\partial}{\partial c_i}+n\gamma_\phi\right]G=0. \tag{19}$$

Here c_i are the couplings, m^2, λ, λ_6, etc., and the β-functions β_i and anomalous dimension γ_ϕ are functions of c_i. The solution of this equation is also standard. Define running couplings which are solutions of the differential equation

$$\Lambda\frac{\partial}{\partial\Lambda}c_i\left(\Lambda\right)=\beta_i\left(c_i\left(\Lambda\right)\right). \tag{20}$$

Then

$$G_n\left(\{x\},c_i\left(\Lambda_1\right),\Lambda_1\right)=e^{-n\int_{\Lambda_1}^{\Lambda_2}\gamma(\Lambda)d\log\Lambda}\,G_n\left(\{x\},c_i\left(\Lambda_2\right),\Lambda_2\right). \tag{21}$$

Equation (18) can be combined with (21) to give

$$G\left(\{sx\},c_i\left(\Lambda\right),\Lambda\right)=s^{n(2-D)/2}e^{-n\int_{\Lambda}^{\Lambda/s}\gamma(\Lambda')d\log\Lambda'}G\left(\{x\},s^{d_i}c_i\left(\Lambda/s\right),\Lambda\right) \tag{22}$$

where d_i is the dimension of coupling c_i. The only difference from (16) is the exponential prefactor, and that c_i is now the running coupling at Λ/s.

It is instructive to look at some examples of renormalization group equations before continuing with our general analysis. In QCD, the renormalization group equation for the dimensionless coupling constant g is

$$\mu\frac{\partial g}{\partial\mu}=-\frac{g^3}{16\pi^2}b_0+\mathcal{O}\left(g^5\right), \tag{23}$$

where $b_0=11N_c/3-2N_f/3$, N_c is the number of colors, and N_f is the number of flavors. Equation (23) is the β-function in any mass independent scheme, such as $\overline{MS}$, and μ is the dimensionful parameter that plays the role of Λ in such a scheme. The anomalous dimension for a field (such as the fermion field ψ) has the form

$$\gamma_\psi=\gamma_\psi^0\frac{g^2}{16\pi^2}+\mathcal{O}\left(g^4\right). \tag{24}$$

Other operators added to the Lagrangian, such as four-Fermi weak decay operators have renormalization group equations of the form

$$\mu\frac{\partial c_i}{\partial\mu}=\gamma_{ij}^0\frac{g^2}{16\pi^2}c_j+\mathcal{O}\left(g^4\right). \tag{25}$$

Let us neglect operator mixing for simplicity, so that γ_{ij} is a diagonal matrix, with elements $\gamma_i\delta_{ij}$. The solutions of the renormalization group equations are[1]

[1] The general case where γ_{ij} is not diagonal can be solved by finding the eigenvalues and eigenvectors of γ_{ij}.

$$
\begin{aligned}
\frac{1}{\alpha_s(\mu_1)}-\frac{1}{\alpha_s(\mu_2)} &= \frac{b_0}{2\pi}\log\frac{\mu_1}{\mu_2},\\
\frac{c_i(\mu_1)}{c_i(\mu_2)} &= \left[\frac{\alpha_s(\mu_1)}{\alpha_s(\mu_1)}\right]^{-\gamma_i^0/2b_0},\\
\exp\left[-\int_{\mu_2}^{\mu_1}\gamma_\psi(\mu)\,d\log\mu\right] &= \left[\frac{\alpha_s(\mu_1)}{\alpha_s(\mu_2)}\right]^{\gamma_\psi^0/2b_0}.
\end{aligned}
\tag{26}
$$

These equations show that the quantum scaling behavior differs from the classical one by logarithms. A more interesting case is a field theory in which the β-function for a dimensionless coupling such as g has the form shown in Fig. 2. The renormalization group equation for g, (23), shows that $g \to g^*$ as $\mu \to 0$ if g starts out in some neighborhood of g^*. For this reason g^* is known as an attractive (or stable) infrared fixed point for g. In this case, the renormalization group scaling in the limit $s \to \infty$ is dominated by $g \approx g^*$, so that

$$
\begin{aligned}
\mu\frac{\partial c_i}{\partial\mu} &= \gamma_{ij}(g^*)\,c_j,\\
\gamma_\phi(\mu) &\to \gamma_\phi(g^*)\,.
\end{aligned}
\tag{27}
$$

Denote the fixed point values $\gamma_{ij}(g^*)$ and $\gamma(g^*)$ by γ_{ij}^* and γ^*, and assume for simplicity that $\gamma_{ij}^* = \gamma_i^*\delta_{ij}$. (As above, the general case is solved by finding the eigenvalues and eigenvectors of γ_{ij}^*.) The solutions of the renormalization group equations in the neighborhood of the fixed point become

$$
\begin{aligned}
\frac{c_i(\mu_1)}{c_i(\mu_2)} &= \left[\frac{\mu_1}{\mu_2}\right]^{\gamma_i^*},\\
\exp\left[-\int_{\mu_2}^{\mu_1}\gamma_\phi(\mu)\,d\log\mu\right] &= \left[\frac{\mu_1}{\mu_2}\right]^{-\gamma^*},
\end{aligned}
\tag{28}
$$

so that (22) becomes

$$
G_n(\{sx\},c_i(\Lambda),\Lambda) = s^{n(2-D)/2}s^{n\gamma^*}G_n\left(\{x\},s^{d_i-\gamma_i^*}c_i(\Lambda/s),\Lambda\right). \tag{29}
$$

This equation shows that scale invariance is recovered in the quantum theory at an infrared stable fixed point, but the quantum dimensions of fields and operators differ from their classical values. Operators now have dimension $D-d_i+\gamma_i^*$, their coefficients have dimension $d_i-\gamma_i^*$, and fields have dimension $(D-2)/2-\gamma^*$. This is the reason why γ, γ_{ij} are called anomalous dimensions. The classification into relevant, irrelevant and marginal operators is the same as before, except that one should use the quantum dimension of the operator which includes the anomalous dimension.

In weakly coupled theories, operator anomalous dimensions can be computed in perturbation theory, and are small. Thus quantum corrections cannot affect which operators are relevant or irrelevant, since the classical dimensions of operators are restricted to be integers or half-integers. The only effect of quantum

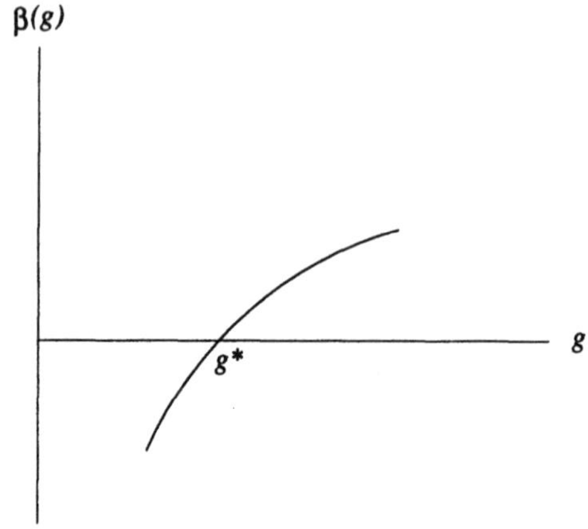

Fig. 2. An infrared stable fixed point of the β function.

corrections is to turn marginal operators into relevant or irrelevant operators, depending on whether their anomalous dimension is negative or positive. In strongly coupled theories, more interesting effects can occur. For example, in walking technicolor theories it is believed that a composite operator $\overline{\psi}\psi$ with classical dimension 3 behaves in the quantum theory as a scalar field with dimension 1, i.e. $\overline{\psi}\psi$ has anomalous dimension -2. A solvable example of this kind exists in two dimensions. The two dimensional Thirring model with a fundamental fermion field

$$\mathcal{L} = \overline{\psi}\left(i\partial\!\!\!/ - m\right)\psi - \frac{1}{2}g\left(\overline{\psi}\gamma^\mu\psi\right)^2, \tag{30}$$

is dual to the sine-Gordon model with a fundamental scalar field

$$\mathcal{L} = \frac{1}{2}\partial_\mu\phi\partial^\mu\phi + \frac{\alpha}{\beta^2}\cos\beta\phi, \tag{31}$$

where the coupling constants g and β are related by

$$\frac{\beta^2}{4\pi} = \frac{1}{1+g/\pi}. \tag{32}$$

The fermion of the Thirring model is the sine-Gordon soliton, and the boson of the sine-Gordon model is a fermion-antifermion bound state in the Thirring model. The mapping (32) shows that the strongly coupled sine-Gordon model with $\beta^2 \approx 4\pi$ can be mapped onto a weakly coupled Thirring model with $g \approx 0$. There are two alternate descriptions of the same theory: (a) A strongly interacting boson theory with large anomalous dimensions[2] (b) A weakly interacting

[2] For example, the operator $\cos\beta\phi$ gets mapped to the Fermion mass term $\overline{\psi}\psi$. In two dimensions, the canonical dimensions of $\cos\beta\phi$ and $\overline{\psi}\psi$ are zero and one, respectively.

fermion theory with small anomalous dimensions. Formally, both descriptions are identical, but clearly (b) is better for doing practical calculations.

The scaling dimensions of fields was determined from the free Lagrangian. That is because one assumes that the effective Lagrangian can be written as a weakly coupled field theory in terms of correctly chosen degrees of freedom at low energies. If the degrees of freedom are strongly coupled, the scaling dimension of the fields may change from their canonical value, as we saw in the sine-Gordon model at $\beta^2 = 4\pi$. Often, the most difficult task in writing down an effective theory is *choosing the right degrees of freedom*. In the sine-Gordon model, it is better to use a weakly coupled soliton field ψ instead of the fundamental field ϕ if $\beta^2 \approx 4\pi$, i.e. the effective Lagrangian for the sine-Gordon model with $\beta^2 \approx 4\pi$ is the Thirring model. Low energy QCD is a weakly coupled theory when written in terms of pion fields, but not when written in terms of quark and gluon fields.[3] At low energies, the Goldstone boson fields scale with canonical dimension zero (they are like angles), which is different from $\overline{q}q$, which has dimension 3 in free field theory. There are many examples of this kind in condensed matter physics. For example, in Landau Fermi liquid theory, the degrees of freedom are weakly interacting quasiparticles, not the strongly interacting electrons.

SUMMARY

We can now summarize the results of this section.

1. Find a good set of variables to describe the dynamics.
2. Write down the effective action as a sum of operators, $\sum_i c_i \mathcal{O}_i$.
3. The scaling rule is that $c_i \to s^{d_i - \gamma_i} c_i$, where d_i is the naive dimension and γ_i is the anomalous dimension. The most important operators are those of lowest dimension. Hopefully, a good choice has been made in (1), so that the anomalous dimensions are small.
4. To include all corrections up to order $1/s^r$, one should include all operators with dimension $\le D + r$, i.e. all terms with coefficients of dimension $\ge -r$.

There are a finite number of operators that contribute to a given order in $1/s$. In four dimensions, the dimensions of scalar, spinor and vector fields is

$$[\phi] = 1, \qquad [\psi] = 3/2, \qquad [A_\mu] = 1. \tag{33}$$

The allowed Lorentz invariant and gauge invariant operators of dimension ≤ 4 are $\phi^n, n \le 4$, $\overline{\psi}\psi$, $\partial_\mu \phi \partial^\mu \phi$, $\overline{\psi} \not{D} \psi$, $\overline{\psi}\psi\phi$, $F_{\mu\nu}F^{\mu\nu}$.

[3] This is obvious in the large N_c limit, where one has a weakly interacting theory of mesons and baryons, with a coupling constant $1/N_c$.

3 Renormalizable Theories vs Effective Theories

Field theory textbooks argue that a quantum field theory should be renormalizable, i.e. that the Lagrangian contains only terms with dimension $\leq D$. Otherwise one needs an infinite number of counterterms, hence an infinite number of unknown parameters, and the theory has no predictive power.

An effective field theory Lagrangian contains an infinite number of terms. Let us write the Lagrangian in the form

$$\mathcal{L}_{\text{eft}} = \mathcal{L}_{\leq D} + \mathcal{L}_{D+1} + \mathcal{L}_{D+2} + \ldots, \tag{34}$$

where $\mathcal{L}_{\leq D}$ contains all terms with dimension $\leq D$, $\mathcal{L}_{D+1}$ contains terms with dimension $D+1$, $\mathcal{L}_{D+2}$ contains terms with dimension $D+2$, and so on. The usual renormalizable Lagrangian is just the first term, $\mathcal{L}_{\leq D}$. There are an infinite number of terms in $\mathcal{L}_{\text{eft}}$, but one still has *approximate predictive power*. The effective Lagrangian is used to compute processes at some scale Λ/s, where Λ is the scale of (possibly unknown) high energy interactions. One can compute with an error of $1/s$ by retaining only $\mathcal{L}_{\leq D}$. Furthermore, one can extend the approximation in a systematic way – to compute with an error of order $1/s^{r+1}$, one needs to retain terms up to $\mathcal{L}_{D+r}$. There are only a finite number of parameters to compute to a given order in $1/s$, so the theory has predictive power.

A non-renormalizable theory is just as good as a renormalizable theory for computations, provided one is satisfied with a finite accuracy

The usual renormalizable field theory result is recovered if one takes the separation of scales $s \to \infty$. In this case, one can compute using a renormalizable Lagrangian $\mathcal{L}_{\leq D}$ with no errors. While exact computations are nice, they are irrelevant. Nobody knows the exact theory up to infinitely high energies. Thus any realistic calculation is done using an effective field theory. The standard "exact" textbook analysis of QED is really an approximate calculation in which terms suppressed by powers of $1/s$ have been neglected.

4 Two Simple Examples

We now consider two simple examples that illustrate the utility of the effective field theory method.

Rayleigh Scattering

The first example is Rayleigh scattering, the scattering of photons off atoms at low energies. Here low energies means energies small enough that one does not excite the internal states of the atom, or cause it to ionize. The atom can be treated as a particle of mass M, interacting with the electromagnetic field. Let $\psi(x)$ denote a field operator that creates an atom at the point x. Then the effective Lagrangian for the atom is

$$\mathcal{L} = \psi^\dagger \left(i\partial_t - \frac{p^2}{2M} \right) \psi + \mathcal{L}_{\text{int}}, \tag{35}$$

where $\mathcal{L}_{\text{int}}$ is the interaction term. Since the atom is neutral, the interaction term is a function of the electromagnetic field strength $F_{\mu\nu} = (\mathbf{E}, \mathbf{B})$. Gauge invariance forbids terms which depend only on the vector potential A_μ. At low energies, the dominant interaction is one which involves the smallest number of derivatives, and the smallest number of photon fields, and has the form

$$\mathcal{L}_{\text{int}} = a_0^3 \; \psi^\dagger \psi \left(c_1 E^2 + c_2 B^2 \right). \tag{36}$$

The electromagnetic field strength has mass dimension two, ψ has mass dimension $3/2$ ($\psi^\dagger i\partial_t \psi$ has dimension four), so that $c_1 a_0^3$ and $c_2 a_0^3$ have mass dimension -3. The typical momentum scale is set by the size of the atom a_0, so one expects $c_{1,2}$ to be of order unity. The interaction (36) gives the scattering amplitude $\mathcal{A} \sim c_i a_0^3 \omega^2$, since the electric and magnetic fields are gradients of the vector potential, so each factor of $\mathbf{E}$ or $\mathbf{B}$ produces a factor of ω. The scattering cross-section is proportional to $|c_i|^2 \, a_0^6 \omega^4$. This has the correct dimensions to be a cross-section, so the phase-space is dimensionless, and one finds that

$$\sigma \propto a_0^6 \; \omega^4. \tag{37}$$

This reproduces the well-know ω^4 dependence of the Rayleigh scattering cross-section, which explains why the sky is blue. One can actually do better, and determine the factors of 4π in (37), but I won't discuss that here. Equation (37) has corrections of order ω/a_0 from higher dimension operators which have been neglected in (36).

The Euler-Heisenberg Lagrangian

The Euler-Heisenberg effective Lagrangian is the effective Lagrangian for photon-photon scattering at energies much lower than the electron mass m_e. The leading order Lagrangian is the free Maxwell theory,

$$\mathcal{L} = -\frac{1}{4} F_{\mu\nu} F^{\mu\nu}. \tag{38}$$

The first interactions that can occur are from higher dimension operators. The lowest non-trivial operators must contain four factors of the field strength $F_{\mu\nu}$ and hence must be of dimension eight,

$$\mathcal{L} = \frac{\alpha^2}{m_e^4} \left[c_1 \left(F_{\mu\nu} F^{\mu\nu} \right)^2 + c_2 \left(F_{\mu\nu} \widetilde{F}^{\mu\nu} \right)^2 \right]. \tag{39}$$

(Terms with only three field strengths are forbidden by charge conjugation symmetry.) The effective interaction (39) is generated from the box diagram of Fig. 3. The box diagram contains four factors of the electric charge e, and one factor of $1/16\pi^2$ for the loop. In addition, the only dimensionful parameter other than the external momenta is the electron mass $m_{\rm e}$. This allows us to write the Lagrangian in the form (39), where $c_{1,2}$ are dimensionless constants. An explicit computation gives

$$c_1 = \frac{1}{90}, \qquad c_2 = \frac{7}{90}. \tag{40}$$

The low energy cross-section for $\gamma\gamma \to \gamma\gamma$ is obtained from the graph in the effective theory, Fig. 3(b). The scattering amplitude is $\mathcal{A} \sim \alpha^2\omega^4/m_{\rm e}^4$, since each gradient of the photon field in (39) produces one factor of ω. This produces a cross-section of order

$$\sigma \sim \left(\frac{\alpha^2\omega^4}{m_{\rm e}^4}\right)^2 \frac{1}{\omega^2}. \tag{41}$$

The phase space factor $1/\omega^2$ is obtained using dimensional analysis. The cross-section must have dimensions of area, so the phase space must have dimension -2. The only dimensionful parameter in the effective theory is the photon energy ω, so the phase space must be proportional to $1/\omega^2$. Thus we find $\sigma \sim \alpha^4\omega^6/m_{\rm e}^8$, with an error of order $\omega^2/m_{\rm e}^2$ from neglected higher order interactions in (39).

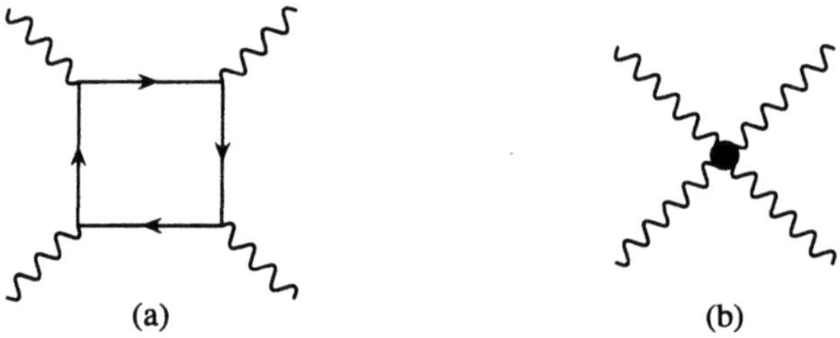

Fig. 3. Light by light scattering in (a) QED and (b) in the Euler-Heisenberg effective theory. The solid dot represents the four-photon interaction from the effective Lagrangian (39).

5 Weak Interactions at Low Energies: Tree Level

The classic example of an effective field theory is the Fermi theory of weak interactions. We first discuss how to obtain the Fermi theory as the low-energy limit of the renormalizable $SU(2) \times U(1)$ electroweak theory at tree level. The use of effective field theories for the tree level weak interactions will seem at first like applying a lot of unnecessary formalism to a trivial problem; the usefulness of the effective field theory method will only become apparent after we study the $\Delta S = 2$ weak interactions, which involve loop corrections in field theory. Finally, we will discuss the weak interactions including the leading logarithmic QCD corrections, for which the effective field theory method is indispensable.

The basic flavor changing vertex in the quark sector is the W coupling to the quark current

$$-\frac{ig}{\sqrt{2}}\, V_{ij}\, \overline{q}_i\, \gamma^\mu\, P_{\rm L}\, q_j, \tag{42}$$

where V_{ij} is the Kobayashi-Maskawa mixing matrix, and $P_{\rm L} = (1-\gamma_5)/2$ is the left-handed projection operator. The lowest order $\Delta S = 1$ amplitude arises from single W exchange (Fig. 4),

$$\mathcal{A} = \left(\frac{ig}{\sqrt{2}}\right)^2 V_{us} V_{ud}^* \,\left(\overline{u}\, \gamma^\mu\, P_{\rm L}\, s\right) \left(\overline{d}\, \gamma^\nu\, P_{\rm L}\, u\right) \left(\frac{-i g_{\mu\nu}}{p^2 - M_{\rm W}^2}\right), \tag{43}$$

where the W boson propagator is in 't Hooft-Feynman gauge, p is the momentum transferred by the W, and u, d, s are quark spinors. The exchange of unphysical scalars $\phi^\pm$ can be neglected, since their Yukawa couplings to the light quarks are very small. The amplitude (43) produces a non-local four-quark interaction, because of the factor of $p^2 - M_{\rm W}^2$ in the denominator. However, if the momentum transfer p is small compared with $M_{\rm W}$, the non-local interaction can be approximated by a local interaction using the Taylor series expansion

$$\frac{1}{p^2 - M_{\rm W}^2} = -\frac{1}{M_{\rm W}^2}\left(1 + \frac{p^2}{M_{\rm W}^2} + \frac{p^4}{M_{\rm W}^4} + \ldots\right), \tag{44}$$

and retaining only a *finite* number of terms. To lowest order, the amplitude is

$$\mathcal{A} = \frac{i}{M_{\rm W}^2}\left(\frac{ig}{\sqrt{2}}\right)^2 V_{us} V_{ud}^* \,\left(\overline{u}\, \gamma^\mu\, P_{\rm L}\, s\right) \left(\overline{d}\, \gamma_\mu\, P_{\rm L}\, u\right) + \mathcal{O}\left(\frac{1}{M_{\rm W}^4}\right). \tag{45}$$

The amplitude (45) can be obtained using the effective Lagrangian

$$\mathcal{L} = -\frac{4 G_{\rm F}}{\sqrt{2}} V_{us} V_{ud}^* \,\left(\overline{u}\, \gamma^\mu\, P_{\rm L}\, s\right) \left(\overline{d}\, \gamma_\mu\, P_{\rm L}\, u\right) + \mathcal{O}\left(\frac{1}{M_{\rm W}^4}\right), \tag{46}$$

where u, d and s are now the quark fields, and we have used the definition

$$\frac{G_{\rm F}}{\sqrt{2}} \equiv \frac{g^2}{8 M_{\rm W}^2}. \tag{47}$$

The effective Lagrangian (46) can be used to study the weak decays of quarks at low energies. The basic interaction is a local four-Fermion vertex, as shown in Fig. 5. To avoid complications with hadronic matrix elements and QCD corrections (which will be discussed later), consider instead the effective Lagrangian for μ decay

$$\mathcal{L} = -\frac{4 G_{\rm F}}{\sqrt{2}} \,\left(\overline{e}\, \gamma^\mu\, P_{\rm L}\, \nu_{\rm e}\right) \left(\overline{\nu}_\mu\, \gamma^\mu\, P_{\rm L}\, \mu\right) + \mathcal{O}\left(\frac{1}{M_{\rm W}^4}\right), \tag{48}$$

whose derivation is almost identical to that of (46). Using (48), neglecting the $1/M_{\rm W}^4$ terms, and integrating over phase space gives the standard result for the muon lifetime at lowest order,

$$\Gamma_\mu = \frac{G_F^2 m_\mu^5}{192\pi^3}. \tag{49}$$

This calculation is well known, and will not be repeated here.

To summarize: at lowest order, the "full theory," which is the $SU(2) \times U(1)$ electroweak theory, can be replaced by the "effective theory," which is QED plus the effective Lagrangian (46) (or (48)), up to corrections of order $1/M_W^4$. The effective theory can be used to compute physical processes such as the muon lifetime. So far, the effective field theory method is a fancy way of saying that we have approximated the W boson propagator in Fig. 4 by $1/M_W^2$. The real advantage of the effective field theory method will be apparent after we have discussed the one-loop $\Delta S = 2$ amplitude including QCD radiative corrections.

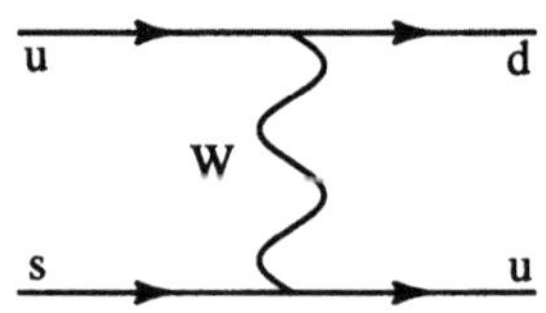

Fig. 4. W exchange diagram for the $\Delta S = 1$ weak interactions.

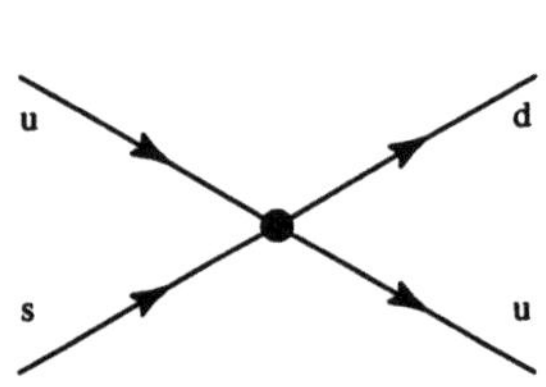

Fig. 5. The effective four-Fermi interaction of (46). This interaction reproduces the results of Fig. 4 to order $1/M_W^2$.

6 Renormalization in Effective Field Theories

In quantum field theory, knowing the Lagrangian is not sufficient to compute results for physical quantities. In addition, one needs to specify a way to get finite, unambiguous answers for physical quantities. In perturbation theory, this corresponds to a choice of renormalization scheme which (i) regulates the integrals and (ii) subtracts the infinities in a systematic way. The effective Lagrangian

(46) that we have constructed is non-renormalizable, since it contains an operator of dimension six, times a coefficient G_F which is of order $1/M_\mathrm{W}^2$. The neglected $1/M_\mathrm{W}^4$ term contains operators of dimension eight, and so on. To use the effective Lagrangian beyond tree level, it is necessary to give a renormalization scheme as part of the definition of the effective field theory. Without this additional information, the effective Lagrangian (46) is meaningless.

It is important to keep in mind that the effective field theory is a *different theory* from the full theory. The full theory of the weak interactions is a renormalizable field theory. The effective field theory is a non-renormalizable field theory, and has a different divergence structure from the full theory. The effective field theory is constructed to correctly reproduce the low-energy effects of the full theory to a given order in $1/M_\mathrm{W}$. The effective Lagrangian includes more terms as one works to higher orders in $1/M_\mathrm{W}$. The effective field theory method is useful only for computing results to a certain order in $1/M_\mathrm{W}$. If one is interested in the answer to all orders in $1/M_\mathrm{W}$, it is obviously much simpler to use the full theory.

The renormalization scheme must be carefully chosen to give a sensible effective field theory. To see what the possible problems might be, consider the flavor diagonal effective Lagrangian from W and Z exchange

$$\mathcal{L} = -\frac{4G_\mathrm{F}}{\sqrt{2}} V_{ui} V_{ui}^* \; (\overline{u}\,\gamma^\mu\, P_\mathrm{L}\, q_i)\,(\overline{q}_i\,\gamma_\mu\, P_\mathrm{L}\, u) + (\mathrm{Z}-\mathrm{exchange}) + \mathcal{O}\left(\frac{1}{M_\mathrm{W}^4}\right), \quad (50)$$

where $i = d, s, b$. At tree level, the W and Z exchange graphs contribute to flavor diagonal parity violating u-quark interactions at order $G_\mathrm{F} \sim 1/M_\mathrm{W}^2$. At one loop, the interaction (50) induces a $Z\overline{u}u$ vertex from the graph in Fig. 6 which is of the form

$$I \sim \frac{1}{M_\mathrm{W}^2} \int d^4k \; \frac{1}{k^2}, \quad (51)$$

neglecting the γ-matrix structure. The $1/k^2$ factor is from the two fermion propagators in the loop, and G_F has been rewritten as $G_\mathrm{F} \sim 1/M_\mathrm{W}^2$. Since the effective field theory is valid up to energies of order M_W, one can estimate the integral using a momentum space cutoff Λ of order M_W,

$$I \sim \frac{1}{M_\mathrm{W}^2} \Lambda^2 \sim \mathcal{O}(1)\,. \quad (52)$$

Thus the interaction (50) produces a one loop correction to the $Z\overline{u}u$ vertex of order one. Similarly, one can show that higher order terms, such as the dimension eight operators, are all equally important. A loop graph of the form Fig. 6 (where the vertex is now a dimension eight operator) is of order

$$I' \sim \frac{1}{M_\mathrm{W}^4} \int d^4k \; \frac{1}{k^2} k^2 \sim \frac{\Lambda^4}{M_\mathrm{W}^4} \sim \mathcal{O}(1)\,, \quad (53)$$

etc. The additional k^2 in the integral (53) is from the extra ∂^2 at the four-quark vertex in the dimension eight operator arising from the order p^2 term in the expansion of (46). The loop graph with an insertion of the dimension eight operator is just as important as the loop graph with an insertion of the dimension six operator; both are of order unity and cannot be neglected. Similarly, all the higher order terms in the effective Lagrangian are equally important, and the entire expansion breaks down. A similar problem also occurs in the flavor changing $\Delta S = 1$ weak interactions that we have been studying, but the analysis is more subtle because of the GIM mechanism, which is why we considered the $Z\overline{u}u$ vertex.

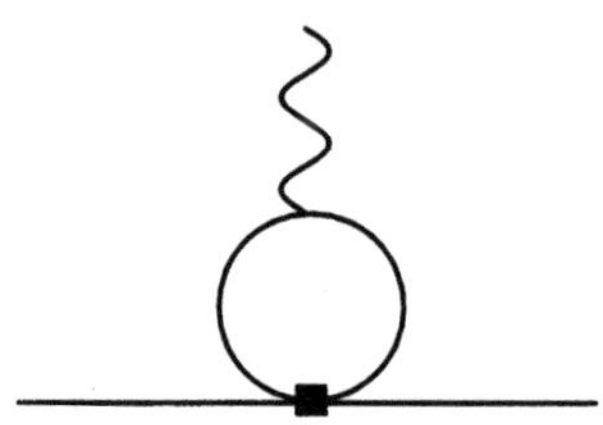

Fig. 6. One loop correction to the $Z\overline{u}u$ vertex. The solid square represents either the dimension six four-quark interaction of eq. (50), or the dimension eight four-quark operator discussed in the text.

The effective field theory expansion breaks down if one introduces a mass-dependent subtraction scheme such as a momentum space cutoff.[4] This problem can be cured if one uses a mass-independent subtraction scheme, such as dimensional regularization and minimal subtraction, in which the dimensional parameter μ only appears in logarithms, and never as explicit powers such as μ^2. In such a subtraction scheme β-functions and anomalous dimensions of composite operators are mass independent. If one estimates the integrals (51) and (53) in a mass-independent subtraction scheme, one finds

$$\begin{aligned} I &= \frac{1}{M_W^2} \int d^4k \, \frac{1}{k^2} \sim \frac{m^2}{M_W^2} \log \mu, \\ I' &= \frac{1}{M_W^4} \int d^4k \, \frac{1}{k^2} k^2 \sim \frac{m^4}{M_W^4} \log \mu, \end{aligned} \tag{54}$$

where m is some dimensionful parameter that is *not* the renormalization scale μ. It must be some other dimensionful scale that enters the loop graph of Fig. 6, such as the quark mass or the external momentum. This completely changes the estimate of the integrals. The integrals are no longer of order one, but are small provided $m \ll M_W$. As a result:

[4] One way to solve this problem is to use a cutoff $\Lambda \ll M_W$. This method does not allow one to easily match between the full and effective theories, or to include QCD corrections.

1. The effective Lagrangian produces a well-defined expansion of the weak amplitudes in powers of m/M_{W}, where m is some low scale such as the quark mass or the external momentum (or Λ_{QCD} when one includes QCD effects). One has a systematic expansion in powers of some low scale over M_{W}. This makes precise what is meant by neglecting $1/M_{\mathrm{W}}^4$ terms in (45).
2. Loop integrals do not have a power law dependence on $\mu \sim M_{\mathrm{W}}$, so one can count powers of $1/M_{\mathrm{W}}$ directly from the effective Lagrangian. Graphs with one insertion of terms in $\mathcal{L}_{\mathrm{eff}}$ of order $1/M_{\mathrm{W}}^2$ produce amplitudes of order $1/M_{\mathrm{W}}^2$. Graphs with one insertion of terms of order $1/M_{\mathrm{W}}^4$ or two insertions of terms of order $1/M_{\mathrm{W}}^2$ produce amplitudes of order $1/M_{\mathrm{W}}^4$, etc.
3. The effective field theory behaves for all practical purposes like a renormalizable field theory if one works to some fixed order in $1/M_{\mathrm{W}}$. This is because there are only a finite number of terms in $\mathcal{L}_{\mathrm{eff}}$ that are allowed to a given order in $1/M_{\mathrm{W}}$. Terms of higher order in $1/M_{\mathrm{W}}$ can be safely neglected because they can never be multiplied by positive powers of M_{W} to produce effects comparable to lower order terms.

It is well-known that different renormalization schemes lead to equivalent answers for all physical quantities. In an effective field theory, a mass-independent subtraction scheme is particularly convenient, since it provides an efficient way of keeping only a few operators in $\mathcal{L}_{\mathrm{eff}}$, and in deciding which Feynman graphs are important. Nevertheless, one must be able to obtain the same results in a mass-dependent scheme such as a momentum space cutoff. This is true in principle: a mass dependent scheme has an infinite number of contributions that are of leading order (from the dimension four, six, eight, ..., operators). If one resums this contribution, then the remaining effects (again from an infinite number of terms) will be of order $1/M_{\mathrm{W}}^2$. Resumming the latter leaves a contribution of $1/M_{\mathrm{W}}^4$, etc. The net result of this procedure is to reproduce the same answer as that obtained much more simply using a mass-independent renormalization scheme. The connection between different renormalization schemes is much more complicated in an effective field theory (which is non-renormalizable), than in a renormalizable field theory.

7 Decoupling of Heavy Particles

There is one important drawback to using a mass-independent subtraction scheme – heavy particles do not decouple. [5] This must obviously be true since the contribution of particles to β-functions does not depend on the particle mass. For example, a 1 TeV charged lepton makes the same contribution as an electron to the QED β-function at 1 GeV.

It is instructive to look at the contribution of a charged fermion to the β-function in QED. Evaluating the diagram of Fig. 7 in dimensional regularization gives

[5] A mass independent subtraction scheme does not satisfy the conditions for the Appelquist-Carazzone theorem.

$$i\frac{e^2}{2\pi^2}\left(p_\mu p_\nu - p^2 g_{\mu\nu}\right)\left[\frac{1}{6\epsilon} - \frac{\gamma}{6} - \int_0^1 dx\ x(1-x)\ \log\frac{m^2 - p^2x(1-x)}{4\pi\mu^2}\right], \tag{55}$$

where p is the external momentum, m is the fermion mass, γ is Euler's constant, and μ is the scale parameter of dimensional regularization.

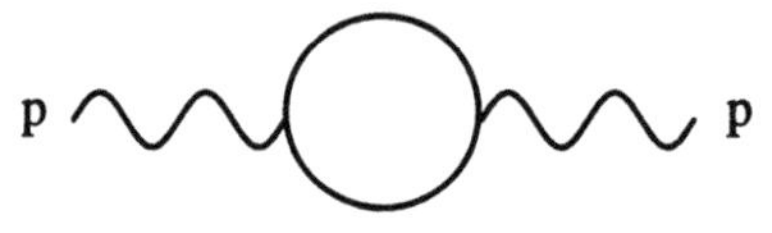

Fig. 7. One loop contribution to the QED β-function from a fermion of mass m.

Mass-Dependent Scheme

In a mass-dependent scheme, such as an off-shell momentum space subtraction scheme, one subtracts the value of the graph at a Euclidean momentum point $p^2 = -M^2$, to get

$$-i\frac{e^2}{2\pi^2}\left(p_\mu p_\nu - p^2 g_{\mu\nu}\right)\left[\int_0^1 dx\ x(1-x)\ \log\frac{m^2 - p^2x(1-x)}{m^2 + M^2x(1-x)}\right]. \tag{56}$$

The fermion contribution to the QED β-function is obtained by acting with $(e/2)Md/dM$ on the coefficient of $i\left(p_\mu p_\nu - p^2 g_{\mu\nu}\right)$,

$$\begin{aligned}\beta(e) &= -\frac{e}{2}M\frac{d}{dM}\frac{e^2}{2\pi^2}\left[\int_0^1 dx\ x(1-x)\ \log\frac{m^2 - p^2x(1-x)}{m^2 + M^2x(1-x)}\right] \\ &= \frac{e^3}{2\pi^2}\int_0^1 dx\ x(1-x)\ \frac{M^2x(1-x)}{m^2 + M^2x(1-x)}.\end{aligned} \tag{57}$$

The fermion contribution to the β-function is plotted in Fig. 8. When the fermion mass m is small compared with the renormalization point M, $m \ll M$, the β-function contribution is

$$\beta(e) \approx \frac{e^3}{2\pi^2}\int_0^1 dx\ x(1-x) = \frac{e^3}{12\pi^2}. \tag{58}$$

As the renormalization point passes through m, the fermion decouples, and for $M \ll m$, its contribution to β vanishes as

$$\beta(e) \approx \frac{e^3}{2\pi^2}\int_0^1 dx\ x(1-x)\frac{M^2x(1-x)}{m^2} = \frac{e^3}{60\pi^2}\frac{M^2}{m^2}. \tag{59}$$

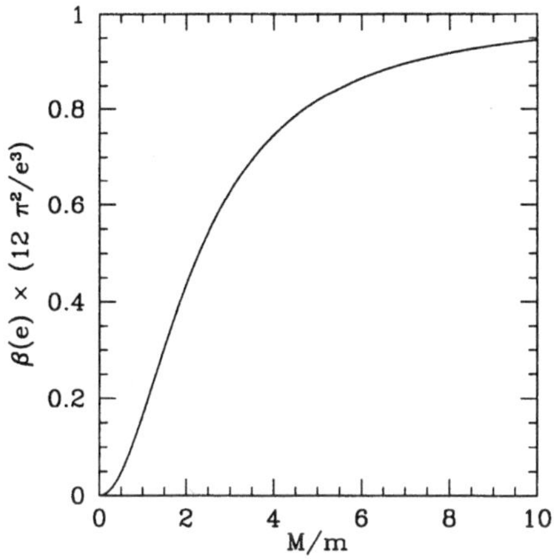

Fig. 8. Contribution of a fermion of mass m to the QED β-function. The result is given for the momentum-space subtraction scheme, with renormalization scale M. The β-function does not attain its limiting value of $e^3/12\pi^2$ until $M \gtrsim 10m$. The fermion decouples for $M \ll m$.

The $\overline{MS}$ Scheme

In the $\overline{MS}$ scheme, one subtracts the $1/\epsilon$ pole and redefines $4\pi\mu^2 e^{-\gamma} \to \mu^2$, to give

$$-i\frac{e^2}{2\pi^2}\left(p_\mu p_\nu - p^2 g_{\mu\nu}\right)\left[\int_0^1 dx\; x(1-x)\log\frac{m^2 - p^2 x(1-x)}{\mu^2}\right]. \tag{60}$$

The fermion contribution to the QED β-function is obtained by acting with $(e/2)\mu d/d\mu$ on the coefficient of $i\left(p_\mu p_\nu - p^2 g_{\mu\nu}\right)$,

$$\begin{aligned}\beta(e) &= -\frac{e}{2}\mu\frac{d}{d\mu}\frac{e^2}{2\pi^2}\left[\int_0^1 dx\; x(1-x)\log\frac{m^2 - p^2 x(1-x)}{\mu^2}\right]\\ &= \frac{e^3}{2\pi^2}\int_0^1 dx\; x(1-x) = \frac{e^3}{12\pi^2},\end{aligned} \tag{61}$$

which is independent of the fermion mass and μ.

The fermion contribution to the β-function in the $\overline{MS}$ scheme does not vanish as $m \gg \mu$, so the fermion does not decouple as it should. There is another problem: the finite part of the Feynman graph in the $\overline{MS}$ scheme at low momentum is

$$-i\frac{e^2}{2\pi^2}\left(p_\mu p_\nu - p^2 g_{\mu\nu}\right)\left[\int_0^1 dx\; x(1-x)\log\frac{m^2}{\mu^2}\right], \tag{62}$$

from (60). For $\mu \ll m$ the logarithm becomes large, and perturbation theory breaks down. These two problems are related. The large finite parts correct for the fact that the value of the running coupling used at low energies is incorrect, because it was obtained using the "wrong" β-function. The two problems can be solved at the same time by integrating out heavy particles. One uses a theory including the fermion when $m < \mu$, and a theory without the fermion when $m > \mu$. Effects of the heavy particle in the low energy theory are included via higher dimension operators, which are suppressed by inverse powers of the heavy particle mass. The matching condition of the two theories at the scale of the fermion mass is that S-matrix elements for light particle scattering in the low-energy theory without the heavy particle must match those in the high-energy theory with the heavy particle.

For the case of a spin-1/2 fermion at one loop, this implies that the running coupling is continuous at $m = \mu$. The β-function is discontinuous at $m = \mu$, since the fermion contributes $e^3/12\pi^2$ to β above m and zero below m. The β-function is a step-function, instead of having a smooth crossover between $e^3/12\pi^2$ and zero, as in the momentum-space subtraction scheme. Decoupling of heavy particles is implemented by hand in the $\overline{MS}$ scheme by integrating out heavy particles at $\mu \sim m$. One calculates using a sequence of effective field theories with fewer and fewer particles. The main reason for using the $\overline{MS}$ scheme and integrating out heavy particles is that it is much easier to use in practice than the momentum-space subtraction scheme. Virtually all radiative corrections beyond one-loop are evaluated in practice using the $\overline{MS}$ scheme.

There are some instances in which heavy particle effects are important in the low energy effective theory. An example of this is the top quark in the standard model. The reason is that the top quark has a mass $m_t = g_t v/\sqrt{2}$, where g_t is the top quark Yukawa coupling, and v is the vacuum expectation value of the Higgs field. Taking m_t large while keeping v fixed is equivalent to taking g_t large. Diagrams involving top quarks and scalars (either the Higgs boson or the longitudinal parts of the W and Z) can be large, because they involve factors of g_t which can cancel any $1/m_t$ suppression. We will see an example of this in the next section, where the $\Delta S = 2$ amplitude is shown to grow with m_t. One can still integrate out the heavy top quark, but the low energy theory contains operators with coefficients which grow with m_t.

8 Weak Interactions at Low Energies: One Loop

The ideas discussed so far can now be applied to the weak interactions at one loop. The amplitude for the $\Delta S = 2$ amplitude for K^0-$\overline{K}^0$ mixing is of order G_F^2. The leading contribution to this amplitude in the standard model is from the box diagram of Fig. 9, where one sums over quarks $i, j = u, c, t$ in the intermediate states. The sum of the W and unphysical scalar exchange graphs is

$$\mathcal{A}^{\text{box}} = \frac{g^4}{128\pi^2 M_W^2} \sum_{i,j} \xi_i \xi_j \, \overline{E}(x_i, x_j) \, \left(\overline{d}\gamma^\mu P_L\, s\right) \left(\overline{d}\gamma_\mu P_L\, s\right), \tag{63}$$

where

$$x_i = \frac{m_i^2}{M_W^2}, \tag{64}$$

$$\xi_i = V_{is} V_{id}^*, \tag{65}$$

$$\begin{aligned}\overline{E}(x,y) = -xy\Bigg\{ & \frac{1}{x-y}\left[\frac{1}{4} - \frac{3}{2}\frac{1}{x-1} - \frac{3}{4}\frac{1}{(x-1)^2}\right]\log x \\ & + \frac{1}{y-x}\left[\frac{1}{4} - \frac{3}{2}\frac{1}{y-1} - \frac{3}{4}\frac{1}{(y-1)^2}\right]\log y - \frac{3}{4}\frac{1}{(x-1)(y-1)}\Bigg\},\end{aligned} \tag{66}$$

and

$$\overline{E}(x,x) = -\frac{3}{2}\left(\frac{x}{x-1}\right)^3 \log x - x\left[\frac{1}{4} - \frac{9}{4}\frac{1}{x-1} - \frac{3}{2}\frac{1}{(x-1)^2}\right]. \tag{67}$$

In the limit $m_u = 0$ and $m_{c,t} \ll M_W$,[6]

$$\mathcal{A}^{\text{box}} = -\frac{G_F^2}{4\pi^2}\left(\overline{d}\gamma^\mu P_L s\right)\left(\overline{d}\gamma_\mu P_L s\right)\left[\xi_c^2 m_c^2 + \xi_t^2 m_t^2 + 2\xi_c\xi_t m_c^2 \log\frac{m_t^2}{m_c^2}\right], \tag{68}$$

using (47). The $\Delta S = 2$ amplitude is of order $1/M_W^4$, rather than $1/M_W^2$ as one might naively expect, because of the GIM mechanism: The quark mass independent piece of the $\Delta S = 2$ amplitude is proportional to

$$\xi_u + \xi_c + \xi_t = \sum_i V_{id} V_{is}^* = 0, \tag{69}$$

which vanishes because the KM matrix is unitary.

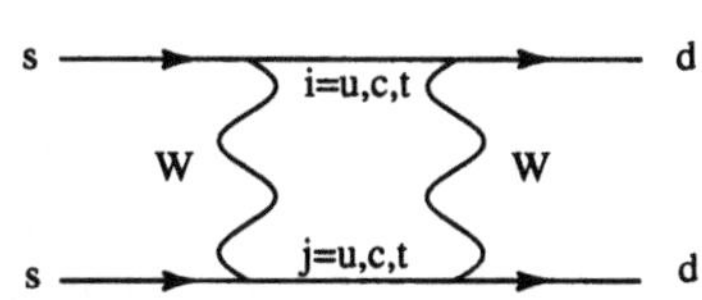

Fig. 9. The box diagram for the $\Delta S = 2$ $K^0 - \overline{K}^0$ mixing amplitude.

[6] The $\Delta S = 2$ amplitude is considered in the limit $m_t \ll M_W$. This was the approximation used in the original calculations, and makes it easier for the reader to compare with the literature. It also simplifies the discussion somewhat, because the t-quark and c-quark can be treated in a similar fashion.

Matching at M_W

We will now reproduce (68) using an effective field theory calculation to one loop. At the scale M_W, the $\Delta S = 2$ amplitude in the full theory is given by a loop graph in the effective theory involving two insertions of the $\Delta S = 1$ interaction, plus a local four-Fermi $\Delta S = 2$ interaction. The sum of the loop graph and the local $\Delta S = 2$ interaction must reproduce the $\Delta S = 2$ interaction in the full theory to order $1/M_W^4$, as shown schematically in Fig. 10. The tree level graphs of Figs. 4 and 5 are chosen to be the same in the full and effective theory to order $1/M_W^2$, but this does not imply that the loop graphs in the full and effective theory are equal to order $1/M_W^4$. The two loop graphs in Fig. 10 would be equal to order $1/M_W^4$ if the loop graphs in the full and effective theory were finite. However, in general, the graphs are infinite, and need subtractions. There is no simple relation between the renormalization prescriptions in the full and effective theories and one needs to add a local $\Delta S = 2$ counterterm at the scale M_W, which is the difference between the loop graphs in the full and effective theories. The graphs in the effective theory are more divergent than in the full theory. In our example, the box diagram in the full theory is convergent by naive power counting.

$$I_{\text{full}} \sim \int d^4k \left(\frac{1}{k}\right)^2 \left(\frac{1}{k^2}\right)^2 , \tag{70}$$

whereas the graph in the effective theory is quadratically divergent,

$$I_{\text{eff}} \sim \int d^4k \left(\frac{1}{k}\right)^2 , \tag{71}$$

where we have used a factor of $1/k$ for each internal fermion line, and $1/k^2$ for each internal boson line. In the case of the standard model, the graph in the effective theory is more convergent than the naive estimate because of the GIM mechanism. As we have seen, the fermion mass-independent part of the diagram is proportional to $\xi_u + \xi_c + \xi_t$, which vanishes. Thus the non-vanishing parts of the graphs in the full and effective theory must involve a factor of the internal fermion mass. In fact, there have to be two factors of the fermion mass because the $\Delta S = 1$ vertex only involves left-handed fields, and a fermion mass changes a left-handed fermion to a right-handed fermion. Thus in the effective theory, the non-zero part of the diagram must have two mass insertions on each of the fermion lines (there is a separate GIM mechanism for each line because of the independent sums over i and j in (63)), as represented in Fig. 10. This increases the degree of convergence of the diagram by two for each internal quark line, and converts it from a diagram that diverges like k^2 to a diagram that converges like $1/k^2$. Since the diagrams in the full and effective theory are both finite, the local $\Delta S = 2$ vertex induced at the scale M_W vanishes.

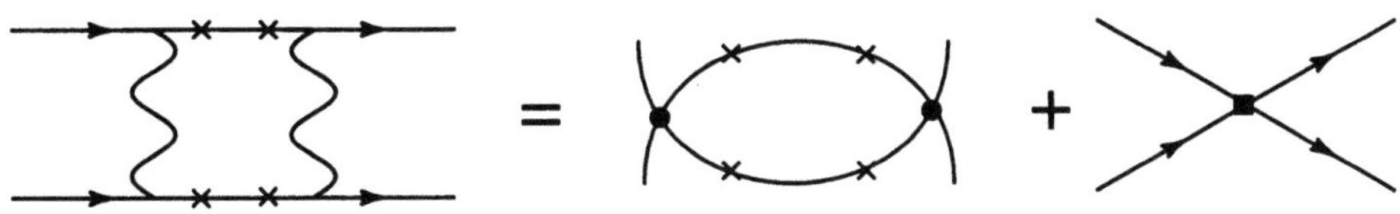

Fig. 10. Box diagram for the $\Delta S = 2$ amplitude in the full and effective theories. The crosses represent fermion mass insertions. The solid circle is a $\Delta S = 1$ vertex, and the solid square is a local $\Delta S = 2$ vertex.

Matching at m_t

The effective Lagrangian remains unchanged down to the scale $\mu = m_\mathrm{t}$, if one neglects QCD radiative corrections. At the scale $\mu = m_\mathrm{t}$, one integrates out the top quark. The "full theory" is now the effective Lagrangian including six quarks, and the "effective theory" is the effective Lagrangian including only five quarks. The $\Delta S = 1$ interactions in the five-quark theory are trivially obtained from those in the six-quark theory, by dropping all terms that contain the t-quark. The $\Delta S = 2$ interactions in the five- and six-quark theories are given in Fig. 11, where the intermediate states in the six-quark theory are the u, c and t quarks, and in the five-quark theory are the u and c quarks. There is no GIM cancellation once the top quark has been integrated out of the theory, so the loop graph in the five-quark theory is divergent, and there will (in principle) be a non-zero counterterm induced at the scale m_t. The value of the counterterm is the difference in the diagrams in the theories above and below m_t, and so is given by the graphs in the theory above m_t that involve at least one t-quark in the loop, as shown in Fig. 12. All other graphs in the six-quark theory are identical to the corresponding graphs in the five-quark theory. The loop graphs in the theory above m_t can be calculated quite simply, and lead to the matching condition

$$c_2(\mu = m_\mathrm{t} - 0) = \frac{G_\mathrm{F}^2}{4\pi^2}\left[\xi_\mathrm{t}^2\, m_\mathrm{t}^2 + 2\xi_\mathrm{c}\xi_\mathrm{t}\left(m_\mathrm{t}^2 + m_\mathrm{c}^2\right) + 2\xi_\mathrm{u}\xi_\mathrm{t}\left(m_\mathrm{t}^2 + m_\mathrm{u}^2\right)\right], \quad (72)$$

where the contributions come from the finite part of Fig. 12, and c_2 is the coefficient of the $\Delta S = 2$ operator $\left(\overline{d}\gamma^\mu P_\mathrm{L}\, s\right)\left(\overline{d}\gamma_\mu P_\mathrm{L}\, s\right)$. Using the relation (69) and neglecting m_u gives

$$c_2(\mu = m_\mathrm{t} - 0) = \frac{G_\mathrm{F}^2}{4\pi^2}\left[-\xi_\mathrm{t}^2\, m_\mathrm{t}^2 + 2\xi_\mathrm{c}\xi_\mathrm{t}\, m_\mathrm{c}^2\right]. \quad (73)$$

Equation (73) is really the difference of two calculations at the scale m_t – one in the full theory and one in the effective theory. Both calculations are sensitive to infrared effects, such as confinement. However, all infrared effects cancel in the difference, and $c_2\left(\mu = m_\mathrm{t} + 0\right) - c_2\left(\mu = m_\mathrm{t} - 0\right)$ is not sensitive to infrared effects. An arbitrary infrared regulator can be used if the diagrams

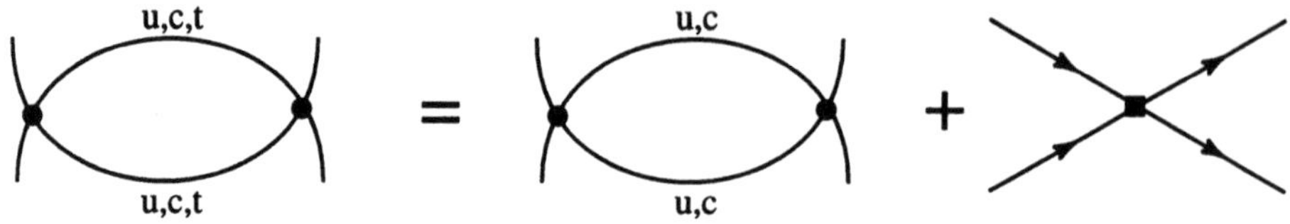

Fig. 11. Matching condition at the t-quark scale. The solid square is the $\Delta S = 2$ counterterm induced at $\mu = m_{\rm t}$.

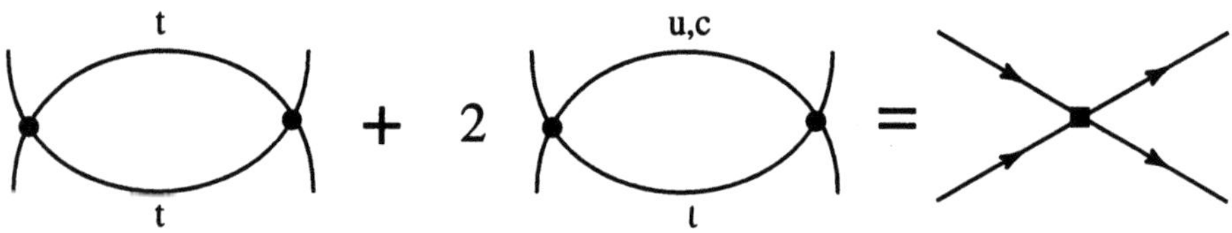

Fig. 12. Graphs to be computed to evaluate the $\Delta S = 2$ counterterm induced at $\mu = m_{\rm t}$.

are infrared divergent. The loop graphs will depend on the choice of regulator, but the matching condition will not. The matching condition is only sensitive to momenta of order $\mu = m_{\rm t}$, so mass parameters such as $m_{\rm u}$ are short distance parameters such as the $\overline{MS}$ mass renormalized at $\mu = m_{\rm t}$.

Scaling from $m_{\rm t}$ to $m_{\rm c}$

The next step is to scale from the scale $m_{\rm t}$ to $m_{\rm c}$. The loop graph Fig. 13 is divergent, because there is no longer a GIM mechanism in the five-quark theory, and c_2 is renormalized proportional to c_1^2, where c_1 is the coefficient of the $\Delta S = 1$ operator. This implies that there is a renormalization group equation for c_2,

$$\mu \frac{d}{d\mu} c_2 = \frac{1}{8\pi^2}\, c_1^2\, m_{\rm c}^2\, \xi_{\rm c}\xi_{\rm t}, \tag{74}$$

where the anomalous dimension is computed using the infinite part of Fig. 13. Integrating this equation from $m_{\rm t}$ to $m_{\rm c}$ gives

$$\begin{aligned} c_2(m_{\rm c}) &= c_2(m_{\rm t}) + \frac{1}{8\pi^2} c_1^2\, m_{\rm c}^2\, \xi_{\rm c}\xi_{\rm t}\, \log\frac{m_{\rm c}}{m_{\rm t}}, \\ &= c_2(m_{\rm t}) - \frac{G_{\rm F}^2}{2\pi^2}\, \xi_{\rm c}\xi_{\rm t}\, m_{\rm c}^2\, \log\frac{m_{\rm t}^2}{m_{\rm c}^2}, \end{aligned} \tag{75}$$

substituting $c_1 = -4G_{\rm F}/\sqrt{2}$.

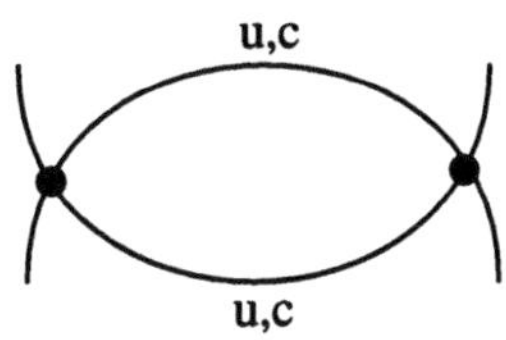

Fig. 13. The infinite part of this graph contributes to the renormalization group scaling of the $\Delta S = 2$ amplitude.

Matching at m_c

Finally, one integrates out the c quark. This is virtually identical to the matching condition at the t-quark scale, and gives

$$c_2(\mu = m_c - 0) = c_2(\mu = m_c + 0) + \frac{G_F^2}{4\pi^2}\left[\xi_c^2 m_c^2 + 2\xi_c\xi_u m_c^2\right] . \tag{76}$$

Combining (72)–(76) reproduces the box diagram computation (68).

There are some important features of the $\Delta S = 2$ computation which are generic to any effective field theory computation. (i) The contributions proportional to the heaviest mass scale m_t arise from matching conditions at that scale. (ii) contributions proportional to lower mass scales (such as m_c) arise from matching at the scale m_c, and also from matching at scales larger than m_c (such as m_t). (iii) Contributions proportional to logarithms of two scales arise from renormalization group evolution between the two scales.

It seems that the effective field theory method is much more complicated than directly computing the original box diagram in Fig. 9. The effective theory method has broken the computation of the box diagram into several steps. The computations involved at each step in the effective field theory are much simpler than the box diagram calculation. The box diagram involves several different mass scales in the internal propagators, which leads to complicated Feynman parameter integrals that must be evaluated. The matching condition computations in the effective field theory each involve only a single mass scale, and are much simpler. One can contrast the full answer (68) with the individual pieces of the effective field theory calculation in (72)–(76). Furthermore, in the effective field theory calculation it is trivial to include the leading logarithmic QCD corrections to the $\Delta S = 2$ amplitude. The corresponding computation in the full theory is far more difficult, and involves computing two loop diagrams such as the one in Fig. 14.

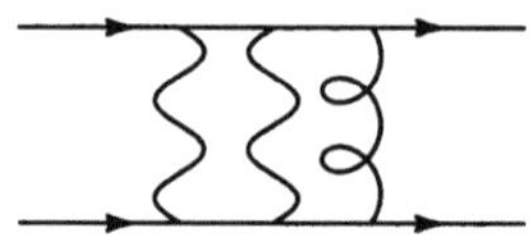

Fig. 14. A QCD radiative correction to the box diagram.

QCD Corrections

The leading logarithmic corrections to the $\Delta S = 2$ amplitude sum all corrections of the form $(\alpha_s \log r)^n$, where r is a large ratio of scales such as M_W/m_c, but neglect corrections of the form $\alpha_s(\alpha_s \log r)^n$. The QCD corrections to the matching condition only involve a single scale, and do not have any large logarithms. For example, the matching condition at the scale m_t only involves corrections that depend on $\alpha_s(\mu)$ and $\log m_t/\mu$. Evaluating these corrections by setting the $\overline{MS}$ parameter $\mu = m_t$ implies that there is no leading logarithmic correction to the matching condition. The only leading logarithmic QCD corrections arise from renormalization group scaling between different scales. This computation is straightforward, and only involves the infinite parts of one loop diagrams. The renormalization group equation (74) is replaced by

$$\mu \frac{d}{d\mu} c_2(\mu) = \frac{1}{8\pi^2}\, m_c^2(\mu)\, c_1^2(\mu)\, \xi_c \xi_t + \gamma_2(\mu)\, c_2(\mu), \tag{77}$$

where m_c has been replaced by the running mass, c_1 has been replaced by the running coupling $c_1(\mu)$, and γ_2 is the anomalous dimension

$$\gamma_2 = \frac{\alpha_s(\mu)}{\pi}, \tag{78}$$

of the $\Delta S = 2$ operator $(\overline{d}\gamma^\mu P_L\, s)(\overline{d}\gamma_\mu P_L\, s)$, which can be obtained from the infinite part of Fig. 15. The running mass $m_c(\mu)$ satisfies the renormalization group equation

$$\mu \frac{d}{d\mu} m_c(\mu) = \gamma_m\, m_c(\mu) = -\frac{2\alpha_s(\mu)}{\pi}\, m_c(\mu). \tag{79}$$

If c_1 satisfies a simple renormalization group equation of the form

$$\mu \frac{d}{d\mu} c_1(\mu) = \gamma_1\, c_1(\mu), \tag{80}$$

one can solve (77)–(80) to obtain the QCD corrected value for $c_2(\mu)$. At one loop, it is convenient to define b and $\hat{\gamma}_i$

$$\mu \frac{d}{d\mu} g = \beta(g) = -b \frac{g^3}{16\pi^2} + \dots, \tag{81}$$

and

$$\gamma_i = \hat{\gamma}_i \, \frac{g^2}{16\pi^2} + \dots, \tag{82}$$

for $i = 1, 2, m$. One can then solve (79) and (80),

$$\begin{aligned} m_c(\mu) &= m_c(\mu') \left[\frac{g(\mu)}{g(\mu')}\right]^{-\hat{\gamma}_m/b} = \left[\frac{\alpha_s(\mu')}{\alpha_s(\mu)}\right]^{\hat{\gamma}_m/2b}, \\ c_1(\mu) &= c_1(\mu') \left[\frac{g(\mu)}{g(\mu')}\right]^{-\hat{\gamma}_1/b} = \left[\frac{\alpha_s(\mu')}{\alpha_s(\mu)}\right]^{\hat{\gamma}_1/2b}. \end{aligned} \tag{83}$$

Substituting (83) into (77) and integrating gives

$$\begin{aligned} c_2(m_c) = {} & c_2(m_t) \left[\frac{\alpha_s(m_t)}{\alpha_s(m_c)}\right]^{\hat{\gamma}_2/2b} \\ & + \frac{m_c^2(m_t)\, c_1^2(m_t)}{g(m_t)^2\, (2 + 2\hat{\gamma}_1/b + 2\hat{\gamma}_m/b - \hat{\gamma}_2/b)} \\ & \times \left[\left(\frac{\alpha_s(m_t)}{\alpha_s(m_c)}\right)^{2+2\hat{\gamma}_1/b+2\hat{\gamma}_m/b} - \left(\frac{\alpha_s(m_t)}{\alpha_s(m_c)}\right)^{\hat{\gamma}_2/2b}\right]. \end{aligned} \tag{84}$$

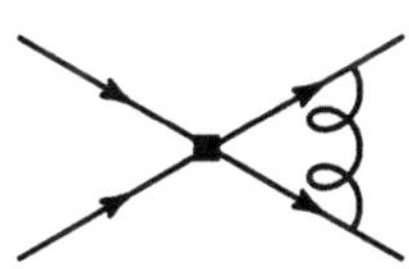

Fig. 15. Graph contributing to the anomalous dimension of the $\Delta S = 2$ operator $(\overline{d}\,\gamma^\mu\, P_L\, s)\,(\overline{d}\,\gamma_\mu\, P_L\, s)$.

The actual computation of these effects in the standard model is more involved, because the $\Delta S = 1$ Lagrangian does not satisfy a simple renormalization group equation of the form (80). There is operator mixing, and (80) is replaced by a matrix equation. Nevertheless, it is possible to compute the results using an effective field theory method, though the final form of the answer is more complicated than (84). The reader is referred to the papers by Gilman and Wise for details. The computation of QCD corrections in the full theory is far more complicated, and has never been done.

To compare the advantages and disadvantages of the full and effective theory computation, let us concentrate only on the m_t part of the $\Delta S = 2$ amplitude. The effective field theory computation gives the $\Delta S = 2$ amplitude as an expansion in powers of m_t/M_W, and we have computed the leading term in (68). The general form of the effective field theory result is

$$\text{answer} = \left(\frac{m_t}{M_W}\right)^2 \left(\frac{\alpha_s(M_W)}{\alpha_s(m_t)}\right)^{\gamma_2/2b} + \left(\frac{m_t}{M_W}\right)^4 \left(\frac{\alpha_s(M_W)}{\alpha_s(m_t)}\right)^{\gamma_4/2b} + \dots \quad (85)$$

where γ_i are the anomalous dimensions of the dimension six, eight, etc. operators. (For example, compare with (84).) Evaluating each of these anomalous dimension is a separate computation. Equation (85) is useful if there is a large ratio of scales, $m_t/M_W \ll 1$, so that one only needs a few terms in the expansion (85). The full theory computation (63) sums up the entire series, and gives an answer of the form

$$\text{answer} = f(m_t/M_W), \quad (86)$$

which is valid for any value of the ratio m_t/M_W. The computations involved in (86) are necessarily more complicated than those for the effective field theory, because one obtains the entire functional form of the answer, rather than the first few terms in a series expansion. However, it is not possible to compute the leading logarithmic QCD corrections to (86), since each term in the expansion has a different anomalous dimension. For the c quark, it is more important to sum the leading QCD corrections, than to include higher order terms in $m_c/M_W \sim 1/50$, and the effective theory method is useful. The recently measured value of the top quark mass indicates that the ratio $m_t/M_W \sim 2$. In this case, it is more important to retain the entire form of the m_t/M_W dependence, than to include the QCD radiative corrections. The way the calculation is done in practice is to integrate out the t-quark and W-boson together at some scale μ which is comparable to both m_t and M_W, and then use an effective theory to scale down to m_c so as to include the QCD corrections between $\{M_W, m_t\}$ and m_c. Clearly, the ideal procedure would be to retain the entire functional form (86), as well as the entire QCD radiative correction. This has been done in a toy model using a non-local effective Lagrangian, but it is not known how to do this in general.

A very different example where an infinite set of anomalous dimensions can be computed is the QCD evolution of parton structure functions. In QCD, the Altarelli-Parisi splitting functions for the parton distribution functions contain the same information as the infinite set of anomalous dimensions of the twist-two operators. The distribution functions can be written as matrix elements of non-local operators, and the one-loop anomalous dimension is a function, whose moments give the anomalous dimensions of the infinite tower of twist two operators.

9 The Non-linear Sigma Model

The previous results discussed effective field theories in the perturbative regime, where one could compute the effective Lagrangian from the full theory in a systematic perturbative expansion. One can also apply effective field theory ideas to situations where one can not derive the effective Lagrangian from the full theory directly. The classic example of this is the use of non-linear sigma models to study spontaneously broken global symmetries, and in particular, the use of chiral Lagrangians to study pion interactions in QCD.

Consider first the linear sigma model with Lagrangian

$$\mathcal{L} = \tfrac{1}{2}\partial_\mu \boldsymbol{\phi} \cdot \partial^\mu \boldsymbol{\phi} - \lambda(\boldsymbol{\phi}\cdot\boldsymbol{\phi} - v^2)^2, \tag{87}$$

where $\boldsymbol{\phi} = (\phi_1, \ldots, \phi_N)$ is a real N-component scalar field. This theory will illustrate some ideas which will be needed for the study of chiral symmetry breaking in QCD. The Lagrangian (87) has a global $O(N)$ symmetry under which $\boldsymbol{\phi}$ transforms as an $O(N)$ vector. The potential has been chosen so that it is minimized for $|\boldsymbol{\phi}| = v$. The set of field configurations where $|\boldsymbol{\phi}| = v$ is known as the vacuum manifold, and in our example, it is the set of points $\boldsymbol{\phi} = (\phi_1, \ldots, \phi_N)$, with $\phi_1^2 + \phi_2^2 + \ldots + \phi_N^2 = v^2$, i.e. it is the $N-1$ dimensional sphere S^{N-1}. The $O(N)$ symmetry can be used to rotate the vector $\langle\boldsymbol{\phi}\rangle$ to a standard direction, which can be chosen to be $(0, 0, \ldots, v)$, the north pole of the sphere. The vacuum of the Lagrangian has spontaneously broken the $O(N)$ symmetry down to the $O(N-1)$ subgroup which acts on the first $N-1$ components. The other generators of $O(N)$ do not leave $(0, 0, \ldots, v)$ invariant. $O(N)$ has $N(N-1)/2$ generators, so the number of Goldstone bosons is equal to the number of broken generators, $N(N-1)/2 - (N-1)(N-2)/2 = N-1$. The $N-1$ Goldstone bosons correspond to rotations of the vector $\boldsymbol{\phi}$, which leave its length unchanged. The potential energy V is unchanged under rotations of $\boldsymbol{\phi}$, so these modes are massless. The remaining mode is a radial excitation which changes the length of $\boldsymbol{\phi}$, and produces a massive excitation, with mass $m_{\mathrm{H}} = \sqrt{8\lambda}\, v$.

It is convenient to switch to "polar coordinates", and define

$$\boldsymbol{\phi} = (\rho + v)\ e^{i\sum_s X^s \cdot \pi^s} \begin{pmatrix} 0 \\ 0 \\ \cdot \\ \cdot \\ \cdot \\ 1 \end{pmatrix}, \tag{88}$$

where $X^s, s = 1, \ldots, N-1$ are $N-1$ broken generators, and π^s and ρ are a new basis for the N fields. This change of variables is only well-defined for small angles π^s. The Lagrangian in terms of the new fields is

$$\mathcal{L} = \frac{1}{2}\partial_\mu \rho \partial^\mu \rho - \lambda\left(\rho^2 + 2\rho v\right)^2 + \frac{1}{2}(\rho + v)^2 \left[\partial_\mu e^{-i\sum_s X^s \cdot \pi^s} \partial^\mu e^{i\sum_s X^s \cdot \pi^s}\right]_{NN}, \tag{89}$$

where $[\,]_{NN}$ is the NN element of the matrix. At energies small compared to the radial excitation mass $\sqrt{8\lambda}\, v$, the ρ field can be neglected, and the Lagrangian reduces to

$$\mathcal{L} = \frac{1}{2}v^2 \left[\partial_\mu e^{-i\sum_s X^s \cdot \pi^s} \partial^\mu e^{i\sum_s X^s \cdot \pi^s}\right]_{NN}, \tag{90}$$

which describes the self-interactions of the Goldstone bosons.

There are some generic features of Goldstone boson interactions that are easy to understand:

1. The Goldstone boson fields are derivatively coupled. The Goldstone bosons describe the local orientation of the ϕ field. A constant Goldstone boson field is a ϕ field that has been rotated by the same angle everywhere in spacetime, and corresponds to a vacuum that is equivalent to the standard vacuum $\langle\phi\rangle = (0, 0, \ldots, 1)$. Thus the Lagrangian must be independent of π^s when π^s is a constant, so only gradients of π^s appear in the Lagrangian.
2. The effective Lagrangian describes a theory of *weakly* interacting Goldstone bosons at low energy. The Goldstone boson couplings are proportional to their momentum, and so vanish for low-momentum Goldstone bosons.
3. The Goldstone boson Lagrangian is non-linear in the Goldstone boson fields. The Goldstone boson Lagrangian describes the dynamics of fields constrained to live on the vacuum manifold. The constraint equation, $\phi_1^2 + \phi_2^2 + \ldots + \phi_N^2 = v^2$, is non-linear, and leads to a non-linear Lagrangian.
4. The vacuum manifold is generically curved (like our sphere S^{N-1}), and does not have a set of global coordinates. The π^s coordinates defined in (88) only make sense for small fluctuations of the Goldstone boson fields about the north pole, which is adequate for perturbation theory. For studying non-perturbative effects or global properties, it is better not to introduce the angular coordinates, but to write the Lagrangian directly in terms of fields that take values on the vacuum manifold, $\pi(x) \in S^{N-1}$.
5. The amplitude for the broken symmetry currents to produce a Goldstone boson from the vacuum is proportional to the symmetry breaking strength v.

10 The CCWZ Formalism

The general formalism for effective Lagrangians for spontaneously broken symmetries was worked out by Callan, Coleman, Wess, and Zumino. Consider a theory in which a global symmetry group G is spontaneously broken to a subgroup H. The vacuum manifold is the coset space G/H. In our example, $G = O(N)$, $H = O(N-1)$, and $G/H = O(N)/O(N-1) = S^{N-1}$.

We would like to choose a set of coordinates which describe the local orientation of the vacuum for small fluctuations about the standard vacuum configuration. Let $\Xi(x) \in G$ be the rotation matrix that transforms the standard vacuum configuration to the local field configuration. The matrix Ξ is not unique: Ξh, where $h \in H$, gives the same field configuration, since the standard vacuum is invariant under H transformations. In our example, one can describe the direction of the vector ϕ by giving the $O(N)$ matrix Ξ, where

$$\phi(x) = \Xi(x) \begin{pmatrix} 0 \\ 0 \\ \cdot \\ \cdot \\ \cdot \\ v \end{pmatrix} . \qquad (91)$$

The same configuration $\boldsymbol{\phi}(x)$ can also be described by $\Xi(x)h(x)$, where $h(x)$ is a matrix of the form

$$h(x) = \begin{pmatrix} h'(x) & 0 \\ 0 & 1 \end{pmatrix}, \tag{92}$$

with $h'(x)$ an arbitrary $O(N-1)$ matrix, since

$$\begin{pmatrix} h'(x) & 0 \\ 0 & 1 \end{pmatrix} \begin{pmatrix} 0 \\ 0 \\ \cdot \\ \cdot \\ \cdot \\ v \end{pmatrix} = \begin{pmatrix} 0 \\ 0 \\ \cdot \\ \cdot \\ \cdot \\ v \end{pmatrix}. \tag{93}$$

The CCWZ prescription is to pick a set of broken generators X, and choose

$$\Xi(x) = e^{iX\cdot\pi(x)}. \tag{94}$$

Consider the $O(N)$ theory for $N=3$, which is the theory of a vector $\boldsymbol{\phi}$ in three-dimensions, and so is easy to visualize. The symmetry group G is the group $G=O(3)$ of rotations in three-space. The standard vacuum configuration $\langle\boldsymbol{\phi}\rangle$ can be chosen to be $\boldsymbol{\phi}$ pointing towards the north pole N, and the unbroken symmetry group $H=O(2)=U(1)$ is rotations about the axis ON, where O is the center of the sphere (see Fig. 16). The group generators are J_1, J_2, J_3, and the unbroken generator is J_3, where J_k generate rotations about the kth axis. The CCWZ prescription is to choose

$$\Xi(x) = e^{i[J_1\pi(x)+J_2\pi_2(x)]} \tag{95}$$

to represent $\boldsymbol{\phi}$ along OA. The matrix Ξ rotates a vector pointing along the ON axis to $\boldsymbol{\phi}=OA$ by rotating along a line of longitude.

Under a global symmetry transformation g, the matrix $\Xi(x)$ is transformed to the new matrix $g\Xi(x)$, since $\boldsymbol{\phi}(x)\to g\boldsymbol{\phi}(x)$. (Note that g is a global transformation, and does not depend on x.) The new matrix $g\Xi(x)$ is no longer in standard form, (94), but can be written as

$$g\,\Xi = \Xi'\,h, \tag{96}$$

since two matrices $g\,\Xi$ and Ξ' which describe the same field configuration differ by an H transformation. That h is non-trivial is a well-known property of rotations in three dimensions. Take an object and rotate it from N to A and then to B. This transformation is not the same as a direct rotation from N to B, but can be written as a rotation about ON, followed by a rotation from N to B. The transformation h in (96) is non-trivial because the Goldstone boson manifold G/H is curved.

The transformation (96) is usually written as

$$\Xi(x) \to g\,\Xi(x)\,h^{-1}(g,\Xi(x)), \tag{97}$$

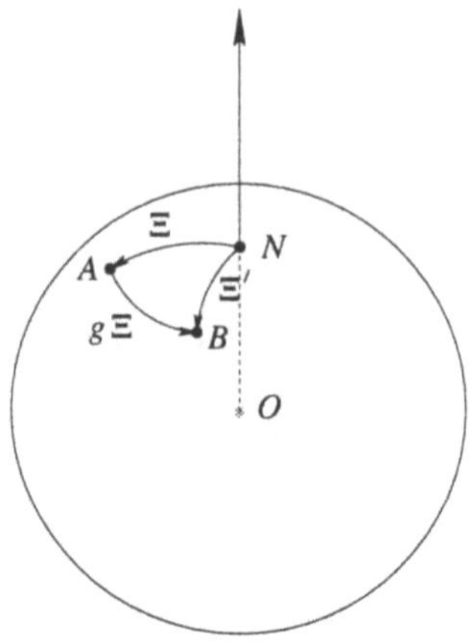

Fig. 16. The vacuum manifold for the $O(3)$ sigma model. The standard configuration $\boldsymbol{\phi}$ is along ON. Under the transformation g, A gets mapped to B.

where we have made clear the implicit dependence of h on x through its dependence on g and $\Xi(x)$. Equations (94) and (97) give the CCWZ choice for the Goldstone boson field, and its transformation law. Any other choice gives the same results for all observables, such as the S-matrix, but does not give the same off-shell Green functions.

11 The QCD Chiral Lagrangian

The CCWZ formalism can now be applied to QCD. In the limit that the u, d and s quark masses are neglected, the QCD Lagrangian has a $SU(3)_{\rm L} \times SU(3)_{\rm R}$ chiral symmetry under which the left- and right-handed quark fields transform independently,

$$\psi_{\rm L}(x) \to L\, \psi_{\rm L}(x), \qquad \psi_{\rm R}(x) \to R\, \psi_{\rm R}(x), \tag{98}$$

where

$$\psi = \begin{pmatrix} u \\ d \\ s \end{pmatrix}. \tag{99}$$

The $SU(3)_{\rm L} \times SU(3)_{\rm R}$ chiral symmetry is spontaneously broken to the vector $SU(3)$ subgroup by the $\langle \overline{\psi}\psi \rangle$ condensate. The symmetry group is $G = SU(3)_{\rm L} \times SU(3)_{\rm R}$, the unbroken group is $H = SU(3)_{\rm V}$, and the Goldstone boson manifold is the coset space $SU(3)_{\rm L} \times SU(3)_{\rm R}/SU(3)_{\rm V}$ which is isomorphic to $SU(3)$. The generators of G are $T^a_{\rm L}$ and $T^a_{\rm R}$ which act on left and right handed quarks respectively, and the generators of H are the flavor generators $T^a = T^a_{\rm L} + T^a_{\rm R}$.

There are two commonly used bases for the QCD chiral Lagrangian, the ξ-basis and the Σ-basis, and we will consider them both. There are many simplifications that occur for QCD because the coset space G/H is isomorphic to a Lie group. This is not true in general; in the $O(N)$ model, the space S^{N-1} is not isomorphic to a Lie group for $N \neq 4$.

The ξ-basis

The unbroken generators of H plus the broken generators X span the space of all symmetry generators of G. One choice of broken generators is to pick $X^a = T^a_{\rm L} - T^a_{\rm R}$. Let the $SU(3)_{\rm L} \times SU(3)_{\rm R}$ transformation be represented in block diagonal form,

$$g = \begin{bmatrix} L & 0 \\ 0 & R \end{bmatrix}, \tag{100}$$

where L and R are the $SU(3)_{\rm L}$ and $SU(3)_{\rm R}$ transformations, respectively. The unbroken transformations have the form (100) with $L = R = U$,

$$h = \begin{bmatrix} U & 0 \\ 0 & U \end{bmatrix}. \tag{101}$$

The Ξ field is then defined using the CCWZ prescription (94)

$$\Xi(x) = e^{iX\cdot\pi(x)} = \exp i \begin{bmatrix} T\cdot\pi & 0 \\ 0 & -T\cdot\pi \end{bmatrix} = \begin{bmatrix} \xi(x) & 0 \\ 0 & \xi^\dagger(x) \end{bmatrix}, \tag{102}$$

where

$$\xi = e^{iT\cdot\pi} \tag{103}$$

denotes the upper block of $\Xi(x)$. The transformation rule (97) gives

$$\begin{bmatrix} \xi(x) & 0 \\ 0 & \xi^\dagger(x) \end{bmatrix} \to \begin{bmatrix} L & 0 \\ 0 & R \end{bmatrix} \begin{bmatrix} \xi(x) & 0 \\ 0 & \xi^\dagger(x) \end{bmatrix} \begin{bmatrix} U^{-1} & 0 \\ 0 & U^{-1} \end{bmatrix}. \tag{104}$$

This gives the transformation law for ξ,

$$\xi(x) \to L\, \xi(x)\, U^{-1}(x) = U(x)\, \xi(x)\, R^\dagger, \tag{105}$$

which defines U in terms of L, R, and ξ.

The Σ basis

The Σ-basis is obtained from the CCWZ prescription using $X^a = T^a_{\rm L}$ for the broken generators. In this case, (94) gives

$$\Xi(x) = e^{iX\cdot\pi(x)} = \exp i \begin{bmatrix} T\cdot\pi & 0 \\ 0 & 0 \end{bmatrix} = \begin{bmatrix} \Sigma(x) & 0 \\ 0 & 1 \end{bmatrix} \tag{106}$$

where

$$\Sigma = e^{iT\cdot\pi} \tag{107}$$

denotes the upper block of $\Xi(x)$. The transformation law (97) is

$$\begin{bmatrix} \Sigma(x) & 0 \\ 0 & 1 \end{bmatrix} \rightarrow \begin{bmatrix} L & 0 \\ 0 & R \end{bmatrix} \begin{bmatrix} \Sigma(x) & 0 \\ 0 & 1 \end{bmatrix} \begin{bmatrix} U^{-1} & 0 \\ 0 & U^{-1} \end{bmatrix}, \tag{108}$$

which gives $U = R$, and

$$\Sigma(x) \rightarrow L\ \Sigma(x)\ R^\dagger. \tag{109}$$

Comparing with (105), one sees that Σ and ξ are related by

$$\Sigma(x) = \xi^2(x). \tag{110}$$

The Lagrangian

The Goldstone boson fields are angular variables, and are dimensionless. When writing down effective Lagrangians in field theory, it is convenient to use fields which have mass dimension one, as for any other spin-zero boson field. The standard choice is to use

$$\xi = e^{iT\cdot\pi/f}, \qquad \Sigma = e^{2iT\cdot\pi/f}, \tag{111}$$

where $f \sim 93$ MeV is the pion decay constant. The π matrix is

$$\boldsymbol{\pi} = \pi^a T^a, \tag{112}$$

where the group generators have the usual normalization $\mathrm{tr}\, T^a T^b = \delta^{ab}/2$,

$$\boldsymbol{\pi} = \frac{1}{\sqrt{2}} \begin{bmatrix} \frac{1}{\sqrt{2}}\pi^0 + \frac{1}{\sqrt{6}}\eta & \pi^+ & \mathrm{K}^+ \\ \pi^- & -\frac{1}{\sqrt{2}}\pi^0 + \frac{1}{\sqrt{6}}\eta & \mathrm{K}^0 \\ \mathrm{K}^- & \overline{\mathrm{K}}^0 & -\frac{2}{\sqrt{6}}\eta \end{bmatrix}. \tag{113}$$

The low energy effective Lagrangian for QCD is the most general possible Lagrangian consistent with spontaneously broken $SU(3) \times SU(3)$ symmetry. Unlike our weak interaction example, one cannot simply compute the effective Lagrangian directly from the original QCD Lagrangian. The connection between the original and effective theories is non-perturbative. The effective Lagrangian has an infinite set of unknown parameters, but we will see that it can still be used to obtain non-trivial predictions for experimentally measured quantities.

It is easy to construct the most general Lagrangian invariant under the transformation $\Sigma \rightarrow L\Sigma R^\dagger$. The most general invariant term with no derivatives must be the product of terms of the form $\mathrm{Tr}\, \Sigma\Sigma^\dagger \ldots \Sigma\Sigma^\dagger$, where Σ and $\Sigma^\dagger$'s alternate. However, $\Sigma\Sigma^\dagger = 1$, so all such terms are constant, and independent of the pion fields. This is just our old result that all Goldstone bosons are derivatively coupled. The only invariant term with two derivatives is

$$\mathcal{L}_2 = \frac{f^2}{4}\, \mathrm{Tr}\ \partial_\mu \Sigma \partial^\mu \Sigma^\dagger. \tag{114}$$

Expanding Σ in a power series in the pion field gives

$$\mathcal{L}_2 = \text{Tr}\ \partial_\mu \pi \partial^\mu \pi + \frac{1}{3f^2}\ \text{Tr}\ [\pi, \partial_\mu \pi]^2 + \ldots\ . \tag{115}$$

The coefficient of the two-derivative term in (114) is fixed by requiring that the kinetic term for the pions in (115) has the standard normalization for scalar fields. The Lagrangian (114) only has terms with an even number of pions, since the pion is a pseudoscalar. The Lagrangian (114) determines all the multi-pion scattering amplitudes to order p^2 in terms of a single constant f. For example, the $\pi - \pi$ scattering amplitude is given by the term $\text{Tr}\,[\pi, \partial_\mu \pi]^2 / 3f^2$, etc.

The Chiral Currents

Noether's theorem can be used to compute the $SU(3)_\text{L}$ and $SU(3)_\text{R}$ currents. If a Lagrangian $\mathcal{L}$ is invariant under an infinitesimal global symmetry transformation with parameter ϵ, the current j_μ is given by computing the change of the Lagrangian when one makes the same transformation, with ϵ a function of x,

$$\delta \mathcal{L} = \partial_\mu \epsilon(x)\ j^\mu(x). \tag{116}$$

The infinitesimal form of the $SU(3)_\text{L}$ transformation $\Sigma \to L\Sigma$ is

$$\Sigma \to \Sigma + i\epsilon_\text{L}^a\, T^a \Sigma, \tag{117}$$

where $L = \exp i\epsilon_\text{L}^a T^a \approx 1 + i\epsilon_\text{L}^a T^a + \ldots$. The change in (114) under (117) is

$$\delta \mathcal{L} = \partial_\mu \epsilon_\text{L}^a\ \ \text{Tr}\ T^a \Sigma \partial^\mu \Sigma^\dagger \tag{118}$$

so that the $SU(3)_\text{L}$ currents are

$$j_\text{L}^\mu = \frac{i}{2} f^2\ \text{Tr}\ T^a \Sigma \partial^\mu \Sigma^\dagger. \tag{119}$$

The right handed currents are obtained by applying the parity transformation, $\pi(x) \to -\pi(-x)$ or by making an infinitesimal $SU(3)_\text{R}$ transformation, so that

$$j_\text{R}^\mu = \frac{i}{2} f^2\ \text{Tr}\ T^a \Sigma^\dagger \partial^\mu \Sigma. \tag{120}$$

The axial current has the expansion

$$j_\text{A}^{\mu a} = j_\text{R}^{\mu a} - j_\text{L}^{\mu a} = -f \partial^\mu \pi^a + \ldots \tag{121}$$

The matrix element $\langle 0 | j_\text{A}^{\mu a} | \pi^b \rangle = i f p^\mu \delta^{ab}$, so that f is the pion decay constant. The experimental value of the π decay rate, $\pi \to \mu \overline{\nu}$ determines $f \approx 93$ MeV.

The low-energy effective theory of the weak interactions is an expansion in some low mass scale (such as m_c or Λ_QCD) over M_W. The QCD chiral Lagrangian is an expansion in derivatives, and so is an expansion in p/Λ_χ. The pion couplings are weak, as long as the pion momentum is small compared with Λ_χ. There are two important questions that have to be answered before one can use the effective Lagrangian: (i) What terms in the effective Lagrangian are required to compute to a given order in p/Λ_χ? (ii) What is the value of Λ_χ? Then one has an estimate of the neglected higher-order terms in the expansion, and the energy at which the effective theory breaks down.

It is useful to eliminate all redundant terms in the effective Lagrangian. One can often eliminate many terms in the effective Lagrangian by making suitable field redefinitions. Field redefinitions are not very useful in renormalizable field theories, because they make renormalizable Lagrangians look superficially non-renormalizable. For example, a field redefinition

$$\phi \to \phi + \epsilon\phi^2, \tag{122}$$

turns

$$\mathcal{L} = \frac{1}{2}\partial_\mu\phi\,\partial^\mu\phi - \frac{1}{2}m^2\phi^2 - \lambda\phi^4 \tag{123}$$

into

$$\mathcal{L} = \frac{1}{2}\partial_\mu\phi\,\partial^\mu\phi - \frac{1}{2}m^2\phi^2 - \lambda\phi^4 + \epsilon\left(2\phi\,\partial_\mu\phi\,\partial^\mu\phi - m^2\phi^3 - 4\lambda\phi^5\right) + \mathcal{O}\left(\epsilon^2\right), \tag{124}$$

which looks superficially like a non-renormalizable interaction. Equations (124) and (123) define identical theories, and the field redefinition (122) has turned a simple Lagrangian into a more complicated one. However, in the case of non-renormalizable theories which contain an infinite number of terms, one can use field redefinitions to eliminate many higher-order terms in the effective Lagrangian (see Ref. 7). The way this is usually done in practice is to use the equations of motion derived from the lowest-order terms in the effective Lagrangian to simplify or eliminate higher order terms.

Weinberg's Power Counting Argument

The QCD chiral Lagrangian is

$$\mathcal{L} = \sum_k \mathcal{L}_k, \tag{125}$$

where $\mathcal{L}_2$, $\mathcal{L}_4$, etc. are the terms in the Lagrangian with two derivatives, four derivatives, and so on. Consider an arbitrary loop graph, such as the one in Fig. 17. It contains m_2 interaction vertices that come from terms in $\mathcal{L}_2$, m_4 interaction vertices from terms in $\mathcal{L}_4$, etc. The general form of the diagram is

$$\mathcal{A} = \int \left(d^4p\right)^L \frac{1}{\left(p^2\right)^I} \prod_k \left(p^k\right)^{m_k}, \tag{126}$$

where L is the number of loops, I is the number of internal lines, and p represents a generic momentum. The factors are easy to understand: there is a d^4p integral for each loop, each internal boson propagator is $1/p^2$, and each vertex in $\mathcal{L}_k$ gives a factor of p^k. In a mass-independent subtraction scheme, the only dimensional parameters are the momenta p. Thus the amplitude $\mathcal{A}$ must have the form $\mathcal{A} \sim p^D$, where

$$D = 4L - 2I + \sum_k k\, m_k, \tag{127}$$

from (126). For any Feynman graph, one can show that

$$V - I + L = 1, \tag{128}$$

where V is the number of vertices, I is the number of internal lines, and L is the number of loops. Combining (127),(128), and using $V = \sum_k m_k$, one gets

$$D = 2 + 2L + \sum_k (k-2)\, m_k. \tag{129}$$

The chiral Lagrangian starts at order p^2, so $k \geq 2$, and all terms in (129) are non-negative. As a result, only a finite number of terms in the effective Lagrangian are needed to work to a fixed order in p, and the chiral Lagrangian acts like a renormalizable field theory. For example, to compute the scattering amplitudes to order p^4, one needs

$$4 = 2 + 2L + \sum_k (k-2)\, m_k, \tag{130}$$

which has the solutions $L = 0$, $m_4 = 1$, $m_{k>4} = 0$, or $L = 1$ and $m_{k>2} = 0$. That is, one only needs to consider tree level diagrams with one insertion of $\mathcal{L}_4$, or one-loop graphs with the lowest order Lagrangian $\mathcal{L}_2$ to compute all scattering amplitudes to order p^4.

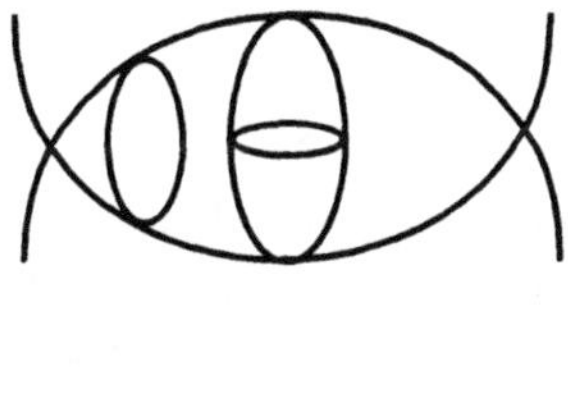

Fig. 17. A loop graph for multipion interactions.

Naive Dimensional Analysis

Consider the $\pi - \pi$ scattering amplitude to order p^4. The power counting argument implies that there are two contributions to this: a tree level graph with one insertion of $\mathcal{L}_4$, and a loop graph using $\mathcal{L}_2$. The loop graph is of order

$$I \sim \int \frac{d^4k}{(2\pi)^4} \frac{k^2}{f^2} \frac{k^2}{f^2} \frac{1}{k^4}, \tag{131}$$

where $1/k^4$ is from the two internal propagators, and each four-pion interaction vertex is of order k^2/f^2, from (115). Here k denotes a generic internal momentum in the Feynman diagram. Estimating this integral gives

$$I \sim \frac{p^4}{16\pi^2} \frac{1}{f^4} \log \mu, \tag{132}$$

where μ is the $\overline{MS}$ renormalization scale, and p represents a generic external momentum. A four derivative operator in the Lagrangian of the form

$$a \, \mathrm{tr}\, \partial_\mu \Sigma \partial^\mu \Sigma^\dagger \partial_\nu \Sigma \partial^\nu \Sigma^\dagger, \tag{133}$$

produces a four-pion interaction of order ap^4/f^4 when one expands the Σ field in a power series in π/f. The total four-pion amplitude, which is the sum of the tree and loop graphs, is μ-independent. A shift in the renormalization scale μ is compensated for by a corresponding shift in a. A change in μ of order one produces a shift in a of order $\delta a \sim 1/16\pi^2$. Generically, a must be at least as big as δa,

$$|a| \gtrsim |\delta a| \sim \frac{1}{16\pi^2}, \tag{134}$$

because a shift in the renormalization point of order one produces a shift in a of this size. Write the effective Lagrangian as

$$\mathcal{L} = \frac{f^2}{4} \left[\mathrm{tr}\, \partial_\mu \Sigma \partial^\mu \Sigma^\dagger + \frac{1}{\Lambda_\chi^2} \mathcal{L}_4 + \frac{1}{\Lambda_\chi^4} \mathcal{L}_6 + \ldots \right] \tag{135}$$

where $1/\Lambda_\chi$ is the expansion parameter of the effective Lagrangian, i.e. (135) gives an expansion for scattering amplitudes in powers of p/Λ_χ. The estimate (134) for the size of the four derivative term implies that

$$\Lambda_\chi \lesssim 4\pi f. \tag{136}$$

One can show that a similar estimate holds for all the higher order terms in $\mathcal{L}$, i.e. the six derivative term has a coefficient of order $1/\Lambda_\chi^4$, etc. Numerous calculations suggest that in QCD, the inequality (136) can be replaced by the estimate

$$\Lambda_\chi \sim 4\pi f \sim 1 \text{ GeV}, \tag{137}$$

for the expansion parameter of the effective Lagrangian. This parameter is large enough that one can apply chiral Lagrangians to low energy processes involving pions and kaons. If the expansion parameter were f instead of $4\pi f$, chiral Lagrangians would not be useful even for pions, since $m_\pi > f$.

The naive dimensional analysis estimate equivalent to (135) is that a term in the Lagrangian has the form

$$f^2 \Lambda_\chi^2 \left(\frac{\pi}{f}\right)^n \left(\frac{\partial}{\Lambda_\chi}\right)^m, \tag{138}$$

as can be seen by expanding (135) in the pion fields. For example, the kinetic term $\mathrm{Tr}\, \partial_\mu \Sigma \, \partial^\mu \Sigma^\dagger$ has a coefficient of order

$$f^2 \Lambda_\chi^2 \left(\frac{\partial}{\Lambda_\chi}\right)^2 \sim f^2, \tag{139}$$

the four derivative term $\mathrm{Tr}\, \partial_\mu \Sigma\, \partial^\mu \Sigma^\dagger \partial_\nu \Sigma\, \partial^\nu \Sigma^\dagger$ has a coefficient of order

$$f^2 \Lambda_\chi^2 \left(\frac{\partial}{\Lambda_\chi}\right)^4 \sim \frac{f^2}{\Lambda_\chi^2} \sim \frac{1}{16\pi^2}, \quad \text{etc.}, \tag{140}$$

which agrees with the earlier estimates.

12 Explicit Symmetry Breaking

The light quark masses explicitly break the chiral $SU(3)_\mathrm{L} \times SU(3)_\mathrm{R}$ symmetry of the QCD Lagrangian. The quark mass term in the QCD Lagrangian is

$$\mathcal{L}_m = -\overline{\psi}_\mathrm{L} M \psi_\mathrm{R} + h.c., \tag{141}$$

where

$$M = \begin{bmatrix} m_\mathrm{u} & 0 & 0 \\ 0 & m_\mathrm{d} & 0 \\ 0 & 0 & m_\mathrm{s} \end{bmatrix}, \tag{142}$$

is the quark mass matrix. The mass term $\mathcal{L}_m$ can be treated as chirally invariant if M is an external field that transforms as

$$M \to L M R^\dagger \tag{143}$$

under chiral $SU(3)_\mathrm{L} \times SU(3)_\mathrm{R}$. The symmetry breaking terms in the chiral Lagrangian are terms that are invariant when M has the transformation rule (143). The symmetry is then explicitly broken when M is fixed to have the value (142). The lowest order term in the effective Lagrangian to first order in M is

$$\mathcal{L}_m = \mu \frac{f^2}{2} \operatorname{tr}\left(\Sigma^\dagger M + M^\dagger \Sigma\right), \tag{144}$$

which breaks the degeneracy of the vacuum and picks out a particular orientation for Σ. All vacua $\Sigma =$ constant are no longer degenerate, and $\Sigma = 1$ is the lowest energy state. Expanding in small fluctuations about $\Sigma = 1$ gives

$$\mathcal{L}_m = -2\mu \operatorname{tr} M \pi^2. \tag{145}$$

Substituting (142) and (113) for M and π and evaluating the trace gives

$$\begin{aligned} M_{\pi^\pm}^2 &= \mu\left(m_\mathrm{u} + m_\mathrm{d}\right) + \Delta M^2, \\ M_{\mathrm{K}^\pm}^2 &= \mu\left(m_\mathrm{u} + m_\mathrm{s}\right) + \Delta M^2, \\ M_{\mathrm{K}^0,\overline{\mathrm{K}}^0}^2 &= \mu\left(m_\mathrm{d} + m_\mathrm{s}\right), \end{aligned} \tag{146}$$

and the π^0, η mass matrix

$$\mu \begin{bmatrix} m_\mathrm{u} + m_\mathrm{d} & \frac{m_\mathrm{u}-m_\mathrm{d}}{\sqrt{3}} \\ \frac{m_\mathrm{u}-m_\mathrm{d}}{\sqrt{3}} & \frac{1}{3}\left(m_\mathrm{u} + m_\mathrm{d} + 4m_\mathrm{s}\right) \end{bmatrix} . \tag{147}$$

To first order in the isospin-breaking parameter $m_\mathrm{u} - m_\mathrm{d}$, the matrix (147) has eigenvalues

$$\begin{aligned} M_{\pi^0}^2 &= \mu\left(m_\mathrm{u} + m_\mathrm{d}\right), \\ M_\eta^2 &= \frac{\mu}{3}\left(m_\mathrm{u} + m_\mathrm{d} + 4m_\mathrm{s}\right). \end{aligned} \tag{148}$$

There is an isospin breaking electromagnetic contribution to the charged Goldstone boson masses ΔM^2 (included in (146)), which is comparable in size to the isospin breaking from $m_\mathrm{u} - m_\mathrm{d}$. To lowest order in $SU(3)$ breaking, ΔM^2 is equal for $\pi^\pm$ and $\mathrm{K}^\pm$, and vanishes for the neutral mesons. The absolute values of the quark masses can not be determined from the meson masses, because they always occur in the combination μm, and μ is an unknown parameter. However, the meson masses can be used to obtain quark mass ratios. From (146)–(147) one gets

$$\frac{m_\mathrm{u}}{m_\mathrm{d}} = \frac{M_{\mathrm{K}^+}^2 - M_{\mathrm{K}^0}^2 + 2M_{\pi^0}^2 - M_{\pi^+}^2}{M_{\mathrm{K}^0}^2 - M_{\mathrm{K}^+}^2 + M_{\pi^+}^2}, \tag{149}$$

$$\frac{m_\mathrm{s}}{m_\mathrm{d}} = \frac{M_{\mathrm{K}^0}^2 + M_{\mathrm{K}^+}^2 - M_{\pi^+}^2}{M_{\mathrm{K}^0}^2 - M_{\mathrm{K}^+}^2 + M_{\pi^+}^2}, \tag{150}$$

and the Gell-Mann–Okubo formula

$$4M_{\mathrm{K}^0}^2 = 3M_\eta^2 + M_\pi^2. \tag{151}$$

Substituting the measured meson masses gives the lowest order values

$$\frac{m_\mathrm{u}}{m_\mathrm{d}} = 0.55, \qquad \frac{m_\mathrm{s}}{m_\mathrm{d}} = 20.1. \tag{152}$$

and $0.99\ \mathrm{GeV}^2 = 0.92\ \mathrm{GeV}^2$ for the Gell-Mann–Okubo formula.

There is an ambiguity in extracting the light quark masses at second order in M. The matrices M and $(\det M)M^{\dagger -1}$ both have the same $SU(3)_\mathrm{L} \times SU(3)_\mathrm{R}$ transformation properties, and are indistinguishable in the chiral Lagrangian. One has an ambiguity of the form

$$M \to M + \lambda\left(\det M\right) M^{\dagger -1} \tag{153}$$

in the quark mass matrix at second order in M. This transformation can be written explicitly as

$$\begin{bmatrix} m_\mathrm{u} & 0 & 0 \\ 0 & m_\mathrm{d} & 0 \\ 0 & 0 & m_\mathrm{s} \end{bmatrix} \to \begin{bmatrix} m_\mathrm{u} + \lambda m_\mathrm{d} m_\mathrm{s} & 0 & 0 \\ 0 & m_\mathrm{d} + \lambda m_\mathrm{u} m_\mathrm{s} & 0 \\ 0 & 0 & m_\mathrm{s} + \lambda m_\mathrm{u} m_\mathrm{d} \end{bmatrix} . \tag{154}$$

One cannot determine the light quark mass ratios to second order using chiral perturbation theory alone, because of the ambiguity (153). This ambiguity can be numerically significant for the ratio $m_\mathrm{u}/m_\mathrm{d}$, since it produces an effective u-quark mass of order $m_d m_s/\Lambda_\chi \sim m_d m_K^2/\Lambda_\chi^2 \sim 0.3 m_d$. The value of m_u is very important, because $m_u = 0$ solves the strong CP problem. The second order term $(\det M)M^{\dagger -1}$ produces an effective u-quark mass that is indistinguishable from m_u in the chiral Lagrangian. Various estimates of the ratio m_u/m_d from different processes (e.g. meson masses, baryon masses $\eta \to 3\pi$) all tend to give the ratio $m_u/m_d \sim 0.56$, so m_u can only be zero if second-order effects in M were of the same size in different processes. One way this could occur is if instanton effects at the scale $\mu \sim 1$ GeV were important. An instanton produces an effective operator of the form $(\det M)M^{\dagger -1}$. If instantons at $\mu \sim 1$ GeV are important, they would lead to an effective mass matrix in the chiral Lagrangian of the form $M_{\rm eff} = M + \lambda \det M M^{-1}$, which would be the same for all processes. This produces $(m_u/m_d)_{\rm eff} \sim 0.56$ in all processes, while still having $m_u = 0$. The only way to distinguish m_u from $(m_u)_{\rm eff}$ is to do a reliable computation that relates the QCD Lagrangian directly to the chiral Lagrangian.

An on-shell particle has $p^2 = M^2$. Since the meson mass-squared is proportional to the quark mass M, the quark matrix M counts as two powers of p for chiral power counting, i.e. terms in $\mathcal{L}_2$ contain two powers of p or one power of M, terms in $\mathcal{L}_4$ contain four powers of p, two powers of p and one power of M, or two powers of M, etc. One can then show that the power counting arguments derived earlier still hold for the effective Lagrangian, including symmetry breaking.

13 π-π Scattering

We now have all the pieces necessary to compute the π-π scattering amplitude near threshold. The full chiral Lagrangian to order p^2 is

$$\mathcal{L}_2 = \frac{f^2}{4} \operatorname{tr} \partial_\mu \Sigma \partial^\mu \Sigma^\dagger + \mu \frac{f^2}{2} \operatorname{tr} \left(\Sigma^\dagger M + M^\dagger \Sigma\right). \tag{155}$$

Expanding this to fourth order in the pion fields gives

$$\mathcal{L}_2 = \frac{1}{3f^2} \operatorname{Tr} \left[\pi, \partial_\mu \pi\right]^2 + \frac{2}{3} \mu \operatorname{Tr} M \pi^4. \tag{156}$$

The π-π scattering amplitude has two contributions, one from the kinetic term and the other from the mass term. Adding the two contributions reproduces the result of Weinberg. The details are left as a homework problem.

The π-π scattering amplitude at order p^4 has two contributions, the one loop diagram Fig. 18(a) involving only the lowest order Lagrangian, and a tree graph Fig. 18(b) with terms from $\mathcal{L}_4$. The answer has the form

$$\frac{A}{16\pi^2}\, p^4 \, \log p^2/\mu^2 + L(\mu)\, p^4, \tag{157}$$

where the first term is from the loop diagram, and the second term is the tree graph contribution from $\mathcal{L}_4$. The coefficient A of the loop graph is completely determined, since there are no unknown parameters in $\mathcal{L}_2$. The loop graph must have a logarithmic term, the so-called chiral logarithm. When $s > 4m_\pi^2$, the π-π scattering amplitude must have an imaginary part from the physical $\pi\pi$ intermediate state, by unitarity. The imaginary part is generated by the chiral logarithm. When $s > 4m_\pi^2$, the argument of the logarithm changes sign, and one gets an imaginary part since $\log(-|r|) = \log|r| + i\pi$. The imaginary part is completely determined by the tree level graph of order p^2, so that the chiral logarithm has a known coefficient. The tree level terms in $\mathcal{L}_4$ are known as low energy constants or counterterms.[7] The total scattering amplitude is μ independent, so the counterterms satisfy the renormalization group equation

$$\mu \frac{d}{d\mu} L(\mu) = \frac{A}{8\pi^2}. \tag{158}$$

The naive dimensional analysis argument discussed earlier is the statement that the counterterm $L(\mu)$ is typically at least as big as the anomalous dimension $A/8\pi^2$.

Fig. 18. Diagrams contributing to $\pi - \pi$ scattering to order p^4. The solid dot represents interaction vertices from $\mathcal{L}_2$, and the solid square represents interactions from $\mathcal{L}_4$.

A generic chiral perturbation theory amplitude has the form (157). There is a chiral logarithm and some counterterms. If one works in a systematic expansion in powers of p, the chiral logarithm is determined completely in terms of lower order terms in the Lagrangian. The counterterms involve additional unknown parameters. There are three main approaches used in the literature to extract useful information from (157):

1. One can hope that the chiral logarithm is numerically more important than the counterterm, when one picks a reasonable renormalization point such as $\mu \sim 1$ GeV. This is formally correct, since $p^4 \log p^2/\mu^2 \gg p^4$ in the limit $p \to 0$. However, in practical examples, p^2 is of order m_π^2 or m_K^2, and the logarithm is -3.9 and -1.4, which is not very large (especially for the K). Nevertheless, the chiral logarithm provides useful information. For example, the correction to f_K/f_π has the form

[7] There are two parts to $\mathcal{L}_4$. There is the infinite part of the coefficient (of order $1/\epsilon$ in dimensional regularization), which is used to cancel divergences from loops using vertices in $\mathcal{L}_2$, and the finite part which affects measurable quantities. The infinite parts are usually never discussed explicitly, and the finite parts are called counterterms.

$$\frac{f_K}{f_\pi} = 1 - \frac{3M_K^2}{64\pi^2 f^2} \log M_K^2/\mu^2 + L(\mu). \tag{159}$$

Setting $\mu \sim \Lambda_\chi$, and neglecting the counterterm gives $f_K/f_\pi = 1.19$, compared with the experimental value of 1.2. The chiral logarithm contribution alone gives a reasonable estimate of the size of the correction (this is just naive dimensional analysis at work), but it also gets the sign correct. The chiral logarithms are also useful in comparing numerical QCD calculations in the quenched approximation, which do not have the full chiral logarithms, with experimental data.

2. The systematic approach which has been used by Gasser and Leutwyler is to write down the most general Lagrangian to order p^4, which contains eight counterterms. This is used to compute $N > 8$ different processes, so that all the counterterms are determined, and one has non-trivial predictions for the remaining $N-8$ amplitudes. This procedure has been more-or-less completed in the meson sector to order p^4, and the results are in good agreement with experiment. At order p^6, there are over 100 terms in the Lagrangian.
3. The third method is to find a process for which there is no counterterm. Typically, this occurs for electromagnetic processes involving neutral particles, such as $K_S^0 \to \gamma\gamma$. Since there is no counterterm, the loop graph must be finite, but it can be non-zero. For example, the leading contribution to $K_S^0 \to \gamma\gamma$ is from the loop graph Fig. 19, and gives an amplitude of order p^2. There are no counterterms for this process at this order. The amplitude at order p^2 is in good agreement with the experimental branching ratio for this process.

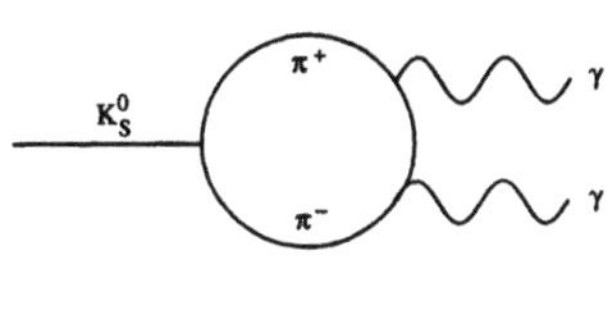

Fig. 19. Leading contribution to $K_S^0 \to \gamma\gamma$.

14 Chiral Perturbation Theory for Matter Fields

Chiral perturbation theory can also be applied to the interactions of the Goldstone bosons with all other particles, which are generically referred to as matter fields. The matter fields (baryon, heavy mesons, etc.) transform as irreducible representations of $SU(3)_V$, but do not form representations of chiral $SU(3)_L \times SU(3)_R$. To discuss the interactions of matter fields, it is more convenient to use the ξ-basis of Sect. 11. We will consider the interactions of the pions with the spin-1/2 baryon octet. The generalization to other matter fields will be obvious. The CCWZ prescription for matter fields such as the baryon is that under a $SU(3)_{\rm L} \times SU(3)_{\rm R}$ transformation, the transformation law is

$$B \to UBU^\dagger, \tag{160}$$

where U is implicitly defined in terms of L and R in (105), and the octet of baryon fields is

$$\begin{bmatrix} \frac{1}{\sqrt{2}}\Sigma^0 + \frac{1}{\sqrt{6}}\Lambda & \Sigma^+ & p \\ \Sigma^- & -\frac{1}{\sqrt{2}}\Sigma^0 + \frac{1}{\sqrt{6}}\Lambda & n \\ \Xi^- & \Xi^0 & -\frac{2}{\sqrt{6}}\Lambda \end{bmatrix}. \tag{161}$$

Under a $SU(3)_{\rm V}$ transformation, $L = R = U$, the baryon transforms as an $SU(3)_{\rm V}$ adjoint. Any transformation that reduces to the adjoint transformation law for $SU(3)_{\rm V}$ transformations is acceptable. For example, one can choose

$$B \to LBL^\dagger, \qquad B \to UBR^\dagger, \qquad etc. \tag{162}$$

The different choices are all equivalent, and correspond to redefining the baryon field. For example, if B has the transformation law (160), then $\xi B \xi^\dagger$ and $B\xi$ transform as $B \to LBL^\dagger$ and $B \to UBR^\dagger$ respectively.

The baryon chiral Lagrangian is the most general invariant Lagrangian written in terms of B and ξ. In writing the Lagrangian, it is convenient to introduce the definitions

$$\begin{aligned} A^\mu &= \tfrac{i}{2}\left(\xi\partial^\mu\xi^\dagger - \xi^\dagger\partial^\mu\xi\right) = \frac{\partial^\mu\pi}{f} + \ldots, \\ V^\mu &= \tfrac{1}{2}\left(\xi\partial^\mu\xi^\dagger + \xi^\dagger\partial^\mu\xi\right) = \frac{1}{2f^2}\left[\pi, \partial^\mu\pi\right] + \ldots, \end{aligned} \tag{163}$$

which transform as

$$A^\mu \to UA^\mu U^\dagger, \tag{164}$$

and

$$V^\mu \to UV^\mu U^\dagger - \partial^\mu U U^\dagger, \tag{165}$$

under $SU(3)_{\rm L} \times SU(3)_{\rm R}$. The covariant derivative on baryons is defined by

$$D^\mu B = \partial^\mu B + \left[V^\mu, B\right], \tag{166}$$

which transforms as

$$D^\mu B \to U\, D^\mu B\, U^\dagger. \tag{167}$$

The most general baryon Lagrangian to order p is

$$\mathcal{L} = \mathrm{tr}\,\overline{B}\left(i\not{D} - m_{\rm B}\right)B + D\,\mathrm{tr}\,\overline{B}\gamma^\mu\gamma_5\left\{A_\mu, B\right\} + F\,\mathrm{tr}\,\overline{B}\gamma^\mu\gamma_5\left[A_\mu, B\right] + \mathcal{L}_\xi, \tag{168}$$

where $\mathcal{L}_\xi$ is the purely meson Lagrangian with $\Sigma = \xi^2$, $m_{\rm B}$ is the baryon mass, $\not{D}$ is the covariant derivative (166) and F and D are the usual axial vector coupling constants, with $g_{\rm A} = F + D$.

The presence of the dimensionful parameter m_B in the Lagrangian ruins the power counting arguments necessary for a sensible effective field theory. Loop graphs in baryon chiral perturbation theory will produce corrections of order $m_B/\Lambda_\chi \sim 1$, so the entire chiral expansion breaks down. There is an alternative formulation of baryon chiral perturbation theory that avoids this problem. The idea is to expand the Lagrangian about nearly on-shell baryons, so that one has a Lagrangian that can be expanded in powers of $1/m_B$, and has no term of order m_B. The method used is similar to that used for heavy quark fields in HQET. Instead of using the Dirac baryon field B, one uses a velocity-dependent baryon field B_v, which is related to the original baryon field B by

$$B_v(x) = \frac{1+\not v}{2} B(x)\, e^{i m_B v\cdot x}, \tag{169}$$

where v is the velocity of the baryon. In the baryon rest frame, $v=(1,0,0,0)$, and

$$B_v(x) = \frac{1+\gamma^0}{2} B(x)\, e^{i m_B t}, \tag{170}$$

which corresponds to keeping only the particle part of the spinor, and subtracting the baryon mass m_B from all energies. In terms of the field B_v, the chiral Lagrangian is

$$\mathcal{L}_v = \operatorname{tr} \overline{B}\,(iv\cdot D)\,B + D \operatorname{tr} \overline{B}\gamma^\mu\gamma_5 \{A_\mu, B\} + F \operatorname{tr} \overline{B}\gamma^\mu\gamma_5 [A_\mu, B] + \mathcal{O}\left(\frac{1}{m_B}\right) + \mathcal{L}_\xi . \tag{171}$$

The baryon mass term is no longer present, and the baryon Lagrangian now has an expansion in powers of $1/m_B$. Note that the baryon chiral Lagrangian starts at order p, whereas the meson Lagrangian starts at order p^2.

A similar procedure can be applied to other matter fields, not just to baryons, provided one can factor the common mass (such as m_B) out of the Feynman graphs. For baryon chiral perturbation theory, this is possible because baryon number is conserved, so one can remove a common mass m_B from all baryons. A similar method also works for hadrons containing a heavy quark, such as the B and B* mesons, because the b-quark number is conserved by the strong interaction, and the B and B* are degenerate in the heavy quark limit. It cannot be used for processes such as $\rho \to \pi\pi$, because the ρ mass m_ρ turns into the pion energy in the final state.

The velocity-dependent Lagrangian $\mathcal{L}_v$ has no dimensionful coefficients in the numerator. This implies that the power counting arguments of an effective field theory are valid. One has two expansion parameters, $1/m_B$ and $1/\Lambda_\chi$. The power counting rule (129) is now

$$D = 1 + 2L + \sum_k m_k\,(k-2) + \sum_k n_k\,(k-1)\,, \tag{172}$$

where m_k is the number of vertices from the p^k terms in the meson Lagrangian, and n_k is the number of vertices from the p^k terms in the baryon Lagrangian. The proof of this result is similar to (129), and will be omitted. The difference between the meson and baryon terms arises because the meson propagator is $1/k^2$, whereas the baryon propagator is $1/k \cdot v$.

The naive dimensional analysis estimate (138) is now

$$f^2 \Lambda_\chi^2 \left(\frac{\pi}{f}\right)^n \left(\frac{\partial}{\Lambda_\chi}\right)^m \left(\frac{B}{f\sqrt{\Lambda_\chi}}\right)^r . \tag{173}$$

For example, the kinetic term $\overline{B}\,(iv \cdot D)\,B$ has a coefficient of order

$$f^2 \Lambda_\chi^2 \left(\frac{\partial}{\Lambda_\chi}\right)^1 \left(\frac{B}{f\sqrt{\Lambda_\chi}}\right)^2 \sim 1, \tag{174}$$

and the four-baryon term $\overline{B}B\,\overline{B}B$ has a coefficient of order

$$f^2 \Lambda_\chi^2 \left(\frac{B}{f\sqrt{\Lambda_\chi}}\right)^4 \sim \frac{1}{f^2}. \tag{175}$$

Similar power counting arguments hold for all strongly interacting gauge theories. For example, in tests for quark and lepton substructure, one uses the operator

$$\frac{4\pi}{\Lambda_{\rm ELP}^2}\,\overline{q}q\,\overline{q}q, \tag{176}$$

and places limits on $\Lambda_{\rm ELP}$. A quark field has the same power counting rules as a baryon field in baryon chiral perturbation theory. Comparing with (175), we see that

$$\Lambda_{\rm ELP} = \Lambda/\sqrt{4\pi}, \tag{177}$$

where Λ is the scale of the composite interactions defined by analogy with the chiral scale Λ_χ: i.e. scattering amplitudes vary on a momentum scale Λ.

πN Scattering

A simple application of the baryon chiral Lagrangian is the computation of the π–N scattering amplitude at threshold, to order p. From (172), the only graphs which contribute are tree graphs which involve terms from the meson Lagrangian at order p^2, and the baryon Lagrangian at order p. The two diagrams which contribute are shown in Fig. 20. The pion-nucleon vertex in Fig. 20(a) vanishes at threshold, since it is proportional to $\boldsymbol{p}$. The two-π nucleon vertex in Fig. 20(b) is

$$\frac{i}{2f^2}\overline{B}\,[\pi, \partial^\mu \pi]\,v^\mu B. \tag{178}$$

The amplitude can be rewritten using $\pi = \pi^a T_{\rm B}^a$, where $T_{\rm B}^a$ are the flavor matrices in the baryon representation. The pions are in the adjoint representation of flavor, so the flavor matrices acting on pions can be written in terms of the structure constants

$$(T_\pi^c)_{ba} = i f_{abc}. \tag{179}$$

Using the commutation relations $\left[T_{\rm B}^a, T_{\rm B}^b\right] = i f_{abc} T_{\rm B}^c$, and evaluating Fig. 20 using the interaction (178) gives the amplitude

$$\mathcal{A} = -\frac{i}{f^2}\, M_\pi\, T_\pi \cdot T_{\rm B}, \tag{180}$$

where we have used $E = M_\pi$ for the energy of the pion. Changing from the non-relativistic normalization of baryon states to the relativistic normalization (where the states are normalized to $2E$) gives the Weinberg-Tomozawa formula for the pion-nucleon scattering amplitude

$$\mathcal{A} = -\frac{2i}{f^2}\, M_{\rm B} M_\pi\ \left(T_\pi \cdot T_{\rm B}\right). \tag{181}$$

$T_\pi \cdot T_{\rm B} = 1/2\left[(T_\pi + T_{\rm B})^2 - T_\pi^2 - T_{\rm B}^2\right] = 1/2\left[I(I+1) - 2 - 3/4\right]$, so that $T_\pi \cdot T_{\rm B}$ is -1 in the isospin-1/2 channel and 1/2 in the isospin-3/2 channel.

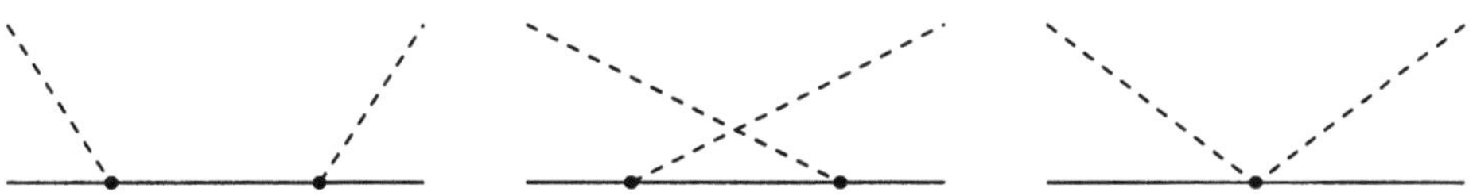

Fig. 20. Contributions to πN scattering at order p.

Non-Analytic Terms

The chiral Lagrangian for matter fields can be used to compute loop corrections. The matter field Lagrangian has an expansion in powers of p, whereas the Goldstone boson Lagrangian had an expansion in powers of p^2. Consequently, loop integrals for matter field interactions can have either even or odd dimension. The even dimensional integrals have the same structure as for the mesons, and lead to non-analytic terms of the form $(M^{2r}/16\pi^2 f^2)\log M^2/\mu^2$, where M is a π, K or η mass, and r is an integer. The μ dependence is cancelled by a corresponding μ dependence in a higher order term in the Lagrangian. The odd dimensional integrals lead to non-analytic terms of the form $(M^{2r+1}/16\pi f^2)$, where r is an integer. These odd-dimensional terms do not have a multiplying logarithm, since the μ dependence cannot be absorbed by a higher dimension operator in the effective Lagrangian. The operator would have to have the form M^{2r+1} which is proportional to $m_{\rm q}^{r+1/2}$, where $m_{\rm q}$ is the quark mass. Such an operator cannot exist in the Lagrangian, since it contains a square-root of the quark mass matrix. Also note that the odd-dimensional integrals have one less power of π in the denominator.

15 Chiral Perturbation Theory for Hadrons Containing a Heavy Quark

Chiral perturbation theory can also be applied to hadrons containing a heavy quark. The hadrons are treated as matter fields, and one writes down the most general possible Lagrangian consistent with the chiral symmetries, as for the spin-1/2 baryons. In addition, one can constrain some of the terms in the Lagrangian using heavy quark symmetry. As a simple example, consider the interaction of the pseudoscalar and vector mesons B and B* (or D and D*) with pions. These mesons can be treated using the velocity dependent formulation, since the b-quark number is conserved by the strong interactions, and the B and B* are degenerate in the heavy quark limit. It is conventional to combine B and B* into a single field H defined by

$$H = \frac{1+\not{v}}{2}\left[B^*_\mu \gamma^\mu - B\gamma_5\right], \tag{182}$$

where B and B^* are column vectors which contain the states $\mathrm{b\overline{u}}$, $\mathrm{b\overline{d}}$, and $\mathrm{b\overline{s}}$. The transformation law for H under heavy quark spin symmetry transformation S_Q and $SU(3)_\mathrm{V}$ flavor symmetry transformation U is

$$H \to S_\mathrm{Q} H U^\dagger. \tag{183}$$

The most general Lagrangian consistent with these symmetries to order p is

$$\mathcal{L} = \mathrm{tr}\,\overline{H}\,(iv\cdot D)\,H + g\,\mathrm{tr}\,\overline{H}H\gamma^\mu\gamma_5 A_\mu, \tag{184}$$

where

$$D^\mu H = \partial^\mu H - HV^\mu. \tag{185}$$

There is only a single coupling constant g which appears to this order, so the $BB^*\pi$ and $B^*B^*\pi$ couplings are related to each other by the heavy quark spin symmetry. The other possible interaction term,

$$\mathrm{tr}\,\overline{H}\gamma^\mu\gamma_5 H A_\mu \tag{186}$$

is forbidden by heavy quark spin symmetry, and is suppressed by one power of $1/m_\mathrm{Q}$. This term splits the $BB^*\pi$ and $B^*B^*\pi$ couplings at order $1/m_\mathrm{Q}$.

The chiral Lagrangian (184) can be used to compute corrections to various quantities for heavy hadrons. For example, one can show that

$$\frac{f_{\mathrm{B_s}}}{f_\mathrm{B}} = 1 - \tfrac{5}{6}\left(1+3g^2\right)\frac{M_\mathrm{K}^2}{16\pi^2 f^2}\log\frac{M_\mathrm{K}^2}{\mu^2}, \tag{187}$$

and

$$\frac{B_{\mathrm{B_s}}}{B_\mathrm{B}} = 1 - \tfrac{2}{3}\left(1-3g^2\right)\frac{M_\mathrm{K}^2}{16\pi^2 f^2}\log\frac{M_\mathrm{K}^2}{\mu^2}, \tag{188}$$

where f_B and B_B are the decay constant for B decay and the bag constant for $\mathrm{B^0}-\overline{\mathrm{B}}^0$ mixing respectively. Further applications can be found in the literature.

Acknowledgments

I would like to thank Gabriela Barenboim, C. Glenn Boyd, Richard F. Lebed, Ira Z. Rothstein, and Oscar Vives for helpful comments on the manuscript. This work was supported in part by a Department of Energy grant DOE-FG03-90ER40546, by a Presidential Young Investigator award from the National Science Foundation, PHY-8958081, and by a "Profesor Visitante IBERDROLA de Ciencia y Tecnología" position at the Departmento de Física Teórica of the University of València.

References

Here are a few references which might be useful for students interested in learning more about the subject.

1. The renormalization group and critical phenomena:
   Ma, S.-K. (1982): *Modern Theory of Critical Phenomena* (Benjamin)
   Wilson, K., Kogut, J. (1974): Phys. Rep. **12**, 75
2. Some references on effective field theories:
   Georgi, H. (1984): *Weak Interactions and Modern Particle Theory* (Benjamin)
   Hall, L.J. (1981): Nucl. Phys. **B178**, 75
   Kaplan, D.B. (1995): e-print `nucl-th/9506035`
   Polchinski, J. (1992): e-print `hep-th/9210046`
   Weinberg, S. (1980): Phys. Lett. **91B**, 51
   Witten, E. (1977): Nucl. Phys. **B122**, 109
3. The sine-Gordon–Thirring duality:
   Coleman, S. (1975): Phys. Rev. **D11**, 2088
4. The Appelquist-Carazzone theorem:
   Appelquist, T., Carazzone, J. (1975): Phys. Rev. **D11**, 2856
5. Weak interactions in the six-quark model:
   Gilman, F.J., Wise, M.B. (1979): Phys. Rev. **D20**, 2392
   Gilman, F.J., Wise, M.B. (1983): Phys. Rev., **D27**, 1128
   Shifman, M., Vainshtein, A., Zakharov, V. (1977): Nucl. Phys. **B120**, 316
   Vainshtein, A., Zakharov, V., Shifman, M. (1975): JETP Lett. **22**, 55
6. Non-local effective actions:
   Bhansali, V., Georgi, H. (1992): e-print `hep-ph/9205242`
7. Simplifying the effective Lagrangian using equations of motion:
   Georgi, H. (1991): Nucl. Phys. **B361**, 339
   Politzer, H.D. (1980): Nucl. Phys. **B172**, 349
8. The CCWZ formalism:
   Callan, C., Coleman, S., Wess, J., Zumino, B. (1969): Phys. Rev. **177**, 2247
   Coleman, S., Wess, J., Zumino, B. (1969): Phys. Rev. **177**, 2239
9. A summary of many of the ideas on chiral perturbation theory:
   Weinberg, S. (1979): Physica **A96**, 327
10. Naive dimensional analysis:

Manohar, A., Georgi, H. (1984): Nucl. Phys **B234**, 189

11. The existence of non-analytic terms in the chiral expansion:
    Li, L.-F., Pagels, H. (1972): Phys. Rev. **D5**, 1509
12. The meson Lagrangian to order p^4:
    Gasser, J., Leutwyler, H. (1984): Ann. Phys. **158**, 142
    Gasser, J., Leutwyler, H. (1985): Nucl. Phys. **B250**, 465
13. Quark masses:
    Gasser, J., Leutwyler, H. (1982): Phys. Rept. **87**, 77
    Kaplan, D.B., Manohar, A.V. (1986): Phys. Rev. Lett. **56**, 2004
    Weinberg, S. (1977): Trans. N.Y. Acad. Sci. **38**, 185
14. Baryon chiral perturbation theory:
    Bijnens, J., Sonoda, H., Wise, M.B. (1985): Nucl. Phys. **B261**, 185
    Gasser, J., Sainio, M.E., Svarc, A. (1988): Nucl. Phys. **B307**, 779
    Jenkins, E., Manohar, A.V. (1991a): Phys. Lett. **B255**, 558
    Jenkins, E., Manohar, A.V. (1991b): Phys. Lett. **B259**, 353
    Langacker, P., Pagels, H. (1974): Phys. Rev. **D10**, 2904
15. A good reference to recent developments in chiral perturbation theory:
    Meißner, U., Ed. (1992): *Effective Field Theories of the Standard Model* (World Scientific, Singapore)
16. Chiral perturbation theory for hadrons containing a heavy quark:
    Burdman, G., Donoghue, J. (1992): Phys. Lett. **B280**, 287
    Wise, M.B. (1992): Phys. Rev. **D45**, 2188
    Yan, T.M. et al. (1992): Phys. Rev. **D46**, 1148
17. Some applications of heavy hadron chiral perturbation theory:
    Grinstein, B. et al. (1992): Nucl. Phys. **B380**, 376

Baryon Structure and the Chiral Symmetry of QCD

Leonid Ya. Glozman

Institut für Theoretische Physik, Universität Graz,
A-8010 Graz, Austria

Abstract. Beyond the spontaneous chiral symmetry breaking scale light and strange baryons should be considered as systems of three constituent quarks with an effective confining interaction and a chiral interaction that is mediated by the octet of Goldstone bosons (pseudoscalar mesons) between the constituent quarks.

1 Introduction

Our aim in physics is not only to calculate some observable and get a correct number but mainly to understand a physical picture responsible for the given phenomenon. It very often happens that a theory formulated in terms of fundamental degrees of freedom cannot answer such a question since it becomes overcomplicated at the related scale. Thus a main task in this case is to select those degrees of freedom which are indeed essential. For instance, the fundamental degrees of freedom in crystals are ions in the lattice, electrons and the electromagnetic field. Nevertheless, in order to understand electric conductivity, heat capacity, etc. we instead work with "heavy electrons" with dynamical mass, phonons and their interaction. In this case a complicated electromagnetic interaction of the electrons with the ions in the lattice is "hidden" in the dynamical mass of the electron and the interactions among ions in the lattice are eventually responsible for the collective excitations of the lattice - phonons, which are Goldstone bosons of the spontaneously broken translational invariance in the lattice of ions. As a result, the theory becomes rather simple - only the electron and phonon degrees of freedom and their interactions are essential for all the properties of crystals mentioned above.

Quite a similar situation takes place in QCD. One hopes that sooner or later one can solve the full nonquenched QCD on the lattice and get the correct nucleon and pion mass in terms of underlying degrees of freedom: current quarks and gluon fields. However, QCD at the scale of 1 GeV becomes too complicated, and hence it is rather difficult to say in this case what kind of physics, inherent in QCD, is relevant to the nucleon mass and its low-energy properties. In this lecture I will try to answer this question. I will show that it is the spontaneous breaking of chiral symmetry which is the most important QCD phenomenon in this case, and that beyond the scale of spontaneous breaking of chiral symmetry light and strange baryons can be viewed as systems of three constituent quarks which interact by the exchange of Goldstone bosons (pseudoscaler mesons) and are subject to confinement.

It is well known that at low temperature and density the approximate $SU(3)_{\mathrm{L}} \times SU(3)_{\mathrm{R}}$ chiral symmetry of QCD is realized in the hidden Nambu-Goldstone mode. The hidden mode of chiral symmetry is revealed by the existence of the octet of pseudoscalar mesons of low mass which represent the associated approximate Goldstone bosons. The η' (the $SU(3)$-singlet) decouples from the original nonet because of the $U(1)_{\mathrm{A}}$ anomaly [1], [2]. Another consequence of the spontaneous breaking of the approximate chiral symmetry of QCD is that the valence quarks acquire their dynamical or constituent mass [3], [4], [5], [6] through their interactions with the collective excitations of the QCD vacuum - the quark-antiquark excitations and the instantons.

We have recently suggested [7], [8] that beyond the scale of spontaneous breaking of chiral symmetry a baryon should be considered as a system of three constituent quarks with an effective quark-quark interaction that is formed by a central confining part and a chiral interaction mediated by the octet of pseudoscalar mesons between the constituent quarks.

The simplest representation of the most important component of the interaction of the constituent quarks mediated by the octet of pseudoscalar bosons in the $SU(3)_{\mathrm{F}}$ invariant limit is

$$H_\chi \sim -\sum_{i<j} V(\mathbf{r}_{ij}) \boldsymbol{\lambda}_i^{\mathrm{F}} \cdot \boldsymbol{\lambda}_j^{\mathrm{F}} \, \boldsymbol{\sigma}_i \cdot \boldsymbol{\sigma}_j. \tag{1}$$

Here $\{\boldsymbol{\lambda}_i^{\mathrm{F}}\}$ represents the flavor $SU(3)$ Gell-Mann matrices and the sums run over the constituent quarks.

Because of the flavor dependent factor $\boldsymbol{\lambda}_i^{\mathrm{F}} \cdot \boldsymbol{\lambda}_j^{\mathrm{F}}$ the chiral boson exchange interaction (1) will lead to orderings of the positive and negative parity states in the baryon spectra, which agree with the observed ones in all sectors. For the spectrum of the nucleon and Δ the strength of the chiral interaction between the constituent quarks is sufficient to shift the lowest positive parity states in the N=2 band (the N(1440) and $\Delta(1600)$) below the negative parity states in the N=1 band (N(1520) – N(1535) and $\Delta(1620) - \Delta(1700)$). In the spectrum of the Λ, on the other hand, it is the negative parity flavor singlet states ($\Lambda(1405) - \Lambda(1520)$) that remain the lowest lying resonances, again in agreement with experiment. The mass splittings between the baryons with different strangeness, and between the Λ and the Σ which have identical flavor, spin and flavor-spin symmetries arise from the explicit breaking of the $SU(3)_{\mathrm{F}}$ symmetry that is caused by the mass splitting of the pseudoscalar meson octet, and the different masses of the u,d and the s quarks.

This lecture has the following structure. Section 2 contains the proof why the commonly used perturbative gluon exchange interaction between the constituent quarks leads to incorrect ordering of positive and negative parity states in the spectra. In Section 3 we outline the importance of the spontaneous breaking of chiral symmetry for low-energy QCD, and in Section 4 we present a short historical sketch of the role of chiral symmetry in quark based models. Section 5 contains a description of the chiral boson exchange interaction (1). In Section 6 we describe the spectra of the nucleon, the Δ resonance and the Λ hyperon as they are predicted by the $SU(3)_{\mathrm{F}}$ symmetric interaction (1). The effect of $SU(3)_{\mathrm{F}}$ breaking in the interaction is considered in Section 7. Section 8 is devoted to an "exact" three-body description of baryons where the interaction (1) is taken into account to all orders. In Section 9 we discuss the role of the exchange current corrections to the baryon magnetic moments that are associated with the pseudoscalar exchange interaction. Finally, in Section 10 some recent lattice - QCD results are discussed.

2 Why the Gluon Exchange Bears No Relation to the Baryon Spectrum

It was accepted by many people (but not by all) that the fine splittings in the baryon spectrum (in analogy to atomic physics they are often called "hyperfine splittings" as they arise from the spin-spin forces) are due to the gluon-exchange interaction between the constituent quarks [9], [10]. Now I shall address myself to a formal consideration why the one gluon exchange interaction cannot be relevant to the baryon spectrum.

The most important component of the one gluon exchange interaction [9] is the so called color-magnetic interaction

$$H_{\mathrm{cm}} \sim -\alpha_{\mathrm{s}} \sum_{i<j} \frac{\pi}{6m_i m_j} \boldsymbol{\lambda}_i^{\mathrm{C}} \cdot \boldsymbol{\lambda}_j^{\mathrm{C}} \boldsymbol{\sigma}_i \cdot \boldsymbol{\sigma}_j \delta(\mathbf{r}_{ij}), \tag{2}$$

where $\{\boldsymbol{\lambda}_i^{\mathrm{C}}\}$ are color $SU(3)$ matrices. It is the permutational color-spin symmetry of the 3q state which is mostly responsible for the contribution of the interaction (2). Indeed, the corresponding two-body matrix element is

$$< [f_{ij}]_{\mathrm{C}} \times [f_{ij}]_{\mathrm{S}} : [f_{ij}]_{\mathrm{CS}} | \boldsymbol{\lambda}_i^{\mathrm{C}} \cdot \boldsymbol{\lambda}_j^{\mathrm{C}} \boldsymbol{\sigma}_i \cdot \boldsymbol{\sigma}_j | [f_{ij}]_{\mathrm{C}} \times [f_{ij}]_{\mathrm{S}} : [f_{ij}]_{\mathrm{CS}} >$$

$$= \left\{ \begin{array}{ll} 8 & [11]_{\mathrm{C}}, [11]_{\mathrm{S}} : [2]_{\mathrm{CS}} \\ -\frac{8}{3} & [11]_{\mathrm{C}}, [2]_{\mathrm{S}} : [11]_{\mathrm{CS}} \end{array} \right. . \tag{3}$$

Thus the symmetrical color-spin pairs (i.e. with a $[2]_{\mathrm{CS}}$ Young pattern) experience an attractive contribution while the antisymmetrical ones ($[11]_{\mathrm{CS}}$) experience a repulsive contribution. Hence the color-magnetic contribution to the Δ state ($[111]_{\mathrm{CS}}$) is more repulsive than to the nucleon ($[21]_{\mathrm{CS}}$) and the Δ becomes heavier than the nucleon. The price is that α_{s} should be larger than unity, which is bad. In addition there is no empirical indication in the spectrum for a large spin-orbit component of the gluon-exchange interaction [9] implied by this big value of α_{s}.

The crucial point is that the interaction (2) feels only colour and spin of the interacting quarks. Thus the structure of the N and Λ spectra has to be the same as these baryons differ only by their flavour structure. If one looks at the Particle Data Group tables, however, one immediately sees a different ordering of the positive and negative parity states in both spectra. In the N spectrum the lowest states are $\frac{1}{2}^+$, N(939); $\frac{1}{2}^+$, N(1440); $\frac{1}{2}^-$, N(1535) – $\frac{3}{2}^-$, N(1520), while in the Λ spectrum the ordering is as follows: $\frac{1}{2}^+$, $\Lambda(1115)$; $\frac{1}{2}^-$, $\Lambda(1405)$ – $\frac{3}{2}^-$, $\Lambda(1520)$; $\frac{1}{2}^+$, $\Lambda(1600)$. A weak flavor dependence via the quark masses in (2) cannot explain this paradox.

The interaction (2) cannot explain why the two-quantum excitations of positive parity N(1440), $\Delta(1600)$, $\Lambda(1600)$ and $\Sigma(1660)$ lie below the one-quantum excitations of negative parity N(1535)–N(1520), $\Delta(1620)-\Delta(1700)$, $\Lambda(1670)-\Lambda(1690)$ and $\Sigma(1750)-\Sigma(?)$, respectively. For instance, the positive parity state N(1440) and the negative parity ones N(1535) – N(1520) have the same mixed ($[21]_{CS}$) color-spin symmetry thus the color-magnetic contribution to these states cannot be very different (a small difference is only due to the different radial structure of the positive and negative parity states). But the N(1440) state belongs to the $N = 2$ shell, while the N(1535) – N(1520) pair is a member of the $N = 1$ band, which means that the N(1440) should lie approximately $\hbar\omega$ above the N(1535)–N(1520). In the Δ spectrum the situation is even more dramatic. The $\Delta(1600)$ positive parity state has a completely antisymmetric CS-Young pattern ($[111]_{CS}$), while the negative parity states $\Delta(1620)-\Delta(1700)$ have a mixed one. Thus the color-magnetic contribution to the $\Delta(1600)$ is much more repulsive than to the $\Delta(1620)-\Delta(1700)$. In addition the $\Delta(1600)$ is the $N = 2$ state, while the pair $\Delta(1620)-\Delta(1700)$ belongs to the $N = 1$ band. As a consequence the $\Delta(1600)$ must lie much higher than the $\Delta(1620)-\Delta(1700)$. All these features are well seen in the explicit 3-body calculations [11].

3 Spontaneous Chiral Symmetry Breaking and Its Consequences for Low-Energy QCD

The QCD Lagrangian with three light flavors has a global symmetry

$$SU(3)_L \times SU(3)_R \times U(1)_V \times U(1)_A, \tag{4}$$

if one neglects the masses of current u,d, and s quarks, which are small compared to a typical low-energy QCD scale of 1 GeV. The $U(1)_A$ is not a symmetry at the quantum level due to the axial anomaly. If the $SU(3)_L \times SU(3)_R$ chiral symmetry of the QCD Lagrangian were intact in the vacuum state we would observe degenerate multiplets in the particle spectrum corresponding to the above chiral group, and all hadrons would have their degenerate partners with opposite parity. Since this does not happen the implication is that the chiral symmetry is spontaneously broken down to $SU(3)_V$ in the QCD vacuum, i.e., realized in the hidden Nambu-Goldstone mode. A direct evidence for the spontaneously broken chiral symmetry is a nonzero value of the quark condensates for the light flavors

$$< \text{vacuum}|\bar{q}q|\text{vacuum} > \approx -(240 - 250\text{MeV})^3, \qquad (5)$$

which represents the order parameter. That this is indeed so, we know from three independent sources: current algebra [12], QCD sum rules [13], and lattice gauge calculations [14]. There are two important generic consequences of the spontaneous chiral symmetry breaking. The first one is an appearance of the octet of pseudoscalar mesons of low mass, π, K, η, which represent the associated approximate Goldstone bosons. The second one is that valence quarks acquire a dynamical or constituent mass. Both these consequences of the spontaneous chiral symmetry breaking are well illustrated by, e.g. the σ-model [15] or the Nambu and Jona-Lasinio model [16]. We cannot say at the moment for sure what the microscopical reason for spontaneous chiral symmetry breaking in the QCD vacuum is. It was suggested that this occurs when quarks propagate through instantons in the QCD vacuum [5], [6].

For the low-energy baryon properties it is only essential that beyond the spontaneous chiral symmetry breaking scale new dynamical degrees of freedom appear - constituent quarks and chiral fields. The low-energy baryon properties are mainly determined by these dynamical degrees of freedom and the confining interaction. This is quite in contrast to pseudoscalar mesons. In the chiral limit, $m_{\mathrm{u}}^0 = m_{\mathrm{d}}^0 = m_{\mathrm{s}}^0 = 0$, all members of the pseudoscalar octet (π, K, η) would have zero mass, which is most clearly seen in the Gell-Mann-Oakes-Renner [12] relations

$$m_{\pi^0}{}^2 = -\frac{1}{f_\pi^2}(m_{\mathrm{u}}^0 < \bar{u}u > + m_{\mathrm{d}}^0 < \bar{d}d >) + O({m_{\mathrm{u,d}}^0}^2),$$

$$m_{\pi^{+,-}}{}^2 = -\frac{1}{f_\pi^2}\frac{m_{\mathrm{u}}^0 + m_{\mathrm{d}}^0}{2}(< \bar{u}u > + < \bar{d}d >) + O({m_{\mathrm{u,d}}^0}^2),$$

$$m_{\mathrm{K}^{+,-}}{}^2 = -\frac{1}{f_\pi^2}\frac{m_{\mathrm{u}}^0 + m_{\mathrm{s}}^0}{2}(< \bar{u}u > + < \bar{s}s >) + O({m_{\mathrm{u,s}}^0}^2),$$

$$m_{\mathrm{K}^0,\overline{\mathrm{K}}^0}{}^2 = -\frac{1}{f_\pi^2}\frac{m_{\mathrm{d}}^0 + m_{\mathrm{s}}^0}{2}(< \bar{d}d > + < \bar{s}s >) + O({m_{\mathrm{d,s}}^0}^2),$$

$$m_\eta{}^2 = -\frac{1}{3f_\pi^2}(m_{\mathrm{u}}^0 < \bar{u}u > + m_{\mathrm{d}}^0 < \bar{d}d > + 4m_{\mathrm{s}}^0 < \bar{s}s >) + O({m_{\mathrm{u,d,s}}^0}^2), \qquad (6)$$

that relate the pseudoscalar meson masses to the quark condensates and current quark masses m^0. Thus the nonzero masses of mesons are determined by the nonzero values of the current quark masses. In the baryon case, even in the chiral limit, baryons would have approximately their actual masses of the order of 1 GeV, as these masses are mostly determined by the dynamical (constituent) masses, the Goldstone boson exchange interaction among them, as well as a confining interaction. The dynamical (constituent) masses are in turn determined mainly by the quark condensates, which is most clearly seen from the gap equations of the Nambu and Jona-Lasinio model, and only weakly dependent on current quark masses. The role of the current quark masses in baryons is to break just the $SU(3)_{\mathrm{F}}$ symmetry in the baryon spectrum.

4 Chiral Symmetry and the Quark Model (Historical Sketch)

The importance of the constraints posed by chiral symmetry for the quark bag [17] and bag-like [18] models for the baryons has been recognized early. In bag models with restored chiral symmetry on the bag surface, or bag-like models, the massless current quarks within the bag were assumed to interact not only by perturbative gluon exchange but also through chiral meson field exchange. In these models the chiral field has the character of a compensating auxiliary field only, rather than a collective low frequency Goldstone quark-antiquark excitation. The possibility of a nonzero quark condensate was also not addressed. As it was discussed in a previous section, it is the quark condensate which is the most important characteristic determining the baryon properties. According to the bag philosophy, however, there is a perturbative QCD phase inside a bag, where all condensates vanish by definition. A general limitation of all bag and bag-like models is, of course, the lack of translational invariance, which is of crucial importance for a description of the excited states.

Common to these models is that the breaking of chiral symmetry arises from the confining interaction. This point of view contrasts with that of Manohar and Georgi [4], who pointed out that there should be two different scales in QCD with 3 flavors. At the first one of these, $\Lambda_{\chi \mathrm{SB}} \simeq 4\pi f_\pi \simeq$ 1 GeV, the spontaneous breaking of the chiral symmetry occurs, and hence at distances beyond $\frac{1}{\Lambda_{\chi \mathrm{SB}}} \simeq 0.2$ fm the valence current quarks acquire their dynamical (constituent) mass (called "chiral quarks" in [4]), and the Goldstone bosons (mesons) appear. The other scale, $\Lambda_{\mathrm{QCD}} \simeq 100 - 300$ MeV, is that which characterizes confinement, and the inverse of this scale roughly coincides with the linear size of a baryon. Between these two scales then the effective Lagrangian should be formed out of the gluon fields that provide a confining mechanism, as well as of the constituent quark and pseudoscalar meson fields. Manohar and Georgi did not, however, specify whether the baryons should be described as bound qqq states or as chiral solitons.

The chiral symmetry breaking scale above fits well with that which appears in the instanton liquid picture of the QCD vacuum [5], [6]. In this model the quark condensates (i.e., equilibrium of virtual quark-antiquark pairs in the vacuum state), as well as the gluon condensate, are supported by instanton fluctuations of a size ~ 0.3 fm. Diakonov and Petrov [6] suggested that at low momenta (i.e., beyond the chiral symmetry breaking scale) QCD should be approximated by an effective chiral Lagrangian of the sigma-model type that contains valence quarks with dynamical (constituent) masses and meson fields. They considered a nucleon as three constituent quarks moving independently of one another in a self-consistent chiral field of the hedgehog form [19]. In this picture the Δ appears as a rotational excitation of the hedgehog, and no explicit confining interaction is included. A very similar description for the nucleon was suggested in [20], [21]. These types of models are now called "quark-soliton models" [22], [23].

The spontaneous breaking of chiral symmetry and its consequences - dynamical quark mass generation, appearance of the quark condensate, and pseudoscalar mesons as Goldstone excitations - are well illustrated by the Nambu and Jona-Lasinio model [16], [24]. This model lacks a confining interaction, which, as argued below, is essential for a realistic description of the properties of baryon physics.

5 The Chiral Boson Exchange Interaction

In an effective chiral description of the baryon structure, based on the constituent quark model, the coupling of the quarks and the pseudoscalar Goldstone bosons will (in the $SU(3)_{\mathrm{F}}$ symmetric approximation) have the form $ig\bar{\psi}\gamma_5\boldsymbol{\lambda}^{\mathrm{F}}\cdot\boldsymbol{\phi}\psi$ (or $g/(2m)\bar{\psi}\gamma_\mu\gamma_5\boldsymbol{\lambda}^{\mathrm{F}}\cdot\psi\partial^\mu\boldsymbol{\phi}$), where ψ is the fermion constituent quark field operator, $\boldsymbol{\phi}$ the octet boson field operator, and g is a coupling constant. A coupling of this form, in a nonrelativistic reduction for the constituent quark spinors, will – to lowest order – give rise to a Yukawa interaction between the constituent quarks, the spin-spin component of which has the form

$$V_{\mathrm{Y}}(r_{ij}) = \frac{g^2}{4\pi}\frac{1}{3}\frac{1}{4m_im_j}\boldsymbol{\sigma}_i\cdot\boldsymbol{\sigma}_j\boldsymbol{\lambda}_i^{\mathrm{F}}\cdot\boldsymbol{\lambda}_j^{\mathrm{F}}\{\mu^2\frac{e^{-\mu r_{ij}}}{r_{ij}} - 4\pi\delta(\mathbf{r}_{ij})\}. \qquad (7)$$

Here m_i and m_j denote the masses of the interacting quarks, and μ that of the meson. There will also be an associated tensor component, which is discussed in Ref. [8].

At short range the simple form (7) of the chiral boson exchange interaction cannot be expected to be realistic and should only be taken to be suggestive. Because of the finite spatial extent of both the constituent quarks and the pseudoscalar mesons the delta function in (7) should be replaced by a finite function, with a range of 0.6-0.7 fm, as suggested by the spatial extent of the mesons. In addition, the radial behaviour of the Yukawa potential (7) is valid only if the boson field satisfies a linear Klein-Gordon equation. The implications of the underlying chiral symmetry of QCD for the effective chiral Lagrangian (which in fact is not known), which contains constituent quarks as well as boson fields, are that these boson fields cannot be described by linear equations near their source. Therefore it is only at large distances, where the amplitude of the boson fields is small, that the quark-quark interaction reduces to the simple Yukawa form.

The latter point is rather important and has to be clarified. The radial dependence in (7) is a direct consequence of a free $(q^2 - \mu^2)^{-1}$ Green function for the boson field, and a pseudoscalar $ig\bar{\psi}\gamma_5\boldsymbol{\lambda}^{\mathrm{F}} \cdot \boldsymbol{\phi}\psi$ or pseudovector $g/(2m)\bar{\psi}\gamma_\mu\gamma_5\boldsymbol{\lambda}^{\mathrm{F}} \cdot \psi\partial^\mu\boldsymbol{\phi}$ coupling in both quark-meson vertices. Making use of the usual "static approximation" and neglecting the recoil corrections at both vertices in a nonrelativistic reduction for the constituent quark spinors, one arrives at (7). The free Green function above comes from the well-known Lagrangian for a free boson field $1/2\partial_\mu\boldsymbol{\phi}\partial^\mu\boldsymbol{\phi} - 1/2\mu^2\boldsymbol{\phi}^2$, which means that the boson field satisfies a linear Klein-Gordon equation.

On the other hand, the underlying $SU(3)_{\mathrm{R}} \times SU(3)_{\mathrm{L}}$ chiral symmetry of QCD tells that the Goldstone boson field cannot be described by such a simple Klein-Gordon Lagrangian. The transformation properties of the Goldstone boson field under $SU(3)_{\mathrm{R}}$ and $SU(3)_{\mathrm{L}}$ chiral rotations, g_{R} and g_{L}, are well defined if the Goldstone boson fields $\boldsymbol{\phi}$ are combined in a unitary matrix [25]

$$U(\boldsymbol{\phi}) = \exp\,(i\frac{\boldsymbol{\lambda}^{\mathrm{F}} \cdot \boldsymbol{\phi}}{f_\pi}), \tag{8}$$

$$U' = g_{\mathrm{R}} U g_{\mathrm{L}}^{-1}. \tag{9}$$

The "kinetic term" in this case is

$$\frac{f_\pi^2}{4} Tr(\partial_\mu U \partial^\mu U^\dagger), \tag{10}$$

which in its expansion in powers of $\boldsymbol{\phi}$ contains not only a free Klein-Gordon kinetic term but also other terms of higher powers. For instance, the terms of fourth order in the meson field give rise to $\pi - \pi$ scattering [26], etc. The full chiral Lagrangian describing the Goldstone boson field is unknown and should contain a quartic term involving derivatives $\partial_\mu U$ and higher order terms. It is clear that a dressed Green function for the field $\boldsymbol{\phi}$ should be very different compared to the free Klein-Gordon Green function since the selfinteraction of the Goldstone boson field is important. Only far away from its source (the fermion current), where the amplitude of the boson field $\boldsymbol{\phi}$ is small, the lowest term in powers of $\boldsymbol{\phi}$ becomes dominant, and therefore only at large distances the quark-quark interaction reduces to its simple Yukawa form.

At this stage the proper procedure should be to avoid further specific assumptions about the short range behavior of $V(r)$ in (1), to extract instead the required matrix elements of it from the baryon spectrum, and to reconstruct by this an approximate radial form of $V(r)$. The overall minus sign in the effective chiral boson interaction in (1) corresponds to that of the short range term in the Yukawa interaction.

The flavor structure of the pseudoscalar octet exchange interaction in (1) between two quarks i and j should be understood as follows :

$$V(r_{ij})\ \boldsymbol{\lambda}_{\rm i}^{\rm F}\cdot\boldsymbol{\lambda}_{\rm j}^{\rm F}\boldsymbol{\sigma}_i\cdot\boldsymbol{\sigma}_j$$
$$=\left(\sum_{a=1}^{3}V_\pi(r_{ij})\lambda_i^a\lambda_j^a+\sum_{a=4}^{7}V_{\rm K}(r_{ij})\lambda_i^a\lambda_j^a+V_\eta(r_{ij})\lambda_i^8\lambda_j^8\right)\boldsymbol{\sigma}_i\cdot\boldsymbol{\sigma}_j. \quad (11)$$

The first term in (11) represents the pion-exchange interaction, which acts only between light quarks. The second term represents the Kaon exchange interaction, which takes place in u-s and d-s pair states. The η-exchange, which is represented by the third term, is allowed in all quark pair states. In the $SU(3)_{\rm F}$ symmetric limit the constituent quark masses would be equal ($m_{\rm u}=m_{\rm d}=m_{\rm s}$), the pseudoscalar octet would be degenerate and the meson-constituent quark coupling constant would be flavor independent. In this limit the form of the pseudoscalar exchange interaction reduces to (1), which does not break the $SU(3)_{\rm F}$ invariance of the baryon spectrum. Beyond this limit the pion, Kaon and η exchange interactions will differ ($V_\pi\neq V_{\rm K}\neq V_\eta$), because of the difference between the strange and u, d quark constituent masses ($m_{\rm u,d}\neq m_{\rm s}$), and because of the mass splitting within the pseudoscalar octet ($\mu_\pi\neq\mu_{\rm K}\neq\mu_\eta$) (and possibly also because of flavor dependence in the meson-quark coupling constant). The source of both the $SU(3)_{\rm F}$ symmetry breaking constituent quark mass differences and the $SU(3)_{\rm F}$ symmetry breaking mass splitting of the pseudoscalar octet is the explicit chiral symmetry breaking in QCD.

6 The Structure of the Baryon Spectrum

The two-quark matrix elements of the interaction (1) are:

$$<[f_{ij}]_{\rm F}\times[f_{ij}]_{\rm S}:[f_{ij}]_{\rm FS}\mid -V(r_{ij})\boldsymbol{\lambda}_i^{\rm F}\cdot\boldsymbol{\lambda}_j^{\rm F}\boldsymbol{\sigma}_i\cdot\boldsymbol{\sigma}_j\mid[f_{ij}]_{\rm F}\times[f_{ij}]_{\rm S}:[f_{ij}]_{\rm FS}>$$
$$=\left\{\begin{array}{ll}-\frac{4}{3}V(r_{ij}) & [2]_{\rm F},[2]_{\rm S}:[2]_{\rm FS}\\ -8V(r_{ij}) & [11]_{\rm F},[11]_{\rm S}:[2]_{\rm FS}\\ 4V(r_{ij}) & [2]_{\rm F},[11]_{\rm S}:[11]_{\rm FS}\\ \frac{8}{3}V(r_{ij}) & [11]_{\rm F},[2]_{\rm S}:[11]_{\rm FS}\end{array}\right. . \quad (12)$$

From these the following important properties may be inferred:

(i) At short range, where $V(r_{ij})$ is positive, the chiral interaction (1) is attractive in the symmetric FS pairs and repulsive in the antisymmetric ones. At large distances the potential function $V(r_{ij})$ becomes negative and the situation is reversed.

(ii) At short range, among the FS-symmetrical pairs, the flavor antisymmetric pairs experience a much larger attractive interaction than the flavor-symmetric ones, and among the FS-antisymmetric pairs the strength of the repulsion in flavor-antisymmetric pairs is considerably weaker than in the symmetric ones.

Given these properties we conclude, that with the given flavor symmetry, the more symmetrical the FS Young pattern is for a baryon the more attractive contribution at short range comes from the interaction (1). For two identical flavor-spin Young patterns $[f]_{\rm FS}$ the attractive contribution at short range is larger for the more antisymmetrical flavor Young pattern $[f]_{\rm F}$.

Consider first, for the purposes of illustration, a schematic model which neglects the radial dependence of the potential function $V(r)$ in (1), and assume a harmonic confinement among quarks as well as $m_{\rm u} = m_{\rm d} = m_{\rm s}$. In this model

$$H_\chi \sim -\sum_{i<j} C_\chi \, \boldsymbol{\lambda}_i^{\rm F} \cdot \boldsymbol{\lambda}_j^{\rm F} \, \boldsymbol{\sigma}_i \cdot \boldsymbol{\sigma}_j . \tag{13}$$

If the only interaction between the quarks were the flavor- and spin-independent harmonic confining interaction, the baryon spectrum would be organized in multiplets of the symmetry group $SU(6)_{\rm FS} \times U(6)_{\rm conf}$. In this case the baryon masses would be determined solely by the orbital structure, and the spectrum would be organized in an *alternative sequence of positive and negative parity states.* The Hamiltonian (13), within a first order perturbation theory, reduces the $SU(6)_{\rm FS} \times U(6)_{\rm conf}$ symmetry down to $SU(3)_{\rm F} \times SU(2)_{\rm S} \times U(6)_{\rm conf}$, which automatically implies a splitting between the octet and decuplet baryons.

For the octet states N, Λ, Σ, Ξ ($N = 0$ shell, N is the number of harmonic oscillator excitations in a 3-quark state) as well as for their first radial excitations of positive parity (breathing modes) N(1440), $\Lambda(1600)$, $\Sigma(1660)$, $\Xi(?)$ ($N = 2$ shell) the flavor and spin symmetries are $[3]_{\rm FS}[21]_{\rm F}[21]_{\rm S}$, and the contribution of the Hamiltonian (13) is $-14C_\chi$. For the decuplet states Δ, $\Sigma(1385)$, $\Xi(1530)$, Ω ($N = 0$ shell) the flavor and spin symmetries, as well as the corresponding matrix element, are $[3]_{\rm FS}[3]_{\rm F}[3]_{\rm S}$ and $-4C_\chi$, respectively. The first negative parity excitations ($N = 1$ shell) in the N and Σ spectra N(1535) - N(1520) and $\Sigma(1750)$ - $\Sigma(?)$ are described by the $[21]_{\rm FS}[21]_{\rm F}[21]_{\rm S}$ symmetries, and the contribution of the interaction (13) in this case is $-2C_\chi$. The first negative parity excitation in the Λ spectrum ($N = 1$ shell) $\Lambda(1405)$ - $\Lambda(1520)$ is flavor singlet $[21]_{\rm FS}[111]_{\rm F}[21]_{\rm S}$, and, in this case, the corresponding matrix element is $-8C_\chi$.

These matrix elements alone suffice to prove that the ordering of the lowest positive and negative parity states in the baryon spectrum will be correctly predicted by the chiral boson exchange interaction (13). The constant C_χ may be determined from the N$-\Delta$ splitting to be 29.3 MeV. The oscillator parameter $\hbar\omega$, which characterizes the effective confining interaction, may be determined as one half of the mass differences between the first excited $\frac{1}{2}^+$ states and the ground states of the baryons, which have the same flavor-spin, flavor and spin symmetries (e.g. N(1440) - N, $\Lambda(1600)$ - Λ, $\Sigma(1660)$ - Σ), to be $\hbar\omega \simeq 250$ MeV. Thus the two free parameters of this simple model are fixed and we can make now predictions. In the N and Σ sectors the mass difference between the lowest excited $\frac{1}{2}^+$ states (N(1440) and $\Sigma(1660)$) and $\frac{1}{2}^- - \frac{3}{2}^-$ negative parity pairs (N(1535) - N(1520) and $\Sigma(1750)$ - $\Sigma(?)$) will then be

$$\mathrm{N}, \Sigma: \quad m(\frac{1}{2}^+) - m(\frac{1}{2}^- - \frac{3}{2}^-) = 250\,\mathrm{MeV} - C_\chi(14-2) = -102\,\mathrm{MeV}, \quad (14)$$

whereas for the Λ system ($\Lambda(1600)$, $\Lambda(1405)$ - $\Lambda(1520)$) it should be

$$\Lambda: \quad m(\frac{1}{2}^+) - m(\frac{1}{2}^- - \frac{3}{2}^-) = 250\,\mathrm{MeV} - C_\chi(14-8) = 74\,\mathrm{MeV}. \quad (15)$$

This simple example shows how the chiral interaction (13) provides different ordering of the lowest positive and negative parity excited states in the spectra of the nucleon and the Λ-hyperon. This is a direct consequence of the symmetry properties of the boson-exchange interaction discussed at the beginning of this section. Namely, the $[3]_{\mathrm{FS}}$ state in the N(1440), $\Delta(1600)$ and $\Sigma(1660)$ positive parity resonances from the $N = 2$ band feels a much stronger attractive interaction than the mixed symmetry state $[21]_{\mathrm{FS}}$ in the N(1535) - N(1520), $\Delta(1620)$ - $\Delta(1700)$ and $\Sigma(1750)$ -$\Sigma(?)$ resonances of negative parity ($N = 1$ shell). Consequently the masses of the positive parity states N(1440), $\Delta(1600)$ and $\Sigma(1660)$ are shifted down relative to the other ones, which explains the reversal of the otherwise expected "normal ordering". The situation is different for $\Lambda(1405)$ - $\Lambda(1520)$ and $\Lambda(1600)$, as the flavor state of $\Lambda(1405)$ - $\Lambda(1520)$ is totally antisymmetric. Because of this the $\Lambda(1405)$ - $\Lambda(1520)$ gains an attractive energy, which is comparable to that of the $\Lambda(1600)$, and thus the ordering suggested by the confining oscillator interaction is maintained.

Consider now, in addition, the radial dependence of the potential with the $SU(3)_{\mathrm{F}}$ invariant version (1) of the chiral boson exchange interaction (i.e., $V_\pi(r) = V_{\mathrm{K}}(r) = V_\eta(r)$). If the confining interaction in each quark pair is taken to have the harmonic oscillator form as above, the exact eigenvalues and eigenstates to the coinfining 3q Hamiltonian are

$$E = (N+3)\hbar\omega + 3V_0, \quad (16)$$

$$\Psi = |N(\lambda\mu)L[f]_{\mathrm{X}}[f]_{\mathrm{FS}}[f]_{\mathrm{F}}[f]_{\mathrm{S}} >, \quad (17)$$

where N is the number of quanta in the state, the Elliott symbol $(\lambda\mu)$ characterizes the $SU(3)$ harmonic oscillator symmetry, and L is the orbital angular momentum. The spatial (X), flavor-spin (FS), flavor (F), and spin (S) permutational symmetries are indicated by corresponding Young patterns (diagrams) $[f]$. All these functions are well known (see, e.g., [27]). Note that the color state $[111]_{\mathrm{C}}$, which is common to all the states, has been suppressed in (17). By the Pauli principle $[f]_{\mathrm{X}} = [f]_{\mathrm{FS}}$.

Table 1. The structure of the nucleon and Δ resonance states up to $N = 2$, including 11 predicted unobserved or nonconfirmed states, indicated by question marks. The predicted energy values (in MeV) are given in the brackets under the empirical ones.

$N(\lambda\mu)L[f]_X[f]_{FS}[f]_F[f]_S$	LS multiplet	average energy	δM_χ
$0(00)0[3]_X[3]_{FS}[21]_F[21]_S$	$\frac{1}{2}^+$, N	939	$-14P_{00}$
$0(00)0[3]_X[3]_{FS}[3]_F[3]_S$	$\frac{3}{2}^+$, Δ	1232 (input)	$-4P_{00}$
$2(20)0[3]_X[3]_{FS}[21]_F[21]_S$	$\frac{1}{2}^+$, N(1440)	1440 (input)	$-7P_{00} - 7P_{20}$
$1(10)1[21]_X[21]_{FS}[21]_F[21]_S$	$\frac{1}{2}^-$, N(1535); $\frac{3}{2}^-$, N(1520)	1527 (input)	$-7P_{00} + 5P_{11}$
$2(20)0[3]_X[3]_{FS}[3]_F[3]_S$	$\frac{3}{2}^+$, Δ(1600)	1600 (input)	$-2P_{00} - 2P_{20}$
$1(10)1[21]_X[21]_{FS}[3]_F[21]_S$	$\frac{1}{2}^-$, Δ(1620); $\frac{3}{2}^-$, Δ(1700)	1660 (1719)	$-2P_{00} + 6P_{11}$
$1(10)1[21]_X[21]_{FS}[21]_F[3]_S$	$\frac{1}{2}^-$, N(1650); $\frac{3}{2}^-$, N(1700) $\frac{5}{2}^-$, N(1675)	1675 (1629)	$-2P_{00} + 4P_{11}$
$2(20)2[3]_X[3]_{FS}[3]_F[3]_S$	$\frac{1}{2}^+$, Δ(1750?); $\frac{3}{2}^+$, Δ(?) $\frac{5}{2}^+$, Δ(?); $\frac{7}{2}^+$, Δ(?)	1750? (1675)	$-2P_{00} - 2P_{22}$
$2(20)2[3]_X[3]_{FS}[21]_F[21]_S$	$\frac{3}{2}^+$, N(1720); $\frac{5}{2}^+$, N(1680)	1700 (input)	$-7P_{00} - 7P_{22}$
$2(20)0[21]_X[21]_{FS}[21]_F[21]_S$	$\frac{1}{2}^+$, N(1710)	1710 (1778)	$-\frac{7}{2}P_{00} - \frac{7}{2}P_{20} + 5P_{11}$
$2(20)0[21]_X[21]_{FS}[21]_F[3]_S$	$\frac{3}{2}^+$, N(?)	? (1813)	$-P_{00} - P_{20} + 4P_{11}$
$2(20)2[21]_X[21]_{FS}[21]_F[21]_S$	$\frac{3}{2}^+$, N(1900?); $\frac{5}{2}^+$, N(2000?);	1950? (1909)	$-\frac{7}{2}P_{00} - \frac{7}{2}P_{22} + 5P_{11}$
$2(20)2[21]_X[21]_{FS}[21]_F[3]_S$	$\frac{1}{2}^+$, N(?); $\frac{3}{2}^+$, N(?) $\frac{5}{2}^+$, N(?); $\frac{7}{2}^+$, N(1990?)	1990? (1850)	$-P_{00} - P_{22} + 4P_{11}$
$2(20)0[21]_X[21]_{FS}[3]_F[21]_S$	$\frac{1}{2}^+$, Δ(1910)	1910 (1903)	$-P_{00} - P_{20} + 6P_{11}$
$2(20)2[21]_X[21]_{FS}[3]_F[21]_S$	$\frac{3}{2}^+$, Δ(1920); $\frac{5}{2}^+$, Δ(1905)	1912 (1940)	$-P_{00} - P_{22} + 6P_{11}$

The full Hamiltonian is the sum of the confining Hamiltonian and the chiral field interaction (1). When the boson exchange interaction (1) is treated in first order perturbation theory, the mass of the baryon states takes the form

$$M = M_0 + N\hbar\omega + \delta M_\chi, \tag{18}$$

where the chiral interaction contribution is $\delta M_\chi =< \Psi|H_\chi|\Psi >$, and $M_0 =$

Table 2. The structure of the Λ-hyperon states up to $N = 2$, including predicted unobserved or nonconfirmed states, indicated by question marks. The predicted energies (in MeV) are given in the brackets under the empirical values.

$N(\lambda\mu)L[f]_X[f]_{FS}[f]_F[f]_S$	LS multiplet	average energy	δM_χ
$0(00)0[3]_X[3]_{FS}[21]_F[21]_S$	$\frac{1}{2}^+, \Lambda$	1115	$-14P_{00}$
$1(10)1[21]_X[21]_{FS}[111]_F[21]_S$	$\frac{1}{2}^-, \Lambda(1405)$; $\frac{3}{2}^-, \Lambda(1520)$	1462 (1512)	$-12P_{00} + 4P_{11}$
$2(20)0[3]_X[3]_{FS}[21]_F[21]_S$	$\frac{1}{2}^+, \Lambda(1600)$	1600 (1616)	$-7P_{00} - 7P_{20}$
$1(10)1[21]_X[21]_{FS}[21]_F[21]_S$	$\frac{1}{2}^-, \Lambda(1670)$; $\frac{3}{2}^-, \Lambda(1690)$	1680 (1703)	$-7P_{00} + 5P_{11}$
$1(10)1[21]_X[21]_{FS}[21]_F[3]_S$	$\frac{1}{2}^-, \Lambda(1800)$; $\frac{3}{2}^-, \Lambda(?)$; $\frac{5}{2}^-, \Lambda(1830)$	1815 (1805)	$-2P_{00} + 4P_{11}$
$2(20)0[21]_X[21]_{FS}[111]_F[21]_S$	$\frac{1}{2}^+, \Lambda(1810)$	1810 (1829)	$-6P_{00} - 6P_{20} + 4P_{11}$
$2(20)2[3]_X[3]_{FS}[21]_F[21]_S$	$\frac{3}{2}^+, \Lambda(1890)$; $\frac{5}{2}^+, \Lambda(1820)$	1855 (1878)	$-7P_{00} - 7P_{22}$
$2(20)0[21]_X[21]_{FS}[21]_F[21]_S$	$\frac{1}{2}^+, \Lambda(?)$	? (1954)	$-\frac{7}{2}P_{00} - \frac{7}{2}P_{20} + 5P_{11}$
$2(20)0[21]_X[21]_{FS}[21]_F[3]_S$	$\frac{3}{2}^+, \Lambda(?)$	? (1989)	$-P_{00} - P_{20} + 4P_{11}$
$2(20)2[21]_X[21]_{FS}[21]_F[3]_S$	$\frac{1}{2}^+, \Lambda(?)$; $\frac{3}{2}^+, \Lambda(?)$; $\frac{5}{2}^+\Lambda(?)$; $\frac{7}{2}^+, \Lambda(2020?)$	2020? (2026)	$-P_{00} - P_{22} + 4P_{11}$
$2(20)2[21]_X[21]_{FS}[111]_F[21]_S$	$\frac{3}{2}^+, \Lambda(?)$; $\frac{5}{2}^+, \Lambda(?)$	? (2053)	$-6P_{00} - 6P_{22} + 4P_{11}$
$2(20)2[21]_X[21]_{FS}[21]_F[21]_S$	$\frac{3}{2}^+, \Lambda(?)$; $\frac{5}{2}^+, \Lambda(2110)$	2110? (2085)	$-\frac{7}{2}P_{00} - \frac{7}{2}P_{22} + 5P_{11}$

$\sum_{i=1}^{3} m_i + 3(V_0 + \hbar\omega)$. The contribution from the chiral interaction to each baryon is a linear combination of the matrix elements of the two-body potential $V(r_{12})$, defined as

$$P_{nl} =< \varphi_{nlm}(\mathbf{r}_{12})|V(r_{12})|\varphi_{nlm}(\mathbf{r}_{12}) > . \tag{19}$$

Here $\varphi_{nlm}(\mathbf{r}_{12})$ represents the oscillator wavefunction with n excited quanta. As we shall only consider the baryon states in the $N \leq 2$ bands, we shall only need the four radial matrix elements P_{00}, P_{11}, P_{20} and P_{22} for the numerical construction of the spectrum.

The contributions to all nucleon, Δ and Λ-hyperon states from the boson exchange interaction, in terms of the matrix elements P_{nl}, are listed in Tables 1 and 2. In this approximate $SU(3)_F$-invariant version of the chiral boson exchange interaction the $\Lambda - \mathrm{N}$ and the $\Xi - \Sigma$ mass differences would solely be ascribed to the mass difference between the s and u,d quarks, since all these baryons have identical orbital structure and permutational symmetries. The states in the Λ-spectrum would be degenerate with the corresponding states in the Σ-spectrum which have equal symmetries.

The oscillator parameter $\hbar\omega$ and the four integrals are extracted from the mass differences between the nucleon and the $\Delta(1232)$, the $\Delta(1600)$ and the N(1440), as well as the splittings between the nucleon and the average mass of the two pairs of states N(1535)–N(1520) and N(1720)–N(1680). This procedure yields the parameter values $\hbar\omega$=157.4 MeV, P_{00}=29.3 MeV, P_{11}=45.2 MeV, P_{20}=2.7 MeV and P_{22}=–34.7 MeV. Given these values, all other excitation energies (i.e., differences between the masses of given resonances and the corresponding ground states) of the nucleon, Δ- and Λ-hyperon spectra are predicted to within $\sim$ 15% of the empirical values where known, and are well within the uncertainty limits of those values. Note that these matrix elements provide a quantitatively satisfactory description of the Λ-spectrum even though they are extracted from the N $-$ Δ spectrum.

The relative magnitudes and signs of the numerical parameter values can be readily understood. If the potential function $V(\mathbf{r})$ is assumed to have the form of a Yukawa function with a smeared δ-function term that is positive at short range $r \leq 0.6 - 0.7$ fm, as suggested by the pion size $\sqrt{< r_\pi^2 >} = 0.66$ fm, one expects P_{20} to be considerably smaller than P_{00} and P_{11}, as the radial wavefunction of the excited S-state has a node, and as it extends further into the region where the potential is negative. The negative value for P_{22} is also natural, since the corresponding wavefunction is suppressed at short range and extends well beyond the expected node in the potential function.

7 The $SU(3)_{\rm F}$ Breaking Chiral Boson Interaction

The model described above has relied on an interaction potential function $V(r)$ in (1) that is flavor independent. A refined version takes into account the explicit flavor dependence of the potential function in (11) ($V_\pi \neq V_{\rm K} \neq V_\eta$). In the following we show how this explicit flavor dependence provides an explanation of the mass splitting between the Λ and the Σ which have the same quark content and the same FS, F and S symmetries, i.e., they are degenerate within the $SU(3)_{\rm F}$ version (1) of the chiral boson exchange interaction.

Beyond the $SU(3)_{\rm F}$ limit the ground state baryons will be determined by the π-exchange radial integral P_{00}^π, the K-exchange one, $P_{00}^{\rm K}$, and by the η-exchange integrals, $P_{00}^{\rm uu} = P_{00}^{\rm ud} = P_{00}^{\rm dd}$, $P_{00}^{\rm us}$ and $P_{00}^{\rm ss}$, where the superscripts indicate quark pairs to which the η-exchange applies. As indicated by the Yukawa interaction (7) these matrix elements should be inversely proportional to the product of the quark masses of the pair state. Thus $P_{nl}^{\rm us} = \frac{m_{\rm u}}{m_{\rm s}} P_{nl}^{\rm uu}$, $P_{nl}^{\rm ss} = (\frac{m_{\rm u}}{m_{\rm s}})^2 P_{nl}^{\rm uu}$. We also assume that $P_{00}^{\rm us} \simeq P_{00}^{\rm K}$, which is suggested by the fact that the quark masses are equal in the states in which these interactions act, and by the near equality of the Kaon and η masses, $\mu_\eta \simeq \mu_{\mathrm{K}}$. Thus we have only two independent radial integrals.

To determine the integrals P^{π}_{00}, P^{K}_{00} and the quark mass difference $\Delta_{\mathrm{q}} = m_{\mathrm{s}} - m_{\mathrm{u}}$ we consider the $\Sigma(1385) - \Sigma$, $\Delta - N$ and $\Lambda - N$ splittings:

$$m_{\Sigma(1385)} - m_{\Sigma} = 4P^{\mathrm{us}}_{00} + 6P^{\mathrm{K}}_{00}, \tag{20}$$

$$m_{\Delta} - m_{\mathrm{N}} = 12P^{\pi}_{00} - 2P^{\mathrm{uu}}_{00}, \tag{21}$$

$$m_{\Lambda} - m_{\mathrm{N}} = 6P^{\pi}_{00} - 6P^{\mathrm{K}}_{00} + \Delta_{\mathrm{q}}, \tag{22}$$

which imply $P^{\mathrm{K}}_{00} = 19.6$ MeV, $\Delta_{\mathrm{q}} = 121$ MeV, if the conventional value of 340 MeV is given to m_{u}, $P^{\pi}_{00} = 28.9$ MeV and the quark mass ratio $m_{\mathrm{s}}/m_{\mathrm{u}} = 1.36$. These values of the matrix elements lead to the values of 65 MeV and 139 MeV for the $\Sigma - \Lambda$ and the $\Xi - \Sigma$ mass differences

$$m_{\Sigma} - m_{\Lambda} = 8P^{\pi}_{00} - 4P^{\mathrm{K}}_{00} - \frac{4}{3}P^{\mathrm{uu}}_{00} - \frac{8}{3}P^{\mathrm{us}}_{00}, \tag{23}$$

$$m_{\Xi} - m_{\Sigma} = P^{\pi}_{00} + \frac{1}{3}P^{\mathrm{uu}}_{00} - \frac{4}{3}P^{\mathrm{ss}}_{00} + \Delta_{\mathrm{q}} \tag{24}$$

in good agreement with the empirical values of 77 MeV and 125 MeV, respectively.

A description of the other parts of the Σ, Ξ and Ω spectra can be found in [8].

8 Three-Body Faddeev Calculations

In the previous sections we have shown how the Goldstone boson exchange (GBE), taken to first order perturbation theory and without explicit parameterizing the radial dependence, can explain the correct level ordering of positive and negative parity states in light and strange baryon spectra, as well as the splittings in those spectra. A question, however, arises about what will happen beyond first order perturbation theory. In order to check this we have numericaly solved three-body Faddeev equations [28]. Besides the confinement potential, which is now taken in linear form, the GBE interaction between the constituent quarks is now included to all orders. These results further support the adequacy of the GBE for baryon spectroscopy.

In addition to the octet-exchange interaction we include here also the flavor-singlet (η') exchange. In the large N_{C} limit the axial anomaly becomes suppressed [29], and the η' becomes the ninth Goldstone boson of the spontaneously broken $U(3)_{\mathrm{L}} \times U(3)_{\mathrm{R}}$ chiral symmetry in addition to the octet of pseudoscalar mesons [30].

For the GBE the spin-spin component of the interaction between the constituent quarks i and j reads:

$$V_\chi(\mathbf{r}_{ij}) = \left\{ \sum_{a=1}^{3} V_\pi(\mathbf{r}_{ij}) \lambda_i^a \lambda_j^a \right.$$
$$\left. + \sum_{a=4}^{7} V_{\mathrm{K}}(\mathbf{r}_{ij}) \lambda_i^a \lambda_j^a + V_\eta(\mathbf{r}_{ij}) \lambda_i^8 \lambda_j^8 + V_{\eta'}(\mathbf{r}_{ij}) \lambda_i^0 \lambda_j^0 \right\} \boldsymbol{\sigma}_i \cdot \boldsymbol{\sigma}_j, \quad (25)$$

where $\lambda^a, a = 1, ..., 8$ are flavor Gell-Mann matrices and $\lambda^0 = \sqrt{2/3}\mathbb{1}$.

In the simplest case, when the boson field satisfies the linear Klein-Gordon equation, one has the following spatial dependence for the meson-exchange potentials in (25):

$$V_\gamma(\mathbf{r}_{ij}) = \frac{g_\gamma^2}{4\pi} \frac{1}{3} \frac{1}{4m_i m_j} \{\mu_\gamma^2 \frac{e^{-\mu_\gamma r_{ij}}}{r_{ij}} - 4\pi\delta(\mathbf{r}_{ij})\}, \quad (26)$$

$$(\gamma = \pi, \mathrm{K}, \eta, \eta'),$$

with quark and meson masses m_i and μ_γ, respectively.

Eq. (26) contains both the traditional long-range Yukawa potential as well as a δ-function term. It is the latter that is of crucial importance for baryon physics. We already discussed in Section 5 that it is strictly valid only for pointlike particles, and that it must be smeared out since the constituent quarks and pseudoscalar mesons have finite size, and in addition the boson fields in a chiral Lagrangian should in fact satisfy a nonlinear equation. Furthermore it is quite natural to assume that at distances $r \ll r_0$, where r_0 can be related to the constituent quark and pseudoscalar meson sizes, there is no chiral boson-exchange interaction, since this is the region of perturbative QCD with the original QCD degrees of freedom. The interactions at these very short distances are not essential for the low-energy properties of baryons. Consequently we use a two-parameter "representation" for the δ-function term in (26)

$$4\pi\delta(\mathbf{r}_{ij}) \Rightarrow \frac{4}{\sqrt{\pi}} \alpha^3 \exp(-\alpha^2(r - r_0)^2). \quad (27)$$

Following the arguments above one should also cut off the Yukawa part of the GBE for $r < r_0$.

The πq coupling constant can be extracted from the phenomenological pion-nucleon coupling [8] as $\frac{g_8^2}{4\pi} = 0.67$. For simplicity (and to avoid any additional free parameter), the same coupling constant is assumed for the coupling between the η- meson and the constituent quark. This is exactly in the spirit of unbroken $SU(3)_{\mathrm{F}}$ symmetry. For the flavor-singlet η', however, we must take a different coupling $\frac{g_0^2}{4\pi}$, as the η' decouples from the pseudoscalar octet due to the $U(1)_{\mathrm{A}}$ anomaly. This fact is illustrated best by the failure of the Gell-Mann–Oakes–Renner relations [12] for the flavor singlet

[1]. Lacking a phenomenological value, we treat $g_0^2/4\pi$ as a free parameter. The constituent masses of the u and d quarks are taken to be 340 MeV, as suggested by the nucleon magnetic moments.

In the present calculation we neglect tensor meson-exchange forces. We expect their role to be of minor importance for the main features of the baryon spectra [8] (mainly due to the absence of the strong δ-function part in this case).

Our full interquark potential is thus given by

$$V(\mathbf{r}_{ij}) = V_\chi^{\text{octet}}(\mathbf{r}_{ij}) + V_\chi^{\text{singlet}}(\mathbf{r}_{ij}) + Cr_{ij}. \tag{28}$$

While all masses and the octet coupling constant are predetermined, we treated r_0, α, $(g_0/g_8)^2$, and C as free parameters and determined their values to be:

$$r_0 = 0.43\,\text{fm},\ \alpha = 2.91\,\text{fm}^{-1},\ (g_0/g_8)^2 = 1.8,\ C = 0.474\,\text{fm}^{-2}.$$

Notice that we do not need any constant V_0, which is usually added to the confining potential. In fact only *four* free parameters suffice to describe all 14 lowest states of the N and Δ spectra, including the absolute value of the nucleon (ground state). At the present stage of determining the qq potential due to GBE we were led by the principle of working with the smallest possible number of free parameters. Therefore we took the octet coupling constant $g_8^2/4\pi$, and likewise the constituent quark mass m, as predetermined. Of course, one may expect that subsequent studies within the GBE model will put further constraints on the parameterization of the qq potential.

The qq potential (28) constitutes the dynamical input into our 3-body Faddeev calculations of the baryon spectra. We show our results in Fig. 1 for the parametrization of the qq interaction as given above. It is well seen that the whole set of lowest N and Δ states is reproduced quite correctly. In the most unfavourable cases deviations from the experimental values do not exceed 3%! In addition all level orderings are correct. In particular, the positive-parity state N(1440) (Roper resonance) lies *below* the pair of negative-parity states N(1535) - N(1520). The same is true in the Δ spectrum with $\Delta(1600)$ and the pair $\Delta(1620)$ - $\Delta(1700)$. We emphasize again that the qq potential (28) is able to predict also the absolute value of the nucleon mass. In previous models an arbitrary constant was usually needed to achieve the correct value of 939 MeV.

At the present stage of our investigation of the baryon spectra with the GBE interaction we have left out the tensor forces. Therefore the fine-structure splittings in the LS-multiplets are not yet introduced. However, it is clear from the observed smallness of these splittings and from the arguments given above, that the tensor component of the GBE can play only a minor role. Here we also note that the Yukawa part of the interaction in (26) is only

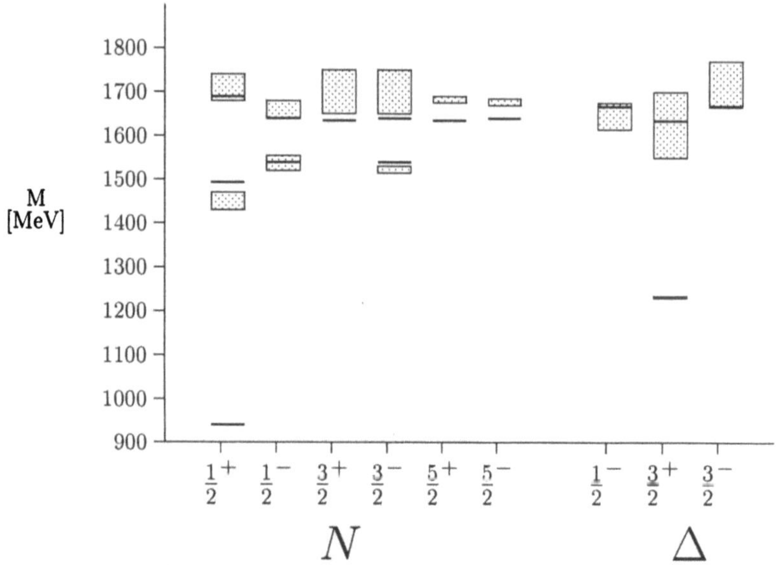

Fig. 1. Energy levels for the 14 lowest non-strange baryons with total angular momentum and parity J^P. The shadowed boxes represent experimental uncertainties.

of secondary importance. In fact, the pattern of Fig. 1 could also be described with the "δ-part" (27) alone (and a slightly modified set of parameters).

It is instructive to learn how the GBE affects the energy levels when it is switched on and its strength (coupling constant) is gradually increased (Fig. 2). Starting out from the case with confinement only, one observes that the degeneracy of states is removed and an inversion of the ordering of positive- and negative-parity states is achieved, both in the N and Δ excitations. From Fig. 2 also the crucial importance of the chiral interaction V_χ becomes evident. Notice that the strength of our confinement, $C = 0.474$ fm^{-2}, is rather small and the confining interaction contributes much less to the splittings than the GBE. The relative "weakness" of our effective confining interaction could be due to a partial cancellation between[1] the much stronger color-electric confinement and the σ-exchange since they are of opposite sign. Due to the same reason one cannot expect that the effective confining interaction between the constituent quarks is strictly of linear form.

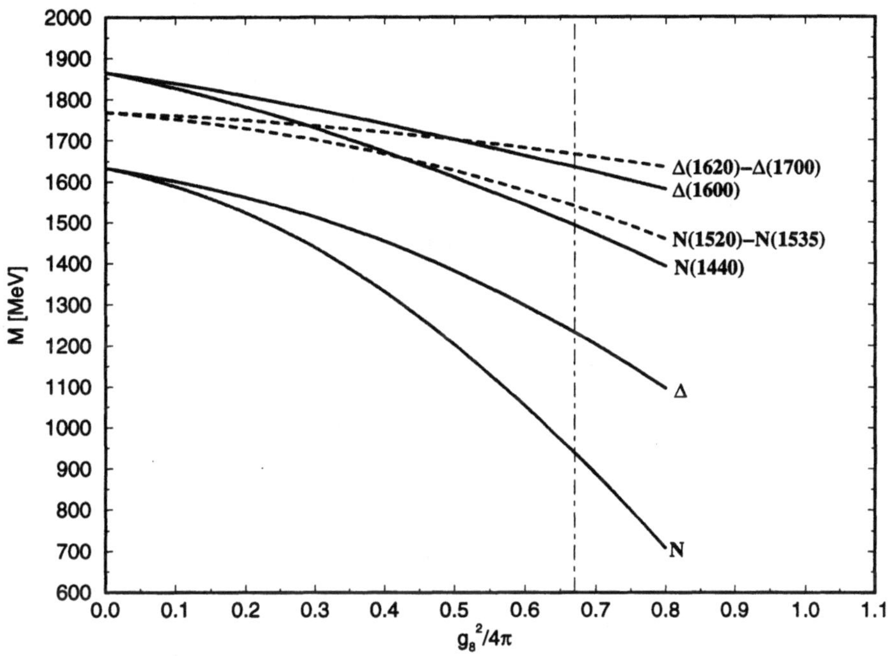

Fig. 2. Level shifts of some lowest baryons as a function of the strength of the GBE. Solid and dashed lines correspond to positive- and negative-parity states, respectively.

9 Exchange Current Corrections to the Magnetic Moments

A flavor dependent interaction of the form (1) will imply the presence of an irreducible two-body exchange current operator, as seen, e.g., directly from the continuity equation, by which the commutator of the interaction and the single particle charge operator is equal to the divergence of the exchange current density [31].

The general form of the octet vector exchange current operator, that is associated with the complete octet mediated interaction (11), is [8]

$$\boldsymbol{\mu}^{\mathrm{ex}} = \mu_{\mathrm{N}}\{\tilde{V}_\pi(r_{ij})(\lambda_i^1\lambda_j^2 - \lambda_i^2\lambda_j^1) + \tilde{V}_{\mathrm{K}}(r_{ij})(\lambda_i^4\lambda_j^5 - \lambda_i^5\lambda_j^4)\}(\boldsymbol{\sigma}_i \times \boldsymbol{\sigma}_j). \quad (29)$$

Table 3. Magnetic moments of the baryon octet (in nuclear magnetons). Column IA contains the quark model impulse approximation expressions, column "exp" the experimental values, column I the impulse approximation predictions, column II the exchange current contribution with $< \varphi_{000}(\mathbf{r}_{12})|\tilde{V}_{\pi}(r_{12})|\varphi_{000}(\mathbf{r}_{12}) >= -0.018$ and $< \varphi_{000}(\mathbf{r}_{12})|\tilde{V}_{\mathrm{K}}(r_{12})|\varphi_{000}(\mathbf{r}_{12}) >= 0.03$, and column III the net predictions.

	IA	exp	I	II	III
p	$\frac{m_{\mathrm{N}}}{m_{\mathrm{u}}}$	+2.79	+2.76	+0.07	+2.83
n	$-\frac{2}{3}\frac{m_{\mathrm{N}}}{m_{\mathrm{u}}}$	-1.91	-1.84	-0.07	-1.91
Λ	$-\frac{1}{3}\frac{m_{\mathrm{N}}}{m_{\mathrm{s}}}$	-0.61	-0.67	+0.06	-0.61
Σ^+	$\frac{8}{9}\frac{m_{\mathrm{N}}}{m_{\mathrm{u}}} + \frac{1}{9}\frac{m_{\mathrm{N}}}{m_{\mathrm{s}}}$	+2.42	+2.68	-0.12	+2.56
Σ^0	$\frac{2}{9}\frac{m_{\mathrm{N}}}{m_{\mathrm{u}}} + \frac{1}{9}\frac{m_{\mathrm{N}}}{m_{\mathrm{s}}}$	?	+0.84	-0.06	+0.72
$\Sigma^0 \to \Lambda$	$-\frac{1}{\sqrt{3}}\frac{m_{\mathrm{N}}}{m_{\mathrm{u}}}$	|1.61|	-1.59	-0.01	-1.60
Σ^-	$-\frac{4}{9}\frac{m_{\mathrm{N}}}{m_{\mathrm{u}}} + \frac{1}{9}\frac{m_{\mathrm{N}}}{m_{\mathrm{s}}}$	-1.16	-1.00	0	-1.00
Ξ^0	$-\frac{2}{9}\frac{m_{\mathrm{N}}}{m_{\mathrm{u}}} - \frac{4}{9}\frac{m_{\mathrm{N}}}{m_{\mathrm{s}}}$	-1.25	-1.51	+0.12	-1.39
Ξ^-	$\frac{1}{9}\frac{m_{\mathrm{N}}}{m_{\mathrm{u}}} - \frac{4}{9}\frac{m_{\mathrm{N}}}{m_{\mathrm{s}}}$	-0.65	-0.59	0	-0.59

Here $\tilde{V}_{\pi}(r)$ and $\tilde{V}_{\mathrm{K}}(r)$ are dimensionless functions that describe π and K exchange, respectively, and which include both the pionic (kaonic) current and the pair current term at long range.

Consider a simplified nonrelativistic constituent quark model. The impulse approximation expressions for the magnetic moments of the ground state octet baryons and their experimental values are listed in Table 3 (columns "IA" and "exp", respectively). A natural approach is to determine the mass ratios $m_{\mathrm{N}}/m_{\mathrm{u}}$ and $m_{\mathrm{N}}/m_{\mathrm{s}}$ to fit the experimental values of the magnetic moments of the Σ^- and Ξ^- octet and the Ω and Δ^{++} ($\mu_{\Omega} = -2.019 \pm 0.054\, \mu_{\mathrm{N}}$, $\mu_{\Delta^{++}} = 4.52 \pm 0.50\, \mu_{\mathrm{N}}$) decuplet baryons, which are unaffected by the exchange current operator (29). While with only two independent variables it is not possible to fit all four experimental magnetic moments exactly, the best overall fit, $\mu_{\Sigma^-} = -1.00\, \mu_{\mathrm{N}}$, $\mu_{\Xi^-} = -0.59\, \mu_{\mathrm{N}}$, $\mu_{\Omega^-} = -2.01\, \mu_{\mathrm{N}}$, $\mu_{\Delta^{++}} = 5.52\, \mu_{\mathrm{N}}$, happens to be obtained with precisely the ratios $m_{\mathrm{N}}/m_{\mathrm{u}} = 2.76$ and $m_{\mathrm{N}}/m_{\mathrm{s}} = 2.01$, which were used for the constituent quark masses to fit ground state baryons ($m_{\mathrm{u}} = 340$ MeV and $m_{\mathrm{s}} = 467$ MeV).

We find (see Table 3) that the meson exchange current contributions systematically improve the predictions of the naive constituent quark model (i.e., with one-body quark currents only) for all known magnetic moments.

As the constituent quarks are not too heavy, both their electromagnetic and axial current operators have significant relativistic correction terms. Their effect is to reduce the magnitude of the predicted values of both the axial coupling constants and the magnetic moments of the baryons that are given by the static quark model. This correction reduces the standard over-prediction of the axial current coupling constant of the nucleon (5/3 vs 1.24) and the strange baryons, but it worsens the mostly satisfactory predictions for the magnetic moments of the baryons that are obtained with the static quark model. In Ref. [32] it is shown that the exchange current corrections associated with the chiral boson exchange interaction between the quarks can compensate for the relativistic correction in the latter case, while leaving it operative in the case of the axial coupling constants. This then makes it possible to obtain at least qualitatively satisfactory simultaneous description of both the magnetic moments and the axial coupling constants.

10 Instead of a Conclusion

Instead of a conclusion we discuss some important recent lattice QCD results in this last section. It was shown already a few years ago that one can obtain a qualitatively correct splitting between Δ and N already within a quenched approximation (for a review and references see [33]). Within the quenched approximation to QCD the sea quark closed loop diagrams generated by gluon lines are neglected. Thus in the quenched approximation for baryons one takes into account only 3 continuous valence quark lines and full gluodynamics. This quenched approximation contains, however, part of antiquark effects related to the Z graphs formed of valence quark lines. One can even construct diagrams within the quenched approximation which correspond to the exchange of the color-singlet isospin 1 or 0 $q\bar{q}$ pairs between valence quark lines [34]. It is also important that these diagrams contribute to the baryon mass to leading order ($\sim N_C$) in a $1/N_C$ expansion [35] (their contribution to the $\Delta - \mathrm{N}$ splitting appears, however, to subleading orders).

From the quenched measurements [33] it is not clear what were the physical reason for the $\Delta - \mathrm{N}$ splitting: gluon exchanges, instantons, or something else. To clarify this question, Liu and Dong have recently measured the $\Delta - \mathrm{N}$ splitting in the quenched and a further so-called "valence approximation" [36]. In the valence approximation the quarks are limited to propagating only forward in time (i.e., Z graphs and related quark-antiquark pairs are removed). The gluon exchange and all other possible gluon configurations, including instantons, are exactly the same in both approximations. The striking result is that the $\Delta - \mathrm{N}$ splitting is observed only in the quenched approximation but not in the valence approximation, in which the N and the

Δ levels are degenerate within error bars. Consequently the $\Delta - \mathrm{N}$ splitting must receive a considerable contribution from the diagrams with $q\bar{q}$ excitations, which correspond to the meson exchanges, but not from the gluon exchange or instanton-induced interaction between quarks (to be precise, the instanton-induced interaction could be rather important for the interactions between quarks and antiquarks).

If the observation of Liu and Dong is confirmed it would be important to measure the relative positions of the lowest excited states of positive and negative parity in the N, Δ, Λ and Σ spectra within both the quenched and the valence approximation. One expects that, if the entire $\Delta-\mathrm{N}$ splitting (i.e., 300 MeV) is due to the antiquark excitations in the quenched approximation, then $1/2^+, \mathrm{N}(1440)$ should be below the negative parity pair $1/2^-, \mathrm{N}(1535) - 3/2^-, \mathrm{N}(1520)$, while in the Λ spectrum the situation should be opposite: the negative parity pair $1/2^-, \Lambda(1405) - 3/2^-, \Lambda(1520)$ should be below the first positive parity excitation $1/2^+, \Lambda(1600)$. In the valence approximation the spin-spin force among quarks, which is due to Goldstone boson exchange, is absent, and the relative position of $1/2^+, \mathrm{N}(1440)$ and $1/2^-, \mathrm{N}(1535) - 3/2^-, \mathrm{N}(1520)$ should be just opposite to the quenched approximation: the $1/2^+, \mathrm{N}(1440)$ should be above the negative parity pair. However, in the Λ spectrum the negative parity pair $1/2^-, \Lambda(1405) - 3/2^-, \Lambda(1520)$ should still be below the $1/2^+, \Lambda(1600)$.

Acknowledgements
It is a pleasure to thank Dan Riska, Zoltan Papp and Willi Plessas for their collaboration that was crucial for the results presented in this lecture.

References

1. S. Weinberg, Phys. Rev. **D11** (1975) 3583.
2. G.'t Hooft, Phys. Rev. **D14** (1976) 3432.
3. S. Weinberg, Physica **96 A** (1979) 327.
4. A. Manohar and H. Georgi, Nucl. Phys. **B234** (1984) 189.
5. E.V. Shuryak, Phys.Rep. **C115** (1984) 152.
6. D. I. Diakonov and V. Yu. Petrov, Phys. Lett. **B147** (1984) 351; Nucl. Phys. **B272** (1986) 457.
7. L. Ya. Glozman and D. O. Riska, The Baryon Spectrum and Chiral Dynamics, Preprint HU-TFT-94-47, [LANL hep-ph 9411279] (1994); Systematics of the Light and Strange Baryons and the Symmetries of QCD, Preprint HU-TFT-94-48, [LANL hep-ph 9412231] (1994).
8. L. Ya. Glozman and D. O. Riska, Physics Reports **268** (1996) 263.
9. A. DeRujula, H. Georgi, and S. L. Glashow, Phys. Rev. **D12** (1975) 147.
10. N. Isgur and G. Karl, Phys. Rev. **D18** (1978) 4187; **D19** (1979) 2653.
11. B. Silvestre-Brac and C. Gignoux, Phys. Rev. **D32** (1985) 743.
12. M. Gell-Mann, R. J. Oakes, and B. Renner, Phys. Rev. **175** (1968) 2195.

13. M. Shifman, A. Vainstein, and V. Zahkarov, Nucl. Phys. **B147** (1979) 385, 448.
14. D. R. Daniel et al, Phys. Rev. **D46** (1992) 3130; M. N. Fukugita et al, Phys. Rev. **D47** (1993) 4739.
15. M. Gell-Mann and M. Levy, Nuovo Cim. **16** (1960) 705.
16. Y. Nambu and G. Jona-Lasinio, Phys. Rev. **122** (1961) 345.
17. A. Chodos and C. B. Thorn, Phys. Rev.**D12** (1975) 2733; G. E. Brown and M. Rho, Phys. Lett. **B82** (1979) 177; A. W. Thomas, Adv. Nucl. Phys. **13** (1984) 1; G. A. Miller, Int. Rev. Nucl. Phys. **1** (1984) 189.
18. W. Weise, International Review of Nucl. Phys., **1** (1984) 116; D. Robson, Topical Conference on Nuclear Chromodynamics, J. Qiu and D. Sivers (eds.), World Scientific, Singapore, 174 (1988).
19. D. I. Diakonov, V. Yu. Petrov, and P. V. Pobylitsa, Nucl. Phys. **B306** (1988) 809.
20. S. Kahana, V. Soni, and G. Ripka, Nucl. Phys. **A415** (1984) 351.
21. M. Birse and M. Banerjee, Phys. Rev. **D31** (1985) 118.
22. C. V. Christov et al., Progr. Part. Nucl. Phys. **37** (1996) 91.
23. R. Alkofer and H. Reinhard, Chiral Quark Dynamics, Springer, 1995.
24. U. Vogl and W. Weise, Progr. Part. Nucl. Phys. **27** (1991) 195.
25. S. Coleman, J. Wess, and B. Zumino, Phys. Rev. **177** (1969) 2239.
26. S. Weinberg, Phys. Rev. Lett. **18** (1967) 188.
27. L. Ya. Glozman and E. I. Kuchina, Phys. Rev. **C49** (1994) 1149.
28. L. Ya. Glozman, Z. Papp, and W. Plessas, Phys. Lett. **B381** (1996) 311.
29. E. Witten, Nucl. Phys. **B156** (1979) 269.
30. S. Coleman and E. Witten, Phys. Rev. Lett. **45** (1980) 100.
31. D. O. Riska, Phys. Reports **181** (1989) 207.
32. K. Dannbom, L. Ya. Glozman, C. Helminen, and D. O. Riska, in preparation.
33. D. Weingarten, Nucl. Phys. B (Proc. Suppl.) **34** (1994) 29.
34. T. D. Cohen and D. B. Leinweber, Comments Nucl. Part. Phys. **21** (1993) 137.
35. G. 't Hooft, Nucl. Phys. **B72** (1974) 461.
36. K. F. Liu and S.-J. Dong, Quark Model from Lattice QCD, Preprint UK 194-03, hep-lat 9411067.

Effective Theory for Heavy Quarks

Thomas Mannel

Institut für Theoretische Teilchenphysik, Universität Karlsruhe,
D-76128 Karlsruhe, Kaiserstraße 12, Germany

Abstract. In this series of lectures the basic ideas of the $1/m_Q$ expansion in QCD (m_Q is the mass of a heavy quark) are outlined. Applications to exclusive and inclusive decays are given.

1 Introduction

With the precise formulation of the $1/m_Q$ expansion in QCD using effective field theory and operator product expansion heavy quark physics has been based on model independent ground, which allows to reduce the uncertainties due to the QCD bound state problem drastically. From the point of view of weak interactions the main interest in processes with heavy quarks lies in the exploration and the test of the CKM sector of the standard model (SM) describing the masses and the mixing of quarks. From the experimental side the running experiments as well as the ones planned in the near future constitute a large effort to explore this part of the SM, which is not yet tested with an accuracy comparable with the one of the coupling of the Z_0 to the fermions.

The main theoretical progress in the description of systems involving a single heavy quark is based on the infinite mass limit of QCD [1], in which two additional symmetries appear that are not present in full QCD [2]. This limit may be regarded as the leading term of a $1/m_Q$ expansion and a systematic approximation to full QCD may be constructed using the methods of effective field theory, the so-called Heavy Quark Effective Theory (HQET) [3]. The heavy mass limit and applications of HQET have been extensively studied, and the development of the field is documented in more or less extensive reviews [4].

The corrections to the heavy mass limit are characterized by two small parameters, namely the strong coupling constant, taken at the scale of the heavy quark $\alpha_s(m_Q)$ and the ratio $\bar{\Lambda}/m_Q$ of the scale of the light degrees of freedom $\bar{\Lambda}$ and the heavy quark mass. While the first kind of corrections may be calculated perturbatively in terms of Feynman diagrams, the second type needs additional non-perturbative input encoded in the matrix elements of higher dimensional operators.

However, the non-perturbative input is to be taken at a small scale, where the symmetries of HQET may be applied and hence the non-perturbative physics is constrained by these symmetries. In particular, for heavy to heavy decays these additional symmetries restrict the number of independent form factors to only a single one, of which the absolute normalization at a specific kinematic point is fixed by the symmetries.

By combining the method of the $1/m_Q$ expansion with the short distance expansion one may obtain a heavy mass expansion also for inclusive decay rates [5]-[7],[8]-[13]. The heavy quark mass sets a scale that is large compared to Λ_{QCD}, and one may use a similar setup as in deep inelastic scattering for the description of inclusive decays. In this way one may not only study total rates, but also differential distributions such as the lepton energy spectra in inclusive semileptonic decays.

This series of three lectures tries to summarize the basic ideas of the heavy mass expansion. In the first lecture the formulation of HQET as an effective field theory is described and the additional symmetries of heavy quarks are introduced. The second and third lecture deal with applications to exclusive and inclusive decays.

2 Effective Field Theory

Effective field theories [14] have become a widely used tool in modern elementary particle physics. An effective theory treatment is convenient if the problem under consideration involves very disparate mass scales such that the physics that is to be described happens at much lower energies than the scale set by some heavy particles in the theory. In such a case it is useful to switch to an effective theory in which the heavy degrees of freedom do not appear explicitly; they only reappear in the effective theory as higher dimensional operators, which are multiplied by coupling constants with negative mass dimension. The scale of the coupling constants is set by the large mass and thus these contributions are small, if the scale of the physics described with the help of the effective theory is small compared to this large mass.

An effective theory is always valid only in a limited region of scales, a natural cut-off is given by the mass of the particle which has been removed by switching from the full to the effective theory. As mentioned above an effective theory involves interactions which would lead to a non-renormalizable theory, if one would consider the theory to all orders in these higher dimensional operators. However, working to a definite order of the expansion in inverse powers of the large scale one does not face any problem concerning renormalization. Starting from the renormalizable dimension-4 piece of the effective theory Lagrangian we may use its renormalization group properties to study the cut-off dependence of the effective theory, which is determined by the short distance properties of the effective theory.

Applying effective theory methods corresponds to an expansion of the Greens functions of the full theory in inverse powers of the large mass scale; such an expansion is only possible up to logarithmic dependencies on this large scale. These logarithms may be accessed using a properly constructed effective theory, where these logarithms correspond to the renormalization group logarithms of the cut off. In this way one may even achieve a resummation of the logarithmic terms using renormalization group methods in the effective theory.

In the case at hand the large scale is the mass m_Q of the heavy quark, and the leading term of this expansion is the static limit. This effective theory (Heavy Quark Effective Theory, HQET) is a a powerful tool, allowing for numerous purely QCD based calculations. Renormalization in this effective theory implies a factorization theorem for the Greens function of full QCD, which means that to any order in the $1/m_Q$ expansion one may factorize the short distance physics from the long distance effects. Explicitly this means for a Greens function G_{full} calculated in full QCD that one may rewrite it as

$$G_{\text{full}}(p_1 \cdots p_n, m_Q, \mu_0 = m_Q) = \sum_j \left(\frac{1}{m_Q}\right)^j Z^{(j)}(m_Q, \mu) G_{\text{eff}}^{(j)}(p_1 \cdots p_n, \mu). \tag{1}$$

where μ_0 is the renormalization point of the full QCD function. Here the constants $Z^{(j)}(m_Q, \mu)$ depend on the factorization scale μ and on logarithms of m_Q; it contains all the short distance effects, which may be calculated perturbatively. Using the renormalization group of the effective theory, one may perform a systematic resummation of logarithmic dependencies on the heavy quark mass.

The Greens functions $G_{\text{eff}}^{(j)}$ are calculated in HQET and do not depend on the heavy mass any more; they contain the long distance effects which are not calculable via perturbation theory. However, as we shall see below, they are constrained by heavy quark symmetries. Consequently one gains complete control over the mass dependence by switching from full QCD to HQET, the effective theory obtained in the heavy mass limit.

In this section we shall discuss the formulation of this effective theory. First we shall review and compare some of the different possibilities to formulate the infinite mass limit. In this limit new symmetries appear which are the key to various model independent statements concerning weak decay matrix elements; these symmetries are reviewed in Sect. 3.1. Finally, we shall consider the systematic approach to the calculation of corrections to the infinite mass limit.

2.1 Infinite Quark Mass Limit as an Effective Field Theory

This issue has been discussed repeatedly and various formulations of the infinite mass limit are available. Of course, as far as physical quantities are concerned, all approaches yield the same result. However, for some special applications one approach may be more convenient than another.

The equivalence of all different approaches is ensured by a theorem well established in the field of effective theories. It has been shown that in an effective theory involving a field ϕ one may perform redefinitions of the fields, such that

$$\phi \to P(\phi, \partial\phi) \tag{2}$$

where P is an arbitrary polynomial function [15], [16].

Such a redefinition will not change the S-matrix, although the Lagrangian (and Greens function) expressed in the redefined fields may look completely different. In this sense the different formulations of HQET are equivalent.

In the following we shall consider two formulations of HQET. The first one nicely exhibits the fact that HQET is an effective theory in the sense that one integrates out a heavy degree of freedom and performs an expansion of the remaining action functional in the large mass scale, which for the case at hand is the heavy quark mass. The process of integrating out the heavy degree of freedom may be performed explicitly as a Gaussian functional integral, and one may construct a formulation of HQET by expanding the result in powers of $1/m_Q$. In this way one obtains a $1/m_Q$ expansion of both the heavy quark field as well as for the Lagrangian.

The second formulation is based on the standard way of separating "upper" and "lower" components of the spinor fields by performing a sequence of Foldy Wouthuysen transformations. These transformations lead to an expansion in $1/m_Q$ for the heavy quark field and for the Lagrangian, which is different for each of these quantities from the other formulation. However, if one calculates a physical quantity, both approaches will yield the same answer, because the $1/m_Q$ expansion has to be unique for an observable.

Integrating out heavy degrees of freedom One may obtain a formulation of the heavy mass limit by integrating out heavy degrees of freedom from the functional integral of QCD Greens functions [17]. This integration may in fact be done explicitly, since for the case at hand it amounts to a Gaussian functional integration. Formulating the heavy mass limit in this way clearly exhibits that it corresponds to an effective theory in the usual sense. We start from the generating functional of the QCD Greens functions

$$Z(\eta,\bar{\eta},\lambda) = \int [dQ][d\bar{Q}][d\phi_\lambda] \exp\left\{ iS + iS_\lambda + i\int d^4x \, (\bar{\eta}Q + \bar{Q}\eta + \phi_\lambda\lambda)\right\}, \tag{3}$$

where $\phi_\lambda = q$, A_μ^a denotes the light degrees of freedom (light quarks q and gluons A_μ) with the action S_λ, while S denotes the piece of the action for the heavy quark Q including its coupling to the gluons

$$S = \int d^4x \, \bar{Q}(i\not{D} - m_Q)Q. \tag{4}$$

where

$$D_\mu = \partial_\mu + igA_\mu, \tag{5}$$

is the covariant derivative of QCD. We have introduced source terms η for the heavy quark and λ for the light degrees of freedom.

We shall consider hadrons containing a single heavy quark, and we assume that this heavy hadron moves with a velocity v

$$v = \frac{p_{\text{hadron}}}{m_{\text{hadron}}}, \quad v^2 = 1, \quad v_0 > 0. \tag{6}$$

This velocity vector may be used to split the heavy quark field Q into an "upper" component ϕ and a "lower" one χ

$$\phi_v = \frac{1}{2}(1+\not{v})Q, \quad \not{v}\phi_v = \phi, \tag{7}$$

$$\chi_v = \frac{1}{2}(1-\not{v})Q, \quad \not{v}\chi_v = -\chi, \tag{8}$$

and to define a decomposition of the covariant derivative into a "longitudinal" and a "transverse" ($\perp$) part

$$D_\mu = v_\mu(v\cdot D) + D^\perp_\mu, \quad D^\perp_\mu = (g_{\mu\nu} - v_\mu v_\nu)D^\nu, \quad \{\not{D}^\perp, \not{v}\} = 0. \tag{9}$$

Using (7-9) the action (4) of the heavy quark field takes the form

$$S = \int d^4x \left[\bar{\phi}\{i(v\cdot D) - m_Q\}\phi - \bar{\chi}\{i(v\cdot D) + m_Q\}\chi + \bar{\phi}i\not{D}^\perp\chi + \bar{\chi}i\not{D}^\perp\phi\right]. \tag{10}$$

The heavy quark in the meson is very close to being on shell, and thus the space time dependence of the heavy quark field is mainly that of a free particle moving with velocity v. This suggests a reparametrization of the fields by removing the space time dependence of a solution of the free Dirac equation. We shall choose the "particle-type" parametrization corresponding to the "positive energy solution" of the Dirac equation

$$\phi_v = e^{-im_Q(v\cdot x)}h_v\,, \qquad \chi_v = e^{-im_Q(v\cdot x)}H_v, \tag{11}$$

such that the space time dependence of the remaining fields h_v and H_v is determined by the residual momentum $k = p - m_Q v$, which is due to binding effects of the heavy quark inside the heavy hadron, and which is a "small" quantity of order Λ_{QCD}.

Expressed in these fields the action of the heavy quark becomes

$$S = \int d^4x \left[\bar{h}_v i(v\cdot D)h_v - \bar{H}_v\{i(v\cdot D) + 2m_Q\}H_v + \bar{h}_v i\not{D}^\perp H_v + \bar{H}_v i\not{D}^\perp h_v\right]. \tag{12}$$

The term containing the sources is also rewritten in terms of the fields h_v and H_v

$$\int d^4x\,(\bar{\eta}\psi + \bar{\psi}\eta) = \int d^4x\,(\bar{\rho}_v h_v + \bar{h}_v\rho_v + \bar{R}_v H_v + \bar{H}_v R_v), \tag{13}$$

where ρ_v and R_v are now source terms for the upper component field h_v and the lower component part H_v respectively.

In terms of the new variables the generating functional reads

$$Z(\rho_v, \bar{\rho}_v, R_v, \bar{R}_v, \lambda) = \int [dh_v][d\bar{h}_v][dH_v][d\bar{H}_v][d\phi_\lambda] \tag{14}$$
$$\times \exp\left\{ iS + S_\lambda + i\int d^4x\, (\bar{\rho}_v h_v + \bar{h}_v \rho_v + \bar{R}_v H_v + \bar{H}_v R_v + \phi_\lambda \lambda) \right\},$$

where the action S for the heavy quark is given in (12).

From (12) it is obvious that the heavy degree of freedom is the lower component field H_v, since it has a mass term $2m_Q$, while the upper component field h_v is a massless field describing the static heavy quark. In the heavy mass limit only the Greens functions involving the field h_v have to be calculated, and hence we integrate over H_v in the functional integral (14) with the sources of the lower component field R_v and $\bar{R}_v$ set to zero. This can be done explicitly, since it is a Gaussian integration

$$Z(\rho_v, \bar{\rho}_v, \lambda) = \int [dh_v][d\bar{h}_v][d\lambda]\Delta \tag{15}$$
$$\times \exp\left\{ iS + S_\lambda + i\int d^4x\, (\bar{\rho}_v^+ h_v^+ + \bar{h}_v^+ \rho_v^+ + \phi_\lambda \lambda) \right\},$$

where now the action functional for the heavy quark becomes a non-local object

$$S = \int d^4x \left[\bar{h}_v^+ i(v\cdot D) h_v^+ - \bar{h}_v^+ \not{D}^\perp \left(\frac{1}{i(v\cdot D) + 2m_Q - i\epsilon} \right) \not{D}^\perp h_v^+ \right]. \tag{16}$$

This Gaussian integration corresponds to the replacement

$$H_v = \left(\frac{1}{2m_Q + ivD} \right) i\not{D}_\perp h_v \tag{17}$$

for the lower component field. Furthermore, the Gaussian integration yields a determinant Δ. In the full theory one may also perform this Gaussian integration, and the determinant obtained contains all the closed loops of heavy quarks. After renormalization of the full theory their contribution starts at order $1/m^2$ with an Uehling potential like term. In the effective theory one may take the determinant Δ to be a constant, if the terms of order $1/m_Q^2$ and higher coming from the closed heavy quark loops are included by matching to the full theory. Since we shall discuss only the leading term of the $1/m_Q$ expansion in this section, we may drop the determinant in what follows.

The non-locality of the action functional is connected to the large scale set by the heavy quark mass, and the non-local terms may be expanded in terms of an infinite series of local operators, which come with increasing powers of $1/m_Q$. In the context of a field theory this corresponds to a short distance expansion and hence these operators have to be renormalized. The tree level relations may be read off from the geometric series expansion of the non-local term in (16). In this way we obtain the expansion of the field and the Lagrangian

$$Q(x) = e^{-im_Q vx}\left[1 + \left(\frac{1}{2m_Q + ivD}\right) i\not{D}_\perp\right] h_v \tag{18}$$

$$= e^{-im_Q vx}\left[1 + \frac{1}{2m_Q}\not{D}_\perp + \left(\frac{1}{2m_Q}\right)^2 (-ivD)\not{D}_\perp + \cdots\right] h_v$$

$$\mathcal{L} = \bar{h}_v(ivD)h_v + \bar{h}_v i\not{D}_\perp \left(\frac{1}{2m_Q + ivD}\right) i\not{D}_\perp h_v \tag{19}$$

$$= \bar{h}_v(ivD)h_v + \frac{1}{2m_Q}\bar{h}_v(i\not{D}_\perp)^2 i h_v$$

$$+ \left(\frac{1}{2m_Q}\right)^2 \bar{h}_v(i\not{D}_\perp)(-ivD)(i\not{D}_\perp)h_v + \cdots$$

A Greens function with an operator insertion is treated in a similar way; the heavy quark fields entering the inserted operator are dealt with in the same way. The net effect of this is that the heavy quark fields in the operator insertion are replaced by the expansion (18).

Foldy Wouthuysen Transformation A second way of formulating the heavy mass limit proceeds along the well known steps performed in deriving the non-relativistic limit of the Dirac equation [18]. The reasoning used here is motivated by quantum mechanics; as a first step one rewrites the equation of motion in a Hamiltonian form

$$iv\partial Q = HQ, \quad H = \not{v}(\not{D}_\perp + m_Q + gvA) \tag{20}$$

where A is the gluon field. Note that in the rest frame we have $v = (1,0,0,0)$, and thus (20) takes the usual form, since $v\partial = \partial_0$ and $vA = A_0$.

In general the Hamiltonian couples the upper and the lower component of the heavy quark field Q, projected out by $P_\pm = (1 \pm \not{v})$. The Foldy Wouthuysen transformation is a transformation of the form

$$Q \to Q' = \exp(iF)Q \quad H \to H' = \exp(iF)\left[H - iv\partial\right]\exp(-iF) \tag{21}$$

where F is a hermitian matrix in the space of the Dirac spinors, such that $\exp(iF)$ is unitary in that space. This requirement is motivated by quantum mechanics where the spinor is interpreted as a wave function, but this interpretation becomes meaningless once we switch to a field theory.

The transformation (21) is designed such that the resulting Hamiltonian H' does not couple upper and lower components of the field Q any more. The generator F of this transformation may be expanded in powers of $1/m_Q$, from which one may construct a $1/m_Q$ expansion of the Hamiltonian and the transformed fields. Removing the mass term of the Hamiltonian by a phase redefinition as in (18), one obtains for the fields and the Lagrangian

$$\begin{aligned} Q(x) &= e^{-im_Q v\cdot x}\left[1+\frac{1}{2m_Q}(i\not{D}_\perp)+\right.\\ &\left.+\frac{1}{4m_Q^2}\left(v\cdot D\not{D}_\perp-\frac{1}{2}\not{D}_\perp^2\right)+\cdots\right]h_v(x)\,, \end{aligned} \tag{22}$$

$$\begin{aligned} \mathcal{L} &= \bar{h}_v\left[iv\cdot D-\frac{1}{2m_Q}\not{D}_\perp^2\right.\\ &\left.+\frac{i}{4m_Q^2}\left(-\frac{1}{2}\not{D}_\perp^2 v\cdot D+\not{D}_\perp v\cdot D\not{D}_\perp-\frac{1}{2}v\cdot D\not{D}_\perp^2\right)+\cdots\right]h_v\,. \end{aligned} \tag{23}$$

We note that the leading order term as well as the terms of order $1/m_Q$ are identical in the two approaches. Differences start to appear at order $1/m_Q^2$, which are terms involving a factor which would vanish by leading order equations of motion. In the Fouldy Wouthuysen formulation the Lagrangian does not contain such terms, while these terms appear in the approach of integrating out the lower components of the heavy quark field. However, subleading terms of a physical matrix element will consist of local contributions originating from the expansion of the field Q as well as of non-local pieces involving time-ordered products of the leading order currents with the subleading terms of the Lagrangian (see below). If the time-ordered products are taken with a term that would vanish by a naive application of the leading-order equation of motion, these will lead to a contact term, i.e. effectively to a local contribution. In this way the terms in the first approach rearrange in such a way that the final result for an observable quantity is the same in both cases. For some practical applications the Foldy Wouthuysen approach has an advantage that all the time-ordered products with the Lagrangian are truly non-local contributions, in other words, none of the contributions will lead to a contact term.

2.2 Corrections to the Heavy Mass Limit

Tree Level Considerations Corrections to the infinite mass limit may be considered in a systematic way. Starting from the tree level expressions given in the last sections one may use the expansion of the Lagrangian and the fields as given in Sect. 2.1. to construct the $1/m_Q$ expansion of full QCD matrix elements. In doing this it will not matter which representation of HQET (e.g. the one that is obtained from integrating out the heavy quark or the one constructed from the Foldy Wouthuysen transformation) is chosen, since the matrix elements have to have a unique $1/m_Q$ expansion.

As an example we shall consider a matrix element of a current $\bar{q}\Gamma Q$ mediating a transition between a heavy meson and some arbitrary state $|A\rangle$. The full QCD Lagrangian $\mathcal{L}$ and the fields Q are expanded in terms of a power series in $1/m_Q$ in the way described in Sect. 2.1, and the matrix element under consideration up to order $1/m_Q$ takes the form:

$$\begin{aligned}\langle A|\bar{q}\Gamma Q|M(v)\rangle &= \langle A|\bar{q}\Gamma h_v|H(v)\rangle \\ &+\frac{1}{2m_Q}\langle A|\bar{q}\Gamma P_- i\not{D} h_v|H(v)\rangle \\ &-i\int d^4x \langle A|T\{L_1(x)\bar{q}\Gamma h_v\}|H(v)\rangle + \mathcal{O}(1/m_Q^2)\end{aligned} \tag{24}$$

where L_1 are the first-order corrections to the Lagrangian as given in (19) or (23). Furthermore, $|M(v)\rangle$ is the state of the heavy meson in full QCD, including all its mass dependence, while $|H(v)\rangle$ is the corresponding state in the infinite mass limit.

Expression (24) displays the generic structure of the higher-order corrections as they appear in any HQET calculation. There will be local contributions coming from the expansion of the full QCD field; these may be interpreted as the corrections to the currents. The non-local contributions, i.e. the time-ordered products, are the corresponding corrections to the states and thus in the r.h.s. of (24) only the states of the infinite-mass limit appear. If one switches to another representation of HQET, one reshuffles terms from the fields into the Lagrangian; in this way the Lagrangian picks up operators which are proportional to the equations of motion.

As an example for the kinds of matrix elements appearing in subleading orders of the $1/m_Q$ expansion we consider the mass of a heavy hadron. In the infinite mass limit this mass is given in terms of the quark mass plus some "binding energy" $\bar{\Lambda}$. The corrections are of order $1/m_Q$ and are given by the matrix elements of the leading correction term of the Lagrangian. One obtains [19]

$$m_H = m_Q\left(1+\frac{\bar{\Lambda}}{m_Q}+\frac{1}{2m_Q^2}(\lambda_1+d_H\lambda_2)+\mathcal{O}(1/m_Q^3)\right) \tag{25}$$

where $d_H = 3$ for the 0^- and $d_H = -1$ for the 1^- meson. Up to this order no non-local terms appear; such terms show up the first time at order $1/m_Q^3$. The parameters $\bar{\Lambda}$, λ_1 and λ_2 correspond to matrix elements involving higher order terms that appear in the effective theory Lagrangian

$$\bar{\Lambda} = \frac{\langle 0|q\, iv\overleftarrow{D}\, \gamma_5 h_v|H(v)\rangle}{\langle 0|q\gamma_5 h_v|H(v)\rangle} \tag{26}$$

$$\lambda_1 = \frac{\langle H(v)|\bar{h}_v(iD)^2 h_v|H(v)\rangle}{2M_H} \tag{27}$$

$$\lambda_2 = \frac{\langle H(v)|\bar{h}_v\sigma_{\mu\nu}iD^\mu iD^\nu h_v|H(v)\rangle}{2M_H} \tag{28}$$

where the normalization of the states is chosen to be $\langle H(v)|\bar{h}_v h_v|H(v)\rangle = 2M_{\rm H} = 2(m_{\rm Q} + \bar{\Lambda})$. These parameters may be interpreted as the binding energy of the heavy meson in the infinite mass limit ($\bar{\Lambda}$), the expectation value of the kinetic energy of the heavy quark (λ_1) and its energy due to the chromomagnetic moment of the heavy quark (λ_2) inside the heavy meson. The latter two parameters play an important role since they parameterize the non-perturbative input needed in the subleading order of the $1/m_{\rm Q}$ expansion.

The only parameter which is easy to access is λ_2, since it is related to the mass splitting between $H(v)$ and $H^*(v,\epsilon)$. From the B-meson system we obtain

$$\lambda_2(m_{\rm b}) = \frac{1}{4}(M_{\rm H^*} - M_{\rm H}) = 0.12 \text{ GeV}^2; \tag{29}$$

from the charm system the same value is obtained. This shows that indeed the spin-symmetry partners are degenerate in the infinite mass limit and the splitting between them scales as $1/m_{\rm Q}$.

The other parameters appearing in (25) are not simply related to the hadron spectrum. Using the pole mass for $m_{\rm Q}$ in (25), QCD sum rules yield for $\bar{\Lambda}$ a value of $\bar{\Lambda} = 570 \pm 70$ MeV [4]. More problematic is the parameter λ_1; from its definition one is led to assume $\lambda_1 < 0$; a more restrictive inequality

$$-\lambda_1 > 3\lambda_2 \tag{30}$$

has been derived in a quantum mechanical framework in [20] and using heavy-flavor sum rules [21]. Furthermore, there exists also a sum rule estimate [22] for this parameter:

$$\lambda_1 = -0.52 \pm 0.12 \text{ GeV}^2. \tag{31}$$

This value is compatible with the bounds; however, it is unexpectedly large since it corresponds to a rms-momentum of the heavy quark inside the meson of

$$\sqrt{\langle \mathbf{p}^2 \rangle} \sim 720 \text{ MeV} \tag{32}$$

which is large compared to the naive guess of $-\lambda_1 \sim (\Lambda_{\rm QCD})^2$ This is the reason why also smaller values of λ_1 have been used in the literature.

Recently there has been an attempt [23] to extract $\bar{\Lambda}$ and λ_1 from the shape of the lepton energy spectrum in inclusive semileptonic B decays (see sections 4.3 and 4.4). The values obtained from this analysis are $\bar{\Lambda} = 0.39 \pm 0.11$ GeV and $-\lambda_1 = 0.19 \pm 0.10$ GeV^2, where the $\overline{\rm MS}$ definition of the mass has been used. The uncertainties quoted are only the 1σ statistical ones; the systematical uncertainties of this approach are difficult to estimate.

Beyond Tree Level Going beyond tree level will induce corrections of order $\alpha_s^n(m_{\rm Q})$, $n = 1, \ldots$. These may be calculated in terms of Feynman diagrams which may be evaluated using the Feynman rules of HQET. Only two Feynman rules are modified compared to full QCD:

	full QCD		HQET	
Propagator of the heavy quark	$\dfrac{i}{\not{p} - m_Q + i\epsilon}$	$\longrightarrow$	$\dfrac{i}{vk + i\epsilon}$,	$p = m_Q v + k$
Heavy quark gluon vertex	$ig\gamma_\mu T^a$	$\longrightarrow$	$igv_\mu T^a$	

For the sake of clarity we shall stick to our example of a heavy light current considered above. To leading order in the $1/m_Q$ expansion one may evaluate the radiative corrections to such a matrix element using the above Feynman rules and finds a divergent result with a divergence related to the short distance behavior. Since HQET is an effective theory, the machinery of effective theory guarantees the factorization of long distance effects from the short distance ones, which are related to the large mass m_Q. Neglecting $1/m_Q$ corrections, this factorization takes the form

$$\langle A|\bar{q}\Gamma Q|M(v)\rangle = Z\left(\frac{m_Q}{\mu}\right)\langle A|\bar{q}\Gamma h_v|H(v)\rangle|_\mu + \mathcal{O}(1/m_Q) \tag{33}$$

From Feynman rule calculations one obtains the perturbative expansion of the renormalization constant Z which generically looks like

$$\begin{aligned} Z\left(\frac{m_Q}{\mu}\right) &= a_{00} \\ &+ a_{11}\left(\alpha_s \ln\left(\frac{m_Q}{\mu}\right)\right) + a_{10}\alpha_s \\ &+ a_{22}\left(\alpha_s \ln\left(\frac{m_Q}{\mu}\right)\right)^2 + a_{21}\alpha_s\left(\alpha_s \ln\left(\frac{m_Q}{\mu}\right)\right) + a_{20}\alpha_s^2 \\ &+ a_{33}\left(\alpha_s \ln\left(\frac{m_Q}{\mu}\right)\right)^3 + a_{32}\alpha_s\left(\alpha_s \ln\left(\frac{m_Q}{\mu}\right)\right)^2 \\ &+ a_{31}\alpha_s^2\left(\alpha_s \ln\left(\frac{m_Q}{\mu}\right)\right) + a_{30}\alpha_s^3 + \cdots \end{aligned} \tag{34}$$

where $\alpha_s = g^2/(4\pi)$.

This factorization theorem corresponds to the statement that the ultraviolet divergences in the effective theory have to match the logarithmic mass dependencies of full QCD. The factorization scale μ is an arbitrary parameter, and the physical quantity $\langle A|\bar{q}\Gamma Q|M(v)\rangle$ does not depend on this parameter. However, calculating the matrix element of this operator in the effective theory and studying its ultraviolet behavior allows us to access the mass dependence of the matrix element $\langle A|\bar{q}\Gamma Q|M(v)\rangle$.

The ultraviolet behavior of the effective theory is investigated by the renormalization group equations. Differentiating (33) with respect to the factorization scale μ yields the renormalization group equation

$$\frac{d}{d\ln\mu}\left\{Z\left(\frac{m_Q}{\mu}\right)\langle A|\bar{q}\Gamma h_v|H(v)\rangle|_\mu\right\} = 0 \tag{35}$$

from which we may obtain an equation which determines the change of the coefficient Z when the scale is changed

$$\left(\frac{d}{d\ln\mu} + \gamma_J(\mu)\right) Z\left(\frac{m_Q}{\mu}\right) = 0 \tag{36}$$
$$\gamma_J(\mu) = \frac{d}{d\ln\mu}\ln(\langle A|\bar{q}\Gamma h_v|H(v)\rangle|_\mu).$$

The quantity γ_J is called the anomalous dimension of the operator $J = \bar{q}\Gamma h_v$ which is universal for all matrix elements of this operator, since it is connected with the short distance behavior of the insertion of J.

Eq.(36) describes the renormalization group scaling in the effective theory. It allows to shift logarithms of the large mass scale from the matrix element of J into the coefficient Z: If the matrix element is renormalized at the large scale m_Q the logarithms of the type $\ln m_Q$ will appear in the matrix element of J while the coefficient Z at this scale will simply be

$$Z(1) = a_{00} + a_{10}\alpha_s(m_Q) + a_{20}\alpha_s^2(m_Q) + a_{30}\alpha_s^3(m_Q) + \cdots \tag{37}$$

The renormalization group equation (36) allows to lower the renormalization point from m_Q to μ; the matrix element renormalized at μ will not contain any logarithms of m_Q any more, they will appear in the coefficient Z in the way shown in (34).

In all cases relevant in the present context the matrix elements will be matrix elements involving hadronic states, which are in most cases impossible to calculate from first principles. However, (36) allows to extract the short distance piece, i.e. the logarithms of the large mass m_Q and to separate it into the Wilson coefficients.

The anomalous dimension may be calculated in perturbation theory in powers of the coupling constant g of the theory. In general, in a renormalizable theory the coupling constant depends on the scale μ at which the theory is renormalized. The scale dependence of the coupling constant is determined by the β function

$$\frac{d}{d\ln\mu}g(\mu) = \beta(\mu). \tag{38}$$

In a mass independent scheme the renormalization group functions γ_O and β will depend on the scale μ only through their dependence on the coupling constant

$$\beta = \beta(g(\mu)) \qquad \gamma_J = \gamma_J(g(\mu)). \tag{39}$$

Hence we may rewrite the renormalization group equation (36) as

$$\left(\mu\frac{\partial}{\partial\mu} + \beta(g)\frac{\partial}{\partial g} + \gamma_J(g)\right) Z\left(\frac{m_Q}{\mu}, g\right) = 0. \tag{40}$$

The renormalization group functions β and γ_O are calculated in perturbation theory; the first term of the β function in QCD is obtained from a one-loop calculation and is given by

$$\beta(g) = -\frac{1}{(4\pi)^2}\left(11 - \frac{2}{3}n_f\right) g^3 + \cdots, \tag{41}$$

where n_f is the number of active flavors, i.e. the number of flavors with a mass less than m_Q.

With this input the renormalization group equation may be solved to yield

$$Z\left(\frac{m_Q}{\mu}\right) = a_{00}\left(\frac{\alpha_s(\mu)}{\alpha_s(m_Q)}\right)^{-\frac{48\pi^2}{33-2n_f}\gamma_1} \tag{42}$$

where γ_1 is the first coefficient in the perturbative expansion of the anomalous dimension $\gamma_J = \gamma_1 g^2 + \cdots$ and $\alpha_s(\mu)$ is the one loop expression for the running coupling constant of QCD

$$\alpha_s(\mu) = \frac{12\pi}{(33 - 2n_f)\ln(\mu^2/\Lambda^2_{QCD})} \tag{43}$$

which is obtained from solving (38) using (41). This expression corresponds to a summation of the leading logarithms $(\alpha_s \ln m_Q)^n$ which is achieved by a one-loop calculation of the renormalization group functions β and γ_Q; in other words, in this way a resummation of the first column of the expansion (34) is obtained.

In a similar way one may also resum the second column of (34), if the renormalization group functions β and γ are calculated to two loops and the finite terms of the one loop expression are included.

Finally, the case we have considered as an example is indeed very simple; in general all operators of a given dimension may mix under renormalization, i.e. instead of a simple anomalous dimension a matrix of anomalous dimensions may occur. For more details on this I refer the reader to a textbook discussion of these issues as given e.g. in [24].

2.3 Heavy Quark Symmetries

The main impact of the heavy quark limit is due to two additional symmetries which are not present in full QCD; the first is a heavy flavor symmetry and the second one is the so-called spin symmetry. The presence of these symmetries implies Wigner-Eckart theorems for transition matrix elements which have far-reaching phenomenological consequences.

We shall first study the heavy flavor symmetry. The interaction of the quarks with the gluons is flavor independent; all flavor dependence in QCD is only due to the different quark masses. In the $1/m_Q$ expansion the leading order Lagrangian is mass independent and hence a flavor symmetry appears relating heavy quarks moving with the same velocity.

For the case of two heavy flavors b and c one has to leading order the Lagrangian [3]

$$\mathcal{L}_{\text{heavy}} = \bar{b}_v(v \cdot D)b_v + \bar{c}_v(v \cdot D)c_v, \tag{44}$$

where b_v (c_v) is the field operator h_v for the b (c) quark moving with velocity v and $D = \partial + igA$ is the QCD covariant derivative. This Lagrangian is obviously invariant under the $SU(2)_{\text{HF}}$ rotations

$$\begin{pmatrix} b_v \\ c_v \end{pmatrix} \to U_v \begin{pmatrix} b_v \\ c_v \end{pmatrix} \quad U_v \in SU(2)_{\text{HF}}. \tag{45}$$

We have put a subscript v for the transformation matrix U, since this symmetry only relates heavy quarks moving with the same velocity.

The second symmetry is the heavy-quark spin symmetry. As is clear from the Lagrangian in the heavy-mass limit, both spin degrees of freedom of the heavy quark couple in the same way to the gluons; we may rewrite the leading-order Lagrangian as

$$\mathcal{L} = \bar{h}_v^{+s}(ivD)h_v^{+s} + \bar{h}_v^{-s}(ivD)h_v^{-s}, \tag{46}$$

where $h_v^{\pm s}$ are the projections of the heavy quark field on a definite spin direction s

$$h_v^{\pm s} = \frac{1}{2}(1 \pm \gamma_5 \not{s})h_v, \quad s \cdot v = 0. \tag{47}$$

This Lagrangian has a symmetry under the rotations of the heavy quark spin and hence all the heavy hadron states moving with the velocity v fall into spin-symmetry doublets as $m_Q \to \infty$. In Hilbert space this symmetry is generated by operators $S_v(\epsilon)$ as

$$[h_v, S_v(\epsilon)] = i\not{\epsilon}\not{v}\gamma_5 h_v \tag{48}$$

where ϵ with $\epsilon^2 = -1$ is the rotation axis. The simplest spin-symmetry doublet in the mesonic case consists of the pseudoscalar meson H(v) and the corresponding vector meson H*(v, ϵ), since a spin rotation yields

$$\exp\left(iS_v(\epsilon)\frac{\pi}{2}\right)|H(v)\rangle = (-i)|H^*(v,\epsilon)\rangle, \tag{49}$$

where we have chosen an arbitrary phase to be $(-i)$.

In the heavy-mass limit the spin symmetry partners have to be degenerate and their splitting has to scale as $1/m_Q$. In other words, the quantity

$$\lambda_2 = \frac{1}{4}(M_{\text{H}^*}^2 - M_{\text{H}}^2) \tag{50}$$

has to be the same for all spin symmetry doublets of heavy ground state mesons. This is well supported by data: For both the (B,B*) and the (D,D*) doublets one finds a value of $\lambda_2 \sim 0.12\ \mathrm{GeV}^2$. This shows that the spin-symmetry partners become degenerate in the infinite mass limit and the splitting between them scales as $1/m_Q$.

In the infinite mass limit the symmetries imply relations between matrix elements involving heavy quarks. For a transition between heavy ground-state mesons H (either pseudoscalar or vector) with heavy flavor f (f') moving with velocities v (v'), one obtains in the heavy-quark limit

$$\langle H^{(f')}(v')|\bar{h}_{v'}^{(f')}\Gamma h_v^{(f)}|H^{(f)}(v)\rangle = \xi(vv')\ \mathrm{Tr}\left\{\overline{\mathcal{H}(v)}\Gamma\mathcal{H}(v)\right\}, \tag{51}$$

where Γ is some arbitrary Dirac matrix and $H(v)$ are the representation matrices for the two possibilities of coupling the heavy quark spin to the spin of the light degrees of freedom, which are in a spin-1/2 state for ground state mesons

$$\mathcal{H}(v) = \frac{\sqrt{M_\mathrm{H}}}{2}\begin{cases} (1+\not{v})\gamma_5 & 0^-,\ (\bar{\mathrm{q}}\mathrm{Q})\ \text{meson} \\ (1+\not{v})\not{\epsilon} & 1^-,\ (\bar{\mathrm{q}}\mathrm{Q})\ \text{meson} \\ & \text{with polarization } \epsilon. \end{cases} \tag{52}$$

Due to the spin and flavor independence of the heavy mass limit the Isgur–Wise function ξ is the only non-perturbative information needed to describe all heavy to heavy transitions within a spin-flavor symmetry multiplet.

Excited mesons have been studied in [25]. They may be classified by the angular momentum of the light degrees of freedom j_l, which is coupled with the heavy quark spin S to the total angular momentum J of the meson. Furthermore, the orbital angular momentum ℓ determines the parity $P = (-1)^{\ell+1}$ of the meson. For a given $\ell > 0$ we can have $j_\mathrm{l} = \ell \pm 1/2$ and the coupling of the heavy quark spin yields two spin symmetry doublets $(J = \ell - 1, J = \ell)$ and $(J = \ell, J = \ell + 1)$. For example, the lowest positive parity $\ell = 1$ mesons are two spin symmetry doublets $(0^+, 1^+)$ and $(1^+, 2^+)$. In the D meson system these states have been observed [26] and behave as predicted by heavy quark symmetry [27]

Similarly as for the mesons heavy-quark symmetries imply that only one form factor is needed to describe heavy to heavy transitions within a spin flavor symmetry multiplet; in other words, there is an Isgur Wise function for each multiplet.

The ground state baryons have been studied in [28], [29], [30]. According to the particle data group they are classified as follows

$$\Lambda_h = [(qq')_0h]_{1/2} \qquad \Xi'_h = [(qs)_0h]_{1/2} \tag{53}$$

$$\Sigma_h = [(qq')_1h]_{1/2} \qquad \Xi_h = [(qs)_1h]_{1/2} \qquad \Omega_h = [(ss)_1h]_{1/2} \tag{54}$$

$$\Sigma_h^* = [(qq')_1h]_{3/2} \qquad \Xi_h^* = [(qs)_1h]_{3/2} \qquad \Omega_h^* = [(ss)_1h]_{3/2}. \tag{55}$$

Here, q, q' refer to u and d quarks, $q \neq q'$ for the Λ_h, but q may be the same as q' for the Σ_h and Σ_h^*. The first subscript (0, 1) is the total spin of the light degrees of freedom, while the second subscript (1/2, 3/2) is the total spin of the baryon.

Spin symmetry forces these baryons into spin symmetry doublets. For the Λ-type baryons (53) the spin rotations are simply a subset of the Lorentz transformations, since the light degrees of freedom are in a spin-0 state. The corresponding spin symmetry doublet is in this case given by the two polarization directions of the heavy baryon. From the point of view of heavy quark symmetries the Λ-type baryons are the simplest hadrons, although from the quark model point of view they are composed of three quarks.

The baryons with the light degrees of freedom in a spin one state may be represented by a pseudovector-spinor object R^μ with $v_\mu R^\mu = 0$ [1]. In general $\gamma_\mu R^\mu \neq 0$ because R^μ contains spin 1/2 contributions as well as spin 3/2 parts. In other words, R^μ contains a Rarita-Schwinger field as well as a Dirac field. Under Lorentz transformations R^μ behaves as

$$R^\mu(v) \to \Lambda^\mu{}_\nu D(\Lambda) R^\nu(\Lambda v), \tag{56}$$

where $\Lambda_{\mu\nu}$ and $D(\Lambda)$ are the Lorentz transformations in the vector and spinor representation, respectively, while under spin rotations we have

$$R^\mu(v) \to -\gamma_5 \not{v} \not{\epsilon} R^\mu(v). \tag{57}$$

The spin-3/2 component of the the pseudovector-spinor object corresponding to the Σ_h^* is projected out by contracting with γ_μ

$$\gamma_\mu R^\mu_{\Sigma_h^*} = 0. \tag{58}$$

The rest of the independent components of R correspond to Σ_h baryon:

$$R^\mu_{\Sigma_h} = \frac{1}{\sqrt{3}}(\gamma^\mu + v^\mu)\gamma_5 u_{\Sigma_h}, \tag{59}$$

where u_{Σ_h} is the Dirac spinor of the Σ_h state. Similar expressions hold for the non-strange baryons $\Xi_h^{(*)}$ and $\Omega_h^{(*)}$.

The spin rotation (57) transforms the Σ-like baryons into the Σ^* states and vice versa. Thus the spin symmetry doublets for the ground state baryons are given by the two polarization directions of the baryons in (53), and by the two states with corresponding light quark flavor numbers in (54) and (55).

Similar to the case of mesons one may derive a Wigner-Eckart theorem for the spin symmetry doublets of the baryons

$$\langle \Lambda_h(v)|\bar{h}\Gamma h'|\Lambda_{h'}(v')\rangle = A(v\cdot v')\bar{u}_{\xi_h}(v)\Gamma u_{\xi_{h'}}(v'), \tag{60}$$

where we have allowed for the possibility of two heavy quark flavors h and h'. In the same way, one obtains two form factors for the $\Sigma_h^{(*)} \to \Sigma_{h'}^{(*)}$ transition

$$\begin{aligned} &\langle \Sigma_h^{(*)}(v)|\bar{h}_v \Gamma h_{v'}|\Sigma_h^{(*)}\rangle \\ &= \bar{R}^\mu_{\Sigma_h^{(*)}}(v)\Gamma R^\nu_{\Sigma_h^{(*)}}(v')\left[B(v\cdot v')g_{\mu\nu} + C(v\cdot v')v'_\mu v_\nu\right]. \end{aligned} \tag{61}$$

[1] One could as well represent the light degrees of freedom by an antisymmetric tensor instead of a pseudovector; this is a completely equivalent formulation [30].

Finally, parity does not allow for transitions between Λ and $\Sigma^{(*)}$ type baryons

$$\langle \Sigma_h^{(*)}(v)|\bar{h}_v \Gamma h_{v'}|\Lambda_h(v)\rangle = 0, \tag{62}$$

and hence these transitions are not only suppressed by the flavor symmetry of the light degrees of freedom, but additionally by heavy quark symmetry.

Excited baryons may be studied along the same lines as for the mesons. The spin symmetry doublets as well as the restrictions on transition matrix elements have been studied in [25].

Heavy quark symmetries thus lead to a strong reduction of the number of independent form factors that describe current induced transitions among heavy hadrons. In addition to that the symmetries even allow us to obtain the normalization of some of these form factors. Since the currents

$$J^{hh'} = \bar{h}_v \gamma_\mu h'_v = v_\mu \bar{h}_v h'_v \tag{63}$$

are the generators of heavy flavor symmetry in the velocity sector v, the normalization of the Wigner-Eckard theorems (51,60,61) is known at the non-recoil point $v = v'$. By standard arguments one obtains for the mesons

$$\xi(vv' = 1) = 1, \tag{64}$$

while the corresponding relation for the baryons is

$$A(vv' = 1) = \sqrt{m_{\Lambda_h} m_{\Lambda_{h'}}} \tag{65}$$

$$B(vv' = 1) = \sqrt{m_{\Sigma_h^{(*)}} m_{\Sigma_{h'}^{(*)}}}\,, \tag{66}$$

where the factor involving the square root of the masses means that the hadron states in (60) are normalized relativistically.

Up to now we have considered only the consequences of heavy quark symmetries for the leading terms of the $1/m_Q$ expansion. However, the additional symmetries also restrict the subleading terms and one of these restrictions is called Lukes theorem [31]. It is a generalization of the Ademollo Gatto theorem [32], which states that in the presence of explicit symmetry breaking the matrix elements of the currents that generate the symmetry are still normalized up to terms which are second order in the symmetry breaking interaction.

For the case at hand the relevant symmetry is the heavy flavor symmetry. This symmetry is an $SU(2)$ symmetry and is generated by three operators $Q_\pm$ and Q_3 with

$$\begin{aligned}
&Q_+ = \int d^3x\, \bar{b}_v(x) c_v(x) \quad Q_- = \int d^3x\, \bar{c}_v(x) b_v(x) \\
&Q_3 = \int d^3x\, (\bar{b}_v(x) b_v(x) - \bar{c}_v(x) c_v(x)) \\
&[Q_+, Q_-] = Q_3 \qquad [Q_+, Q_3] = -2Q_+ \qquad (Q_+)^\dagger = Q_-
\end{aligned} \tag{67}$$

Let us denote the ground state flavor symmetry multiplet as $|B\rangle$ and $|D\rangle$. Then the operators act in the following way

$$\begin{aligned} &Q_3|B\rangle = |B\rangle \qquad Q_3|D\rangle = -|D\rangle \\ &Q_+|D\rangle = |B\rangle \qquad Q_-|B\rangle = |D\rangle \\ &Q_+|B\rangle = Q_-|D\rangle = 0. \end{aligned} \tag{68}$$

The Hamiltonian of this system has a $1/m_{\mathrm{Q}}$ expansion of the form

$$\begin{aligned} H &= H_0^{(\mathrm{b})} + H_0^{(\mathrm{c})} + \frac{1}{2m_{\mathrm{b}}} H_1^{(\mathrm{b})} + \frac{1}{2m_{\mathrm{c}}} H_1^{(\mathrm{c})} + \cdots \\ &= H_0^{(\mathrm{b})} + H_0^{(\mathrm{c})} + \frac{1}{2}\left(\frac{1}{2m_{\mathrm{b}}} + \frac{1}{2m_{\mathrm{c}}}\right)(H_1^{(\mathrm{b})} + H_1^{(\mathrm{c})}) \\ &\quad + \frac{1}{2}\left(\frac{1}{2m_{\mathrm{b}}} - \frac{1}{2m_{\mathrm{c}}}\right)(H_1^{(\mathrm{b})} - H_1^{(\mathrm{c})}) + \cdots \\ &= H_{\mathrm{symm}} + H_{\mathrm{break}}. \end{aligned} \tag{69}$$

In the second equation, the first line is still symmetric under heavy flavor $SU(2)$ while the term in the second line does not commute any more with $Q_\pm$, but it still commutes with Q_3. In other words, to order $1/m_{\mathrm{Q}}$ we still have common eigenstates of H and Q_3, which we shall denote as $|\tilde{B}\rangle$ and $|\tilde{D}\rangle$. Sandwiching the commutation relation we get

$$\begin{aligned} 1 &= \langle\tilde{B}|Q_3|\tilde{B}\rangle = \langle\tilde{B}|[Q_+, Q_-]|\tilde{B}\rangle \\ &= \sum_n \left[\langle\tilde{B}|Q_+|\tilde{n}\rangle\langle\tilde{n}|Q_-|\tilde{B}\rangle - \langle\tilde{B}|Q_-|\tilde{n}\rangle\langle\tilde{n}|Q_+|\tilde{B}\rangle\right] \\ &= \sum_n \left[|\langle\tilde{B}|Q_+|\tilde{n}\rangle|^2 - |\langle\tilde{B}|Q_-|\tilde{n}\rangle|^2\right] \end{aligned} \tag{70}$$

where $|\tilde{n}\rangle$ form a complete set of states of the Hamiltonian $H_{\mathrm{symm}} + H_{\mathrm{break}}$. The matrix elements may be written as

$$\langle\tilde{B}|Q_\pm|\tilde{n}\rangle = \frac{1}{E_B - E_n}\langle\tilde{B}|[H_{\mathrm{break}}, Q_\pm]|\tilde{n}\rangle \tag{71}$$

where E_B and E_n are the energies of the states $|\tilde{B}\rangle$ and $|\tilde{n}\rangle$ respectively. In the case $|\tilde{n}\rangle = |\tilde{D}\rangle$ the matrix element will be of order unity, since both the numerator as well as the energy difference in the denominator are of the order of the symmetry breaking. For all other states the energy difference in the denominator is non-vanishing in the symmetry limit, and hence this difference is of order unity; thus the matrix element for these states will be of the order of the symmetry breaking. From this we conclude

$$\langle\tilde{B}|Q_+|\tilde{D}\rangle = 1 + \mathcal{O}\left[\left(\frac{1}{2m_{\mathrm{b}}} - \frac{1}{2m_{\mathrm{c}}}\right)^2\right]. \tag{72}$$

In particular, the weak transition currents at the non-recoil point $v = v'$ are proportional to these symmetry generators and hence we may conclude that for some of these matrix elements we only have corrections of the order $1/m_Q^2$.

Another restriction on the $1/m_Q$ expansion is imposed by the so-called reparametrization invariance [33] which is basically the remnant of the original Lorentz covariance of full QCD. The full theory depends only on the momentum of the heavy quark P, and the splitting of this momentum in an on-shell part $m_Q v$ and a residual momentum k corresponding to the covariant derivative acting on the heavy static field h_v is arbitrary. Formally this means that the Lagrangian does not depend on the velocity v, if all orders of the $1/m_Q$ expansion are included; the v dependence only enters once the expansion is truncated.

Using the representation (19) and (18) the Lagrangian is invariant under the transformation

$$
\begin{aligned}
&v \to v + \delta v \qquad v \cdot \delta v = 0 \\
&h_v \to h_v + \frac{\delta \not{v}}{2}\left(1 + P_- \frac{1}{2m + ivD} i\not{D}\right) h_v \\
&iD \to -m\,\delta v.
\end{aligned}
\tag{73}
$$

This invariance is the so-called reparametrization invariance, which has non-trivial consequences, since it relates terms of different orders of the $1/m$ expansion.

3 Application to Exclusive Decays

The heavy mass limit and the resulting additional symmetries allow us to restrict the matrix elements which occur in weak transitions of heavy hadrons. We shall consider in the following in some detail the semileptonic $\mathrm{b} \to \mathrm{c}$ transition, which we shall treat as a heavy to heavy decay. In section 3.2 we investigate the consequences of heavy quark symmetries for transitions of the heavy to light type.

3.1 Transitions of the Type Heavy → Heavy

For the case of a heavy to heavy transition the Wigner Eckart theorem (51) implies that there is only a single form factor which describes the weak decays of heavy hadrons; furthermore, the heavy mass limit yields the normalization of this form factor at the kinematic point $v = v'$.

Treating both the b and the c quark as heavy, the semileptonic decays $\mathrm{B} \to \mathrm{D}^{(*)} \ell \nu$ are the phenomenologically relevant examples. The matrix elements for these transitions are in general parameterized in terms of six form factors

$$
\langle D(v')|\bar{c}\gamma_\mu b|B(v)\rangle = \sqrt{m_\mathrm{B} m_\mathrm{D}}\left[\xi_+(y)(v_\mu + v'_\mu) + \xi_-(y)(v_\mu - v'_\mu)\right] \tag{74}
$$

$$\langle D^*(v',\epsilon)|\bar{c}\gamma_\mu b|B(v)\rangle = i\sqrt{m_{\rm B}m_{\rm D^*}}\xi_V(y)\varepsilon_{\mu\alpha\beta\rho}\epsilon^{*\alpha}v'^\beta v^\rho \tag{75}$$

$$\begin{aligned}\langle D^*(v',\epsilon)|\bar{c}\gamma_\mu\gamma_5 b|B(v)\rangle &= \sqrt{m_{\rm B}m_{\rm D^*}}\left[\xi_{A1}(y)(vv'+1)\epsilon^*_\mu - \xi_{A2}(y)(\epsilon^* v)v_\mu\right.\\ &\left.\quad -\xi_{A2}(y)(\epsilon^* v)v'_\mu\right],\end{aligned} \tag{76}$$

where we have defined $y = vv'$. Due to the Wigner Eckart theorem (51) these six form factors are related to the Isgur Wise function by

$$\xi_i(y) = \xi(y) \text{ for } i = +, V, A1, A3, \qquad \xi_i(y) = 0 \text{ for } i = -, A2. \tag{77}$$

Since heavy quark symmetries also yield the normalization of the Isgur Wise function, we know the absolute value of the differential rate at the point $v = v'$ in terms of the meson masses and V_{cb}. Hence we may use this to extract V_{cb} from these decays in a model independent way by extrapolating the lepton spectrum to the kinematic endpoint $v = v'$. Using the mode $\mathrm{B} \to \mathrm{D}^{(*)}\ell\nu$ one obtains the relation

$$\lim_{v\to v'} \frac{1}{\sqrt{(vv')^2-1}}\frac{d\Gamma}{d(vv')} = \frac{G_{\rm F}^2}{4\pi^3}|V_{cb}|^2(m_{\rm B}-m_{\rm D^*})^2 m_{\rm D^*}^3|\xi_{A1}(1)|^2, \tag{78}$$

where ξ_{A1} is equal to the Isgur Wise function in the heavy mass limit, and hence $\xi_{A1}(1) = 1$.

Corrections to this relation have been calculated along the lines outlined above in leading and subleading order. A complete discussion may be found in more extensive review articles (see e.g. Neubert's review [4]), including reference to the original papers. Here we only state the final result

$$\begin{aligned}\xi_{A1}(1) = x^{6/25}&\left[1 + 1.561\frac{\alpha_s(m_c)-\alpha_s(m_b)}{\pi} - \frac{8\alpha_s(m_c)}{3\pi}\right.\\ &\left.+z\left\{\frac{25}{54} - \frac{14}{27}x^{-9/25} + \frac{1}{18}x^{-12/25} + \frac{8}{25}\ln x\right\} - \frac{\alpha_s(\bar{m})}{\pi}\frac{z^2}{1-z}\ln z\right]\\ &+\delta_{1/m^2},\end{aligned} \tag{79}$$

where we use the abbreviations

$$x = \frac{\alpha_s(m_c)}{\alpha_s(m_b)}, \qquad z = \frac{m_c}{m_b}$$

and $\bar{m}$ is a scale somewhere between m_b and m_c.

Up to the term δ_{1/m^2} all these contributions may be calculated perturbatively, including the dependence on z. The quantity δ_{1/m^2} parameterizes the non-perturbative contributions, which enter here at order $1/m^2$. These corrections may be expressed in terms of the kinetic energy λ_1, the chromomagnetic moment λ_2, which are given in (27) and (28) respectively, and matrix elements involving time-ordered products between the current and the corrections of the Lagrangian

$$\begin{aligned}\delta_{1/m^2} &= -\left(\frac{1}{2m_c}\right)^2 \frac{1}{2}\left(-\lambda_1 + \lambda_2 + (-i)^2 \frac{1}{2\sqrt{M_B M_D}}\right. \\ &\times \left.\int d^4x\, d^4y\, \langle B^*(v,\epsilon)| T\left[\mathcal{L}_b^{(1)}(x) \bar{b}_v c_v \mathcal{L}_c^{(1)}(y)\right] |D^*(v,\epsilon)\rangle\right) \\ &+ \mathcal{O}(1/m_c^3, 1/m_b^2, 1/(m_c m_b)),\end{aligned} \tag{80}$$

where $\mathcal{L}_Q^{(1)}$ is the first order Lagangian for the quark Q as given in (19) or (23) and M_B (M_D) are the masses of the B (D) meson in the heavy quark limit. Here we display only the largest contribution of order $1/m_c^2$; the complete expression, including the $1/m_b^2$ and $1/(m_c m_b)$ terms, may be found in [34], [35].

Thus the correction δ_{1/m^2} is given in terms of λ_1 defined in (27), λ_2 given in (28) and a non-local matrix element involving a time-ordered product. The problem concerning the determination of λ_1 has been considered already above; similarly it is not easy to obtain information on the matrix element involving the time-ordered product, and thus the corrections of order $1/m^2$ will finally limit our ability to determine the CKM matrix element V_{cb} in a model independent way, at least using the approach described here.

Various estimates for δ_{1/m^2} have been given in the literature. The first estimate of this correction has been given in [34] using the GISW model [36], which is based on a wave function for the light quark. In this work $\delta_{m^2} = -2\% \ldots -3\%$ has been obtained. Another estimate with weaker assumptions yields $\delta_{m^2} = 0 \ldots -5\%$ [35], but both estimates have been criticized recently as being too small. Based on heavy flavor sum rules it has been argued in [37] that the $1/m^2$ corrections can be quite large $\delta_{m^2} = 0\% \ldots -8\%$ [37]. These various estimates indicate the size of the theoretical error involved in the determination of V_{cb} from the exclusive channel $B \to D^* \ell \bar{\nu}_\ell$; a generally accepted value for these corrections has been given recently [38]

$$\delta_{m^2} = -0.055 \pm 0.025 \tag{81}$$

from which one obtains

$$\xi_{A1}(1) = 0.91 \pm 0.03\,. \tag{82}$$

This result has been used to extract V_{cb} from CLEO [39] as well as from LEP data [40]. The values obtained are

$$|V_{cb}| = 0.0386 \pm 0.0019 \pm 0.0020 \pm 0.0014 \qquad \text{CLEO} \tag{83}$$

$$|V_{cb}| = 0.0392 \pm 0.0025 \pm 0.0027 \pm 0.0015 \qquad \text{ALEPH} \tag{84}$$

where (82) has been used. Note that the third error in $|V_{cb}|$ is due to the theoretical uncertainties, which by now almost match the experimental ones.

3.2 Transitions of the Type Heavy → Light

Heavy quark symmetries may also be used to restrict the independent form factors appearing in heavy to light decays. For the decays of heavy mesons into light 0^- and 1^- particles heavy quark symmetries restrict the number of independent form factors to six, which is just the number needed to parameterize the semileptonic decays of this type. Furthermore, no absolute normalization of form factors may be obtained from heavy quark symmetries in the heavy to light case; only the relative normalization of B meson decays heavy to light transitions may be obtained from the corresponding D decays.

In general we shall discuss matrix elements of a heavy to light current which have the following structure

$$J = \langle A|\bar{\ell}\Gamma h_v|H(v)\rangle, \tag{85}$$

where Γ is an arbitrary Dirac matrix, ℓ is a light quark (u, d or s) and A is a state involving only light degrees of freedom.

Spin symmetry implies that the heavy quark index hooks directly to the heavy quark index of the Dirac matrix of the current. Thus one may write for the transition matrix element (91)

$$\langle A|\bar{\ell}\Gamma h_v|H(v)\rangle = \text{ Tr } (\mathcal{M}_A \Gamma H(v)) \tag{86}$$

where the matrix $H(v)$ representing the heavy meson has been given in (52). The matrix $\mathcal{M}_A$ describes the light degrees of freedom and is the most general matrix which may be formed from the kinematical variables involved. Furthermore, if the energies of the particles in the state A are small, i.e. of the order of $\Lambda_{\rm QCD}$, the matrix $\mathcal{M}_A$ does not depend on the heavy quark; in particular it does not depend on the heavy mass $m_{\rm H}$. In the following we shall discuss some examples.

The first example is the heavy meson decay constant, where the state A is simply the vacuum state. The heavy meson decay constant is defined by

$$\langle 0|\bar{\ell}\gamma_\mu\gamma_5 h_v|H(v)\rangle = f_{\rm H} m_{\rm H} v_\mu, \tag{87}$$

and since $|A\rangle = |0\rangle$ the matrix $\mathcal{M}_0$ is simply the unit matrix times a dimensionful constant[2] and one has, using (86)

$$\langle 0|\bar{\ell}\gamma_\mu\gamma_5 h_v|H(v)\rangle = \kappa \text{ Tr } (\gamma\gamma_5 H(v)) = 2\kappa\sqrt{m_{\rm H}} v_\mu. \tag{88}$$

As discussed above the constant κ does not depend on the heavy mass and thus one infers the well-known scaling law for the heavy meson decay constant from the last two equation

$$f_{\rm H} \propto \frac{1}{\sqrt{m_{\rm H}}}. \tag{89}$$

[2] Note that contributions proportional to $\not{v}$ may be eliminated using

$$H(v)\not{v} = -H(v).$$

Including the leading and subleading QCD radiative corrections one obtains a relation between f_B and f_D

$$f_\mathrm{B} = \sqrt{\frac{m_\mathrm{c}}{m_\mathrm{b}}} \left(\frac{\alpha_\mathrm{s}(m_\mathrm{b})}{\alpha_\mathrm{s}(m_\mathrm{c})}\right)^{-6/25} \left[1 + 0.894 \frac{\alpha_\mathrm{s}(m_\mathrm{c}) - \alpha_\mathrm{s}(m_\mathrm{b})}{\pi}\right] f_\mathrm{D} \sim 0.69 f_\mathrm{D}. \tag{90}$$

The second example are transitions of a heavy meson into a light pseudoscalar meson, which we shall denote as π. The matrix element corresponding to (85) is

$$J_\mathrm{P} = \langle \pi(p) | \bar{\ell} \Gamma h_v | H(v) \rangle, \tag{91}$$

where p is the momentum of the light quark,

The Dirac matrix $\mathcal{M}_\mathrm{P}$ for the light degrees of freedom appearing now in (86) depends on p and v. It may be expanded in terms of the sixteen independent Dirac matrices 1, γ_5, γ_μ, $\gamma_5\gamma_\mu$, and $\sigma_{\mu\nu}$ taking into account that it has to behave like a pseudoscalar. The form factors appearing in the decomposition of $\mathcal{M}_\mathrm{P}$ depend on the variable $v \cdot p$, the energy of the light meson in the rest frame of the heavy one. In order to compare different heavy to light transitions by employing heavy flavor symmetry this energy must be sufficiently small, since the typical scale for the light degrees of freedom has to be of the order of Λ_QCD to apply heavy quark symmetry[3]. For the case of a light pseudoscalar meson the most general decomposition of $\mathcal{M}_P$ is

$$\mathcal{M}_P = \sqrt{v \cdot p} A(\eta) \gamma_5 + \frac{1}{\sqrt{v \cdot p}} B(\eta) \gamma_5 \not{p}, \tag{92}$$

where we have defined the dimensionless variable

$$\eta = \frac{v \cdot p}{\Lambda_\mathrm{QCD}}. \tag{93}$$

The form factors A and B are universal in the kinematic range of small energy of the light meson, i.e. where the momentum transfer to the light degrees of freedom is of the order Λ_QCD; in this region η is of order unity. This universality of the form factors may be used to relate various kinds of heavy to light transitions, e.g. the semileptonic decays like $\mathrm{D} \to \pi e \nu$, $\mathrm{D} \to \mathrm{K} e \nu$ or $\mathrm{B} \to \pi e \nu$ and also the rare decays like $\mathrm{B} \to \mathrm{K} \ell^+ \ell^-$ or $\mathrm{B} \to \pi \ell^+ \ell^-$ where ℓ denotes an electron or a muon.

As an example we give the relations between exclusive semileptonic heavy to light decays. The relevant hadronic current for this case may be expressed in terms of two form factors

$$\begin{aligned} \langle \pi(p) | \bar{\ell} \gamma (1 - \gamma_5) h_v | H(v) \rangle &= F_1(v \cdot p) m_\mathrm{H} v_\mu + F_2(v \cdot p) p_\mu \\ &= F_+(v \cdot p)(m_\mathrm{H} v_\mu + p_\mu) + F_-(v \cdot p) q_\mu \end{aligned} \tag{94}$$

[3] Note that in this case the variable $v \cdot p$ ranges between 0 and $m_\mathrm{H}/2$ where we have neglected the pion mass. Thus at the upper end of phase space the variable $v \cdot p$ scales with the heavy mass and heavy quark symmetries are not applicable any more.

where

$$F_\pm(v\cdot p)=\frac{1}{2}\left(F_1(v\cdot p)\pm F_2(v\cdot p)\right)\,. \tag{95}$$

Inserting this into (91) one may express $F_\pm$ in terms of the universal form factors A and B

$$F_1(v\cdot p) = F_+(v\cdot p)+F_-(v\cdot p)=-2\sqrt{\frac{v\cdot p}{m_{\mathrm{H}}}}A(\eta)\,, \tag{96}$$

$$F_2(v\cdot p) = F_+(v\cdot p)-F_-(v\cdot p)=-2\sqrt{\frac{m_{\mathrm{H}}}{v\cdot p}}B(\eta)\,. \tag{97}$$

From these relations one may read off the scaling of the form factors with the heavy mass which was already derived in [41].

This may be used to normalize the semileptonic B decays into light mesons relative to the semileptonic D decays. One obtains

$$F_\pm^{\mathrm{B}}(v\cdot p)=\frac{1}{2}\left(\sqrt{\frac{m_{\mathrm{D}}}{m_{\mathrm{B}}}}\pm\sqrt{\frac{m_{\mathrm{B}}}{m_{\mathrm{D}}}}\right)F_+^{\mathrm{D}}(v\cdot p)+\frac{1}{2}\left(\sqrt{\frac{m_{\mathrm{D}}}{m_{\mathrm{B}}}}\mp\sqrt{\frac{m_{\mathrm{B}}}{m_{\mathrm{D}}}}\right)F_-^{\mathrm{D}}(v\cdot p) \tag{98}$$

Note that F_+ for the B decay is expressed in terms of F_+ *and* F_- for the D decays. In the limit of vanishing fermion masses only F_+ contributes, which means that the F_- contribution to the rate is of the order of $m_{\mathrm{lepton}}/m_{\mathrm{H}}$. Thus it will be extremely difficult to determine experimentally.

The case of a heavy meson decaying into a light vector meson may be treated similarly. The matrix element for the transition of a heavy meson into a light vector meson (denoted generically as ρ in the following) is given again by (85) and is in this case

$$J_{\mathrm{V}}=\langle\rho(p,\epsilon)|\bar{\ell}\Gamma h_v|H(v)\rangle. \tag{99}$$

Using (86) one has

$$\langle\rho(p,\epsilon)|\bar{\ell}\Gamma h_v|H(v)\rangle = \mathrm{Tr}\ \left(\mathcal{M}_{\mathrm{V}}\Gamma H(v)\right), \tag{100}$$

where now the Dirac matrix $\mathcal{M}_{\mathrm{V}}$ has to be a linear function of the polarization of the light vector meson.

The most general decomposition is given in terms of four dimensionless form factors

$$\mathcal{M}_{\mathrm{V}}=\sqrt{v\cdot p}\,C(\eta)(v\cdot\epsilon)+\frac{1}{\sqrt{v\cdot p}}D(\eta)(v\cdot\epsilon)\not{p}+\sqrt{v\cdot p}\,E(\eta)\not{\epsilon}+\frac{1}{\sqrt{v\cdot p}}F(\eta)\not{p}\not{\epsilon} \tag{101}$$

where the variable η has been defined in (93).

Similar to the case of the decays into a light pseudoscalar meson (100) may be used to relate various exclusive heavy to light processes in the kinematic range where the energy of the outgoing vector meson is small. For example, the semileptonic decays $\mathrm{D}\to\rho e\nu$, $\mathrm{D}\to\mathrm{K}^*e\nu$ and $\mathrm{B}\to\rho e\nu$ are related among themselves and all of them may be related to the rare heavy to light decays $B \to K^* \ell^+ \ell^-$ and $B \to \rho \ell^+ \ell^-$ with $\ell = e, \mu$.

Data on these decays are still very sparse; there are first measurements of the decays $B \to \pi \ell \nu$ and $B \to \rho \ell \nu$ from CLEO [42], from which total rates may be obtained. From this one may extract a value of V_{ub} by employing form factor models, and the value given by CLEO is

$$|V_{ub}| = (3.3 \pm 0.2^{+0.3}_{-0.4} \pm 0.7) \times 10^{-3} \tag{102}$$

where the last uncertainty represents the variation of the result between different models. In order to perform a model independent determination along the lines discussed above a good measurement of the lepton energy spectra in these decays is needed.

Finally we comment on the heavy to light transitions of baryons. For the Λ-type heavy baryons (53) spin symmetry relates different polarizations of the same particle and thus imposes interesting constraints. Consider for example the matrix element of an operator $\bar{\ell} \Gamma h_v$ between a heavy Λ_Q and a light spin-1/2 baryon B_ℓ. It is described by only two form factors,

$$\langle B_\ell(p) | \bar{\ell} \Gamma h_v | \Lambda_Q(v) \rangle = \bar{u}_\ell(p) \{ F_1(v \cdot p) + \not{v} F_2(v \cdot p) \} \Gamma u_{\Lambda_Q}(v). \tag{103}$$

Thus in this particular case spin symmetry greatly reduces the number of independent Lorentz-invariant amplitudes which describe the heavy to light transitions.

This has some interesting implications for exclusive semileptonic Λ_c decays. For the case of a left handed current $\Gamma = \gamma_\mu (1 - \gamma_5)$, the semileptonic decay $\Lambda_c \to \Lambda \ell \bar{\nu}_\ell$ is in general parameterized in terms of six form factors

$$\begin{aligned} \langle \Lambda(p) | \bar{q} \gamma_\mu (1 - \gamma_5) c | \Lambda_c(v) \rangle &= \bar{u}(p) \left[f_1 \gamma_\mu + i f_2 \sigma_{\mu\nu} q^\nu + f_3 q^\mu \right] u(p') \\ &+ \bar{u}(p) \left[g_1 \gamma_\mu + i g_2 \sigma_{\mu\nu} q^\nu + g_3 q^\mu \right] \gamma_5 u(p'), \end{aligned} \tag{104}$$

where $p' = m_{\Lambda_c} v$ is the momentum of the Λ_c whereas $q = m_{\Lambda_c} v - p$ is the momentum transfer. From this one defines the ratio G_A/G_V by

$$\frac{G_A}{G_V} = \frac{g_1(q^2 = 0)}{f_1(q^2 = 0)}. \tag{105}$$

In the heavy c quark limit one may relate the six form factors f_i and g_i $(i = 1, 2, 3)$ to the two form factors F_j $(j = 1, 2)$

$$f_1 = -g_1 = F_1 + \frac{m_\Lambda}{m_{\Lambda_c}} F_2 \tag{106}$$

$$f_2 = f_3 = -g_2 = -g_3 = \frac{1}{m_{\Lambda_c}} F_2 \tag{107}$$

from which one reads off $G_\mathrm{A}/G_\mathrm{V} = -1$. This ratio is accessible by measuring in semileptonic decays $\Lambda_\mathrm{c} \to \lambda \ell \bar{\nu}_\ell$ the polarization variable α

$$\alpha = \frac{2G_\mathrm{A}G_\mathrm{V}}{G_\mathrm{A}^2 + G_\mathrm{V}^2} \tag{108}$$

which is predicted to be $\alpha = -1$ in the heavy c quark limit. The subleading corrections to the heavy c quark limit have been estimated and found to be small [43]

$$\alpha < -0.95, \tag{109}$$

and recent measurements yield

$$\alpha = -0.91 \pm 0.49 \quad \text{ARGUS [44]} \tag{110}$$
$$\alpha = -0.89^{+0.17+0.09}_{-0.11-0.05} \quad \text{CLEO[45]} \tag{111}$$

and are in satisfactory agreement with the theoretical predictions.

Recently the CLEO collaboration also measured the ratio of the form factors F_1 and F_2, averaged over phase space. Heavy quark symmetries do not fix this form factor ratio, at least not for a heavy to light decay, while for a heavy to heavy decay the form factor F_2 vanishes in the heavy mass limit for the final state quark. CLEO measures [46]

$$\left\langle \frac{F_2}{F_2} \right\rangle_{\text{phase space}} = -0.25 \pm 0.14 \pm 0.08 \tag{112}$$

which is in good agreement with model estimates [47].

4 The $1/m_\mathrm{Q}$ Expansion in Inclusive Decays

For inclusive decays a $1/m_\mathrm{Q}$ expansion is obtained for the rates by an approach similar to the one known from deep inelastic scattering [5]-[13]. The first step consists of an operator product expansion (OPE) which yields an infinite sum of operators with increasing dimension. The dimensions of the operators are compensated by inverse powers of a large scale, which is in general of the order of the heavy mass scale. The decay probability is then given as forward matrix elements of these operators between the state of the decaying heavy hadron; these matrix elements still have a mass dependence, which then may be extracted in terms of a $1/m_\mathrm{Q}$ expansion using HQET as for exclusive decays.

The method described below also allows us to deal with inclusive non-leptonic processes and hence in principle opens the possibility for a calculation of lifetimes and branching fractions in the framework of the $1/m_\mathrm{Q}$ expansion. This is remarkable, since non-leptonic processes are usually very hard to deal with, in particular the $1/m_\mathrm{Q}$ expansion has not (yet ?) brought any success in the field of exclusive non-leptonic decays.

Applying the OPE to the energy spectra of the charged lepton in inclusive semileptonic decays of heavy mesons, the relevant expansion parameter is not $1/m_Q$, but rather $1/(m_Q - 2E_\ell)$; the denominator is thus the energy release of the decay. In almost all phase space the energy release is of the order of the heavy mass; it is only in the endpoint region that it becomes small and hence the expansion breaks down. This problem may be fixed by a resummation of terms in the operator product expansion, which strongly resembles the summation corresponding to leading twist in deep inelastic scattering. Analogously to the parton-distribution function, a universal function appears, which determines all inclusive heavy-to-light decays.

4.1 Operator Product Expansion for Inclusive Decays

The effective Hamiltonian for a decay of a heavy (down-type) quark is in general linear in the decaying heavy flavored quark

$$\mathcal{H}_{\rm eff} = \bar{Q}R \tag{113}$$

where the operator R describes the decay products. In the following we shall consider semileptonic decays, for which

$$R_{\rm sl} = \frac{G_{\rm F}}{\sqrt{2}} V_{Qq}\ \gamma_\mu(1-\gamma_5)q\ (\bar{\nu}_\ell\gamma^\mu(1-\gamma_5)\ell), \tag{114}$$

where q is an up-type quark (c or u, since we shall consider b decays). Similarly, for non-leptonic decays the Cabbibo allowed contribution corresponds to

$$\begin{aligned} R_{\rm nl} = \frac{G_{\rm F}}{2\sqrt{2}} V_{Qq} V^*_{q'q''}\ &[(C_+(m_{\rm b}) + C_-(m_{\rm b}))\gamma_\mu(1-\gamma_5)q\ (\bar{q}'\gamma^\mu(1-\gamma_5)q'') \\ &+ (C_+(m_{\rm b}) - C_-(m_{\rm b}))\gamma_\mu(1-\gamma_5)q''\ (\bar{q}'\gamma^\mu(1-\gamma_5)q)]\,, \end{aligned} \tag{115}$$

where q' (q'') is a down-type (up-type) quark and V_{Qq} the corresponding CKM matrix element. The coefficients $C_\pm(m_{\rm b})$ are the QCD corrections obtained from the renormalization group running between $M_{\rm W}$ and $m_{\rm b}$; in leading logarithmic approximation these coefficients are [48]

$$C_\pm(m_{\rm b}) = [\frac{\alpha_{\rm s}(M_{\rm W}^2)}{\alpha_{\rm s}(m_{\rm b}^2)}]^{\gamma_\pm}, \text{ with } \gamma_+ = \frac{6}{33-2n_{\rm f}} = -\frac{1}{2}\gamma_- \tag{116}$$

where $\alpha_{\rm s}(\mu)$ is the one-loop expression (43) for the running coupling coupling constant of QCD.

Finally, for radiative rare decays we have

$$R_{\rm rare} = \frac{G_{\rm F}}{\sqrt{2}} V_{tb}V^*_{ts}C_7(m_{\rm b})\frac{e}{16\pi^2} m_{\rm b}\sigma_{\mu\nu}(1+\gamma_5)sF^{\mu\nu} \tag{117}$$

where $C_7(m_b)$ is again a coefficient obtained from running between M_W and m_b. Its value is $C_7(m_b) \sim 0.3$, the corresponding analytical expression may be found in [49].

The inclusive decay rate for a heavy hadron H containing the quark Q may be related to a forward matrix element by

$$\begin{aligned} \Gamma \propto & \sum_X (2\pi)^4 \delta^4(P_H - P_X) |\langle X | \mathcal{H}_{\text{eff}} | H(v) \rangle|^2 = \\ & \int d^4x \, \langle H(v) | \mathcal{H}_{\text{eff}}(x) \mathcal{H}_{\text{eff}}^\dagger(0) | H(v) \rangle \\ & = 2 \, \text{Im} \int d^4x \, \langle H(v) | T\{\mathcal{H}_{\text{eff}}(x) \mathcal{H}_{\text{eff}}^\dagger(0)\} | H(v) \rangle . \end{aligned} \tag{118}$$

where $|X\rangle$ is the final state, which is summed over to obtain the inclusive rate.

The matrix element appearing in (119) contains a large scale, namely the mass of the heavy quark. The first step towards a $1/m_Q$ expansion is to make this large scale explicit. This may be done by a phase redefinition. This leads to

$$\Gamma \propto 2 \, \text{Im} \int d^4x \, e^{-im_Q vx} \, \langle H(v) | T\{\widetilde{\mathcal{H}}_{\text{eff}}(x) \widetilde{\mathcal{H}}_{\text{eff}}^\dagger(0)\} | H(v) \rangle , \tag{119}$$

where

$$\widetilde{\mathcal{H}}_{\text{eff}} = \bar{Q}_v R \qquad Q_v = e^{-im_Q vx} Q . \tag{120}$$

This relation exhibits the similarity between the cross-section calculation in deep inelastic scattering and the present approach to total rates. In deep inelastic scattering there appears a large scale, which is the momentum transfer to the leptons, while here the mass of the heavy quark appears as a large scale.

The next step is to perform an operator product expansion of the product of the two Hamiltonians. After the phase redefinition the remaining matrix element does not involve large momenta of the order of the heavy quark mass any more and hence a short-distance expansion becomes useful, if the mass m_Q is large compared to the scale $\bar{\Lambda}$ determining the matrix element. The next step is thus to perform an operator-product expansion, which has the general form

$$\begin{aligned} & \int d^4x \, e^{im_Q vx} \, \langle H(v) | T\{\widetilde{\mathcal{H}}_{\text{eff}}(x) \widetilde{\mathcal{H}}_{\text{eff}}^\dagger(0)\} | H(v) \rangle \\ & = \sum_{n=0}^{\infty} \left(\frac{1}{2m_Q} \right)^n \hat{C}_{n+3}(\mu) \langle H(v) | \mathcal{O}_{n+3} | H(v) \rangle_\mu , \end{aligned}$$

where $\mathcal{O}_n$ are operators of dimension n, with their matrix elements renormalized at scale μ, and $\hat{C}_n$ are the corresponding Wilson coefficients. These coefficients encode the short distance physics related to the heavy quark mass scale and may be calculated in perturbation theory. All long distance contributions connected to the hadronic scale $\bar{\Lambda}$ are contained in the matrix elements of the operators $\mathcal{O}_{n+3}$.

Still the matrix elements of $\mathcal{O}_{n+3}$ are not independent of the heavy quark mass scale, but this mass dependence may be expanded in powers of $1/m_{\mathrm{Q}}$ by means of heavy quark effective theory. This is achieved by expanding the heavy quark fields appearing in the operators $\mathcal{O}_n$ using (18) (or, equivalently, (22)) as well as the states by including the corrections to the Lagrangian given in (19) (or (23)) as time-ordered products. In this way the mass dependence of the total decay rate may be accessed completely within an expansion in $1/m_{\mathrm{Q}}$.

The lowest-order terms of the operator product expansion are the dimension-3 operators. Due to Lorentz invariance and parity there are only two combinations which may appear, namely $\bar{Q}_v \not{v} Q_v$ or $\bar{Q}_v Q_v$. Note that the Q_v operators differ from the full QCD operators only by a phase redefinition, and hence $\bar{Q}_v \not{v} Q_v = \bar{Q} \not{v} Q$ and $\bar{Q}_v Q_v = \bar{Q} Q$. The first combination is proportional to the Q-number current $\bar{Q}\gamma_\mu Q$, which is normalized even in full QCD, while the second differs from the first one only by terms of order $1/m_{\mathrm{Q}}^2$

$$\bar{Q}_v Q_v = v_\mu \bar{Q}_v \gamma_\mu Q_v + \frac{1}{2m_{\mathrm{Q}}^2} \bar{h}_v \left[(iD)^2 - (ivD)^2 + \frac{i}{2}\sigma_{\mu\nu} G^{\mu\nu} \right] h_v + \mathcal{O}(1/m_{\mathrm{Q}}^3), \tag{121}$$

where $G_{\mu\nu}$ is the gluon field strength.

Thus the matrix elements of the dimension-3 contribution is known to be normalized; in the standard normalization of the states this implies

$$\langle H(v)|\mathcal{O}_3|H(v)\rangle = \langle H(v)|\bar{Q}_v \not{v} H_v|B(v)\rangle = 2m_{\mathrm{H}} \tag{122}$$

where m_{H} is the mass of the heavy hadron. To lowest order in the heavy mass expansion we may furthermore replace $m_{\mathrm{B}} = m_{\mathrm{Q}}$ and hence we may evaluate the leading term in the $1/m_{\mathrm{Q}}$ expansion without any hadronic uncertainty. Generically the dimension-3 contribution yields the free quark decay rate. This has been previously used as a model for inclusive decays, but now it turns out to be the first term in a systematic $1/m_{\mathrm{Q}}$ expansion of total rates.

A dimension-four operator contains an additional covariant derivative, and thus one has matrix elements of the type

$$\langle H(v)|\mathcal{O}_4|H(v)\rangle \propto \langle H(v)|\bar{Q}_v \Gamma D_\mu Q_v|H(v)\rangle = \mathcal{A}_\Gamma v_\mu \,. \tag{123}$$

Since the equations of motion apply for this tree level matrix element, one finds that the constant $\mathcal{A}_\Gamma$ has to vanish, and thus there are no dimension-four contributions. This statement is completely equivalent to Lukes theorem [31], since we are considering a forward matrix element, i.e. a matrix element at zero recoil [35].

The first non-trivial non-perturbative contribution comes from dimension-5 operators and is of order $1/m_{\mathrm{Q}}^2$. For mesonic decays there are only the two parameters λ_1 and λ_2 given in (27) and (28), which correspond to matrix elements of the subleading terms of the Lagrangian. They parameterize the non-perturbative input in the order $1/m_Q^2$. For Λ_Q-type baryons the parameter λ_2 vanishes due to heavy quark spin symmetry, while the kinetic energy parameter λ_1 is non-zero as well. In the framework of the $1/m_Q$ expansion this leads to a difference in lifetimes between mesons and baryons.

4.2 Calculation of Total Decay Rates

In this subsection we shall collect the results for the total rates including the first non-trivial non-perturbative correction.

Inserting R_{sl} as given in (114) one obtains for the total inclusive semileptonic decay rate $B \to X_c \ell \nu$

$$\Gamma(B \to X_c \ell \nu) = \frac{G_F^2 m_b^5}{192\pi^3} |V_{cb}|^2 \left[\left(1 + \frac{\lambda_1}{2m_c^2}\right) f_1\left(\frac{m_c}{m_b}\right) - \frac{9\lambda_2}{2m_c^2} f_2\left(\frac{m_c}{m_b}\right)\right], \tag{124}$$

where the two f_j are phase-space functions

$$\begin{aligned} f_1(x) &= 1 - 8x^2 + 8x^6 - x^8 - 24x^4 \log x, \\ f_2(x) &= 1 - \frac{8}{3}x^2 - 8x^4 + 8x^6 + \frac{5}{3}x^8 + 8x^4 \log x. \end{aligned} \tag{125}$$

The result for $B \to X_u \ell \nu_\ell$ is obtained from (124) as the limit $m_c \to 0$ and the replacement $V_{cb} \to V_{ub}$

$$\Gamma(B \to X_u \ell \nu) = \frac{G_F^2 m_b^5}{192\pi^3} |V_{ub}|^2 \left[1 + \frac{\lambda_1 - 9\lambda_2}{2m_b^2}\right]. \tag{126}$$

As it has been discussed above, the leading non-perturbative corrections in (124) and (126) are parameterized by λ_1 and λ_2. Estimates for these parameters have been discussed in Sect. 2; in order to estimate the total effect of the non-perturbative effects we shall insert a range of values $-0.3 > \lambda_1 > -0.6$ GeV^2; from this we obtain

$$\frac{\lambda_1 - 9\lambda_2}{2m_b^2} \sim -(3 \cdots 4)\% \tag{127}$$

This means that the non-perturbative contributions are small, in particular compared to the perturbative ones, which have been calculated some time ago [50], [51]. For the decay $B \to X_u \ell \bar{\nu}_\ell$ the lowest order QCD corrections are given by

$$\Gamma(B \to X_u \ell \bar{\nu}_\ell) = \frac{G_F^2 m_b^5}{192\pi^3} |V_{ub}|^2 \left[1 + \frac{2\alpha}{3\pi}\left(\frac{25}{4} - \pi^2\right)\right] = 0.85 |V_{ub}|^2 \Gamma_b, \tag{128}$$

and thus the typical size of QCD radiative corrections is of the order of ten to twenty percent.

Similarly, one obtains the result for non-leptonic decays as

$$\begin{aligned}\Gamma(\mathrm{B}\to X_\mathrm{c}) &= 3\frac{G_\mathrm{F}^2 m_\mathrm{b}^5}{192\pi^3}\left\{A_1 f_1\left(\frac{m_\mathrm{c}}{m_\mathrm{b}}\right)\left[1+\frac{1}{2m_\mathrm{b}^2}(\lambda_1-9\lambda_2)\right]\right. \\ &\left. -48A_2 f_3\left(\frac{m_\mathrm{c}}{m_\mathrm{b}}\right)\frac{1}{2m_\mathrm{b}^2}\lambda_2\right\}\end{aligned} \tag{129}$$

where the coefficients A_i are given by a combination of the Wilson coefficients $C_\pm(m_\mathrm{b})$ (116)

$$A_1 = \frac{1}{3}[C_-^2(m_\mathrm{b}) + 2C_+^2(m_\mathrm{b})], \qquad A_2 = \frac{1}{6}[C_+^2(m_\mathrm{b}) - C_-^2(m_\mathrm{b})], \tag{130}$$

and $f_3(x) = (1-x^2)^3$ is another phase space function. Again the non-perturbative corrections turn out to be small, in the region of a few percent compared to the leading term, and the perturbative corrections turn out to be much larger than this.

Finally, for the rare decay $\mathrm{B}\to X_s\gamma$ one may as well calculate the non-perturbative contribution in terms of λ_1 and λ_2. One obtains

$$\Gamma(\mathrm{B}\to X_s\gamma) = \frac{\alpha G_\mathrm{F}^2}{16\pi^4} m_\mathrm{b}^5 |V_{ts}V_{td}^*|^2 |C_7(m_\mathrm{b})|^2 \left[1+\frac{1}{2m_\mathrm{b}^2}(\lambda_1-9\lambda_2)\right], \tag{131}$$

and the relative size of the non-perturbative corrections is the same as in the $\mathrm{B}\to X_\mathrm{u}\ell\bar{\nu}_\ell$ decays.

Typically the non-pertubative corrections are much smaller than the radiative corrections. The only exception is the endpoint region of lepton energy spectra which receives both large perturbative as well as non-perturbative corrections. However, this is only a small region in phase space and the corrections to the total rates remain moderate.

4.3 Lifetimes of Heavy Hadrons

The subject of heavy hadron lifetimes is strongly related to non-leptonic processes, which have been considered already some time ago [52]-[55]; however the application of the $1/m_\mathrm{Q}$ expansion has turned many assumptions into quantitative arguments. As outlined in the last section the $1/m_\mathrm{Q}$ expansion allows to calculate total rates, even for non-leptonic processes, and hence a QCD-based calculation of lifetimes becomes possible. A recent review of this subject is given in [56].

Studying the formulae obtained in the $1/m_\mathrm{Q}$ expansion up to order $1/m_\mathrm{Q}^2$ one finds that lifetime differences between B mesons do not occur up to this level of the expansion; in other words, any difference between the $\mathrm{B}^\pm$ and the B^0 or the B_s lifetimes are induced by effects of the order $1/m_\mathrm{b}^3$. These effects are due to dimension six operators of the four-quark form

$$\mathcal{O}_6 = (\bar{b}\Gamma q)(\bar{q}\Gamma b). \tag{132}$$

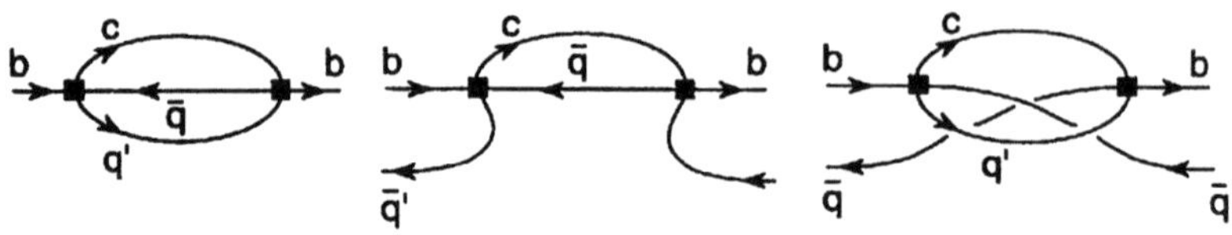

Fig. 1. Diagrams for non-leptonic decays of B mesons. The first diagram corresponds to the leading term in the $1/m_Q$ expansion, the second one to weak annihilation, and the third one to Pauli interference. Diagrams taken from [56].

In Fig. 1 the relevant diagrams are shown. The second and the third ones yield dimension six operators of the form (132); comparing to the terminology used in phenomenological models the second diagram is called weak annihilation (WA) and the third one Pauli interference (PI).

The four quark operators appearing at order $1/m_Q^3$ are usually estimated applying the vacuum insertion assumption, although this procedure has been criticized recently [57]. It amounts to replace

$$\langle B|(\bar{b}\Gamma q)(\bar{q}\Gamma b)|B\rangle \to \langle B|(\bar{b}\Gamma q)|0\rangle\langle 0|(\bar{q}\Gamma b)|B\rangle$$

and hence the parameter entering the estimates of the lifetime differences is the decay constant f_B.

The WA piece has been considered in [53], [58] and has been found to be small, of the order of one percent. The dominant contribution comes from the PI diagram. For the lifetime of the B^- one obtains [56]

$$\Gamma(B^-) = \Gamma_{1/m_Q^2}(B) + \Delta\Gamma_{PI}(B^-) \tag{133}$$

where the PI contribution is

$$\begin{aligned}\Delta\Gamma_{PI}(B^-) &= \frac{G_F^2 m_b^5}{192\pi^3}|V_{cb}|^2 \cdot 24\pi^2 \frac{f_B^2}{M_B^2}[C_+^2(m_b) \\ &\quad -C_-^2(m_b) + \frac{1}{N_C}(C_+^2(m_b) + C_-^2(m_b))]\end{aligned} \tag{134}$$

where $C_\pm$ have been given in (116). QCD radiative corrections change the sign of the PI contribution; the constructive interference at the scale M_W is turned into a destructive one at the scale m_b prolonging the lifetime of the B^- relative to the B_0. Using HQET the running of the coefficients below m_b

has also been calculated; it has been found that the destructive interference is amplified by the running below m_b, since

$$\Delta\Gamma_{PI}(B^-) = \frac{G_F^2 m_b^5}{192\pi^3}|V_{cb}|^2 \cdot 24\pi^2 \frac{f_B^2}{M_B^2}\kappa^{-4}\left[(C_+^2(m_b) - C_-^2(m_b))\kappa^{9/2}\right.$$
$$\left. + \frac{C_+^2(m_b) + C_-^2(m_b)}{3} - \frac{1}{9}(\kappa^{9/2} - 1)(C_+^2(m_b) - C_-^2(m_b))\right], \qquad (135)$$

where

$$\kappa = [\frac{\alpha_s(\mu_{had}^2)}{\alpha_s(m_b^2)}]^{1/b}, \text{ with } b = 11 - \frac{2}{3}n_f \qquad (136)$$

and μ_{had} is some small hadronic scale, which may be defined, for instance, by $\alpha_s(\mu_{had}) = 1$. Putting in numbers the result is [56]

$$\frac{\tau(B^-)}{\tau(B_d)} = 1 + 0.05 \cdot \frac{f_B^2}{(200\text{ MeV})^2} \qquad (137)$$

Hence the $1/m_Q$ expansion (in combination with vacuum insertion) predicts a slightly longer lifetime for the charged B meson compared to the neutral one. However, this is based on the vacuum insertion assumption which has been reconsidered recently in [57]; it is claimed that possible non-factorizable contributions may be enhanced by large prefactors thereby invalidating (137).

However, the estimate (137) is compatible with data; the latest compilation yields [59]

$$\frac{\tau(B^+)}{\tau(B^0)} = 1.019 \pm 0.048\,. \qquad (138)$$

Applying the $1/m_Q$ expansion for the lifetimes also to the D system one finds large corrections. From the experimental side it is known that the lifetime differences are very large and that the naive parton model expectation fails by a large margin. This may be taken as an indication that the c quark mass is indeed not large enough to justify a $1/m_Q$ expansion, at least for the non-leptonic processes.

The lifetimes of the two neutral B mesons are equal up to terms of order $1/m_b^3$, hence one expects

$$\bar{\tau}(B_d) = \bar{\tau}(B_s) \qquad (139)$$

to a good accuracy. Here $\bar{\tau}$ denotes the average lifetime of the two mass eigenstates of the $B^0 - \bar{B}^0$ system. While the lifetimes in the B_d system are practically the same such that the difference may be neglected, it has been pointed out that this is not necessarily the case in the B_s system; the lifetime difference has been estimated in [60] to be

$$\frac{\Delta\Gamma(B_s)}{\bar{\Gamma}(B_s)} = \frac{\Gamma(B_{s,\text{short}}) - \Gamma(B_{s,\text{long}})}{\bar{\Gamma}(B_s)} = 0.18 \cdot \frac{f_{B_s}^2}{(200\text{ MeV})^2} \qquad (140)$$

Hence this difference may be as large as twenty percent and thus it cannot be neglected any more e.g. in an analysis of B_s-$\bar{B}_s$ mixing.

Finally we shall consider the b baryon lifetimes. Here we expect the differences between the meson and the baryon lifetimes to be of order $1/m^2$, since the matrix elements of the kinetic energy operator as well as of the chromomagnetic moment operator are different between baryons and mesons; in particular, due to spin symmetry the matrix element of the chromomagnetic moment operator vanishes for Λ-type baryons, since the light degrees of freedom are in a spin-0 state. Probably only the Λ-type baryons will be stable against electromagnetic and strong decays, and hence their lifetime will be determined by the weak decay of the heavy quark.

Theoretical estimates of the Λ_{b} lifetime have been attempted in [56], [57] and yield the expectation that

$$\frac{\tau(\Lambda_{\mathrm{b}})}{\tau(\mathrm{B_d})} \gtrsim 0.9. \tag{141}$$

This has to be confronted with the recent data [59]

$$\frac{\tau(\Lambda_{\mathrm{b}})}{\tau(\mathrm{B_d})} = 0.80 \pm 0.05. \tag{142}$$

Thus the lifetime is slightly below the expectation, although this is not yet a significant deviation from the prediction. In particular, older data indicated that this lifetime ratio could have been as low as 0.7; such a low value would clearly indicate a theoretical problem in the $1/m_{\mathrm{Q}}$ expansion of the lifetimes, but the new data are in better agreement with the theoretical expectations.

4.4 Lepton Energy Spectra

The method of the operator-product expansion may also be used to obtain the non-perturbative corrections to the charged lepton energy spectrum [8]. In this case the operator product expansion is applied not to the full effective Hamiltonian, but rather only to the hadronic currents. The rate is written as a product of the hadronic and leptonic tensor

$$d\Gamma = \frac{G_{\mathrm{F}}^2}{4m_{\mathrm{B}}} |V_{Qq}|^2 W_{\mu\nu} \Lambda^{\mu\nu} d(\mathrm{PS}), \tag{143}$$

where $d(\mathrm{PS})$ is the phase-space differential. The short-distance expansion is then performed for the two currents appearing in the hadronic tensor. Redefining the phase of the heavy-quark fields as in (11) one finds that the momentum transfer variable relevant for the short-distance expansion is $m_{\mathrm{Q}}v - q$, where q is the momentum transfer to the leptons.

The structure of the expansion for the spectrum is identical to the one of the total rate. The contribution of the dimension-3 operators yields the free-quark decay spectrum, there are no contributions from dimension-4 operators, and the $1/m_b^2$ corrections are parameterized in terms of λ_1 and λ_2. Calculating the spectrum for $B \to X_c \ell \nu$ yields [9]-[12]

$$\begin{aligned}\frac{d\Gamma}{dy} = \frac{G_F^2 \mid V_{cb}\mid^2 m_b^5}{192\pi^3}&\Theta(1-y-\rho)y^2 \left[\{3(1-\rho)(1-R^2)-2y(1-R^3)\}\right. \\ &+\frac{\lambda_1}{[m_b(1-y)]^2}(3R^2-4R^3)-\frac{\lambda_1}{m_b^2(1-y)}(R^2-2R^3) \\ &-\frac{3\lambda_2}{m_b^2(1-y)}(2R+3R^2-5R^3)+\frac{\lambda_1}{3m_b^2}[5y-2(3-\rho)R^2+4R^3] \\ &\left.+\frac{\lambda_2}{m_b^2}[(6+5y)-12R-(9-5\rho)R^2+10R^3]\right]+\mathcal{O}\left[(\Lambda/[m_b(1-y)])^3\right]\end{aligned} \tag{144}$$

where we have defined

$$\rho = \left(\frac{m_c}{m_b}\right)^2 \quad R = \frac{\rho}{1-y} \tag{145}$$

and

$$y = 2E_\ell/m_b \tag{146}$$

is the rescaled energy of the charged lepton.

This expression is somewhat complicated, but it simplifies for the decay $B \to X_u \ell \nu$ since then the mass of the quark in the final state may be neglected. One finds

$$\begin{aligned}\frac{d\Gamma}{dy} = \frac{G_F^2 \mid V_{ub}\mid^2 m_b^5}{192\pi^3}&\left[\left(2y^2(3-2y)+\frac{10y^2}{3}\frac{\lambda_1}{m_b^2}+2y(6+5y)\frac{\lambda_2}{m_b^2}\right)\Theta(1-y)\right. \\ &\left.-\frac{\lambda_1+33\lambda_2}{3m_b^2}\delta(1-y)-\frac{\lambda_1}{3m_b^2}\delta'(1-y)\right] .\end{aligned} \tag{147}$$

Figure 2 shows the distributions for inclusive semileptonic decays of B mesons. The spectrum close to the endpoint, where the lepton energy becomes maximal, exhibits a sharp spike as $y \to y_{max}$. In this region we have

$$\frac{d\Gamma}{dy} \propto \Theta(1-y-\rho)\left[2+\frac{\lambda_1}{(m_Q(1-y))^2}\left(\frac{\rho}{1-\rho}\right)^2\left\{3-4\left(\frac{\rho}{1-\rho}\right)\right\}\right], \tag{148}$$

which behaves like δ-functions and its derivatives as $\rho \to 0$, which can be seen in (147). This behavior indicates a breakdown of the operator product expansion close to the endpoint, since for the spectra the expansion parameter is not $1/m_Q$, but rather $1/(m_Q - qv)$, which becomes $1/(m_Q[1-y])$ after the integration over the neutrino momentum. In order to obtain a description of the endpoint region, one has to perform some resummation of the operator product expansion.

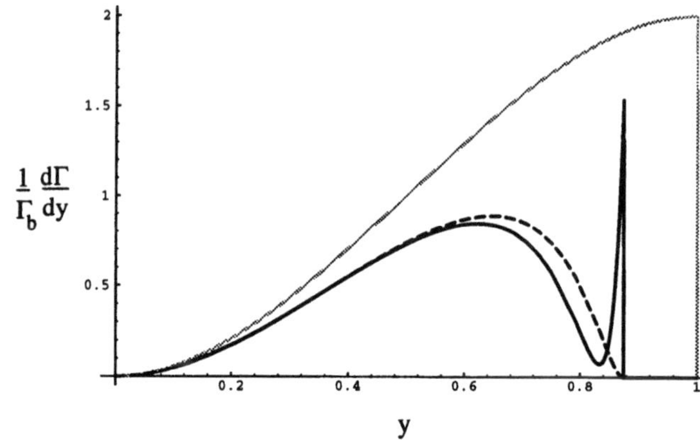

Fig. 2. The electron spectrum for free quark $b \to c$ decay (dashed line), free quark $b \to u$ decay (grey line), and $B \to X_c e \bar{\nu}_e$ decay including $1/m_b^2$ corrections (solid line) with $\lambda_1 = -0.5$ GeV^2 and $\lambda_2 = 0.12$ GeV^2. The figure is from [11].

4.5 Resummation in the Endpoint Region

Very close to the endpoint of the inclusive semileptonic decay spectra only a few resonances contribute. In this resonance region one may not expect to have a good description of the spectrum using an approach based on parton-hadron duality; here a sum over a few resonances will be appropriate.

In the variable y the size of this resonance region is however of the order of $(\bar{\Lambda}/m_Q)^2$ and thus small. In a larger region of the order $\bar{\Lambda}/m_Q$, which we shall call the endpoint region, many resonances contribute and one may hope to describe the spectrum in this region using parton-hadron duality.

It has been argued in [61] that the δ-function-like singularities appearing in (147) may be reinterpreted as the expansion of a non-perturbative function describing the spectrum in the endpoint region. Keeping only the singular terms of (147) we write

$$\frac{1}{\Gamma_b}\frac{d\Gamma}{dy} = 2y^2(3-2y)S(y), \tag{149}$$

where

$$S(y) = \Theta(1-y) + \sum_{n=0}^{\infty} a_n \delta^{(n)}(1-y) \tag{150}$$

is a non-perturbative function given in terms of the moments a_n of the spectrum, taken over the endpoint region. These moments themselves have an expansion in $1/m_Q$ such that $a_n \sim 1/m_Q^{n+1}$, and we shall consider only the leading term in the expansion of the moments, corresponding to the most singular contribution to the endpoint region.

Comparing (147) with (149) and (150) one obtains that

$$a_0 = \int dy(S(y) - \Theta(1-y)) = \mathcal{O}(1/m_Q^2) \tag{151}$$

$$a_1 = \int y(S(y) - \Theta(1-y)) = -\frac{\lambda_1}{3m_Q^2} \tag{152}$$

where the integral extends over the endpoint region.

The non-perturbative function implements a resummation of the most singular terms contributing to the endpoint and, in the language of deep inelastic scattering, corresponds to the leading twist contribution. This resummation has been studied in QCD [63], [62] and the function $S(y)$ may be related to the distribution of the light cone component of the heavy quark residual momentum inside the heavy meson. The latter is a fundamental function for inclusive heavy-to-light transitions, which has been defined in [62]

$$f(k_+) = \frac{1}{2M_H}\langle H(v)|\, \bar{h}_v\, \delta(k_+ - iD_+)\, h_v\, |H(v)\rangle, \tag{153}$$

where $k_+ = k_0 + k_3$ is the positive light cone component of the residual momentum k. The relation between the two functions S and f is given by

$$S(y) = \frac{1}{m_Q} \int\limits_{-m_Q(1-y)}^{\bar{\Lambda}} dk_+ f(k_+) \tag{154}$$

from which we infer that the nth moment of the endpoint region is given in terms of the matrix element $\langle H(v)|\bar{h}_v(iD_+)^n h_v|H(v)\rangle$.

The function f is a universal distribution function, which appears in all heavy-to-light inclusive decays; another example is the decay $B \to X_s\gamma$ [64], [62], where this function determines the photon-energy spectrum in a region of order $1/m_Q$ around the K^* peak.

In principle f has to be determined by other methods than the $1/m_Q$ expansion, e.g. from lattice calculations or from a model, or it has to be determined from experiment by measuring the photon spectrum in $B \to X_s\gamma$ or the lepton spectrum in $B \to X_u \ell\bar{\nu}$. In the context of the ACCMM model [51] f has been calculated in [65].

Some of the properties of f are known. Its support is $-\infty < k_+ < \bar{\Lambda}$, it is normalized to unity, and its first moment vanishes. Its second moment is given by a_1, and its third moment has been estimated [62], [35]. A one-parameter model for f has been suggested in [63], which incorporates the known features of f

$$f(k_+) = \frac{32}{\pi^2 \bar{\Lambda}}(1-x)^2 \exp\left\{ -\frac{4}{\pi}(1-x)^2 \right\} \Theta(1-x), \tag{155}$$

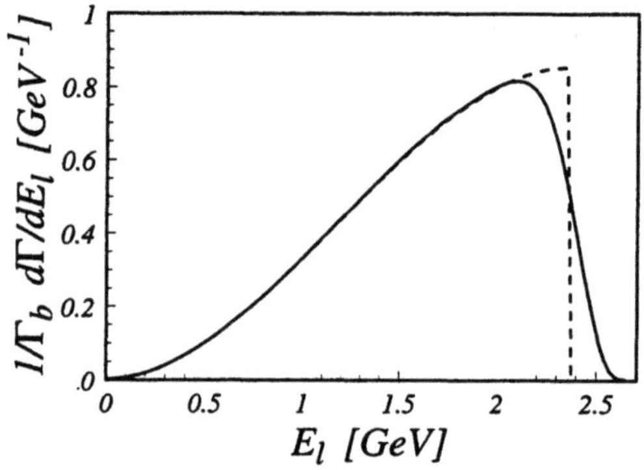

Fig. 3. Charged-lepton spectrum in $\mathrm{B} \to X_u \ell \bar{\nu}$ decays. The solid line is (149) with the ansatz (155), the dashed line shows the prediction of the free-quark decay model. The figure is from [63].

where $x = k_+/\bar{\Lambda}$, and the choice $\bar{\Lambda} = 570$ MeV yields reasonable values for the moments. In Fig. 3 we show the spectrum for $\mathrm{B} \to X_u \ell \nu_\ell$ using the ansatz (155).

Including the non-perturbative effects yields a reasonably behaved spectrum in the endpoint region and the δ-function-like singularities have disappeared. Furthermore, the spectrum now extends beyond the parton model endpoint; it is shifted from $E_\ell^{\max} = m_Q/2$ to the physical endpoint $E_\ell^{\max} = M_H/2$, since f is non-vanishing for positive values of $k_+ < \bar{\Lambda} = M_H - m_Q$.

5 Conclusions

The development of the field of heavy quark physics has been indeed remarkable over the last few years, experimentally as well as theoretically. From the experimental side, the progress in the technology of detectors (e.g. silicon vertex detectors) opened the possibility to study b physics even at machines which originally were not designed for this kind of research. In this way also the high energy colliders (in particular LEP and TEVATRON) could contribute substantially in this area, since they allow to measure states (such as the $\mathrm{B_s}$ and the b flavored baryons) which lie above the threshold of the $\Upsilon(4s)$-B-factories.

From the theoretical side the heavy quark limit and HQET brought an important success, since it provides a model independent and QCD based framework for the description of processes involving heavy quarks. The effective theory approach has originally been formulated for exclusive decays but in the past few years a heavy mass expansion has been set up also for inclusive transitions.

As far as exclusive heavy to heavy decays are concerned, the additional symmetries of the heavy mass limit restrict the number of non-perturbative functions in a model independent way; furthermore, heavy quark symmetries fix the absolute normalization of some of the transition amplitudes at the point of maximum momentum transfer. Phenomenologically this has improved our knowledge on the CKM matrix element V_{cb} dramatically; with the value $|V_{cb}| = (39.5 \pm 2.0) \times 10^{-3}$ the relative precision of this CKM matrix element is now about 5% and thus at a level of the precision with which the Cabbibo angle is known.

In heavy to light decays heavy quark symmetries do not work as efficiently; in this case only the relative normalization of B decays versus the corresponding D decays may be obtained. From the experimental side there are first measurements of $B \to \pi \ell \nu$ and $B \to \rho \ell \nu$ from the CLEO collaboration and an extraction of the CKM matrix element V_{ub} from these processes is still to some extent model dependent. The latest value for this CKM matrix element is $V_{ub} = (3.3 \pm 0.2^{+0.3}_{-0.4} \pm 0.7) \times 10^{-3}$ where the last error is due to the model dependence.

HQET does not yet have much to say about exclusive non-leptonic decays; even for the decays $B \to D^{(*)} D_s^{(*)}$, which involves three heavy quarks, heavy quark symmetries are not sufficient to yield useful relations between the decay rates [66]. Of course, with additional assumptions such as factorization one can go ahead and relate the non-leptonic decays to the semileptonic ones; however, this is a very strong assumption and it is not clear in what sense factorization is an approximation. On the other side, the data on the non-leptonic B decays support factorization, and first attempts to understand this from QCD and HQET have been undertaken [67]; however, the problem of the exclusive non-leptonic decays still needs clarification and hopefully the heavy mass expansion will also be useful here.

The $1/m_Q$ expansion obtained from the OPE and HQET offers the unique possibility to calculate the transition rates for inclusive decays in a QCD based and model independent framework. The leading term of this expansion is always the free quark decay, and the first non-trivial corrections are in general given by the mean kinetic energy of the heavy quark inside the heavy hadron λ_1 and the matrix element λ_2 of the chromomagnetic moment operator.

The method also allows us to calculate differential distributions, such as the charged lepton energy spectrum in inclusive semileptonic decays of heavy hadrons. For this case, the expansion parameter is the inverse of the energy release $m_b - 2E_\ell$, where E_ℓ is the lepton energy. Close to the endpoint, the energy release is small and thus the expansion in its inverse powers becomes useless. In this kinematic region one may partially resum the $1/m_Q$ expansion, obtaining a result closely analogous to the leading twist term in deep inelastic scattering. Particularly in the endpoint region a non-perturbative function is needed which corresponds to the parton distributions parameterizing the deep inelastic scattering.

The leading term of the $1/m_Q$ expansion is the free quark decay, and the result for the semileptonic branching fraction in this approximation has been well known for some time. The first non-perturbative corrections turn out to be quite small and hence the main corrections are the perturbative QCD corrections, where in a recent calculation also the effects of finite charm quark mass have been taken into account [68], [69]. The radiative corrections lower the semileptonic branching fraction somewhat compared to the parton model.

There has been some discussion on the issue of the semileptonic branching fraction triggered by the fact that the data used to be as low as $\mathrm{Br}(B \to X\ell\nu) \sim 10.5\%$ with a relative error of about ten percent. Such a low value for the semileptonic branching fraction in combination with the charm counting in B decays would indicate some theoretical problem; however, the recent LEP data yield a value of $\mathrm{Br}(B \to X\ell\nu) = (11.5 \pm 0.3)\%$ which is compatible with the theoretical expectations.

The expansion in powers of the inverse quark mass has become the standard tool in heavy quark physics and with the forthcoming experiments one may expect a strong improvement in our knowledge of the CKM sector of the SM, in particular a test of CP violation as it is encoded in the CKM matrix of the SM.

Acknowledgment
I thank the organizers of the Schladming School for the invitation and for providing such a beautiful environment for this conference.

References

1. M. Voloshin and M. Shifman, Sov. J. Nucl. Phys. **45** (1987) 292 and **47** (1988) 511; E. Eichten and B. Hill, Phys. Lett. **B234** (1990) 511; a more complete set of references can be found in one of the reviews [4]
2. N. Isgur and M. Wise, Phys. Lett. **B232** (1989) 113 and **B237** (1990) 527
3. B. Grinstein, Nucl. Phys. **B339** (1990) 253; H. Georgi, Phys. Lett. **B240** (1990) 447; A. Falk, H. Georgi, B. Grinstein and M. Wise, Nucl. Phys. **B343** (1990) 1
4. H. Georgi: contribution to the *Proceedings of TASI–91*, by R. K. Ellis et al. (eds.) (World Scientific, Singapore, 1991); B. Grinstein: contribution to *High Energy Phenomenology*, R. Huerta and M. A. Peres (eds.) (World Scientific, Singapore, 1991); N. Isgur and M. Wise: contribution to *Heavy Flavors*, A. Buras and M. Lindner (eds.) (World Scientific, Singapore, 1992); M. Neubert, Phys. Rept. **245** (1994) 259; T. Mannel, contribution to *QCD–20 years later*, P. Zerwas and H. Kastrup (eds.) (World Scientific, Singapore 1993); T. Mannel, J. Phys. **G 21** (1995) 1007; M. Neubert, CERN-TH/96-55, to appear in Int. J. Mod. Phys.
5. I. Bigi et al., preprint TPI-MINN-92-67-T, Talk given at *Particles & Fields 92*: 7th Meeting of the Division of Particles Fields of the APS (DPF 92), Batavia, IL, 10-14 Nov. 1992
6. I. Bigi, preprint UND-HEP-92-BIG06, Talk given at *26th International Conference on High Energy Physics (ICHEP 92)*, Dallas, TX, 6-12 Aug. 1992

7. I. Bigi et al., Phys. Lett. **B293** (1992) 430, (E) Phys. Lett. **B297** (1993) 477
8. J. Chay, H. Georgi and B. Grinstein, Phys. Lett. **B247** (1990) 399
9. I. Bigi, N. Uraltsev and A. Vainshtein, Phys. Lett. **B293** (1992) 430; I. Bigi et al., Minnesota TPI-MINN-92/67-T (1992) and Phys. Rev. Lett. **71** (1993) 496
10. B. Blok et al., Phys. Rev. **D49** (1994) 3356
11. A. Manohar and M. Wise, Phys. Rev. **D49** (1994) 1310
12. T. Mannel, Nucl. Phys. **B423** (1994) 396
13. A. Falk, M. Luke and M. Savage, Phys. Rev. **D49** (1994) 3367
14. E. Witten, Nucl. Phys. **B122** (1977) 109; S. Weinberg, Physica **96A** (1979) 327; J. Polchinski, Nucl. Phys. **B231** (1984) 269; H. Georgi, *Weak Interactions and Modern Particle Theory*, Benjamin/Cummings, Menlo Park (1984)
15. S. Coleman, J. Wess and B. Zumino, Phys. Rev. **177** (1969) 2239; C. G. Callan, S. Coleman, J. Wess and B. Zumino, Phys. Rev. **177** (1969) 2247
16. W. Kilian and T. Ohl, Phys. Rev. **D50** (1994) 4649
17. T. Mannel, W. Roberts and Z. Ryzak, Nucl. Phys. **B 368** (1992) 204
18. J. Körner and G. Thompson, Phys. Lett. **B264** (1991) 185
19. A. Falk, B. Grinstein and M. Luke, Nucl. Phys. **B357** (1991) 185
20. I. Bigi et al., Int. J. Mod. Phys. **A9** (1994) 2467
21. I. Bigi et al., preprint CERN-TH-7250/94 (1994), hep-ph/9405410
22. P. Ball and V. Braun, Phys. Rev. **D49** (1994) 2472
23. M. Gremm et al., Preprint CALT-68-2043, hep-ph/9603314
24. J. Collins, *Renormalization*, Cambridge Monographs on Mathematical Physics, Cambridge University Press 1984
25. A. Falk, Nucl. Phys. **B 378** (1992) 79
26. T. Bergfeld et al. (CLEO Collaboration), Phys. Lett. **B340** (1994) 194
27. E. Eichten, C. Hill and C. Quigg, Phys. Rev. Lett. **71** (1993) 4116
28. N. Isgur and M. Wise, Nucl. Phys. **B348** (1991) 276
29. H. Georgi, Nucl. Phys. **B348** (1991) 293
30. T. Mannel, W. Roberts and Z. Ryzak, Nucl. Phys. **B355** (1991) 38
31. M. Luke, Phys. Lett. **B252** (1990) 447
32. M. Ademollo and R. Gatto, Phys. Rev. Lett. **13** (1964) 264
33. M. Luke and A. Manohar, Phys. Lett. **B286** (1992) 348; Y. Chen, Phys. Lett. **B317** (1993) 421
34. A. Falk and M. Neubert, Phys. Rev. **D47** (1993) 2965
35. T. Mannel, Phys. Rev. **D50** (1994) 428
36. N. Isgur, D. Scora, B. Grinstein and M. Wise, Phys. Rev. **D39** (1989) 799
37. M. Shifman, N.G. Uraltsev and A. Vainshtein, Minnesota Preprint TPI-MINN-94-13-T (1994)
38. M. Neubert, Preprint CERN-TH-95-107, Talk given at 30th Rencontres de Moriond: Electroweak Interactions and Unified Theories, Meribel les Allues, France, 11-18 Mar 1995
39. B. Barish et al., (CLEO Collaboration), Phys. Rev. **D 51** (1995) 1014
40. D. Buskulic et al., (ALEPH Collaboration), Phys. Lett. **B 359** (1995) 236
41. N. Isgur and M. Wise, Phys. Rev. **D42** (1990) 2388
42. K. Berkelman (CLEO COllaboration), talk given at the III. German Russian Workshop on Heavy Quark Physics, Dubna, Russia, 20-22 May 1996
43. G. Lin and T. Mannel, Phys. Lett. **B321** (1994) 417

44. H. Albrecht, et al. (ARGUS Collaboration) Phys. Lett. **B326** (1994) 320
45. T. Bergfeld, et al. (CLEO Collaboration) Phys. Lett. **B323** (1994) 219
46. G. Crawford et al, (CLEO COllaboration), Cornell preprint CLNS 94/1306, CLEO 94/24 (1994)
47. J. Körner and M. Krämer, Phys. Lett. **B275** (1992) 495
48. F. Gilman and M. Wise, Phys. Rev. **D20** (1979) 2392; A. Buras et al., Nucl. Phys. **B370** (1992) 69
49. B. Grinstein, R. Springer and M. Wise, Nucl. Phys. **B319** (1988) 271; M. Misiak, Nucl. Phys. **B393** (1993) 23
50. A. Ali and E. Pietarinen, Nucl. Phys. **B154** (1979) 519; N. Cabibbo, G. Corbo and L. Maiani, Nucl. Phys. **B155** (1979) 83; G. Corbo, Nucl. Phys. **B212** (1983) 99; M. Jezabek and J. H. Kühn, Nucl. Phys. **B320** (1989) 20; A. Falk et al., Phys. Rev. **D49** (1994) 3367
51. G. Altarelli et al., Nucl. Phys. **B208** (1982) 365
52. R. Rückl, CERN preprint Print-83-1063 1983 (unpublished)
53. B. Guberina et al., Phys. Lett. **89B** (1979) 111
54. J. Ellis, M. Gaillard and D. Nanopoulos, Nucl. Phys. **B100** (1975) 313
55. N. Cabibbo and L. Maiani, Phys. Lett. **89B** (1979) 111
56. I. Bigi et al., preprint CERN-TH.7132/94, to appear in *B Decays*, S. Stone (ed.), World Scientific
57. M. Neubert and C. Sachrajda, Preprint CERN-TH-96/19, hep-ph/9603202
58. I. Bigi and N. Uraltsev, Phys. Lett. **B280** (1992) 120
59. G. Sciolla, talk given at the III. German Russian Workshop on Heavy Quark Physics, Dubna, Russia, 20-22 May 1996
60. M. Voloshin et al., Sov. J. Phys. **46** (1987) 112
61. M. Neubert, Phys. Rev. **D49** (1994) 2472
62. I. Bigi et al., Int. J. Mod. Phys. **A9** (1994) 2467
63. T. Mannel and M. Neubert, Phys. Rev. **D 50** (1994) 2037
64. M. Neubert, Phys. Rev. **D49** (1994) 4623
65. I. Bigi et al, preprint CERN-TH.7159/94
66. T. Mannel, W. Roberts and Z. Ryzak, Phys. Lett. **B248** (1990) 392
67. M. Dugan and B. Grinstein, Phys. Lett. **B255** (1991) 583
68. E. Bagan et al., Phys. Lett. **B342** (1995) 362
69. E. Bagan et al., Phys. Lett. **B 351** (1995) 546

Seminars

Several participants gave interesting seminar talks fitting to the general theme of the School. It was only possible to include a few of these talks in the present volume. In what follows, we give a complete list of the seminar speakers (alphabetically ordered) and the titles of their contributions.

E. Avakyan (Gomel Polytech. Inst.):
The description of Kaon physics in the quark confinement model

B. Bajc (J. Stefan Inst. Ljubljana):
$c \to u\ g$ in Cabbibo suppressed D meson radiative weak decays

N. Brambilla (Univ. Milano):
Wilson loop approach and dual QCD

A. Denner (Univ. Würzburg):
The Goldstone-boson equivalence theorem in higher orders in the background-field method

St. Dittmaier (Univ. Bielefeld):
Integrating out heavy fields in the path integral

M. Fabre de la Ripelle (Univ. Paris-Sud):
Simple calculations with quark potential models

M. Feurstein (TU Wien): Topological aspects of lattice QCD field theoretic versus geometric definitions

Ch. Gattringer (MPI München):
Chiral lattice fermions - results in 2 dimensions

B. Geyer (Univ. Leipzig):
Evolution kernels for non-singlet structure functions in polarized deep inelastic scattering

L. Glozman (Univ. Graz):
Baryon structure and the chiral symmetry of QCD

Ch. Gutsfeld (RWTH Aachen):
Gross-Neveu and Schwinger model on the lattice

Th. Heinzl (Univ. Regensburg):
Hamiltonian approach to the Gribov problem

P. Labelle (Mc Gill Univ. Montreal):
An explicit and powerful application of effective field theories to QED bound states

A. Laser (Univ. Heidelberg):
Bound states in the hot electroweak phase

A. Mihaly (TU Wien):
Meson-meson potentials with improved action in lattice QCD

K. Passek (Rudjer Boskovic Inst. Zagreb):
Exclusive photoproduction of large momentum-transfer K and K* mesons

V. Pershin (Univ. Tomsk):
Massive background fields equations of motion in bosonic string theory in polarized deep inelastic scattering

G. Prosperi (Univ. Milano):
Bethe-Salpeter like equation in QCD in a Wilson-loop context

E. Remiddi (Univ. Bologna):
The analytical value of the 6th order electron (g-2) in QED

A. Ruffing (MPI Mnchen):
Quantum mechanics in discretized phase space

M.G. Schmidt (Univ. Heidelberg):
Worldline formulation of QFT

R. Shulyakovski (Acad. Sci. Minsk):
Correlation function for two gluons produced by the instanton field

P. Skala (TU Wien):
Magnetic monopoles and the dual superconductor picture in lattice QCD

D. Vaman (Inst. Atom. Phys. Bukarest):
Supersymmetries and motions in Taub-NUT spinning space

MIX
Papier aus verantwortungsvollen Quellen
Paper from responsible sources
FSC® C105338

If you have any concerns about our products,
you can contact us on
ProductSafety@springernature.com

In case Publisher is established outside the EU,
the EU authorized representative is:
Springer Nature Customer Service Center GmbH
Europaplatz 3, 69115 Heidelberg, Germany

Printed by Libri Plureos GmbH
in Hamburg, Germany